Advances in Intelligent Systems and Computing

Volume 833

Series editor

Janusz Kacprzyk, Polish Academy of Sciences, Warsaw, Poland
e-mail: kacprzyk@ibspan.waw.pl

The series "Advances in Intelligent Systems and Computing" contains publications on theory, applications, and design methods of Intelligent Systems and Intelligent Computing. Virtually all disciplines such as engineering, natural sciences, computer and information science, ICT, economics, business, e-commerce, environment, healthcare, life science are covered. The list of topics spans all the areas of modern intelligent systems and computing such as: computational intelligence, soft computing including neural networks, fuzzy systems, evolutionary computing and the fusion of these paradigms, social intelligence, ambient intelligence, computational neuroscience, artificial life, virtual worlds and society, cognitive science and systems, Perception and Vision, DNA and immune based systems, self-organizing and adaptive systems, e-Learning and teaching, human-centered and human-centric computing, recommender systems, intelligent control, robotics and mechatronics including human-machine teaming, knowledge-based paradigms, learning paradigms, machine ethics, intelligent data analysis, knowledge management, intelligent agents, intelligent decision making and support, intelligent network security, trust management, interactive entertainment, Web intelligence and multimedia.

The publications within "Advances in Intelligent Systems and Computing" are primarily proceedings of important conferences, symposia and congresses. They cover significant recent developments in the field, both of a foundational and applicable character. An important characteristic feature of the series is the short publication time and world-wide distribution. This permits a rapid and broad dissemination of research results.

Advisory Board

More information about this series at http://www.springer.com/series/11156

Kazimierz Choroś · Marek Kopel
Elżbieta Kukla · Andrzej Siemiński
Editors

Multimedia and Network Information Systems

Proceedings of the 11th International
Conference MISSI 2018

 Springer

Editors
Kazimierz Choroś
Faculty of Computer Science
 and Management
Wrocław University of Science
 and Technology
Wrocław, Poland

Elżbieta Kukla
Faculty of Computer Science
 and Management
Wrocław University of Science
 and Technology
Wrocław, Poland

Marek Kopel
Faculty of Computer Science
 and Management
Wrocław University of Science
 and Technology
Wrocław, Poland

Andrzej Siemiński
Faculty of Computer Science
 and Management
Wrocław University of Science
 and Technology
Wrocław, Poland

ISSN 2194-5357   ISSN 2194-5365   (electronic)
Advances in Intelligent Systems and Computing
ISBN 978-3-319-98677-7   ISBN 978-3-319-98678-4   (eBook)
https://doi.org/10.1007/978-3-319-98678-4

Library of Congress Control Number: 2018950645

This Springer imprint is published by the registered company Springer Nature Switzerland AG
The registered company address is: Gewerbestrasse 11, 6330 Cham, Switzerland

Preface

The volume consists of the Proceedings of the 11th International Conference on Multimedia & Network Information Systems (MISSI 2018) held during September 12–14, 2018, in Wrocław, Poland. The Conference was organized by the Wrocław University of Science and Technology. The Conference ran under the patronage of the Committee of Informatics of the Polish Academy of Sciences, Wrocław Scientific Society, Forum of IT Companies, and Polish Information Processing Society.

The MISSI Conferences have already obtained the status of a well-recognized and respected biennial event. They span now over the period of more than 20 years. Many participants are frequent guests of the Conference. They have the opportunity to see how the research area evolves, new topics are raised, and some old are much advanced as it is exemplified by tasks like speech, face, or handwriting recognition. Not long ago they were hot research topics, and now they are embedded in commercial products.

The 11th edition marks a new opening for the Conference. For the first time, it included the 6th International Workshop on Computational Intelligence for Multimedia Understanding, and additionally four special sessions were organized.

The keynote lectures were given by three outstanding scientists, and the summaries of their lectures are included in the Proceedings. The organizers succeeded once more in gathering a great number of scientists from all across Europe and well beyond.

From all submissions, the Editorial Board has selected 58 papers with the highest quality for oral presentation and publication in the Proceedings. Most of the papers focus on applying different AI methods to multimedia or natural language processing. Many papers have the distinct practical trait. They describe and evaluate some experimental systems. There is no shortage of papers of a theoretical nature.

The volume consists of eight chapters:

1) Multimedia Systems
2) Web Systems and Network Technologies

3) Natural Language Processing and Information Retrieval
4) Computational Intelligence for Multimedia Understanding
5) Accessing Multilingual Information and Opinions
6) Intelligent Audio Processing
7) Innovations in Web Technologies
8) Video Game Development Methods and Technologies.

The first chapter includes papers on the broad topic of multimedia systems. It is presented at various levels of processing. The papers on low-level processing address topics like linear compression of blurry photographs or estimation of noise in digital images. At the other end of the spectrum, we have got a paper that proposes a concept of visual knowledge representation. In between, there are papers on such up-to-date areas as video summarization, using the virtual assistant as a communication channel for road traffic situation, and analyzing the accuracy of the methods of fuzzy clustering of images. A group of papers deals with human-related issues: sign language recognition and tracking human upper limb motion.

The second chapter offers a large number of papers that explore the data available on the Internet and the novel technologies used to process them. It goes without saying they apply recent techniques originating from big data and artificial intelligence.

A picture may be worth a thousand words, but a natural language is unbeatable when it goes about precise conveying of information. The Internet spans the whole world and is therefore inherently multilingual. The papers collected in the third chapter discuss various aspects of natural language processing such as word sense disambiguation, measuring the efficiency of machine translation, or stylometry.

The next chapter is devoted to the computational intelligence for multimedia understanding. This is a diverse area of study, and it has papers, e.g., on deep learning approach to human osteosarcoma cell detection and classification, a hybrid CNN approach for single image depth estimation, or detection of suspicious accounts on Twitter using Word2Vec and sentiment analysis.

The accessing multilingual information and opinions chapter concentrates upon the different aspects of summarization problem. It does not stop at the more traditional work such as speech-to-cross-lingual text summarization but goes well beyond that and evaluates, e.g., multimedia content summarization algorithms.

The intelligent audio processing chapter is very much technically oriented. It presents among others advanced techniques for spectral and cepstral feature extraction techniques for phoneme modeling or the selection of features for multimodal vocalic segment classification.

The papers from the innovations in Web technologies part have a distinctive human touch. They propose a multimedia platform for the Mexican cultural heritage, dealing with the visualization of big data from Polish digital libraries or recognition of the pathology of the human movement with the use of mobile technology and machine learning.

The video game section is not only about gaming per se as it contains such topics as accident prevention during immersion in virtual reality or wide field of view projection using rasterization.

Our special thanks go to the Program Chairs, Special Session Chair, Organizing Chairs, Publicity Chairs for their work for the Conference. We sincerely thank all the members of the International Program Committee for their valuable efforts in the review process, which helped us to guarantee the highest quality of the selected papers for the MISSI Conference. We cordially thank the organizers and chairs of special sessions who contributed to the success of the Conference.

We would like to express our thanks to the keynote speakers: Alberto del Bimbo from University of Florence in Italy, Bogusław Cyganek from AGH University of Science and Technology in Poland, and Andrzej Czyżewski from Gdańsk University of Technology in Poland for their world-class plenary speeches.

We wish to thank the members of the Organizing Committee for their excellent work and for their considerable effort.

We cordially thank all the authors for their valuable contributions and the other participants of this Conference. The Conference would not have been possible without their support.

The pace of progress in the field of Multimedia and Internet Information Systems ever increases and so does the need for an up-to-date account of the current work in the area. The editors of the volume will be pleased if the Proceedings will provide in the researches with the results of current work done at other universities. At the same time, we hope that they would be a source of an inspiration for a new generation of scientific works and it will prompt them to devote their time and effort to uphold the progress of research on the area.

These were our goals. If we achieve them, it will be the best reward for our efforts.

Kazimierz Choroś
Marek Kopel
Elżbieta Kukla
Andrzej Siemiński

Organization

General Chair

Kazimierz Choroś   Wrocław University of Science and Technology, Poland

Steering Committee

Kazimierz Choroś   Wrocław University of Science and Technology, Poland

Grzegorz Dobrowolski   AGH University of Science and Technology, Poland

Dosam Hwang   Yeungnam University, Republic of Korea

Czesław Jędrzejek   Poznań University of Technology, Poland

Janusz Kacprzyk   Polish Academy of Sciences, Poland

Ngoc Thanh Nguyen   Wrocław University of Science and Technology, Poland

Toyoaki Nishida   Kyoto University, Japan

Maria Pietruszka   University of Łódź, Poland

Gottfried Vossen   University of Münster, Germany

Special Session Chair

Andrzej Siemiński   Wrocław University of Science and Technology, Poland

Organizing Chair

Elżbieta Kukla   Wrocław University of Science and Technology,
                                Poland

Publicity Chairs

Bernadetta Maleszka   Wrocław University of Science and Technology,
                                Poland
Marek Kopel   Wrocław University of Science and Technology,
                                Poland

Keynote Speakers

Alberto del Bimbo   University of Florence, Italy
Bogusław Cyganek   AGH University of Science and Technology,
                                Poland
Andrzej Czyżewski   Gdańsk University of Technology, Poland

Main Track and Special Session International Program Committee

Reza Andrea   STMIK Widya Cipta Dharma, Indonesia
Edoardo Ardizzone   University of Palermo, Italy
Remigiusz Baran   Kielce University of Technology, Poland
László Bokor   Budapest University of Technology and
                                Economics, Hungary
Tiago Jose de Carvalho   Federal Institute of Education, Science
                                and Technology of Sao Paulo, Brazil
Yusra Chabchoub   ISEP, BILab, France
Raja Chiky   ISEP, BILab, France
Kazimierz Choroś   Wrocław University of Science and Technology,
                                Poland
Rozenn Dahyot   Trinity College Dublin, Ireland
Ivanoe De Falco   ICAR – CNR, Italy
Dominique Fohr   University of Lorraine, France
Michal Grega   AGH University of Science and Technology,
                                Poland
Michal Haindl   Institute of Information Theory and Automation
                                of the CAS, Czech Republic
Zbigniew Huzar   Wrocław University of Science and Technology,
                                Poland

Gabriele Pieri	CNR—Istituto di Scienza e Tecnologie dell'Informazione, Italy
Marcin Pietranik	Wrocław University of Science and Technology, Poland
Marco Reggiannini	CNR—Istituto di Scienza e Tecnologie dell'Informazione, Italy
Cristina Ribeiro	INESC TEC—Faculty of Engineering, University of Porto, Portugal
Jorge Rodas-Osollo	LaNTI-UACJ, Mexico
Filip Rudziński	University of Lorraine, France
Emanuele Salerno	CNR—Istituto di Scienza e Tecnologie dell'Informazione, Italy
Bala Sebastian	Opole University, Poland
Syed Afaq Shah	The University of Western Australia, Australia
Andrzej Siemiński	Wrocław University of Science and Technology, Poland
Leszek Sliwko	University of Westminster, UK
Kamel Smaïli	University of Lorraine, France
Piotr Sobolewski	Wrocław University of Science and Technology, Poland
Piotr Szczuko	Gdańsk University of Technology, Poland
Salvatore-Antoine Tabbone	University of Lorraine, France
Behçet Uğur Töreyin	Informatics Institute, Istanbul Technical University (ITU), Turkey
Juan-Manuel Torres	University of Avignon, France
Bogdan Trawiński	Wrocław University of Science and Technology, Poland
Maria Trocan	Institut Supérieur d'Électronique de Paris, France
Jonathan Weber	University of Lorraine, France
Aleksander Zgrzywa	Wrocław University of Science and Technology, Poland
Amaia Mendez Zorrilla	University of Deusto, Spain

Additional Reviewers

Indah Astuti	Janusz Rafałko
Rayane El Sibai	Gintautas Tamulevičius
Muhammad Hanif	Pin Zhang
Attila Nagy	Sławomir Zieliński

Special Session Organizers

Satellite Workshop IWCIM—6th International Workshop on Computational Intelligence for Multimedia Understanding

Behçet Uğur Töreyin	Informatics Institute, Istanbul Technical University (ITU), Turkey
Maria Trocan	Institut Supérieur d'Électronique de Paris (ISEP), France
Davide Moroni	Institute of Information Science and Technologies ISTI—CNR, Italy

AMIS—Special Session on Accessing Multilingual Information and Opinions

Kamel Smaïli	University of Lorraine, France
Mikołaj Leszczuk	AGH University of Science and Technology, Poland
Begoña García-Zapirain	University of Deusto, Spain
Juan Manuel Torres Moreno	University of Avignon, France

IAP—Special Session on Intelligent Audio Processing

Bożena Kostek	Gdańsk University of Technology, Poland

INWEBTECH—Special Session on Innovations in Web Technologies

Jolanta Mizera-Pietraszko	Opole University, Poland
Ricardo Rodriguez Jorge	Universidad Autónoma de Ciudad Juárez, Chihuahua, Mexico
Edgar Alonso Martinez Garcia	Universidad Autónoma de Ciudad Juárez, Chihuahua, Mexico

VGDMT 2018—Special Session on Video Game Development Methods and Technologies

Reza Andrea	STMIK Widya Cipta Dharma, Indonesia
Marek Kopel	Wrocław University of Science and Technology, Poland
Paweł Rohleder	Techland, Poland
Piotr Sobolewski	Wrocław University of Science and Technology, Poland

Contents

Invited Papers

Identity Recognition by Incremental Learning

Alberto del Bimbo$^{(\boxtimes)}$, Federico Pernici, Matteo Bruni,
and Federico Bartoli

Dipartimento di Ingegneria dell'Informazione, Università degli Studi di Firenze,
Via Santa Marta, 3, 50137 Florence, Italy
alberto.delbimbo@unifi.it

Face recognition systems nowadays benefit from the improved performance of new classification models combined with the availability of large datasets of face images and the increase of computational power. Effective performance were achieved for both the *closed-set* [1] and *open-set* [2] face recognition scenarios, where the system should return the best ranked gallery identity or – in the *open-set* case – eventually reject the probe. However, in real world contexts, a different *open-world* scenario [3] is also frequent. In this case, the face recognition system must address unseen-before subjects and include the novel identities in the learning system so to be able to recognize such identities in the next future. Differently from the *closed-set* and *open-set* scenarios, in this case the gallery dataset is dynamic, and the subject identities must be learned incrementally from the observations. Parametric learning methods like deep networks that are trained to learn mappings by adjusting connection weights to minimize the error in the output, while are well suited to perform *closed-set* recognition - and can be adapted to the *open-set* case as well - are not able to support incremental learning as required in the *open-world* case, without catastrophic forgetting.

In this talk, we present and discuss a novel approach to unsupervised on-line incremental learning of face identities in video streams that provides an effective solution to the *open-world* case. In our approach, we collect deep face descriptors into a memory module and build identity representations incrementally, distilling the most relevant descriptors that are collected with a smart controller mechanism that takes into account, dynamically, the relevance of each descriptor to be learned as representative of its identity. Identity matching between the observations and the face descriptors in the memory is performed using Reverse Nearest Neighbor with distance ratio criterion [4]. To avoid that descriptors of the same identity accumulate in memory, we associate to each descriptor-identity pair a dimensionless quantity referred to as *eligibility-to-be-learned* that is set to 1 (one) when the descriptor is loaded into the memory and is decreased at each match. Highly matched descriptors are clearly redundant and can be removed from memory with no sensible effect on the identity representation. We also exploit the temporal coherence of the video sequence as a sort of supervisor to decide when instantiate a new identity, in order to reduce the proliferation of new identities. We show that this incremental learning procedure can operate at real-time pace with GPU support and stabilizes asymptotically around the probability density function of the descriptors of each identity as time progresses. It also scales gracefully as the number of identities that are observed grows. In support to our findings, we will present experimental evidence that shows the effectiveness of the method for *open-world* face recognition on large video datasets [5].

© Springer Nature Switzerland AG 2019
K. Choroś et al. (Eds.): MISSI 2018, AISC 833, pp. 3–4, 2019.
https://doi.org/10.1007/978-3-319-98678-4_1

References

1. Parkhi, O.M., Vedaldi, A., Zisserman, A.: Deep face recognition. In: Proceedings of the British Machine Vision Conference 2015, Swansea, UK (2015)
2. Scheirer, W.J., de Rezende Rocha, A., Sapkota, A., Boult, T.E.: Toward open set recognition. IEEE Trans. Pattern Anal. Mach. Intell. **35**(7), 1757–1772 (2013)
3. Bendale, A., Boult, T.: Towards open world recognition. In: Proceedings of the IEEE Conference on Computer Vision and Pattern Recognition (2015)
4. Korn, F., Muthukrishnan, S.: Influence sets based on reverse nearest neighbor queries. In: Proceedings of the 2000 ACM SIGMOD International Conference on Management of Data, SIGMOD 2000, New York, NY, USA (2000)
5. Pernici, F., Bartoli, F., Bruni, M., Del Bimbo, A.: Memory based online learning of deep representations from video streams. In: IEEE Conference on Computer Vision and Pattern Recognition, CVPR 2018, Salt Lake City, Utah, USA (2018)

Modern Approaches to Multi-dimensional Visual Signals Analysis

Bogusław Cyganek[✉]

AGH University of Science and Technology,
Al. Mickiewicza 30, 30-059 Kraków, Poland
cyganek@agh.edu.pl

Each day enormous amounts of visual data is produces (Internet, Youtube, etc.). For their efficient processing a synergy between theoretical achievements in multi-dimensional data classification and computational technologies has to be reached. In this talk the focus will be laid upon modern approaches to multi-dimensional visual signal analysis with recently developed deep learning and tensor based methods, on the one hand, and their practical realizations, on the other. Special part will be devoted to modern classification ensembles, joining multi-dimensional approach as well as deep neural architectures.

Deep learning revolutionized machine learning world. Especially convolutional neural networks allowed previously attainable results in such domains as image classification, object detection, tracking, speech recognition, natural text processing, and many more. There are various widely available computer platforms, such as TensorFlow, which allow for development of new deep applications by the machine learning community. On the other hand, processing of multi-dimensional signals was greatly facilitated with recent achievements in tensor theory, adjusted to the problems of multi-dimensional data representation and analysis. In this talk we will try to shed more light on the two domains, with special emphasis on their common aspects. First, basic facts behind data representation and analysis with tensors will be presented. In this respect, an overview of various tensor decomposition methods, such as the higher order singular value decomposition as well as the best rank-($R1, R2, ..., RN$) algorithms, will be presented. A brief introduction to various deep neural network architectures, especially suited for processing of multi-dimensional signals, will follow.

Last but not least, the talk will be endowed with description of numerous applications and novel processing platforms. Further information can be found in the following references [1–22].

References

1. Bader, W.B., Kolda, T.G.: MATLAB tensor classes for fast algorithm prototyping. ACM Trans. Math. Softw. (2004)
2. Cichocki, A., Zdunek, R., Amari, S.: Nonnegative matrix and tensor factorization. IEEE Sig. Process. Mag. **25**(1), 142–145 (2008)
3. Cyganek, B., Krawczyk, B., Woźniak, M.: Multidimensional data classification with chordal distance based kernel and support vector machines. J. Eng. Appl. Artif. Intell. **46**(Part A), 10–22 (2015)

© Springer Nature Switzerland AG 2019
K. Choroś et al. (Eds.): MISSI 2018, AISC 833, pp. 5–6, 2019.
https://doi.org/10.1007/978-3-319-98678-4_2

4. Cyganek, B.: An analysis of the road signs classification based on the higher-order singular value decomposition of the deformable pattern tensors. In: Advanced Concepts for Intelligent Vision Systems, ACIVS 2010. LNCS, vol. 6475, pp. 191–202. Springer (2010)
5. Cyganek, B.: Object Detection and Recognition in Digital Images: Theory and Practice. Wiley, Hoboken (2013)
6. Cyganek, B.: Object recognition with the higher-order singular value decomposition of the multi-dimensional prototype tensors. In: 3rd Computer Science On line Conference (CSOC 2014). Advances in Intelligent Systems and Computing, pp. 395–405. Springer (2014)
7. Cyganek, B., Woźniak, M.: On robust computation of tensor classifiers based on the higher-order singular value decomposition. In: The 5th Computer Science On-line Conference on Software Engineering Perspectives and Application in Intelligent Systems 2016 (CSOC2016), Advances in Intelligent Systems and Computing, vol. 465, pp. 193–201. Springer (2016)
8. Jain, A., Tompson, J., LeCun, Y., Bregler, C.: Modeep: a deep learning framework using motion features for human pose estimation. In: Proceedings of the Asian Conference on Computer Vision (ACCV), pp. 302–315 (2014)
9. Kolda, T.G., Bader, B.W.: Tensor decompositions and applications. SIAM Rev. 51, 455–500 (2008)
10. Koziarski, M., Cyganek, B.: Image recognition with deep neural networks in presence of noise – dealing with and taking advantage of distortions. Integr. Comput. Aided Eng. 24(4), 1–13 (2017)
11. Krizhevsky, A., Sutskever, I., Hinton, G.E.: Imagenet classification with deep convolutional neural networks. In: Advances in Neural Information Processing Systems, pp. 1097–1105 (2012)
12. Lathauwer, de L.: Signal Processing Based on Multilinear Algebra. Ph.D. dissertation, Katholieke Universiteit Leuven (1997)
13. de Lathauwer, L., de Moor, B., Vandewalle, J.: A multilinear singular value decomposition. SIAM J. Matrix Anal. Appl. 21(4), 1253–1278 (2000)
14. de Lathauwer, L., de Moor, B., Vandewalle, J.: On the best Rank-1 and Rank-(R1, R2, ..., RN) approximation of higher-order tensors. SIAM J. Matrix Anal. Appl. 21(4), 1324–1342 (2000)
15. LeCun, Y.A., Bottou, L., Orr, G.B., Müller, K.-R.: Efficient backprop. In: Neural Networks: Tricks of the Trade, 2nd edn., pp. 9–48 (2012)
16. Muti, D., Bourennane, S.: Survey on tensor signal algebraic filtering. Sig. Process. 87, 237–249 (2007)
17. Novikov, A., Podoprikhin, D., Osokin, A., Vetrov, D.P.: Tensorizing neural networks. In: Proceedings of the Advances in Neural Information Processing Systems (NIPS), pp. 442–450 (2015)
18. Savas, B., Eldén, L.: Handwritten digit classification using higher order singular value decomposition. Pattern Recogn. 40, 993–1003 (2007)
19. Simonyan, K., Zisserman, A.: Very deep convolutional networks for large-scale image recognition. In: Proceedings of the International Conference on Learning Representations (ICLR) (2015)
20. Sun, J., Tao, D., Faloutsos, C.: Incremental tensor analysis: theory and applications. ACM Trans. Knowl. Discovery Data 2(3), 11 (2008)
21. Tucker, L.R.: Some mathematical notes on three-mode factor analysis. Psychometrika 31, 279–311 (1966)
22. Vasilescu, M.A., Terzopoulos, D.: Multilinear analysis of image ensembles: TensorFaces. In: Proceedings of Eurpoean Conference on Computer Vision, pp. 447–460 (2002)

New Applications of Sound and Vision Engineering

Andrzej Czyżewski[✉]

Multimedia Systems Department, Faculty of Electronics,
Telecommunications and Informatics, Gdansk University of Technology,
ul. Narutowicza 11/12, 80-233 Gdańsk, Poland
ac@pg.edu.pl

Multimedia, Sound & Vision Engineering are relatively new fields within the area of science and technology, but teaching and research in this area has been carried out at Gdansk University of Technology (Gdansk, Poland) for nearly 5 decades. The scope of scientific interests of the department covers areas such as: multimedia technology, digital signal and image processing, particularly based on methods pertaining to the field of artificial intelligence and telecommunications, speech acoustics, the psychophysiology of perception, advanced image processing with applications in biomedical engineering, biometrics, public safety, and also cultural heritage restoration.

In recent years, increasingly complex methods and algorithms for automated analysis and processing of signals, image and video data are being developed. The Multimedia Systems Department team contributed to the rapid growth of research results in above domains by participating in recent years in 5 international and more than 10 domestic projects which brought more than 500 papers published by the department's specialists. The topics selected to be covered in the keynote paper illustrate the scope of the currently carried-out research in the department, mainly in the domain of video and sound processing and their applications.

The aim of the project ALOFON (Audiovisual speech transcription method) is to conduct research aimed at developing a methodology of automatic speech phonetic transcription (in English), based on the use of a combination of information derived from the analysis of audio and video signals. In particular, basic research is conducted on the relationship between allophonic variation in speech, i.e. differences in the nature of the same sounds resulting from different arrangements of articulators depending on the phonetic environment (i.e. neighboring phones or prosodic features) and objective signal parameters. Phonetic transcription of speech employing this method can be carried out with greater accuracy than using only the acoustic modality.

The project HCIBRAIN (Human-computer communication methods for diagnosis and stimulation of patients with severe brain injuries) develops in accordance with the assumption regarding the implementation of basic research and experiments in the field of diagnosis and therapy of non-communicating patients. The main objective of the project is to develop the concept and solutions of an integrated multimodal system together with a diagnostic and therapeutic procedure validated in the course of tests, which will provide a more effective and wider approach to the diagnosis and rehabilitation of severely impaired patients, in particular those remaining in a coma.

© Springer Nature Switzerland AG 2019
K. Choroś et al. (Eds.): MISSI 2018, AISC 833, pp. 7–9, 2019.
https://doi.org/10.1007/978-3-319-98678-4_3

8   A. Czyżewski

On the basis of the project IDENT (Multimodal biometric system for bank client identity verification) the team has built a scientific synergy with the biggest Polish Bank (PKO BP), both in terms of technical cooperation, as well as in the domain of assessing the feasibility of implemented biometric solutions which provide the subject of joint research and development work lasting 3 years, already. It turned out that the biometric system developed within the project IDENT uses both: the original proposals based e.g. on multivariate analysis of dynamic electronic signature submitted with a special developed pen (installed in bank tellers), applications of image analysis, as well as more widely known in biometrics: speaker recognition, or the analysis of the topology of blood vessels in the palm (employing a commercially available sensor). In effect of the project IDENT, comprehensive and valuable information was obtained concerning the effectiveness of the biometric authentication methods developed as a result of the large test series carried-out in real bank environment conditions. Owing to the project, carried-out together with the largest Polish Bank PKO BP, a study test was carried-out in 60 bank outlets (on 100 teller stands) employing of a group of 10.000 bank customers.

The objective of the project INZNAK (Intelligent Road Signs with V2X Interface for Adaptive Traffic Controlling) is to develop a conceptual design and research tests of a new kind of intelligent road signs which will enable the prevention of the most common collisions on highways, resulting from the rapid stacking of vehicles resulting from accidental heavy braking. A range of products will be developed, including intelligent road signs: standing, hanging and mobile ones, displaying dynamically updated driving speed limit, determined automatically, by embedded electronic module, enabling multimodal measurement of traffic conditions (video, sound, and analysis of meteorological conditions). The intelligent road sign will communicate the speed calculated in relation to the information received from a row of similar signs placed along a stretch of highway that will communicate with each other via a wireless network remaining optionally adjustable remotely. Its development requires addressing a number of issues of research and technology, such as: effective, and independent of weather conditions, traffic estimation made on the basis of the simultaneous analysis of several types of data representation, the method of calculating the velocity gradient for various traffic situations considering the road topology, creating a platform for self-organizing and reliable wireless connection and performing scheduled on an adequate scale tests of prototypes. The planned implementation will lead to the development of products that increase road safety for which there exists worldwide market demand. The solution also fits into, in an original way, in the rapidly growing trend of development of communication of cars with the road infrastructure, enabling an access to digital road infrastructure for all drivers.

The project INUSER (Integrated Systems for Managing Wind Farms) assumes a development of set of solutions for the monitoring and for the diagnosing the condition of selected parts of power wind turbines. Critical structural elements of a typical wind turbine construction, which require special attention due to the of maintaining good technical condition are: the main gearbox mechanism with the generator, turbine blades together with the angle adjusting mechanism, turbine tower and its connection to foundations. A damage or a malfunction of any of the items listed above detected too late may cause a damage to the components or whole installation. Therefore, solutions for audio-video monitoring of the technical state of components of the wild turbine provides the vital task of the project. The spatial distribution of vibroacoustic energy and its propagation based on the measurement of parameters of sound intensity within given gridpoints, employing a special probe developed by the department's research & engineering team is the main challenge of this newest project contracted by the department specialists.

Multimedia Systems

A Concept of Visual Knowledge Representation

Tatiana Jaworska[(✉)]

Systems Research Institute, Polish Academy of Sciences,
01-447, 6 Newelska Street, Warsaw, Poland
Tatiana.Jaworska@ibspan.waw.pl

Abstract. The image semantic representation is a very challenging task. This article presents a concept of using visual analysis to represent knowledge based on large amounts of massive, dynamic, ambiguous multimedia. This concept is based on the semantic representation of these visual resources. We argue that the most important factor in building a semantic representation is defining the ordered and hierarchical structure, as well as the relationships among entities. This concept has stemmed from the content-based image retrieval analysis.

Keywords: Image semantics · Knowledge representation · Order · Similarity

1 Introduction

For many years researchers have been intensively striving to describe image semantics. It is an element of a widely understood knowledge representation for further knowledge retrieval.

So far, all knowledge has been represented in language form, in the beginning artificial, and now more or less natural which, at the same time, is the biggest obstacle in the proliferation of the knowledge repository. The best example is Wikipedia, without detracting from its merit, where articles differ depending on the national versions.

Figure 1 represents the location of visual knowledge representation in the whole pyramid of decision making support. So far, we have developed the data and information retrieval systems. It means that the decision maker has received raw, or slightly processed, mainly aggregate data. Recently, content-based image retrieval systems have caused a great breakthrough in information analysis, becoming the front-end element in the domain of knowledge retrieval systems [1].

With a deluge of images and photos, and the development of graphical interfaces in computers, mobiles, etc., the new generation is more and more dependent on visual information rather than textual. Producers and programmers steadily multiply icons, emoticons and other graphical symbols of all kinds. It concerns not only human-machine interaction systems but, first of all, pattern recognition and machine learning, as well as artificial intelligence. All this suggests that we should construct a visual knowledge representation system, rather than textual ones, e.g. domain ontologies. Our objective is the creation of a visual knowledge representation as the first step to a visual knowledge retrieval system because effective retrieval is possible only when a proper representation has been prepared.

© Springer Nature Switzerland AG 2019
K. Choroś et al. (Eds.): MISSI 2018, AISC 833, pp. 13–22, 2019.
https://doi.org/10.1007/978-3-319-98678-4_4

14   T. Jaworska

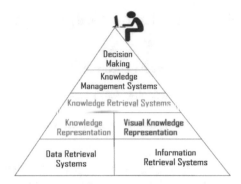

Fig. 1. The decision making support pyramid.

We are aware of the fact that we cannot totally avoid description in knowledge representation, but in this paper we present a concept of knowledge representation, focused on images as much as possible. We will demonstrate that images and, broadly understood, multimedia have such a large information potential that we can reduce the use of a natural language to nearly zero and, thanks to this, make our system much more universal.

1.1 Categorical Perception

We claim that the most important observation is the way in which the human brain perceives images and what conclusions it draws from visual information. Let us think for a moment about the task of identifying an animal – a potentially dangerous one. First of all, we recognize the general posture. If we know this kind of animal because of our earlier experience, we can recognize it based on the fragment of its silhouette, for example a head. Then we decide whether to escape because of the threat the animal poses, or to catch it because it might promise a tasty meal. This knowledge is deduced without the name of the animal [2].

In this example two aspects are important: (i) assignation to category: danger/safe and edible/inedible; (ii) the kind of categories which are orthogonal to each other.

Thus categorization is fundamental in prediction, inference, classification, and further, in knowledge representation and decision making. Categorical perception is the phenomenon of perception of distinct categories when there is a gradual change in a variable along a continuum. Our perception is based on the different aspects of reality, whereas the understanding of images is inextricably connected with the human experience and the reality we live in. That is why we started preparing the framework of our system from defining the most universal and mutually separable areas of abstraction in order to cover the widest possible range of semantics. Moreover, to organize the knowledge in each area of abstraction, we had to define the way of ordering according to which we would arrange images to enable the user to navigate the system.

The mathematical framework for human perception problems, involving ambiguous image perception, was provided by Tim Poston in 1978 in [3]. These problem shall be ignored here, as they reach beyond the scope of our interests.

2   The Concept of Image Wiki

Visual knowledge representation and, later, visual knowledge retrieval systems offer quite new capabilities to the decision makers. A unified measure reflecting visual semantic similarity has not been developed yet, although we claim that semantic information can be described in context. In each context each image can be treated as an information granule, classified by a vector of categories. We can go even further: each object in an image can also be an information granule [4]. For each seed of information which constitutes a visual object, we define a coverage and specificity between which the trade-off must remain. Information granules of type-2 give most abstractive notions, e.g. a set of many people and buildings defines a city. Information granules are characterised by:

- The semantic similarity, which means that feature vectors for them should be more or less similar;
- The functional similarity, understood as similarity of classes, assigned to similar visual objects;
- The semantic unambiguity - the communication impact is defined by semantic classes;
- Representation of objects from the real world.

Information granularity connects strictly with the notion of scale.

Thinking globally about all the sets of existing images, we can tentatively select the following most common areas of abstraction which we understand as orthogonal dimensions by which we can characterize each image:

- Scale[1] – an image can present information about objects of different size: from galactic clusters to atoms. The scale should change linearly or at intervals, depending on the structure of data.
- Time – changes linearly in our system, according to our common intuition. However, in the system there is information about the photo acquisition date, but this same photo appears in a domain chronology, which means that if a photo presents a geological mesozoic formation, it is presented in a geological chronology.
- Hierarchy – our algorithm organizes images in a general-to-specific relationship of content (depth of abstraction) which is deliberately less strict in comparison with formal linguistic relations.
- Content domain – taxonomy – covers different areas of abstraction. It is the most voluminous because it has to address as many domains as humans are interested in. It shall connect to existing ontologies [5].
- Geographical location – connects the location where a photo has been taken with GIS or Google maps.
- Image author – information about the image author, can link to the author's website or images of the author's masterpieces.

---

[1] There are two notions called *scale*. A s*cale* here means the size of the object represented by an image. In Subsect. 2.2 we use the notion *scale,* in fact, as a **scale of measure** which is different from the one described here.

- Action – presents some actions and movies in videos.
- Information granularity – images in their nature contain information in the form of visual information granules. Semantic objects are the best candidates for such granules, with the bounding box as coverage and the centroid as location. Objects have an assigned vector of classes or prototypes;
- Exemplification – at the level of abstraction where detailed exemplification is possible. At the lowest level images will be organized according to some similarity measures to make browsing among them easier.

As we can see from the above list, each of these dimensions represents an important aspect of each image and has to be organized by means of a different, immanent order (as mentioned in the introduction). All the above-listed parameters will be elaborated on below, beginning from the basic notions

The assumptions in our project are very wide, however, the framework of the system is still under construction. In order to create the system, obviously gradually, we use the existing CBIR systems [6, 7] in different domains to obtain a sufficient number of images and order them based on their similarity in different metrics. The help of experts is needed in the proper organization of domain knowledge. The latest achievements, namely CNNs [8, 9], allow a quick selection of objects in an image in order to add to each selected object some more detailed images at the lower levels (see Fig. 4). The object images which will be attached on a lower level to the more general image shall be accepted by experts if the images are relevant in terms of the taxonomy of the particular branch of domain knowledge. At each level of browsing, the user will be able to jump to another aspect. For example, the user finds a satellite image of a fragment of their town and wants to see the changes in urban development in the last 30 years. Then they switch to the time search and can see the photos and archive plans of this selected area.

Obviously, we assume the availability of these data. The system shall offer free access to any user willing to edit the content similarly to the rules in the textual Wikipedia, thanks to which the amount of images will increase very fast and will be revised by experts to maintain semantic correction.

Each new added image contains all the information required to be localized in a proper place in the whole structure. At each level, many exemplifications will be accessible and each object in them will be connected to the lower level images in a domain hierarchy.

The system also has to contain drawings, schemes, slides and other graphical materials in order to help the user to understand the presented knowledge. In many cases we understand an image not as a photo, but as a specially prepared photographic or animated illustration to visualize how a scheme or process works. We would like to emphasize that the system in its assumption is much wider than GIS, CBIR systems and ontologies, but connects them together and, because of this fact, it partially uses the mechanism and algorithms implemented in these systems.

2.1 Preliminaries

In order to unify our concept, we have applied the notion of a total order and a partial order, according to which we build relations over a set of visual entities X. By an entity of the system we understand any visual information granules, such as: images, videos, 3D graphics, depth maps, etc.

A (non-strict) partial order $P = (X, \leq)$ is a binary relation $\leq$ over a set X, satisfying particular axioms. The axioms for a non-strict partial order state that the relation $\leq$ is reflexive, antisymmetric, and transitive [10]. That is, for all a, b, and c in X, it must satisfy:

- $a \leq a$ (reflexivity: every element is related to itself).
- if $a \leq b$ and $b \leq a$, then $a = b$ (antisymmetry: two distinct elements cannot be related in both directions).
- if $a \leq b$ and $b \leq c$, then $a \leq c$ (transitivity: if a first element is related to a second element, and, in turn, that element is related to a third element, then the first element is related to the third element).

In other words, a partial order is an antisymmetric pre-order. A set with a partial order is called a partially ordered set (also called a *poset*). A well-known example of a partial order is a linear order, particularly one based on real numbers.

Graphically, a poset can be represented in the form of a Hasse diagram (see Fig. 2) [11] where each element of X is a vertex and the upward line symbolizes the binary relation which holds between comparable elements that are immediate neighbours. Such a diagram, with labelled vertices, uniquely determines its partial order.

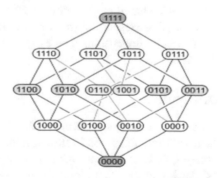

Fig. 2. An example of a Hasse diagram for a 4-element set ordered by inclusion $\subseteq$. (Follow Wikipedia: https://en.wikipedia.org/wiki/Hasse_diagram).

In the most common sense of the term, a graph is an ordered pair $G = (X, E)$ comprising a set X of vertices, nodes or points together with a set E of edges, arcs or lines, which are 2-element subsets of X (i.e. an edge is associated with two vertices, and that association takes the form of an unordered pair comprising those two vertices) [12].

We have introduced all these notions in order to be able to organize images semantically. Some aspects, such as time or scale, are linear, so the linear order is used naturally, but others, such as hierarchy or taxonomy, can be defined only in a different way [13].

Then, we have to use a different scale of measure which describes the multidimensional nature of information. Here, we follow Stevens [14], being aware that there are some other approaches to the problem of scale and scaling. In our case the division of scales into four types: nominal, ordinal, intervallic, and linear is appropriate for the dimensions we have proposed.

The nominal scale is apparently the simplest one, but it is formally described by the category theory, from the mathematical point of view, and it is basic to all kinds of classification, which is one of the most important parameters when we discuss semantic image analysis, especially when the value of the quality of the information is difficult to assign. As Rozeboom [15] claims, a semantic scale is equivalent to a formal scale. In such a situation, classes or categories and properties are coded by the values of a natural variable, which are equivalent to a formal scale. Strictly speaking, a scale-name A and set $\mathbf{a}$ of symbols compose a semantic scale $\langle A, \mathbf{a} \rangle$ where $\mathbf{a} \in \{x_1, \ldots, x_n\}$ for natural variables $\Delta \in \{d_1, \ldots, d_n\}$ in a given coding system. Then, we can always find a formal scale ϕ of $\Delta \langle \phi, \Delta \rangle$, such that A and ϕ have the same argument domain $\mathbf{D} = 1, \ldots, n$ and there exists a function f from $\mathbf{a}$ into Δ, such that: $f \colon \mathbf{a} \to \Delta$, $d = f(x)$. Then, f is called a scaling transformation of $\mathbf{a}$ into Δ and it must be one-one.

This scheme has to be extended in the image case because image information is much more complex. Semantic variables are represented by a vector of attributes for each of the above-mentioned dimensions.

The linear and ordinal scales are based on an order described above.

The interval scale will be used to present information for which it is impossible to assign an exact value of a particular dimension, such as geological eras, for instance, for images of geological layers.

2.2   How to Build Such a System?

Data structure should be similar to a Geographic Information System (GIS) data structure for raster or grid data, with the difference in the attribute representation. As we mentioned in the preliminaries, a Hasse diagram for a poset $P = (X, \leq)$ represents an acyclic graph $G = (X, \prec)$, such that maximal elements are situated at the top of the diagram, and for two vertices $a \prec b$, where a vertex a is over b. As we know, an acyclic graph represents a tree as a data structure. Hence, as a basic data structure, we use a tree and because of the fact that images have rectangular structure and objects selected from them are polygons we apply the R-tree structure. There are no pure R-tree structures because there is an option of changing aspects which forces the correlations between R-tree structures [16]. The layer structure, being a characteristic feature of a GIS, in our case, is equivalent to levels in trees. We use Oracle DBMS for a set of all images, data, attributes, etc.

It is also important that the user has access to an image in most possible semantic correlations. Hence, once more we relate to the image of a geological mesozoic formation. Not only will this image appear in a domain geologic context, but the user will also see it analyzing the work(s) of the geologist who took this photo, and/or they will referred to the mineralogical domain.

3   The System Navigation and the User Interface

Below, in Fig. 3, we present the main screen of our system where icons symbolizing the main search parameters/aspects (from the top left time down to the bottom right information granularity and sequence of the example of the same kind of entity). Each of them opens further windows, enabling the user to surf down a particular dimension to find the image he/she wants to learn something. It is not an attempt to present similar images, it is a system which organizes images in a semantic order which depends on the aspect the user is browsing at a particular moment.

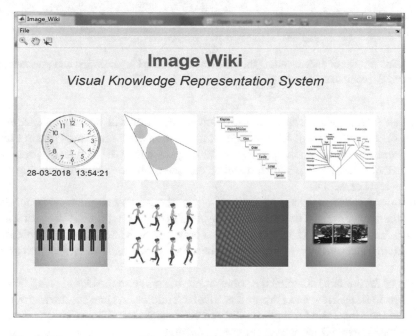

Fig. 3.   The main screen of the Image Wiki system. Each square is an active icon.

For example, the user is in an aspect *hierarchy* and then they go through the anatomy of an animal (see Fig. 4). When the user wants to look at different photos of a fruit fly (drosophila), they can switch into an enumeration, then look at other photographs of the same species (a fruit fly in this particular example). Browsing examples of objects at the same level is available after selecting ![icon] and by clicking on the left/right grey arrow situated on both sides of the screen (see Fig. 4). But when the user wants to continue analysing the fruit fly anatomy details she\he clicks on the eyes, legs, or wings and moves lower to images from a microscope, presenting in detail particular organs.

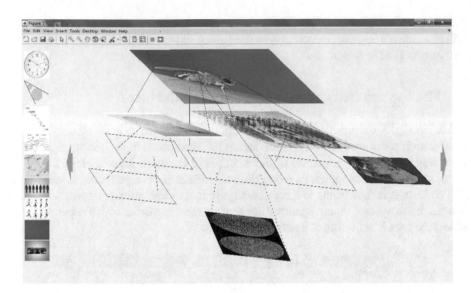

Fig. 4. The manner of understanding the mechanism of visual organization and presentation of taxonomy. All component images (CC).

A fast search for interesting information by the user will be available through a system of icons which will facilitate quick navigation across the most general levels. The system will start from the previous browsing stage to save the time of going down to a particular level and it will remember the search history to aid returning to the previous level of search.

In order to change a domain, the user clicks the icon symbolizing taxonomy and selects the proper domain of interest. The system is based on different kinds of images, which means not only photographs, but also drawings, schemes, sketches, videos, etc. The domains can overlap. At each level, there is a list of tools to switch into another aspect.

One of the domains contains the zoom-in maps, similar to the Google map. The user can move to street view and later to a particular building. At present, there are photographs of buildings, and through satellite stereo images their 3D models are calculated more often. When 3D models of different building interiors are available, we will be able to incorporate them into our system in order to enable the user to visit these buildings inside.

Navigation at a particular level will be intuitive or rather similar to the navigation of present graphical programs, where the scroll wheel zooms in and out the image, a click on an object with the left button presents this object in more detail on a more precise level, a click with the right button anywhere in the image moves the user to the upper level.

All visual objects in an image are active. For example, Fig. 5 presents a quick selection of the scale in which the user wants to operate. It means that a click on a left nebula moves the user to astronomical knowledge, a click on the Earth moves them to the Earth map and in the next step at once to the level of the Google map, a click on the atom moves to chemistry and the atomic scale, etc.

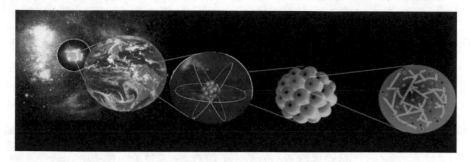

Fig. 5. An example of space distances of image content, enabling the user to select the scale of interest, from light years for astronomical images, through kilometers and meters related to the Earth, down to atom diameters and even further, to the subatomic scale.

4   Conclusions and Further Works

In this paper, we present the framework which is being created in order to elaborate the visual knowledge representation for further visual knowledge retrieval. Taking into consideration the state-of-the-art semantic analysis methods, we can safely assert that the construction of the above-described system is fully technically feasible, though time consuming. However, the fact that the semantic analysis of image by machines is still far from human abilities remains a considerable challenge. Additionally, effective management of such huge sets of visual information will require intensive research in terms of both hardware resources and new algorithms for sharing and managing visual information.

References

1. Belongie, S., Perona, P.: Visipedia circa 2015. Pattern Recogn. Lett. **72**, 15–24 (2016). https://doi.org/10.1016/j.patrec.2015.11.023
2. Hawking, J., Blakeslee, J.: On Intelligence: How a New Understanding of the Brain Will Lead to the Creation of Truly Intelligent Machines, p. 262. Henry Holt and Company, New York (2004)
3. Poston, T., Stewart, I.: Nonlinear modeling of multistable perception. Syst. Res. Behav. Sci. **4**(23), 318–334 (1978). https://doi.org/10.1002/bs.3830230403
4. Yu, F., Pedrycz, W.: The design of fuzzy information granules: tradeoffs between specificity and experimental evidence. Appl. Soft Comput. **9**, 264–273 (2009). https://doi.org/10.1016/j.asoc.2007.10.026
5. Chmiel, P., Ganzha, M., Jaworska, T., Paprzycki, M.: Combining semantic technologies with a content-based image retrieval system – preliminary considerations. In: Application of Mathematics in Technical and Natural Sciences, Conference Proceedings, Albena, Bulgaria, pp. 100001–1000012 (2017). https://doi.org/10.1063/1.5007405

6. Jaworska, T.: The concept of a multi-step search-engine for the content-based image retrieval systems. In: Information Systems Architecture and Technology. Web Information Systems Engineering, Knowledge Discovery and Hybrid Computing, Wrocław, pp. 189–200 (2011)
7. Jaworska, T.: A search-engine concept based on multi-feature vectors and spatial relationship. In: Christiansen, H., De Tré, G., Yazici, A., Zadrożny, S., Larsen, H.L. (eds.) Flexible Query Answering Systems, vol. 7022, pp. 137–148. Springer, Heidelberg (2011). https://doi.org/10.1007/978-3-642-24764-4_13
8. Krizhevsky, A., Sutskever, I., Hinton, G.E.: ImageNet classification with deep convolutional neural networks. In: Proceedings of the 25th International Conference on Neural Information Processing Systems – NIPS 2012, Lake Tahoe, Nevada, USA, 03–06 December 2012, pp. 1097–1105 (2012)
9. Zhang, N., Donahue, J., Girshick, R., Darrell, T.: Part-based R-CNNs for fine-grained category detection. In: 13th European Conference on Computer Vision - ECCV. Proceedings, Part I, Zurich, Switzerland, 6–12 September 2014, pp. 834–849 (2014). https://doi.org/10.1007/978-3-319-10590-1_54
10. Simovici, D.A., Djeraba, C.: Partially ordered sets. In: Mathematical Tools for Data Mining: Set Theory, Partial Orders, Combinatorics, p. 615. Springer, London (2008). https://doi.org/10.1007/978-1-4471-6407-4
11. Freese, R.: Automated lattice drawing. In: Second International Conference on Formal Concept Analysis, Sydney, Australia, 23–26 February 2004, pp. 580–596 (2004). https://doi.org/10.1007/978-3-540-24651-0_12
12. Bang-Jensen, J., Gutin, G.Z.: Basic terminology, notation and results. In: Digraphs Theory, Algorithms and Applications, pp. 1–30. Springer, London (2009). https://doi.org/10.1007/978-1-84800-998-1
13. Bart, E., Porteous, I., Perona, P., Welling, M.: Unsupervised learning of visual taxonomies. In: IEEE Conference on Computer Vision and Pattern Recognition, Anchorage, USA, 23–28 June 2008, pp. 1–8, (2008)
14. Stevens, S.S.: On the theory of scales of measurement. Science **103**(2684), 677–680 (1946). https://doi.org/10.1126/science.103.2684.677
15. Rozeboom, W.W.: Scaling theory and the nature of measurement. Synthese **16**(2), 170–233 (1966). https://doi.org/10.1007/bf00485356
16. Manolopoulos, Y., Nanopoulos, A., Papadopoulos, A.N., Theodoridis, Y.: R-Trees: Theory and Applications, 3rd edn., p. 194. Springer, London (2010). https://doi.org/10.1007/978-1-84628-293-5

Three-Dimensional Multi-resolution and Hardware Accelerated Compression for the Efficient Interactive Web Applications

Daniel Dworak[1]([⊠]) and Maria Pietruszka[2]

[1] Institute of Information Technology, Lodz University of Technology, Łódź, Poland
`daniel.dworak@edu.p.lodz.pl`
[2] Faculty of Economics and Sociology, University of Lodz, Łódź, Poland
`maria.pietruszka@uni.lodz.pl`

Abstract. This paper focuses on methods of preparation, compression, coding and decoding the 3D triangle meshes for the Web purposes. There has been proposed a multi-resolution analysis method for reduction of redundant 3D data with split, predict and update schemes. We introduced the criteria of vertex deletion and quality measurements for optimum surface curvature and proper triangulation without model's artifacts. Due to a long processing time of large 3D data sets, we decided to elaborate parallel compression algorithm for the GPU in CUDA technology. Described method of compression is a first step of complete workflow that we proposed to achieve smaller files for optimized transferring through the Web.

Keywords: 3D graphics for the Web · 3D geometry compression
Multi-resolution analysis · CUDA · Parallel computing · GPGPU

1 Introduction

The modern 3D Web applications consists of large amount of data that should be stored, sent, received, decoded and displayed in the real time. The most popular way of storing 3D data in computer graphics is based on mesh of triangles, which consist of vertices, normals and texture coordinates.

Chun [1] noticed that four aspects should be concerned while preparing 3D container for the interactive Web applications: *pipeline* (how to create and prepare 3D models), *serving* (how to send data efficiently and fast), *loading* (how to reduce time of decoding data) and *rendering* (which structures are better for hardware and how to optimize data for displaying).

Preparing 3D data for efficient transmission through the Web can be solved in many ways. First of all, the 3D objects should be modeled according to basic requirements of computer graphics (thoroughly described in [5]) like: maximum number of 65536 vertices per model's segment, normalized texture coordinates

© Springer Nature Switzerland AG 2019
K. Choroś et al. (Eds.): MISSI 2018, AISC 833, pp. 23–34, 2019.
https://doi.org/10.1007/978-3-319-98678-4_5

and normal vectors, squared textures with dimensions of power of 2 etc. Secondly, there are variety of Web browsers used via different Internet bandwidth, operating systems and GPU's, therefore 3D models should be interchangeable and possibly simple. Last, but not least aspect concerns the interactivity of Web applications [1].

There are few propositions of transmission formats for the 3D Web, but only glTF (gl Transmission Format) is a standardized one. We proposed efficient **3dPNG** format (Three-dimensional Portable Network Geometry) [4] capable of storing compressed 3D data encoded in the channels of two-dimensional PNG format. There has been proposed to use a GPU rendering pipeline in WebGL technology by Vertex and Fragment Shaders [6] for decoding files and rendering them. As the result, the 3dPNG format is able to reduce files' size and decoding time as well.

The compression of 3D models is the crucial stage for preparing data for transmission via the Web. This paper focuses on the three-dimensional data compression that performs hardware accelerated calculations on GPU. At early phase, we elaborated CPU multi-resolution analysis algorithm with prediction scheme. However, due to complexity of calculations it had been taking long time to determine final result. It has been decided to implement parallel algorithm for CUDA architecture what allowed to reduce compression time greatly.

2   Related Work

The **geometry compression**, as Deering [2] defined first time, is a general time-space compromise, that offers as much advantages between memory space and transmission time as possible. The solution for increasing the interactivity of Web applications, Maglo et al. [10] called **three-dimensional compression**, that reduces object complexity and applies 3D optimization techniques.

The biggest development of 3D compression methods took place in late 90s. The first algorithms of processing 3D data based on **lossy compression**, that causes an irretrievable loss of data parts. Deering [2] has adopted algorithms of quantization and Huffman coding, that primarily were designed for 2D data retrieval. Rossignac and Taubin [17] have developed first algorithm for conversion 3D mesh to the 2D representation of graph structure. They have defined specialized language for saving such data with usage of basic symbols $\{C, L, E, R, S\}$. Then, first attempt for progressive transmission and retrieval of meshes has been proposed by Hoppe [7]. It allowed to render 3D objects on demand and at any level of detail without sending whole meshes.

In computer graphics, for three-dimensional compression are often applied **clustering** methods [9]. There are hierarchical methods like *k-means* or *k-medoids* [14] and also modifications or improvements of such methods. *PAM (Partitioning Around Medoids)* clustering unlike k-medoids trends to intragroup and intergroup variability minimalization. There are also efficient methods that allows to analyse a density of large input data (like vertices) called *CATS (Clustering After Transformation and Smoothing)* [13] and to calculate the similarities between polygons like *CLARA(NS) (Clustering Large Applications based* *on RANdomized Search)* [12]. The main disadvantage of such methods is noticeable deformation of meshes due to influence of so called peculiar vertices or faces that are treated like a noise. The long time of clustering is also a weak point of those methods.

Multi-resolution mesh analysis allows to produce detailed and raw object representation at different levels of detail (LoD). Sweldens [8, 16] used **second generation wavelet** transform for optimal multi-resolution deconstruction of irregular meshes of triangles. There has been used lifting scheme with distribution, prediction and update schemes. It also allows to analyse meshes with additional attributes stored in vertices like normal vectors, texture coordinates or colors. Szczęsna [15] has also used wavelet based transform, enriched with single quality measurement for better triangulation.

Most of three-dimensional compression methods need a lot of time to perform many complicated calculations. Authors of these algorithms do not even mention about processing time, because they compare only the quality of output meshes. It is really important to create efficient environment, when processing a lot of input data. Parallel computing with usage of GPU (Graphics Processing Unit) is a modern trend in huge data sets processing, which is called GPGPU (General Purpose computation on Graphics Processing Units) [18]. At the beginning, it was possible to perform general purpose calculations in programmable rendering pipeline via Vertex and Fragment Shaders. CUDA technology, developed by the Nvidia corporation, allows to perform general purpose parallel computations through API CUDA and kernels. However, due to technological differences between CPU and GPU (i.e. number of cores, access to memory) it is hard (sometimes even impossible) to redesign or develop algorithm that could have been working on traditional processor CPU (i.e. because of data type or data processing type).

3   Three-Dimensional Compression Using Multi-resolution Analysis and Transform

General Scheme of multi-resolution analysis and transform for 3D compression mesh looks as follows:

1. **Split** vertices of mesh into two subsets with odd c_{2k+1}^{j+1} and even c_{2k}^{j+1} indices (where j is a resolution level of a mesh, k is a vertex index).
2. **Predict** the change of local curvature when deleting odd vertices that are farthest away from the regression surfaces.
3. **Update** the mesh by half edge collapse retriangulation based on even vertices applying the measurements of triangulation quality.

Split Scheme. Odd elements determine factors d while even elements are output values of representation at lower resolution level. The main aim of the split scheme is extraction of two subsets with odd c_{2k+1}^{j+1} and even c_{2k}^{j+1} indices. Odd vertices, after fulfilling appropriate conditions, are removed from input mesh. The selection of odd elements needs to perform analysis of every vertex of input mesh, by determining its neighbourhood and detail value d (the Euclidean distance between vertex and its regression surface). We decided to use a local membership criteria of every vertex to a surface determined by vertices from local neighbourhood. In every step, every vertex is analysed and there is created a regression surface that occurs as close to every neighbouring vertex as possible (Fig. 1A). This surface is determined by minimization of the sum of orthogonal squares of the distances between neighbouring vertices to that plane [11]. The algorithm for determining the regression surface looks as follows:

1. Find eigenvector that corresponds with the lowest eigenvalue of adjacency matrix $F^T F$ and defines normal vector of the sought surface.
2. Find eigenvectors that correspond with the highest eigenvalue of adjacency matrix $F^T F$ using power method (we have proposed to use vector iteration).
3. Determine coefficients of the regression surface that correspond with the eigenvectors of the adjacency matrix $F^T F$.

Predict Scheme. Then, there is determined the d distance between vertex and its regression surface for every vertex. The list of such as distances is sorted in ascending order. The lowest value means the lowest influence to the curvature of the local neighbourhood. Those values that are close to zero value lies almost on the regression surface and can be treated like a redundant data, therefore they can be removed. The arbitrary t vertices with lowest values d are deleted from the mesh. However, vertex cannot be the first-level neighbour of one of the deleted ones. It means the considered vertex cannot be deleted when at least one of the previously deleted vertices has been its direct neighbour. Moreover, there is built a bounding sphere with its middle point (x_s, x_y, x_z) at considered vertex and radius (Eq. 1).

$$r_f = max(\{d_f^0, d_f^1, ..., d_f^n\}) \tag{1}$$

This radius equals to the distance to the farthest first-level neighbouring vertex (Fig. 1B). Vertex (v_i^x, v_i^y, v_i^z) of the neighbouring level greater than 1 that fulfills equation (Eq. 2) cannot be deleted in the current iteration.

$$(v_i^x - x_s)^2 + (v_i^y - y_s)^2 + (v_i^z - z_s)^2 \leq r_f^2 \tag{2}$$

We introduced such as requirement to prevent vertices deleting from the same flat surface (i.e. table). The algorithm deletes vertices as long as t vertices (or t percent of whole input data) are deleted from the mesh.

Update Scheme. After removing t vertices, there appear many holes due to modification of triangulation. There are many ways to retriangulate area with the holes (i.e. Delaunay triangulation [3]). It has been decided to use *half edge collapse* that removes all edges and collapses edges into one end chosen point (Fig. 2). Methods described here allows to separate global optimization from local optimization. The newly created triangulation is determined by minimization of chosen criteria.

Before collapsing edges, for every possible triangulation, there are applied two criteria of triangulation to provide optimal and proper triangulation.

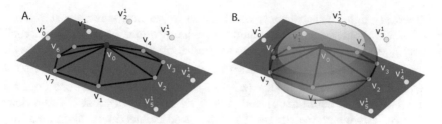

Fig. 1. Predict scheme. A: Regression surface of considered vertex v_0 (red) based on neighbouring vertices (green) v_1, v_2, v_3 etc. B: Vertices inside the bounding sphere (yellow dots: v_1^1 and v_2^1) cannot be deleted in present iteration.

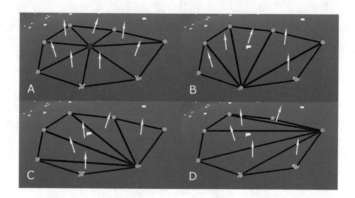

Fig. 2. The criteria of triangulation quality: A. quality of red vertex neighbouring in original mesh is examined; B, C and D quality examination of every possible triangulation after half edge collapse into blue vertex.

The criteria of *equilateral triangle* (Eq. 4) prevents from creating skinny triangles, which are not preferred for 3D rendering. If the value of $Q_s^q = 1$ for triangulation variant s, then it is the optimum, which means the triangles are equilateral and value $Q_s^q \approx 0$ means skinny and long triangles of such as triangulation. The triangulation with the highest Q_t value (Eq. 3) is the best variant from all possible triangulations.

$$Q_t = \max_{v_s \in N(v_i)} Q_q^s \qquad (3)$$

$$Q_s^q = \frac{4\sqrt{3}}{(m-2)} \sum_{j=1}^{m-2} \frac{A_j}{a_j^2 + b_j^2 + c_j^2} \qquad (4)$$

where $N(v_i)$ vertices from local neighbourhood of the vertex v_i, s - variant of triangulation, m - number of vertices in $N(v_i)$, A - area of the triangle, a, b, c - lengths of triangle's sides.

The second criteria was introduced to create triangles, which produce the *least modification of local surface curvature* (Eq. 5). There has been introduced the dihedral angle measurement, which is an angle between normal vectors of triangles with joint edges. It is equal to scalar product of two normal vectors $(n_j^- \cdot n_j^+)$, that are perpendicular to triangles on the both sides of j edge. If $(1 - (n_j^- \cdot n_j^+)) = 0$ it means the both triangles are placed on the same surface (i.e. countertop). However if $(1 - (n_j^- \cdot n_j^+)) = 1$ the angle between normals is $90°$ (i.e. the edge of countertop) and cannot be deleted, because it could deform the surface. The mean measurement is calculated for neighbourhood area before triangulation (Q_δ') and every possible variant after triangulation (Q_δ^s):

$$Q_\delta^s = \frac{1}{m-3} \sum_{j=1}^{m-3} (1 - (n_j^- \cdot n_j^+)) \tag{5}$$

The lowest difference between before and after triangulation's curvature has the lowest Q_c value (Eq. 6).

$$Q_c = \min_{v_s \in N(v_i)} |Q_\delta' - Q_\delta^s| \tag{6}$$

Finally, the most optimum triangulation is determined by two criteria measurements (Eq. 7). When Q is close to an value of one, then it is the most optimum triangulation and the vertex v_s can be chosen to half edge collapse, because such as triangulation is possibly similar to the pre-triangulation surface and the most proper one.

$$Q = \max_{v_s \in N(v_i)} |Q_q^s - (Q_\delta' - Q_\delta^s)| \tag{7}$$

3.1 Hardware Accelerated Multi-resolution Analysis and Compression

Preparing Data. It has been decided to use CUDAfy.NET API that allows to develop high performance GPGPU applications easily and directly from C# project. It was necessary to change the structures that we used before due to lack of lists or dynamic arrays. At the beginning, there is created a CUDAfy module with determined types of target device (Cuda), id (which GPU) or language (Cuda or OpenCL). Then, any data that should be processed is transferred to the memory of selected device by *gpu.CopyToDevice(data)* function. The launch of GPU context is performed by function *gpu.Launch(grid_size, block_size, kernel_name, fields, arg0, arg1)* where *grid_size* is a number of width and height elements, *block_size* is a block size that depends on device parameters, *kernel_name* is a name of kernel function, *fields* is a number of input data sets and *arg0, arg1* etc. are previously transferred data sets.

The kernel part with function called *Compress()* is the main part for performing calculations. The described compression method is a complicated and time-consuming algorithm, especially for huge 3D data sets. The greatest speed up of GPGPU calculations takes place for so called atomic (i.e. addition, subtraction, finding min or max value) and matrix operations. The described compression algorithm consists of many atomic and matrix computations, like

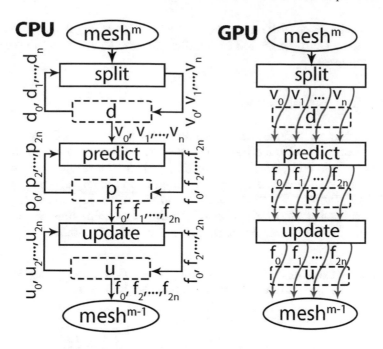

Fig. 3. The comparison of multi-resolution analysis and compression for CPU and GPU. Three main blocks: split, predict and update compute new data values of d, p and u based on vertices (v) and faces (f) of the mesh.

determining regression surface or quality of triangulation. The specific architecture of GPU with thousands of cores allows to perform many parallel operations.

It has been noticed that, there are many iterations that can be performed in parallel way, therefore there are distinguished three main blocks of code: split, predict and update (Fig. 3). These blocks are separated, so values can be calculated independently and locally in every block. It led us to create parallel algorithm in CUDA technology based on dynamic parallelism technique, which allows us to create and sync nested operations called child processing blocks. It makes possible to configure and create grids and then control the flow of a data in particular blocks.

First of all, the *split* block determines the detail values d, based on regression surfaces which are created for every vertex and its neighbouring. To determine vertices with the lowest values of d it's necessary to *wait* for computing all d values. There is used CUDA built-in function *__syncthreads()*, which works like a barrier (waiting point) to ensure that all threads in the same block have done their calculations. It is common to synchronize threads after blocks of code that perform write-read operations in GPU memory. Secondly, *predict* block determines distances between vertices and their regression surfaces. Then, simple parallel odd-even sort algorithm is performed to determine top t vertices with lowest values d. These vertices, after fulfilling the same conditions like CPU version does, are removed. The final independent *update* block creates new triangulation on areas with removed odd vertices. The triangulation and its quality calculations can be executed in parallel, because there are no dependencies between odd areas.

4   Quality Benchmarks

The comparison of output meshes quality was performed with two types of criteria - Hausdorff distance (Q_H, the maximum distance between vertices of input and output) and equilateral triangle (Q_E). As the result, we have got the two types of measurements for vertices, triangles and triangulation quality. The tables below (Tables 1 and 2) compare the quality of output models for author's compression (called DD), 3ds Max (Optimize algorithm) and Maya (Reduce algorithm). The best values are highlighted, the worst are italic.

Table 1. The comparison of *Stanford Bunny* (2500 vertices) mesh quality for author's compression, 3ds Max and Maya software at three levels of detail.

	DD compression		3ds Max		Maya	
	Q_E	Q_H	Q_E	Q_H	Q_M	Q_H
80%	**0.759**	**198.024**	*0.688*	198.909	0.699	*198.992*
50%	**0.734**	**198.212**	0.710	*199.479*	*0.681*	199.102
20%	**0.718**	**199.062**	0.713	199.875	*0.676*	*199.944*

Table 2. The comparison of *Chair* (6309 vertices) mesh quality for author's compression, 3ds Max and Maya software at three levels of detail.

	DD compression		3ds Max		Maya	
	Q_E	Q_H	Q_E	Q_H	Q_M	Q_H
80%	**0.646**	**145.495**	0.631	*148.487*	0.615	147.601
50%	**0.642**	**147.557**	*0.598*	*149.133*	0.601	148.021
20%	**0.644**	**148.141**	0.614	*149.568*	*0.612*	149.012

Comparing visual results (Fig. 4), in case of *Bunny* model there are no noticeable differences up to 50% reduction. However, at 20% reduction the 3ds Max produces the worst output mesh (compare bunny's ears). There are also some artifacts in ears of Maya output model. The visual comparison of *Chair* model (Fig. 5) shows interesting differences. The flat and slightly curved surfaces (seating, backrest and armrests) are getting creased at 80% level and getting even worse at 50%, while comparing 3ds Max and Maya models. Moreover, the parts of chair's legs of Maya mesh at 50% and 20% reduction level are lacking (it started at 80% level).

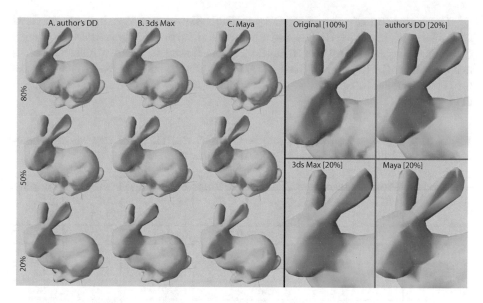

Fig. 4. The comparison of *Bunny* mesh quality after processing: A. author's DD compression, B. 3ds Max and C. Maya software with zoomed areas on the right.

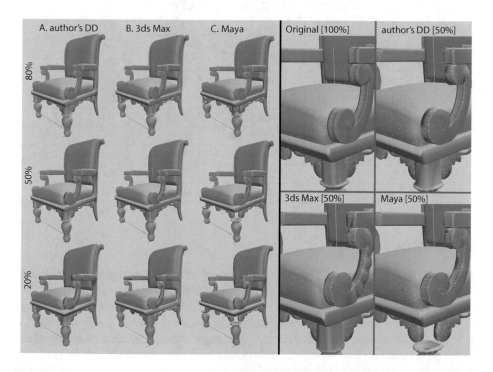

Fig. 5. The comparison of *Chair* mesh quality after processing: A. author's DD compression, B. 3ds Max and C. Maya software with zoomed areas on the right.

The reduction of compression time (Table 3) was possible because of GPU computing realized in CUDA technology. The type of input data and non-dependent stages of multi-resolution analysis allowed to perform parallel calculations efficiently (up to 87 times faster) with non-linear growth of time, unlike CPU. The output 3dPNG files are 2÷4 times smaller than transmission standard glTF.

Table 3. The comparison of time and ratio (CPU and GPU) of 80% author's compression for 3D models at different levels of complexity.

3D model	Input OBJ [KB]	Input vertices	CPU [s]	GPU [s]	Ratio[$\frac{CPU}{GPU}$]	Output 3dPNG [KB]	Output glTF [KB]
Teddybear	172	1598	47	0,7	62,84	4	6
Bunny	296	2500	64	0,9	64,07	9	16
Sphere1	1126	4092	144	2	73,39	23	52
Chair	1275	6309	202	3	72,37	32	90
Table1	1126	7872	191	2,7	70,74	30	89
Portrait	1698	10072	436	6	71,33	42	131
Sphere2	3305	19802	1 079	17	62,53	66	200
Chandelier	3382	23039	1 585	20	77,73	67	235
Room112	6550	39202	4 379	56	78,55	123	467
Stairways1	7981	43000	5 128	67	76,55	156	626
Chair	8583	44501	5 379	72	74,96	165	681
Table2	10130	53296	8 415	106	79,11	189	799
Fireplace	10951	62947	11 820	155	76,36	199	840
Roof	15328	64140	12 439	178	69,82	271	1152
Room002	17299	77407	18 159	235	77,32	260	1225
Wardrobe	21381	115561	27 259	339	80,41	361	1588
Stairways2	31928	159888	76 026	875	86,91	528	2587

5 Conclusions

The main aim of our researches is development of complete workflow for modeling, automatic compression, encoding, storing, transferring and decoding 3D geometry optimized for the Web. The introduced algorithm for multi-resolution compression with author's enhancements allows to reduce 3D mesh complexity at any level of detail. The tests showed our algorithm is noticeably better than the popular commercial software (Tables 1 and 2). The results show the criteria of equilateral triangle (Q_E), which allows to create more proper triangulation with author's compression. The criteria of Hausdorff distance shows the differences between input and output mesh vertices. The author's algorithm creates triangulation based on existing vertices, therefore the Hausdorff distance criteria (Q_H) is also lower (better).

Previously developed 3dPNG format in combination with described here multi-resolution and hardware accelerated compression allows to create efficient solution for 3D Web applications with proper reduction of mesh complexity and high encoding compression ratio of 3D data.

References

1. Chun, W.: WebGL models: end-to-end. In: Cozzi, P., Riccio, C. (eds.) OpenGL Insights, pp. 431–454. CRC Press, July 2012
2. Deering, M.: Geometry compression. In: SIGGRAPH 1995 Proceedings of the 22nd Annual Conference on Computer Graphics and Interactive Techniques, pp. 13–20 (1995)
3. Delaunay, B.: Sur la sphère vide. In: Izvestia Akademii Nauk SSSR, Otdelenie Matematicheskikh i Estestvennykh Nauk 7, pp. 793–800 (1934)
4. Dworak, D., Pietruszka, M.: Fast encoding of huge 3D data sets in lossless PNG format. In: New Research in Multimedia and Internet Systems, Advances in Intelligent Systems and Computing Series 314, pp. 15–24. Springer (2015)
5. Dworak, D., Kuroczyński, P.: Virtual reconstruction 3.0: new approach of web-based visualisation and documentation of lost Cultural Heritage. In: Digital Heritage. Progress in Cultural Heritage: Documentation, Preservation and Protection. Lecture Notes in Computer Science 10058, pp. 292–306. Springer (2017)
6. Dworak, D., Pietruszka, M.: PNG as fast transmission format for 3D computer graphics in the web. In: Proceedings of the 10th International Conference on Multimedia & Network Information Systems 2016, Advances in Intelligent Systems and Computing Series 506, pp. 15–26. Springer, Wrocław (2017)
7. Hoppe, H.: Progressive meshes. In: ACM SIGGRAPH 1996 Proceedings, pp. 99–108
8. Khodakovsky, A., Schröder, P., Sweldens, W.: Progressive geometry compression. In: SIGGRAPH 2000 Proceedings of the 27th Annual Conference on Computer Graphics and Interactive Techniques, pp. 271–278 (2000)
9. Liu, X., Cheng, S.Y., Zhang, X.W., Yang, X.R., Nguyen, T.B., Sukhan, L.: Unsupervised segmentation in 3D planar object maps based on fuzzy clustering. In: 8th International Conference on Computational Intelligence and Security, pp. 364–368 (2013)
10. Maglo, A., Lavoue, G., Dupont, F., Hudelot, C.: 3D mesh compression: survey, comparisons and emerging trends. ACM Comput. Surv. 9(4) (2013). Art. 39
11. Mathews, J.H., Fink, K.K.: Numerical Methods Using Matlab, 4th Edn. (2004). pp. 599–607
12. Ng, R.T., Han, J.L.: CLARANS: a method for clustering objects for spatial data mining. IEEE Trans. Knowl. Data Eng. 14(5), 1003–1016 (2002)
13. Serban, N., Wasserman, L.: CATS: clustering after transformation and smoothing. J. Am. Stat. Assoc. 100(471), 990–999 (2005)
14. Sim, K., Yap, G.E., Hardoon, D.R., Gopalkrishnan, V., Cong, G., Lukman, S.: Centroid-based actionable 3D subspace clustering. IEEE Trans. Knowl. Data Eng. 25(6), 1213–1226 (2013)
15. Szczęsna, A.: The lifting scheme for multiresolution wavelet-based transformation of surface meshes with additional attributes. In: Computer Vision and Graphics, Proceedings of ICCVG 2008, LNCS 5337, pp. 487–495. Springer (2009)

16. Sweldens, W.: The lifting scheme: a construction of second generation wavelets. SIAM J. Math. Anal. Madrid, 511–546 (1995)
17. Taubin, G., Rossignac, J.: Geometric compression through topological surgery. ACM Trans. Graph. **17**(2), 84–115 (1998)
18. Wawrzonowski, M., Szajerman, D., Daszuta, M., Napieralski, P.: Mobile devices' GPUs in cloth dynamics simulation. In: Proceedings of the 2017 Federated Conference on Computer Science and Information Systems, pp. 1283–1290 (2017)

Using the Virtual Assistant Alexa as a Communication Channel for Road Traffic Situation

Viktória Kerekešová, František Babič[✉], and Vladimír Gašpar

Department of Cybernetics and Artificial Intelligence, Faculty of Electrical Engineering and Informatics, Technical University of Košice, Letná 9, 042 00 Košice, Slovakia
{viktoria.kerekesova,frantisek.babic,vladimir.gaspar}@tuke.sk

Abstract. A virtual assistant is a software agent that can perform tasks or services for an individual. In 2017, the capabilities and usage of virtual assistants are expanding rapidly. In our work, we focused on virtual assistants integrated into smart devices like Amazon Echo, Google Home or Apple HomePod. Amazon virtual assistant Alexa has a leading position around the world and thanks to the opening the development platform, it offers an increasing number of available skills. The skill represents a functionality implemented by any independent developer and submitted for certification. After this skill passes certification, it is published in the Alexa Skills Store and available for anyone to use. We choose this virtual assistant as a new communication channel for traditional announcement about road traffic situation. We created two own skills to cover the typical questions about the current traffic situations like "how is the traffic in the city" or "how is the traffic on the way from one point to the second", etc. For this purpose, we designed and implemented middleware services to get relevant data based on user request from the road traffic database. Both skills were submitted for the certification, and the first one is already available in the Alexa Skills Store.

Keywords: Virtual assistant · Traffic announcement · Alexa

1 Introduction

Typical traffic announcement (TA) refers to the broadcasting of a specific type of traffic report on the Radio Data System (RDS). It is based on data provided by drivers based on their own real-time experiences. This data is stored in databased and published through different multimedia channels, such as radio or web pages. However, both approaches have their advantages or disadvantages. When we are in the car, it is good to know about possible traffic accidents or barriers on the road, but sometimes is the TA too long and we received not only useful information for us. Also, we can acquire important information through relevant web page with traffic information, but sometimes it is available only as a list of accidents and we need to make more searching, such as where the particular street is and whether it is on our route.

The virtual assistant available within smart speaker can improve this user experience based on simpler obtaining relevant information. The general rule for speech giving is

© Springer Nature Switzerland AG 2019
K. Choroś et al. (Eds.): MISSI 2018, AISC 833, pp. 35–44, 2019.
https://doi.org/10.1007/978-3-319-98678-4_6

100 to 200 words per minute, but we can write only about 40 words per minute. The popularity of this type of equipment is steadily increasing, so it is important to use this trend, for example, to build a digital product. In our case, the digital product is represented by a community of people who use of contribute data about the traffic situation every day. This type of digital product has also a marketing potential; it allows us to better target ad campaigns, to personalize offered services, to customer loyalty.

The whole paper is organized as follow. We present briefly our motivation and the current situation in the area of voice assistants. Next sections describe the whole life cycle containing design, implementation, and testing. The conclusion summarizes our work and outlines some future work.

1.1 Virtual Assistants

In the past, the voice-controlled virtual assistants were limited to mobile devices. Hauswald et al. defines a virtual assistant as "an application that uses inputs such as the user's voice, vision (images), and contextual information to provide assistance by answering questions in natural language, making recommendations, and performing actions" [7]. Myers et al. understand it as personalized application on which users may have preferences over a wide range of functions within the system, including how tasks are performed, how and when meetings are scheduled, and how the system interacts with the user [8].

The voice-controlled virtual assistants represent a significant change in information access, not only by introducing voice control and touch gestures, but also by enabling dialogues with the user. Authors in [9] focused on user satisfaction with different scenarios of use. They concluded that in some scenarios it is the most important a task completion. In other cases, the amount of effort spent is in the first place. Cowan et al. [10] studied an issue of infrequent users' experiences of intelligent personal assistants. The results covered, for example, a frustration at limitations on fully hands-free interaction or significant concerns around privacy and transparency.

This situation changed in November 2014 when Amazon launched a new product including smart personal assistant technology on a speaker connected to a Wi-Fi home network equipped with a microphone. The user can use it to provide voice guidance. Two years later, Google introduced Google's smart home speaker with the voice assistant. In 2017, Apple's HomePod was launched, with already well-known virtual assistant Siri. Finally, the Cortana voice assistant from Microsoft available on the market was linked to the Harman Kardon speaker.

A still untapped potential of the virtual assistant is confirmed by selected results of various reports, such as Tractica estimates 2017 market size for virtual digital assistants (VDA) software was 816 million USD and forecast 7.7 billion USD in 2025[1]. The Voicebot Smart Speaker Consumer Adoption Report found that 19.7% of U.S. adults

---

[1] https://www.tractica.com/newsroom/press-releases/enterprise-virtual-digital-assistant-users-to-surpass-1-billion-by-2025/.

have access to smart speakers today[2]. Some other interesting findings are visualized in the Fig. 1.

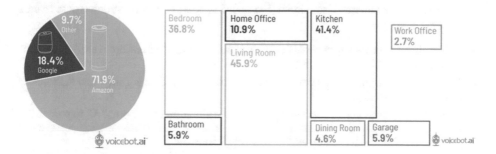

Fig. 1. The Voicebot Smart Speaker Consumer Adoption Report January 2018: left - smart speaker market share installed base 2017, right - primary household location for smart speakers.

Based on available information source, such as [3–6], we performed a simple comparison of the top four virtual assistants available on the market, see Table 1.

Table 1. A simple comparison of the selected virtual assistant.

	Alexa	Google assistant	Siri	Cortana
Developed by	Amazon	Google	Apple	Microsoft
Number of available skills (January 2018)	25 784	1 719	–	235
Speaker	Amazon Echo	Google Home	Apple HomePod	Harman Kardon
Number of offered language	3	10	21	8
Slovak language	No	No	No	No
Development support	Yes	Yes	Yes	Yes

2   Building the Skills for Alexa

Our motivation is to provide the traffic information through home smart device Amazon Echo [2] using virtual assistant Alexa. For this purpose, we designed and implemented two new skills. Finally, both skills and their implementation was tested by unit, integration, system and user-acceptance level.

---

[2] https://www.voicebot.ai.

2.1 Questions Proposal

At first, it was necessary to design a set of questions for communication with Alexa on this topic. Our list emerged from identified user requirements and available data from a pilot study, such as:

- The skill user needs to get an overview of today's traffic situation overall as well as sub-categories.
- The skill user should get an overview of the traffic situation within a few kilometers of the specified location.
- The skill user must get an overview of the traffic situation from point "A" to point "B".

    Alexa will be able to answer to the following questions:

- What "category" is in "city"? How is the traffic in "city"?
- What "category" is in last "x" hours? How is the traffic in last "x" hours?
- What "category" is in "x" km from "street", "city"? How is the traffic in "x" km from "street", "city"?
- What "category" is on the way from "street", "city" to "street", "city"? How is the traffic on the way from "street", "city" to "street", "city"?
- What "category" is on the way from home to work? How is the traffic on the way from home to work?

    Each question contains variables that need to be changed for a specific desired value. The *category* can acquire restrictions, accidents, patrols or radars; x is an integer value and *street* represents the name of street.

2.2 Architecture

We proposed three-level architecture (Fig. 2): presentation, application and data layers. Each component in this architecture is important to provide the right information at the right time to the right user. The presentation layer is represented by home assistant Amazon Echo which is responsible for handling the user requests and providing the answers to them. This device connects to the voice-controlled virtual assistant called Alexa, which responds to the name "Alexa".

    When the assistant registers a new question, it is sent to the application layer containing Amazon services like Alexa Voice Services (AVS), Alexa Skills Kit and AWS lambda. The AVS is Amazon's suite of services built around its voice-controlled intelligent virtual assistant for the home and other environments. In the AVS environment, services are known as skills. M. Taylor published an article presenting 50 most useful Alexa skills[3]; we have selected several for illustration:

---

[3] https://www.cnet.com/how-to/amazon-echo-most-useful-alexa-skills/.

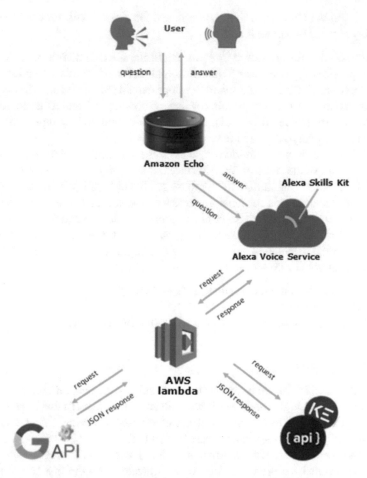

Fig. 2. The proposed architecture containing Amazon and newly implemented middleware services.

- The Capital One skill allows users to check the credit card balance or make a payment when one is due.
- The Harmony skill by Logitech will allow users to control their entertainment system using their voice though a Harmony hub-based remote, for example: "Alexa, turn on the TV".
- With the skill Anova Culinary, users can look up cooking guides and begin cooking using their voice, for example: "Alexa, ask Anova to help me cook a steak".
- To double-check what internal temperature is considered safe when cooking different meats, users can use the Meat Thermometer: Alexa, ask Meat Thermometer what is the best temperature for steak".
- For tracking the food, users can use the Track by Nutritionix skill, for example: "Alexa, ask Food Tracker how many calories are in two eggs and three slices of bacon".

- If a user wants to find a future trip, he can use the Kayak skill, for example: "Alexa, ask Kayak where I can go for $500".

The proposed skills are stored in the Alexa Skills Kit as a definition of the relevant metadata, possible types of user's questions, and a list of related variables with their values. The Alexa Skills Kit, a software development kit (SDK), is made freely available to developers and skills are available for instant download from Amazon.com. It is a collection of self-service APIs, tools, documentation, and code samples that makes it fast and easy for everyone to add new skills to Alexa.

AWS Lambda is an event-driven, serverless computing platform provided by Amazon as a part of the Amazon Web Services. It is a computing service that runs code in response to events and automatically manages the computing resources required by that code. In our case it is responsible for data processing. It means that component communicates with the data layers containing primary services to obtain the right data from traffic database ("api"). The secondary services on this layer are responsible for translation, a route calculation and geocoding ("Google API"). Data about traffic situation can be available in JSON format containing at least:

- a category to which the traffic situation is assigned,
- a date of record's creation,
- coordinates like a latitude, longitude and a name of the street.

2.3   First Alexa Skill

With the Alexa Skills Kit, we were able to create skills with a custom interaction model. We implemented the logic for the skills with the voice interface. In this case, we mapped the users' spoken input to the intents of our cloud-based service. Intent represents an action that fulfills a user's spoken request (see Fig. 3).

Once the user asks something Amazon Echo or Echo DOT integrated with Alexa, Alexa Service translates the voice form of the question into the text and sent it to the Alexa Skill Kit within a specified skill. Based on the context and the words used in the question, Alexa Skills Kit identifies the intent code to run. As part of the implementation, we used open source Node.js platform, which allows the creation of fast and scalable network applications using JavaScript. We have also used several APIs offered by Google. We've used the Google Maps Directions API to calculate and find a route between two locations. For geocoding and reverse geocoding, we used the Google Maps geocoding API. The translation of traffic situation description in English has performed dynamically thanks to the Google Cloud Translation API.

Fig. 3. A sample of intent "GetTrafficBetweenLocations".

2.4 Second Alexa Skill

The second skill provides the user the latest information from the stored traffic news. If this input is available in the mp3 format, a user can listen to it as well as on the radio in the Slovak language. This skill is a part of the "Flash Briefing" skills, which are already available in the Alexa Skills Store. In this case, it is possible to link own skill to the predefined Alexa questions like "Alexa, what's my Flash Briefing?" or "Alexa, what's in the news?". This type of skill can be triggered regularly at some specific time, for example, early morning after wakeup.

2.5 Pilot Study

Testing within simple pilot study helped us to verify the code quality, a reliability of the outputs and a level of user acceptance. For this purpose, we used a traditional V-model [1]. The V-Model demonstrates the relationships between each phase of the development lifecycle and its associated phase of testing, such as unit, integration, system and user acceptance testing.

As a pilot study, we choose the traffic announcements services provided by web application Košice Online and Radio Košice. We tested examples like:

- How is the traffic in Košice?
- What restrictions are in last 3 h?
- What radars are in 2 km from Ondavská, Košice?
- How is the traffic on the way from Štúrova, Košice to Jesenná, Prešov?
- What accidents are on the way from home to work?

For unit testing, we used the framework called "Alexa Conversation," developed by Expedia, and freely provided for developers of new skills who need to quickly and easily test the proper functionality of code blocks. It is built on the Mocha JavaScript testing framework, which runs on Node.js, and after that generates reports. An example of the unit test report is shown in the Fig. 4. Each of the blocks was tested with multiple tests that varied with different inputs. They verified what the function block returns as a response if the correct input values are entered but also if some values are incorrect or even missing. For example, a user would ask about a category that does not exist or would not tell the street, on the route from which he wants to hear the traffic situation.

```
Conversation: Test GetTrafficBetweenLocationsIntent
    User triggers: GetTrafficBetweenLocationsIntent SLOTS:
    {category: traffic, origin: ,originCity: ,
    destination: Hlavna, destinationCity: Presov}
    √ Alexa's plain text response should equal:  I could not find the
      origin street or city you are looking for. Try to ask me again.

PASSED
```

Fig. 4. One of the performed unit tests.

The cooperation between all components of the proposed architecture was tested by the integration testing. For this purpose, we used the Alexa Skills Kit. We were able to insert the expected questions in both text and voice formats. After entering a question, we could verify whether the received response contained all the details of that type of

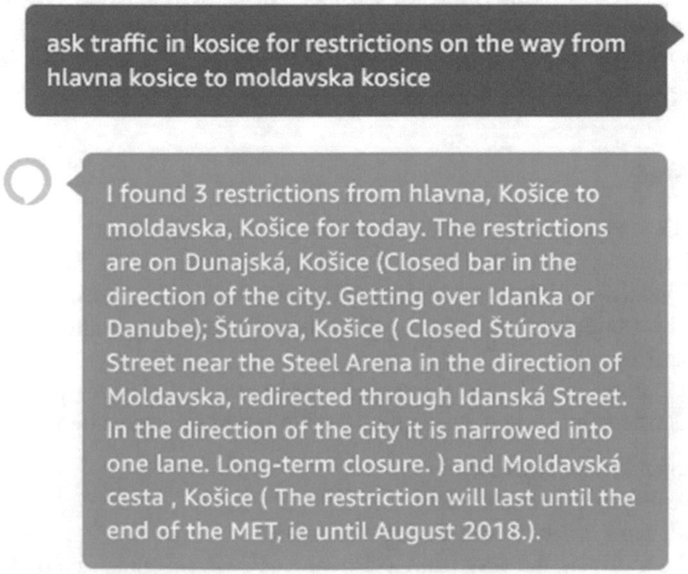

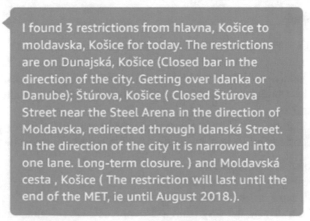

Fig. 5. An example of the integration testing.

question. So if the user asks for traffic accidents, then he or she will really hear an answer about traffic accidents. The example of a question and relevant answer used in this testing is shown in the Fig. 5.

System testing is testing conducted on a complete, integrated system to evaluate the system's compliance with its specified requirements. We tested different combination to evaluate it the right answer was returned to the user. All tests passed, so no further code interventions were needed. We also investigate the performed of our skills and related services. The performance of the software product belongs to the most important quality metrics. For voice assistant skills, the expected response speed should be in milliseconds to seconds. The first skill was tested in three ways:

1. How long it will take until the user returns an answer that no records have been found. We got a response like "No accidents for today yet." approximately after 200 ms.
2. We focused on responses that return traffic accidents only in a certain category. This is a more complicated response, so we expected longer execution time. We had 4 categories (restrictions, accidents, radars, and patrols) and 5 different questions; we tested the execution time of 20 questions. We received the responses on average after 600 ms.
3. Finally, we tested the response speed for the most complex answers. In this case, the user is interested in the overall traffic situation, so multiple categories are included in the responses. The answer started with "I found" x "traffic records" - x specified the number of found records and continued by listing all records found in different categories. The average response time, in this case, was 1 s and a few milliseconds (Fig. 6).

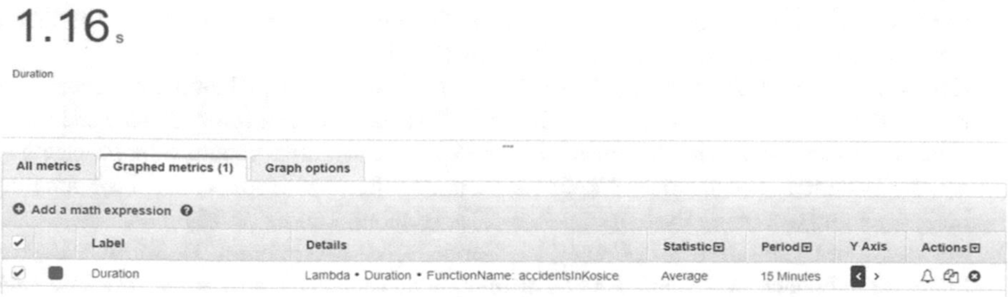

1.16 s

Duration

Fig. 6.   An example of the performance testing.

3   Conclusion

This paper presents a new opportunity how to present the information about road traffic situation. We created two new skills for Alexa virtual assistant that will be available in Alexa Skills Store after their successful certification. Currently, this type of skill is not very popular in the store. We found two similar skills, such as Got Jam (Singapore's traffic application to check for traffic jams and roadblocks) and UK Roads (the status of UK motorways, listing any incidents, accidents, closures, maintenance work, etc.)

We evaluated our skills with simple pilot study - all tests passed. Proposed approach aims to the community of people who like new technology and daily use their car. Since a price of Amazon Echo DOT in Slovakia starts from 70 EUR; designed skills represent the interesting digital product for marketing purposes.

Acknowledgment. This work was supported by the Slovak Research and Development Agency under the contract No. APVV-16-0213 and the Cultural and Educational Grant Agency of the Ministry of Education and Academy of Science of the Slovak Republic under grant No. 005TUKE-4/2017.

References

1. Forsberg, K., Mooz, H.: The relationship of system engineering to the project cycle. In: Proceedings of the First Annual Symposium of National Council on System Engineering, pp. 57–65 (1991)
2. Weber, A.: Amazon Echo: The Best User Guide to Master Amazon Echo Fast, USA (2016)
3. López, G., Quesada, L., Guerrero, L.A.: Alexa vs. Siri vs. Cortana vs. Google assistant: a comparison of speech-based natural user interfaces. In: Nunes, I. (ed.) Advances in Human Factors and Systems Interaction, AHFE 2017, Advances in Intelligent Systems and Computing, vol. 592. Springer, Cham (2018)
4. Sathi, A.: Cognitive devices as human assistants. In: Cognitive (Internet of) Things. Palgrave Macmillan, New York (2016)
5. Chung, H., Iorga, M., Voas, J., Lee, S.: Alexa, Can I Trust You? Computer **50**, 100–104 (2017)
6. Canbek, N.G., Mutlu, M.E.: On the track of artificial intelligence: learning with intelligent personal assistants. J. Hum. Sci. **13**(1), 592–601 (2016)
7. Hauswald, J., Laurenzano, M.A., Zhang, Y., Li, C., Rovinski, A., Khurana, A., Dreslinski, R. G., Mudge, T., Petrucci, V., Tang, L., Mars, J.: Sirius: an open end-to-end voice and vision personal assistant and its implications for future warehouse scale computers. In: Proceedings of the Twentieth International Conference on Architectural Support for Programming Languages and Operating Systems, pp. 223–238. Istanbul, Turkey (2015)
8. Myers, K., Berry, P., Blythe, J., Conley, K., Gervasio, M., McGuinness, D., Morley, D., Avi Pfeffer, A., Pollack, M., Tambe, M.: An intelligent personal assistant for task and time management. AI Mag. **28**(2), 47–61 (2007)
9. Kiseleva, J., Williams, K., Jiang, J., Awadallah, A.H., Crook, A.C., Zitouni, I., Anastasakos, T.: Understanding user satisfaction with intelligent assistants. In: Proceedings of the 2016 ACM on Conference on Human Information Interaction and Retrieval, pp. 121–130. ACM, Carrboro, North Carolina, USA (2016)
10. Cowan, B.R., Pantidi, N., Coyle, D., Morrissey, K., Clarke, P., Al-Shehri, S., Earley, D., Bandeira, N.: "What can I help you with?": infrequent users' experiences of intelligent personal assistants. In: Proceedings of the 19th International Conference on Human-Computer Interaction with Mobile Devices and Services, Vienna, Austria (2017)

Application of the Sensor Fusion for the System to Track Human Upper Limb Motion

Łukasz Leśniczek[1] and Krzysztof Brzostowski[2]

[1] IT Kontrakt Sp. z o.o., ul. Gwiaździsta 66, 53-413 Wrocław, Poland
lukaszlesniczek@gmail.com
[2] Faculty of Computer Science and Management,
Wrocław University of Science and Technology,
Wyb. Wyspiańskiego 27, 50-370 Wrocław, Poland
krzysztof.brzostowski@pwr.edu.pl

Abstract. The paper presents the system to track and visualize the human movement. The system is equipped with inertial measurement units, consisting triaxial accelerometers, gyroscopes and magnetometers. The estimates of the relative position and orientation in the proposed system are obtained using the data from the sensors and a biomechanical model. To estimate the orientation of the body segments we applied Madgewick algorithm. The biomechanical model applied in our studies is based on twists and exponential maps. As a proof-of-concept, we applied our approach to an upper limb to illustrate its ability to track human motion.

Keywords: State estimation · Wearable technology
Human kinematics

1 Introduction

Rapid development in the field of wireless sensor networks (WSN), wearable devices and signal processing techniques paved the way to build modern systems for tracking the human status and performance in everyday life. The applications incorporate many low-cost, energy-efficient sensors, which acquire thousands of data per second. Nowadays, we are surrounded by smart devices. Almost every modern smartphone or tablet has built-in sensors, that can provide valuable information about their user [9]. Different variations of such equipment create powerful systems that operate in following areas: health care [3], sport [2], industry [7], robotics [16] and entertainment [4].

Current trends lean towards adding more and more electronics to our clothes, like watches, shoes or glasses. Moreover, progress in micro-electromechanical devices leads to more comfortable use of such devices. The idea of applying micro-electromechanical devices (MEMS for short) in human motion analysis is not new, but it is becoming successful because of improved performance and

© Springer Nature Switzerland AG 2019
K. Choroś et al. (Eds.): MISSI 2018, AISC 833, pp. 45–55, 2019.
https://doi.org/10.1007/978-3-319-98678-4_7

affordable prices [20]. All of that helps to focus our thoughts on the new ways
to make use of the collected data. One of the fields that have benefited from
the research on wireless technologies and sensing technologies is human motion
tracking.

Based on the technologies it is possible to build the accurate, non-invasive
and portable systems for human motion tracking. The interest in these technologies is motivated by the fact that they help to overcome many issues raised by
classical solutions such as the optical systems. For example, the systems based
on wearable sensors do not suffer from occlusions. Moreover, their usage is not
limited to the laboratory setting.

Wearable systems for the analysis of human motion are usually multi-sensor
systems composed of triaxial inertial sensors such as the accelerometer and the
gyroscope. Some systems have an additional sensor, i.e., the triaxial magnetometer. The magnetometers can measure the strength of the local magnetic field. The
measurements of local magnetic field help to improve the overall performance of
the MEM devices [19]. These sensors can be used to reconstruct the pose or at
least either the position or the orientation of the body they are attached to.

The main disadvantage of the inertial sensors is their susceptibility to noise.
The main sources of noise relate to time-varying bias, and magnetic field distortion [4,25]. The bias drift corresponds to the higher temperature of gyroscope
during the measurements while magnetic field distortion, caused by electronic
and magnetic devices, affects the measurements acquired from the magnetometer. The noise in inertial measurements makes it difficult to apply the sensors in
human motion tracking. For example, the time-varying bias results in a drift of
the estimation that is unreliable after a few seconds.

Amongst the existing approach, to process signals in multi-sensor systems, we
propose sensor fusion. Sensor fusion offers various methods, techniques, and architectures to process signals acquired from different sources. The result of sensor
fusion is the coherent and consistent format of processed information [10].

To verify the accuracy of the proposed approach, we designed and implemented
the system based on the principle of 3D modeling. The problem of accuracy verification was solved by a 3D graphical representation of the user [24].

2   Related Works

The inertial sensors are commonly used in wide range of applications as an alternative to conventional approaches. The aim is to present the short introduction
to instrumental and computational aspects of tracking human motion. The main
problem in human motion tracking is an estimation of orientation, position, and
relation between body's segments (human kinematics).

One of the widespread algorithm to estimate the orientation of the human
segment is Kalman filter (KF) [24,25]. Since the Kalman Filter is suitable for linear problems, in the case of orientation computation the Extended Kalman Filter
(EKF) is commonly used. The Kalman filter and EKF have some weak points
such as the high computational cost. Moreover, to obtain satisfactory results,

the high sampling frequency of data is required. Unfortunately, the sampling frequency in popular MEMS-based sensors is limited.

An alternative to KF and EKF are the complementary filters. In [15] a filter based on this approach is presented. The main steps in the method are object's orientation computations for low frequencies based on acceleration and magnetometer measurements, and the parallel orientation calculation based on gyroscopic data. Then, these results are fused.

The other example of the filter based on the complementary filter is the Madgwick filter [12–14]. The method gives the accurate results at the low sampling frequency. The filter incorporates gradient descent algorithm to calculate the direction of gyroscope bias drift and to fuse it with data gathered from the accelerometer and the magnetometers to mitigate their inaccuracies. Low computation cost is another unquestionable advantage of this approach which makes it adequate for real-time modeling.

Another fundamental problem in human motion tracking is the representation of body's segment orientation in three-dimensional space. The main approaches are Euler angles, rotation matrix, quaternions, and exponential notation. The most popular way to represent the orientation is based on the quaternions [17]. The method is free of gimbal locks. The drawback of using quaternions is a lack of an immediate way to interpret the orientation. An alternative to the quaternions is an exponential map. The method needs only one 4×4 matrix for the object with three degrees of freedom.

In the human kinematics modeling yet another problem is the parametrization of the kinematic chain in three-dimensional space. For the objects connected by joints, we have to take into account the relations between the connected joints. It is because the movements of the one object affect the orientation of the others. The problem is well-known in the theory of kinematic chains [1]. Kinematic chain describes a group of linked objects in such way that it is possible to determine the orientation and position of each object relating to other objects in the chain [5]. One of the methods to describe the kinematic chain is twists and exponential maps convention [23]. Twists describe the movement composed of rotation and translation in the same plane. It is the main advantage of twists. They incorporate two kinds of movements in one short notation. Exponential maps help to reduce the complexity of human movements by mapping similar movements to the same groups. The other approach to parameterize the kinematic chain is the Denavit-Hartenberg convention [6].

3   Materials and Methods

3.1   Experimental Design

To assess the effectiveness of the proposed approach we used the self-developed sensing and processing systems. The system (Fig. 1) is composed of the connection manager module, data acquisition module, and calibration module. The system was developed with the use of the *Shimmer Connect Library* and *32feet.NET*

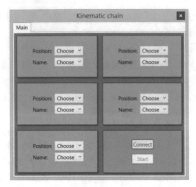

Fig. 1. Data acquisition module of the designed system.

libraries. The sensing part of the system can transfer data from up to five wearable inertial sensors attached to the human body. In our studies, we use inertial sensors from the sensing platform Shimmer2r [21]. The sensor is composed of 12-bit the tri-axis gyroscope, accelerometer, and magnetometer. Raw signals were sampled at 1000 [Hz] and transferred to the processing system through Bluetooth protocol. All inertial sensor used in our studies are calibrated by the *Shimmer 9DoF Calibration Application* [21].

3.2   Data Acquisition

In our experiment, we used three Shimmer sensors. We place two sensors next to the corresponding joints (Fig. 3), and one of the sensors have been placed on the chest to be a point of reference (Fig. 2). The attachment of the sensors guarantees to map full range of the movements. We connect all devices to the server computer through Bluetooth protocol.

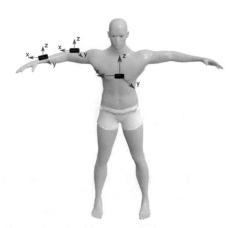

Fig. 2. The sensor placement and the point of reference.

To test the model performance, we select a set of typical postures for the human elbow (Fig. 3). As the ground truth, we used the goniometer measurements.

Fig. 3. The sensor placement on the arm and the forearm.

3.3   The System to Visualize Human Upper Limb Motion

To test the reliability of tracking of the kinematic chain, a full 3D live system to visualize the human body has been developed (Fig. 4). To build a 3D human model we used *MakeHuman* application.

The *MakeHuman*-based 3D human model was transformed to *T-pose* by a *Blender* application. In the last step we used 3D graphics *Mogre* engine to render final 3D model. 3D graphics *Mogre* engine is a C# port of *Ogre3D*. The visualization module is responsible for providing direct feedback to the user and for visual comparison which can be helpful to determine model correctness.

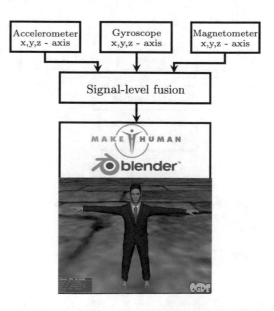

Fig. 4. The system to visualize the human joint tracking.

At this stage, the user was asked to perform any upper limb exercises. Then, we can compare the movements to the generated 3D model visualization.

3.4   Introduction to Sensor Fusion

Signals acquired from the accelerometer, gyroscopes, and magnetometers need a suitable signal processing technique. In the field of the human motion, the sensor fusion is a standard approach. Sensor fusion methods help to determine the orientations and poses of the human body parts based on the data acquired from inertial sensors.

The main concepts of sensor fusion are the levels of abstraction. In general, we can divide sensor fusion into three categories: signal, feature, and decision. Because our approach is based on the signal level fusion, below we provide the introduction to the first level sensor fusion.

Signal-Fusion Algorithms. Signal-level fusion is used to combine raw signal acquired from different sensors. The signals at this level are commensurate, i.e., acquired from direct measurements of the same features. The example is the signals from accelerometer and gyroscope applied to measure the motion of the human body.

The methods at the signal level allow obtaining the coherent description of the object of interest. It is important that new data acquired from original data processing can be useful to get the new description of the object which it is not possible with the use of existed sensors. In Fig. 5 we show the model of data fusion at the first level.

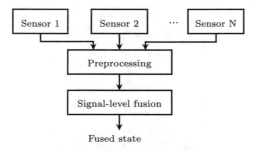

Fig. 5. Signal-level fusion (based on [11]).

4   Algorithm for Estimating the Upper Limb Kinematic Chain

The section introduces the methodology to estimate human joints based on the inertial data and sensor fusion techniques (Fig. 6). The first step in our approach is the signal preprocessing. The step is important because the data gathered from

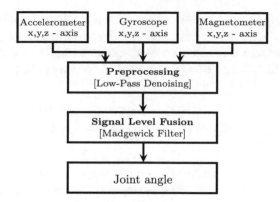

Fig. 6. The proposed methodology to estimate human joints.

inertial sensors are noised [8]. To this end, we applied the low-pass filtering. The next step in proposed approach is signal-level fusion to determine the limb orientation. In our work, we used the Madgwick filter [13]. It has been proven that the Madgwick is an efficient and computationally inexpensive algorithm designed to fuse the signals from inertial sensors [14].

Based on the orientations of human limbs, the next step is to create a model of the kinematic chain for the human upper limb. The proposed model has low computational cost by applying the high efficient twists and exponential maps techniques to build a kinematic chain of human upper limbs.

4.1 Twists and Exponential Maps

Let us define the movements of the upper limb as a set of rotations and translations. The important factor is to describe the root point of these transformations. The sensors are attached to the different areas of the human body. Additionally, every device has its coordinate system, which in fact make it impossible to compare values from different sensors directly. That is why there is a necessity of transformation from the one coordinate system to another [23]. When we describe the position of the joint S_1 about another S_2 we can define the relationship as follows

$$p_{S_2 S_1} = \mathbf{p}_{S_1 S_2} + \mathbf{R}_{S_1 S_2} \cdot p_{S_1 S_2} \tag{1}$$

where $\mathbf{p}_{S_1 S_2} \in \mathbb{R}^3$ is the vector position from the centre of one joint to another and $\mathbf{R}_{S_1 S_2} \in SO(3)$ (special orthogonal group) is the orientation of joint S_2 in relation to joint S_1. Then we can define joint configuration $\mathbf{G}$ as:

$$p_{S_2 S_1} = \begin{bmatrix} \mathbf{p}_{S_2 S_1} \\ 1 \end{bmatrix} = \begin{bmatrix} \mathbf{R}_{S_1 S_2} & \mathbf{p}_{S_1 S_2} \\ 0 & 1 \end{bmatrix} \begin{bmatrix} p_{S_1 S_2} \\ 1 \end{bmatrix} = \mathbf{G}_{S_1 S_2} p_{S_1 S_2} \tag{2}$$

The configuration matrix $\mathbf{G}_{S_1 S_2}$ is used to transform orientation of the joint from the one coordinate system to another. The main problem of this task is the computational complexity of matrices multiplication. Moreover, for three axes we would have to define three separate matrices to transform the point in space. However, when we use twist technique, we need only one matrix for each joint [22]. For each of the configuration group $\mathbf{G}_{S_1 S_2}$ there is a corresponding twist in a tangent space $se(3)$ which we define as:

$$
\begin{aligned}
se(3) &= \{(\mathbf{v}, \hat{\boldsymbol{\Omega}}) : \mathbf{v} \in \mathbb{R}^3, \hat{\boldsymbol{\Omega}} \in so(3)\} \\
so(3) &= \{\mathbf{M} \in \mathbb{R}^{n \times n} : \mathbf{M}^T = -\mathbf{M}\}
\end{aligned}
\tag{3}
$$

$$
\hat{\boldsymbol{\Omega}} = \begin{bmatrix} 0 & -\omega_3 & \omega_2 \\ \omega_3 & 0 & -\omega_1 \\ -\omega_2 & \omega_1 & 0 \end{bmatrix}
\tag{4}
$$

where $\boldsymbol{\omega}$ represents the rotation of the joint. Now we can formulate the twist as

$$
\boldsymbol{\Xi} = \begin{bmatrix} \mathbf{v} \\ \boldsymbol{\omega} \end{bmatrix}^{\wedge} = \begin{bmatrix} \hat{\boldsymbol{\Omega}} & \mathbf{v} \\ \mathbf{0} & 0 \end{bmatrix} \in \mathbb{R}^{4 \times 4}
\tag{5}
$$

where $\mathbf{v}$ is a position vector of the centers of the joints. As we can see, the twist combines rotation and translation in one matrix improving algorithm performance. Next step is to map the resulting twist back to the orientation of the sensor. To do that we used exponential map

$$
\begin{aligned}
e^{\theta \boldsymbol{\Xi}} &= \begin{bmatrix} e^{\theta \hat{\boldsymbol{\Omega}}} & (\mathbf{I} - e^{\theta \hat{\boldsymbol{\Omega}}})(\hat{\boldsymbol{\Omega}} \mathbf{v} + \boldsymbol{\omega} \boldsymbol{\omega}^T \mathbf{v} \theta) \\ \mathbf{0} & 1 \end{bmatrix} \quad \boldsymbol{\omega} \neq \mathbf{0} \\
e^{\theta \boldsymbol{\Xi}} &= \begin{bmatrix} \mathbf{I} & \mathbf{v}\theta \\ 0 & 1 \end{bmatrix} \quad \boldsymbol{\omega} = \mathbf{0}
\end{aligned}
\tag{6}
$$

To calculate exponential on the twist the Rodrigues's formula was used:

$$
e^{\theta \hat{\boldsymbol{\Omega}}} = \mathbf{I} + \hat{\boldsymbol{\Omega}} \sin \theta + \hat{\boldsymbol{\Omega}}^2 (1 - \cos \theta)
\tag{7}
$$

4.2   Kinematic Chain

The next step in proposed approach is to model the kinematic chain of the human upper limb, i.e., an approach to link all of the sensors together based on the estimated orientation of the sensor and the method of its transformation to other coordinates systems. The joints in the human body are connected to each other, i.e., moving one joint often changes the position and orientation of others. To cope with this complex problem, we defined a kinematic chain for upper limb in the form

$$
\begin{aligned}
\mathbf{G}_{chest,wrist}(\Theta) &= \exp(\theta_{chest} \boldsymbol{\Xi}_{chest} + \theta_{chest} \boldsymbol{\Xi}_{shoulder} \\
&\quad + \theta_{shoulder} \boldsymbol{\Xi}_{elbow} + \theta_{elbow} \boldsymbol{\Xi}_{wrist}) \mathbf{G}_{elbow,wrist}(0)
\end{aligned}
\tag{8}
$$

Above equation represents the upper limb kinematic chain transforming coordinates of wrist joint to the base coordination system, which in our case is the chest [18]. For other joints, the equation should be simplified according to the position of the joint in the chain.

5   Results and Disscussion

The section demonstrates the results of upper limb motion tracking. In the presented experiments we focus on the angle tracking in elbow joints. During experiment, we were acquiring the data from the inertial sensors attached to the human arm (Fig. 3). To estimate the orientations of the sensors in the kinematic chain we applied Madgwick filter. To assess the quality of the results the user was instructed to perform a couple of static and dynamic movements.

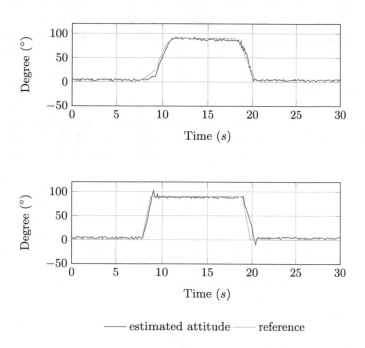

Fig. 7. Results of elbow joint angle tracking in quasi-static condition (top), and dynamic condition (bottom).

Experiments were performed for quasi-static and dynamic motions. Figure 7 (top) presents the elbow joint angle in a quasi-static motion. Quasi-static motion is a slow, uniform motion, which because of the small changes of the orientation in the following time frame can be recognized as a pseudo-static. In this case, the proposed solution works well. Achieved results do not differ significantly from the expected results. The highest difference between measured and calculated results was 8°. In Fig. 7 (bottom) we show the performance of the algorithm in a dynamic motion. In this case, the rate of the orientation change was high which increased the distortions influencing the results. It is especially easy to notice in the last part of the measurements when the response to the algorithm was too slow compared to actual measurements. It is clear that model tends to overestimate fast movements, but works well in a quasi-static case.

We also evaluate the overall model accuracy by using the self-developed system (Fig. 4). To do that a 3D live visualization module was used. The visual comparison of the movements between an application user and his computer model showed that proposed solution correctly estimates user behavior. Additionally, results of the computations are presented to the user almost immediately, which makes it perfect for the scenarios when instant feedback is necessary. Experiments showed the high quality of the results, especially in quasi-static motion. Quality of the estimated angles and orientations decrease with fast, dynamic movements. In the case of dynamic conditions, we have to apply more sophisticated algorithms. It is worth to mention that proposed solution is not strictly restricted to tracking upper limb movements and can be easily extended to create whole body kinematic chain.

6 Conclusions

An inertial sensors-based approach to tracking human movements is presented. The algorithm can estimate the position and orientation of the body's segments. The proposed algorithm is the part of the system to capture and visualize the human motion.

Experimental results show that the proposed algorithm works well. The quality of the position and orientation estimates is satisfactory. Some differences occur when the quasi-static and dynamic movements are compared. The estimation quality for the first kind of movement is slightly better than for the second one. This is due to the additional acceleration terms existed in measurements acquired from the accelerometer.

In further works we plan to extend our approach to the more body segments.

References

1. Abdel-Malek, K., Arora, J.: Human Motion Simulation: Predictive Dynamics. Academic Press (2013)
2. Ahmadi, A., Rowlands, D., James, D.: Towards a wearable device for skill assessment and skill acquisition of a tennis player during the first serve. Sports Technol. **2**(3–4), 129–136 (2009)
3. Au, A., Kirsch, R.: EMG-based prediction of shoulder and elbow kinematics in able-bodied and spinal cord injured individuals. IEEE Trans. Rehabil. Eng. **8**(4), 471–480 (2000)
4. Bao, L., Intille, S.: Activity recognition from user-annotated acceleration data. In: Pervasive Computing, pp. 1–17. Springer (2004)
5. Craig, J.: Wprowadzenie do robotyki: mechanika i sterowanie, pp. 35–128. Wydawnictwa Naukowo-Techniczne (1995)
6. Denavit, J.: A kinematic notation for lower-pair mechanisms based on matrices. ASME J. Appl. Mech, 215–221 (1955)
7. Fougner, A., Scheme, E., Chan, A., Englehart, K., Stavdahl, O.: A multi-modal approach for hand motion classification using surface emg and accelerometers. In: Engineering in Medicine and Biology Society, EMBC, 2011 Annual International Conference of the IEEE, pp. 4247–4250. IEEE (2011)

8. Guraliuc, A., Barsocchi, P., Potortì, F., Nepa, P.: Limb movements classification using wearable wireless transceivers. IEEE Trans. Inf Technol. Biomed. **15**(3), 474–480 (2011)

9. Jaitner, T., Gawin, W.: A mobile measure device for the analysis of highly dynamic movement techniques. Procedia Eng. **2**(2), 3005–3010 (2010)

10. Khaleghi, B., Khamis, A., Karray, F.: Multisensor data fusion. In: Multisensor Data Fusion: From Algorithms and Architectural Design to Applications, p. 15 (2015)

11. Llinas, J., Hall, D., Liggins, M.: Handbook of Multisensor Data Fusion: Theory and Practice. CRC Press, Boca Raton (2009)

12. Madgwick, S.: An efficient orientation filter for inertial and inertial/magnetic sensor arrays. Report x-io and University of Bristol (UK) (2010)

13. Madgwick, S., Harrison, A., Vaidyanathan, R.: Estimation of IMU and MARG orientation using a gradient descent algorithm. In: 2011 IEEE International Conference on Rehabilitation Robotics (ICORR), pp. 1–7. IEEE (2011)

14. Madgwick, S., Vaidyanathan, R., Harrison, A.: An efficient orientation filter for inertial measurement units (IMUS) and magnetic angular rate and gravity (MARG) sensor arrays. Technical report, Department of Mechanical Engineering, April 2010

15. Mahony, R., Hamel, T., Pflimlin, J.: Nonlinear complementary filters on the special orthogonal group. IEEE Trans. Autom. Control **53**(5), 1203–1218 (2008)

16. Murray, R., Li, Z., Sastry, S., Sastry, S.: A Mathematical Introduction to Robotic Manipulation. CRC Press, Boca Raton (1994)

17. Parent, R.: Animacja komputerowa: algorytmy i techniki, pp. 40–71, 181–209, 459–505. Wydawnictwo Naukowe PWN (2012)

18. Roetenberg, D., Luinge, H., Slycke, P.: Xsens MVN: full 6DOF human motion tracking using miniature inertial sensors. Technical report, Xsens Motion Technologies BV (2009)

19. Roetenberg, D., Luinge, H., Baten, C., Veltink, P.: Compensation of magnetic disturbances improves inertial and magnetic sensing of human body segment orientation. IEEE Trans. Neural Syst. Rehabil. Eng. **13**(3), 395–405 (2005)

20. Shaeffer, D.: Mems inertial sensors: a tutorial overview. IEEE Commun. Mag. **51**(4), 100–109 (2013)

21. Shimmer: Shimmer, October 2015. http://www.shimmersensing.com

22. Tchoń, K., Mazur, A., Hossa, R., Duleba, I., Muszyński, R.: Manipulatory i roboty mobilne. Modele, planowanie ruchu, sterowanie, pp. 23–90. Akademicka Oficyna Wydawnicza PLJ (2000)

23. Xi, C.: Human motion analysis with wearable inertial sensors. Ph.D. thesis, University of Tennessee, Knoxville, August 2013

24. Zhang, Z., Wong, W.C., Wu, J.: Wearable sensors for 3D upper limb motion modeling and ubiquitous estimation. J. Control Theory Appl. **9**(1), 10–17 (2011)

25. Zhou, H., Hu, H., Harris, N., Hammerton, J.: Applications of wearable inertial sensors in estimation of upper limb movements. Biomed. Signal Process. Control **1**(1), 22–32 (2006)

Estimation of the Amount of Noise in Digital Images Using Finite Differences

Janusz Kowalski[1], Grzegorz Mikołajczak[2], and Jakub Pęksiński[2(✉)]

[1] Faculty of Medicine, Pomeranian Medical University, Szczecin, Poland
janus@pum.edu.pl
[2] Faculty of Electrical Engineering, West Pomeranian University of Technology,
Szczecin, Poland
jpeksinski@zut.edu.pl

Abstract. The paper presents a novel method for estimating the variance of noise in a digital image. This method is based on filtering corrupted image using finite differences. As a result an original signal is removed while noise remains so its variance can be calculated. In the paper several differential filter's masks of various sizes were derived. The proposed method was verified on test images corrupted by the additive Gaussian distributed noise. The method was compared with several previously published estimation methods. It was shown that the novel method performs well for a large range of noise variance values.

Keywords: Digital filter · Finite differences · Noise estimation

1 Introduction

Estimation of the amount of noise is an essential part of image processing, especially in the noise reduction field. Many algorithms have been proposed that use information about the noise to process the image optimally. Application of noise estimation in signal processing methods eliminates the necessity to provide the value of the amount of noise by the user, and makes it much more automated. Image restoration, edge detection and adaptive filtration are typical examples of such processes.

A significant practical issue in signal digital processing algorithms is searching for the methods of improving the image through removal of deformations being the result of noise. The effectiveness of filtration is the function of many factors, including: the selected filtration algorithm, certain information about noise features as well as the standard image. The information of particular significance refers to noise features - random or determined, its spectral density distribution, variance, etc. In most cases, complete data about the noise is not available and, thus, attempts are made at estimation - assessment of the level of noise expressed with variance. Availability of the information about the level of noise enables obtaining of the optimum quality of filtration, especially in such areas as: removal of noises from astronomical images [2], reconstruction of images [3], detection of edges [4], segmentation of images [9], movement estimation [7], reduction of noises in MRI images [13], smoothing of images [15] and other.

© Springer Nature Switzerland AG 2019
K. Choroś et al. (Eds.): MISSI 2018, AISC 833, pp. 56–66, 2019.
https://doi.org/10.1007/978-3-319-98678-4_8

The issue of noise level estimation involves, most frequently, determination of its variance σ_n^2 (or standard deviation σ_n) from a digital image, with the assumption that the noise is an additive, stationary process, not correlated with the useful signal with the average value equal to zero.

The methods of estimating the noise level may, basically, be divided into two groups: smoothing and block ones [8, 12]. The smoothing methods are based on filtration of the corrupted image and assume that the image obtained after filtration is the original image. The difference between the corrupted and refined image is the noise, from which the noise variance is determined [8, 13]. The problem in these methods is that the process of filtration not only influences the noise, but also the useful signal. In block methods, the value of noise variance is estimated based on the analysis of the value of so called local variances determined from the so called "flat areas". For, it is assumed that the level of signal variability in these areas is only determined by the noise. In these methods, a significant difficulty is automation of searching for a suitable area, i.e. determination of the criteria according to which such a region is included in the "flat areas" [6].

Except for these two groups of estimation methods, many other may be found in literature [5, 10, 14]. Among them, there are methods based on: analysis of histograms [12] or combination of block methods with smoothing ones [15].

2 Noise Estimation Using Finite Differences

Let us consider noisy data y(n):

$$y(n) = s(n) + n(n) \tag{1}$$

where $s(n)$ is the original (noiseless) data and $n(n)$ is the additive white Gaussian noise with zero mean and variance σ_n^2. The goal is to estimate the standard deviation σ_n. The method proposed in this paper consists in processing the signal $y(n)$ in a way that the original signal $s(n)$ is removed from the Signal $y(n)$ but the noise $n(n)$ is retained. In such a case the variance of the processed signal is the estimate of the noise variance σ_n^2. It can be obtained through use of the characteristics of finite differences.

The first order finite difference is defined as:

$$\Delta y(n) = y(n+1) - y(n) \tag{2}$$

the second order finite difference is:

$$\Delta^2 y(n) = y(n+2) - 2y(n+1) + y(n) \tag{3}$$

Generally, the k order finite difference is:

$$\Delta^k y(n) = \sum_{i=0}^{k} c_{i,k} y(n+k=i) \tag{4}$$

$$c_{i,k} = (-1)^i \binom{k}{i} \tag{5}$$

By differentiating any input signal y(n) by means of k-order finite difference a new signal w(n) is obtained:

$$w(n) = \Delta^k y(n) \tag{6}$$

If the input signal $y(n)$ consists only of the Gaussian distributed noise n(n) with zero mean and its variance $V(n(n))$ equals σ_n^2, the variance of the signal $w(n)$ is:

$$V(w(n)) = V\left(\sum_{i=0}^{k} c_{i,k} n(n+k-i)\right) = \sum_{i=0}^{k} \left(c_{i,k}^2\right)\sigma_n^2 \tag{7}$$

thus the variance σ_n^2 equals:

$$\sigma_n^2 = \frac{V(w(n))}{\sum\limits_{i=0}^{k} c_{i,k}^2} \tag{8}$$

Therefore, the noise variance can be obtained if the variance of the differentiated noise $V(w(n))$ and the sum of coefficients c for a given differentiation order are given. This sum can be expressed using formula [1]:

$$\sum_{i=0}^{k} c_{i,k}^2 = \sum_{i=0}^{k} \binom{k}{i}^2 = \frac{(2k)!}{(k!)^2} \tag{9}$$

Yet if the input signal is a polynomial $x(t_n)$ of the form:

$$s(t_n) = \sum_{k=0}^{m} a_k t_n^k = a_0 + a_1 t_n + \cdots + a_m t_n^m \tag{10}$$

and assuming that $\Delta t = t_{n+1} - t_n$, the first order difference of such a signal equals:

$$\Delta s(t_n) = a_1 \Delta t + a_2 (2t_n \Delta t + \Delta t^2) + \cdots + a_m (t_n^{m-1} + m t_n^{m-1} + \cdots powers < m - 1) \tag{11}$$

The first order difference of the m degree polynomial is the polynomial of the $m - 1$ degree, hence the m order difference is a constant independent from the variable t, and the $m + 1$ order difference equals zero. Therefore, if the input signal is a signal y (n) given by (1), where s(n) is a data signal which can be expressed in form of the m degree polynomial (10), and if n(n) is a Gaussian distributed uncorrected noise with zero mean, the $m + 1$ order finite difference of such a signal includes only this noise and the data signal is removed. Then, the estimate of the noise variance σ_n^2 can be obtained by applying Eq. (8).

3   Maximally Flat Frequency Characteristic Filter

Signal differentiating can be treated as filtration by means of a maximally flat frequency characteristic filter, from now on referred to as a differentiating filter. In the case of one-dimensional signal, the filter mask is presented in Fig. 1. Assuming the symmetry of coefficients a_i to a_0, a general filter equation is given by:

$$\bar{f}_x = a_0 f_x + \sum_{k=1}^{N} a_k (f_{x+k} + f_{x-k}) \tag{12}$$

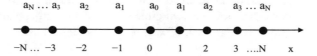

Fig. 1.   One-dimensional filter mask of the size N

The coefficients in (12) were obtained through the test function method - *ftest* == 1, $x, x^2, \dots, x^{N-1}$. For each test function the left side of (12) is zero. Apart from that, it is assumed that $a_N = 1$. By assuming $x = 0$ and substituting *ftest* = 1 to (12), we get:

$$a_0 = -2 - 2 \sum_{k-1}^{N} a_k \tag{13}$$

By analogy, with *ftest* = x^m, where $m = 2, 4, 6, \dots$, we get:

$$\sum_{k=1}^{N} a_k k^m = -N^m \tag{14}$$

For *ftest* = x^m, where $m = 1, 3, 5\dots$, owing to the symmetry of the coefficients Eq. (12) gets nullified. Out of (13) and (14) a matrix equation is obtained, the solution of which is given by the following formulas for filter coefficients:

$$a_0 = 2(-1)^N \binom{2N-1}{N} \; ; \; a_k = (-1)^{N-1} \binom{2N}{N-k} \tag{15}$$

The comparison of the coefficients of the differentiating filter (15) and the coefficients of the finite differences (5) leads to the following conclusion: the filter coefficients for the mask of the size N are equal to the $2N$ order finite difference coefficients.

Hence, the filter (12) conducts the signal differentiation. The transmittance of this filter is given by:

$$H(\omega) = a_0 + 2 \sum_{k=1}^{N} a_k \cos(k\omega) \tag{16}$$

The filter will remove a data signal only if the condition of maximally flat frequency characteristic is satisfied:

$$\frac{\delta^k H(\omega)}{\delta \omega^k} = 0 \; for \; \omega = 0, \; k = 0, 1, 2 \ldots \tag{17}$$

Out of (16) and (17) a set of Eqs. (13) and (14) can be obtained, thus the condition of the maximally flat characteristic of the differentiating filter for $\omega = 0$ is equivalent to nullification of the filter Eq. (12) using the following test functions, hence it ensures the removal of the data signal of an assumed form of a polynomial.

The mask's coefficients for the two-dimensional signal are obtained similarly. A general filter equation for the mask of size $N \times M$ is defined as:

$$\bar{f}_{x,y} = \sum_{i=0}^{N-1} \sum_{j=0}^{M-1} a_{i,j} \cdot f_{x-\frac{N-1}{2}+i, y-\frac{N-1}{2}+j} \tag{18}$$

Let a mask be a square mask $N \times N$ and let it be characterized by the symmetry to the middle coefficient that ensures, similarly as in the one-dimensional case, the nullification of (18) for all test functions with even power. An example of the mask of the desired filter of the size 3×3 is presented in Fig. 2.

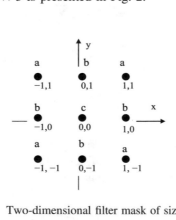

Fig. 2.   Two-dimensional filter mask of size 3×3

The filter equation for above mask takes the form:

$$\bar{f}_{x,y} = cf_{x,y} + b(f_{x-1,y} + f_{x+1,y} + f_{x,y-1} + f_{x,y+1}) + a(f_{x-1,y-1} + f_{x-1,y+1} + f_{x+1,y-1} + f_{x+1,y+1})$$

(19)

Through substitution of the following test functions ftest to (19) we obtain equations as follow: - for ftest = 1

$$4a + 4b + c = 0 \tag{20}$$

- for ftest = x^2 and ftest = y^2

$$4a + 2b = 0 \tag{21}$$

- for *ftest* = $x^2 y^2$

$$4a = 0 \tag{22}$$

When *ftest* = x^m, y^m or $x^m y^m$, where $m = 1, 3, 5, ...$, Eq. (19) gets nullified owing to the symmetry between the coefficients. Equations (20), (21) and (22) lead to $a = b = c = 0$. Assuming that the coefficient b = 1 and taking (20) and (22) into account, we obtain:

$$a = 0, \ b = 1, \ c = -4 \tag{23}$$

yet taking (20) and (21) into account, we obtain:

$$a = -\frac{1}{2}, \ b = 1, \ c = -4 \tag{24}$$

Therefore, in the case of two-dimensional signal the differentiating filter mask cannot be unequivocally defined.

The values of the coefficients (23) and (24) can be obtained by employing transmittance form corresponding to the filter (19):

$$H(\omega 1, \omega 2) = c + 2b(\cos(\omega 1) + \cos(\omega 2)) + 4a \cos(\omega 1) \cos(\omega 2) \tag{25}$$

and taking account of the maximally flat characteristic condition:

$$\frac{\delta^k H(\omega 1, \omega 2)}{\delta \omega 1^k} = 0 \ for \ \omega 1 = 0, \ \omega 2 = 0, \ k = 0, 1, 2$$
$$\frac{\delta^k H(\omega 1, \omega 2)}{\delta \omega 2^k} = 0 \ for \ \omega 1 = 0, \ \omega 2 = 0, \ k = 0, 1, 2 \tag{26}$$

The masks D1 and D2 corresponding to the obtained solutions (23) and (24) are presented in Fig. 3. The mask D1 describes the Laplace filter used in digital signal

$$D1 = \begin{bmatrix} 0 & 1 & 0 \\ 1 & -4 & 1 \\ 0 & 1 & 0 \end{bmatrix} \quad D2 = \begin{bmatrix} -1 & 2 & -1 \\ 2 & -4 & 2 \\ -1 & 2 & -1 \end{bmatrix}$$

Fig. 3. Differentiating filter masks of size 3×3

processing. By analogy, masks for any N can be obtained. Figures 4, 5 and 6 show masks obtained for N = 5, 7, 9.

$$A0 = \begin{bmatrix} 1 & -14 & 26 & -14 & 1 \\ -14 & 96 & -164 & 96 & -14 \\ 26 & -164 & 276 & -164 & 26 \\ -14 & 96 & -164 & 96 & -14 \\ 1 & -14 & 26 & -14 & 1 \end{bmatrix} \quad A1 = \begin{bmatrix} 1 & -10 & 18 & -10 & 1 \\ -10 & 64 & -108 & 64 & -10 \\ 18 & -108 & 180 & -108 & 18 \\ -10 & 64 & -108 & 64 & -10 \\ 1 & -10 & 18 & -10 & 1 \end{bmatrix}$$

Fig. 4. Differentiating filter masks of size 5×5

$$B0 = \begin{bmatrix} 1 & -6 & 15 & -20 & 15 & -6 & 1 \\ -6 & 36 & -90 & 120 & -90 & 36 & -6 \\ 15 & -90 & 225 & -300 & 225 & -90 & 15 \\ -20 & 120 & -300 & 400 & -300 & 120 & -20 \\ 15 & -90 & 225 & -300 & 225 & -90 & 15 \\ -6 & 36 & -90 & 120 & -90 & 36 & -6 \\ 1 & -6 & 15 & -20 & 15 & -6 & 1 \end{bmatrix} \quad B1 = \begin{bmatrix} 1 & -9 & 27 & -38 & 27 & -9 & 1 \\ -9 & 72 & -207 & 288 & -207 & 72 & -9 \\ 27 & -207 & 585 & -810 & 585 & -207 & 27 \\ -38 & 288 & -810 & 1120 & -810 & 288 & -38 \\ 27 & -207 & 585 & -810 & 585 & -207 & 27 \\ -9 & 72 & -207 & 288 & -207 & 72 & -9 \\ 1 & -9 & 27 & -38 & 27 & -9 & 1 \end{bmatrix}$$

Fig. 5. Differentiating filter masks of size 7×7

$$C = \begin{bmatrix} 1 & -8 & 28 & -56 & 70 & -56 & 28 & -8 & 1 \\ -8 & 64 & -224 & 448 & -560 & 448 & -224 & 64 & -8 \\ 28 & -224 & 784 & -1568 & 1960 & -1568 & 784 & -224 & 28 \\ -56 & 448 & -1568 & 3136 & -3920 & 3136 & -1568 & 448 & -56 \\ 70 & -560 & 1960 & -3920 & 4900 & -3920 & 1960 & -560 & 70 \\ -56 & 448 & -1568 & 3136 & -3920 & 3136 & -1568 & 448 & -56 \\ 28 & -224 & 784 & -1568 & 1960 & -1568 & 784 & -224 & 28 \\ -8 & 64 & -224 & 448 & -560 & 448 & -224 & 64 & -8 \\ 1 & -8 & 28 & -56 & 70 & -56 & 28 & -8 & 1 \end{bmatrix}$$

Fig. 6. Differentiating filter masks of size 9×9

In the ensuing study, the differentiating filter will be applied to the estimation of noise level in images. As it was shown in (8) noise variance equals the variance of the filtered signal divided by the sum of differentiating coefficient, i.e. the sum of coefficients of the filter mask.

4   Results

For noise level estimation three well known, real test images were used - Lenna, Mandrill and Pentagon, and one synthetic image with the constant gray level 128. The images were corrupted by the Gaussian distributed additive noise whose level was incremented within the range $\sigma_n = 1 - 20$. For each value σ_n ten independent noises were generated. For each filtration process the estimation error E_n, i.e. the difference between the true σ_n and the estimated σ_{est} standard deviation of noise, was calculated:

$$E_n = |\sigma_n - \sigma_{est}| \tag{27}$$

Then, for each value σ_n average μE_n and the standard deviation σE_n of the estimation error were calculated:

$$\mu E_n = \frac{\sum_{i=1}^{N} En(i)}{N} \quad \sigma E_n = \sqrt{\frac{\sum_{i=1}^{N} (En(i) - \mu E_n)^2}{N}} \tag{28}$$

where N is a number of images multiplied by a number of noise generations for a given an. The calculations were conducted with the use of Mathcad 2014.

Figure 7 serve as a comparison of values μE_n for different mask sizes shown in Figs. 3, 4, 5 and 6. The smallest estimation errors were obtained for the mask C of the size 9×9, and the largest for the mask Dl of the size 3×3 (Table 1).

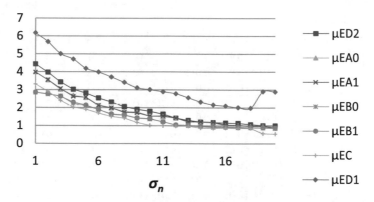

Fig. 7. Average of estimation error μE_n for different mask sizes

Similarly, the smallest values of the standard deviation of the estimation error were obtained for the mask C. In general, the larger the mask size is, the more exact the results are, but at the same time, the less efficient the estimation process is. The results for masks A0 and Al, as well as for B0 and Bl are similar (almost identical), hence it makes no difference which of them is used for noise estimation (Table 2).

The results for the masks Dl, D2 (worst results) and C (best results) were compared to the results of the noise level estimation obtained for other methods well known in the

Table 1. The results of average estimation error μE_n for different mask sizes and methods

σ_n	μED1	μED2	μEA0	μEA1	μEB0	μEB1	μEC	μEAV	μEMED	μELAP	μEBLK
1	6.17	4.43	3.99	3.97	2.84	2.85	3.31	5.45	6.74	3.76	2.64
2	5.68	3.97	3.54	3.55	2.75	2.77	2.91	4.90	6.01	3.35	1.81
3	5.01	3.43	3.02	3.06	2.65	2.63	2.40	4.26	5.54	2.98	1.73
4	4.71	3.02	2.68	2.62	2.23	2.30	2.02	3.90	5.01	2.76	1.54
5	4.19	2.81	2.56	2.58	2.10	2.14	1.91	3.56	4.70	2.56	1.05
6	3.99	2.54	2.12	2.13	1.90	1.89	1.70	3.10	4.32	2.13	1.00
7	3.71	2.31	1.98	1.99	1.62	1.65	1.51	2.96	4.02	2.02	1.00
8	3.41	2.05	1.78	1.74	1.58	1.59	1.42	2.83	3.79	1.98	1.03
9	3.11	1.91	1.76	1.72	1.42	1.43	1.18	2.76	3.54	1.87	1.21
10	3.01	1.80	1.54	1.53	1.37	1.39	1.02	2.30	3.23	1.71	1.38
11	2.89	1.64	1.51	1.54	1.22	1.21	0.99	2.04	3.10	1.68	1.41
12	2.78	1.42	1.42	1.43	1.03	1.04	0.97	1.98	2.96	1.54	1.45
13	2.56	1.30	1.21	1.23	1.00	1.03	0.97	1.84	2.76	1.45	1.67
14	2.32	1.21	1.20	1.22	0.95	0.96	0.87	1.81	2.69	1.34	1.91
15	2.16	1.18	1.16	1.20	0.91	0.94	0.86	1.76	2.51	1.24	1.92
16	2.10	1.16	1.01	1.05	0.92	0.93	0.87	1.71	2.34	1.16	2.01
17	2.01	1.14	1.01	1.03	0.89	0.92	0.86	1.70	2.07	1.15	2.13
18	1.98	1.07	0.97	0.98	0.89	0.90	0.86	1.68	2.01	1.17	2.13
19	2.90	1.01	0.96	0.97	0.88	0.89	0.56	1.67	1.98	1.18	2.16
20	2.89	1.00	0.87	0.89	0.86	0.87	0.54	1.67	2.00	1.17	2.18

Table 2. The results of standard deviation of estimation error σE_n for different mask sizes and methods

σ_n	σED1	σED2	σEA0	σEA1	σEB0	σEB1	σEC	σEAV	σEMED	σELAP	σEBLK
1	0.01	0.01	0.01	0.01	0.01	0.01	0.01	0.01	0.01	0.01	0.05
2	0.01	0.01	0.01	0.01	0.01	0.012	0.01	0.015	0.013	0.02	0.049
3	0.01	0.01	0.01	0.02	0.015	0.01	0.015	0.01	0.12	0.04	0.074
4	0.01	0.02	−.012	0.016	0.017	0.012	0.014	0.02	0.16	0.01	0.1
5	0.01	0.02	0.01	0.017	0.02	0.015	0.015	0.02	0.02	0.02	0.08
6	0.011	0.018	0.015	0.020	0.013	0.02	0.02	0.034	0.021	0.03	0.1
7	0.01	0.018	0.015	0.021	0.02	0.023	0.02	0.04	0.018	0.02	0.09
8	0.01	0.017	0.018	0.018	0.02	0.025	0.03	0.032	0.016	0.021	0.11
9	0.01	0.02	0.02	0.03	0.025	0.03	0.034	0.04	0.02	0.03	0.1
10	0.012	0.024	−.025	0.034	0.03	0.03	0.042	0.034	0.021	0.02	0.12
11	0.015	0.023	0.03	0.037	0.04	0.034	0.04	0.05	0.023	0.04	0.13
12	0.02	0.021	0.04	0.05	0.043	0.034	0.04	0.05	0.021	0.02	0.15
13	0.023	0.03	0.05	0.057	0.047	0.04	0.055	0.04	0.034	0.03	0.12
14	0.02	0.035	0.063	0.062	0.051	0.035	0.05	0.05	0.032	0.023	0.09
15	0.019	0.036	0.06	0.07	0.046	0.04	0.05	0.06	0.025	0.021	0.15
16	0.02	0.04	0.076	0.08	0.05	0.05	0.06	0.09	0.03	0.031	0.14
17	0.035	0,045	0.068	0.08	0.06	0.052	0.055	0.1	0.026	0.03	0.12
18	0.03	0.045	0.08	0.1	0.065	0.055	0.10	0.07	0.03	0.04	0.14
19	0.04	0.05	0.09	0.12	0.063	0.05	0.14	0.09	0.02	0.05	0.18
20	0.04	0.055	0.1	0.11	0.07	0.06	0.08	0.1	0.025	0.05	0.2

literature, and illustrated in form of curves shown in Fig. 8. The methods are referred to by using the following abbreviations:

AV - average-filter estimation method [12],
MED - median-filter estimation method [12],
LAP - Laplace-filter estimation method [9],
BLK - estimation method based on averaging of 10 percent of variances measured in a 7×7 image blocks [10].

The only weakness of the method proposed in this paper is a considerable time-consumption of calculations. The larger the filter mask is, the more time-consuming the calculations are. Therefore, it is reasonable to select only a certain region of an image for calculations, and it could be a field for further investigations.

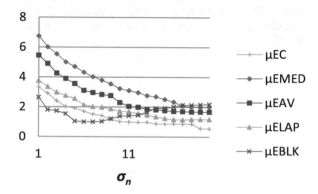

Fig. 8. Average of estimation error μE_n for different estimation methods

References

1. Gradshteyn, I.S., Ryzhik, I.M.: Table of Integrals, Series, and Products. Elsevier (2007)
2. Murtagh, F., Starck, J.L.: Image processing through multiscale analysis and measurement noise modelling. Stat. Comput. J. **10**, 95–103 (2000)
3. Al-Ghaib, H., Adhami, R.: On the digital image additive white Gaussian noise estimation. In: Industrial Automation, Information and Communications Technology (IAICT), pp. 90–96 (2014)
4. Canny, J.: A computational approach to edge detection. IEEE Trans. Pattern Anal. Mach. Intell. **9**, 679–698 (1986)
5. Narayanan, N.R., Ponnappan, S., Reichenbach, S.: Effects of noise on information content of remote sensing images. Geocarto Int. **18**(2), 15–26 (2003)
6. Ghazal, M., Amer, A.: Homogeneity localization using particle filters with application to noise estimation. IEEE Trans. Image Process. **20**(7), 1788–1798 (2011)
7. de Hann, G., Kwaaitaal-Spassova, T.G., Larragy, M.M., Schutten, R.J.: Television noise reduction IC. IEEE Trans. Cons. Electron. **44**(1), 143–154 (1998)
8. Pyatykh, S., Hesser, J., Zheng, L.: Image noise level estimation by principal component analysis. IEEE Trans. Image Process. **22**(2), 687–699 (2013)

9. Immerkaer, J.: Fast noise variance estimation. Comput. Vis. Image Underst. **64**(2), 300–302 (1996)
10. Mastin, G.A.: Adaptive filters for digital noise smoothing: an evaluation. Comput. Vis. Graph. Image Process. **31**, 103–121 (1985)
11. Meer, P., Jolion, J., Rosenfeld, A.: A fast parallel algorithm for blind estimation of noise variance. IEEE Trans. Pattern Anal. Mach. Intell. **12**(2), 216–223 (1990)
12. Olsen, S.I.: Estimation of noise in images: an evaluation. Graph. Models Image Process. **55** (4), 319–323 (1993)
13. Aja-Fernández, S., Pieciak, T., Vegas-Sánchez-Ferrero, G.: Spatially variant noise estimation in MRI: a homomorphic approach. Med. Image Anal. **20**(1), 184–197 (2015)
14. Zhu, S., Yuille, A.: Region competition: unifying snakes, region growing, and Bayes MDL for multiband image segmentation. IEEE Trans. Pattern Anal. Mach. Intell. **18**, 884–900 (1996)
15. Kowalski, J., Peksinski, J., Mikolajczak, G.: Detection of noise in digital images by using the averaging filter name COV. In: Intelligent Information and Database Systems, 5th Asian Conference, ACIIDS 2013, vol. 7803, pp. 1–8. Springer, Berlin (2013)
16. Lee, J.S.: Refined filtering of image noise using local statistics. Comput. Vis. Graph. Image Process. **15**, 380–389 (1981)

Linear Filtering of Blurry Photos

Jerzy Kisilewicz[✉]

Chair of Computer Systems and Networks, Wroclaw University of Science
and Technology, Wybrzeże Wyspiańskiego 27, 50-370 Wrocław, Poland
jerzy.kisilewicz@pwr.edu.pl

Abstract. The paper considers the possibility of using linear two-dimensional
filters to improve the sharpness of images. It is assumed that the point spread
function (PSF) dissipates point in a circle of radius r. For given r, the linear
filters are calculated to minimize the mean square error between the point image
and the result of the convolution of the filter and the image of the blurred point.
The results of filtration of Cameraman and Lena_color images previously dispersed using the assumed PSF are shown.

Keywords: Two-dimensional filters · Image processing · Image equalization

1 Introduction

The equalization of blurry or stirred photos is one of the most important and interesting
problems in image processing. These defects are very popular and they are difficult to
equalize. They are much more difficult than the noise reduction or equalization of
improper exposure.

Although the mathematical theory of equalization of blurry or stirred images has
been developed for about 80 years, there are no effective algorithms that give good
results. The obtained results, although they achieve a significant improvement in the
clarity of the image, usually are burdened with visible distortions, and obtaining the
satisfactory results often requires many hours or several days of calculations. [3, 5, 6].

Consider a monochrome image with dimensions $A \times B$ points whose intensity is
determined by the function $f(x, y)$, where $0 \leq x \leq A - 1$, $0 \leq y \leq B - 1$. The
spread function of $(2a + 1) \times (2b + 1)$ points affects on this image. The points
intensity of that function is determined by the $h(x, y)$, where $-a \leq x \leq a$, $-b \leq x$
b, where $(2a + 1) < A$ and $(2a + 1) < B$. Also the additive noise $n(x,y)$ has influence
on this image. The points intensity of the resulted blurred image is described by the
function $g(x, y)$ which is given by the formula [2, 9]

$$g(x,y) = h(x,y) * f(x,y) + n(x,y) \tag{1}$$

where * is a two-dimensional convolution operator

$$h(x,y) * f(x,y) = \sum_{i=-a}^{a}\sum_{j=-b}^{b} h(i,j)f(x-i,y-j). \tag{2}$$

© Springer Nature Switzerland AG 2019
K. Choroś et al. (Eds.): MISSI 2018, AISC 833, pp. 67–76, 2019.
https://doi.org/10.1007/978-3-319-98678-4_9

The problem of image g defocus equalization is to find the image $\tilde{f}$ which is as close as possible to the image f. However, the function $h(x, y)$ may or may not be known [1, 4, 6]. In further considerations, we will assume that the function $h(x, y)$ is known and we will omit the noise. Therefore, the equalization problem will be reduced to the pure deconvolution problem.

Many of known methods use the Fourier transformation [6, 9]. Using the two-dimensional Fourier transform to the images g, h, f and n we transform (1) into

$$G(u, v) = H(u, v)F(u, v) + N(u, v) \tag{3}$$

where G, H, F and N represent images g, h, f and n in the two-dimensional space of frequencies u and v, and the product $H(u, v)F(u, v)$ is the ordinary product of "element times element", not the matrix product.

Theoretically dividing (3) by $H(u, v)$ we obtain

$$\tilde{F}(u, v) = \frac{G(u, v)}{H(u, v)} = F(u, v) + \frac{N(u, v)}{H(u, v)}. \tag{4}$$

So we have to construct a $\tilde{c}(x, y)$ filter with a two-dimensional transform

$$\tilde{C}(u, v) = \frac{1}{H(u, v)}. \tag{5}$$

Unfortunately, the practical implementation of the formula (4) and the filter (5) is difficult because the function $H(u, v)$ for some u, v has values close to zero or even equal to zero [9].

Another, iterative approach, called the Lucky-Richardson method, was proposed by Richardson [7] in 1972 and by Lucy [5] in 1974, and it has been improved by many authors [3]. The weak point of this method is the problem of the iteration stop criterion. This method is based on the formula

$$\tilde{f}_{k+1}(x, y) = \tilde{f}_k(x, y)\left[h(-x, -y) * \frac{g(x, y)}{h(x, y) * \tilde{f}_k(x, y)}\right], \tag{6}$$

where k is the iteration number, and $*$ is a two-dimensional convolution.

This method is often used in programs that process astronomical images, and calculations can take up to several days [9].

In this work, we will calculate two-dimensional linear filters that minimize the mean square error between the corrected image and the reference image, and we will show the use of these filters for blurry images equalization.

2   Linear Equalization Filter

In the vast majority of practical cases, the image becomes out of focus when each of its points is transformed into a circle with a defined radius [1, 9]. If the radius of this circle is larger, then the picture is more out of focus. Therefore, if the original image is one white point in the center of a black background, the blurred (dispersed) picture presents a gray circle. This blurred image of a single point is called the point spread function (PSF). In formulas (1) to (9) this is the function $h(x, y)$. PSF characterizes the dispersion of original image [1, 4, 6, 9]. Finding the PSF is a separate problem and will not be considered here. In this paper, we assume the same brightness of all points of the circle in PSF. Such an assumption well approximates the practically existing PSFs [1, 9].

If the circle radius of a dissipated point is equal to r pixels, then we assume that the PSF is given by

$$h(x, y) = \begin{cases} s & \text{when} \quad x^2 + y^2 \leq r^2 \\ 0 & \text{when} \quad x^2 + y^2 > r^2 \end{cases}, \tag{7}$$

where $s = \frac{1}{\pi r^2}$.

We will look for such a function $c(x, y)$, which minimizes the sum of squares of differences between the points of the convolution $c(x,y)*h(x,y)$ and the points of the picture $\underline{u}(x,y)$ of a single white point. So we will solve the equation

$$\operatorname*{grad}_{c(x,y)} (c(x, y) * h(x, y) - \underline{u}(x, y))^2 = 0. \tag{8}$$

To provide (8) in a matrix form, let us represent the convolution $c*h$ as a product of matrices. Assuming that c is an image with dimensions $n \times m$ points, and h is an image with dimensions $N \times M$ points and also indexing rows and columns from zero, according to (2) we get

$$u(x, y) = c(x, y) * h(x, y) = \sum_{i=0}^{n-1} \sum_{j=0}^{m-1} c(i,j) h(x - i, y - j), \tag{9}$$

where $0 \leq x < \underline{N} = N + n - 1$, $0 \leq y < \underline{M} = M + m - 1$.

Let the images u and $\underline{u}$ will be represented by columnar vectors, each of which has $\underline{N} \times \underline{M}$ elements

$$U = \left[u_{x\underline{M}+y} = u(x, y) \right], \tag{10a}$$

$$\underline{U} = \left[\underline{u}_{x\underline{M}+y} = \underline{u}(x, y) \right]. \tag{10b}$$

Similarly, the filter c will be represented by columnar vector with $n \times m$ elements

$$C = \left[c_{im+j} = c(i,j) \right].$$ (11)

Let's define the matrix H with $\underline{N} \times \underline{M}$ rows and $n \times m$ columns

$$H = \left[h_{x\underline{M}+y,\, im+j} = \begin{cases} h(x-i, y-j) & \text{when} \quad 0 \leq x-i < N, \quad 0 \leq y-j < M \\ 0 & \text{otherwise} \end{cases} \right].$$ (12)

Now, the convolution (9) can be written as the product of matrix H by vector C, so $U = HC$. The condition (8) reduces to the calculation of C from the system of $n \times m$ linear equations

$$H^{\mathrm{T}}HC = H^{\mathrm{T}}U,$$ (13)

where T represents a matrix transposition.

3   Examples of Filters

We will examine the possibility of correction of less blurry and more blurry images. Assume three cases in which one bright pixel is blurred on the circle of radius r equal to 5, 10 and 14 pixels. It was assumed that the side of the square filter would be k times larger than the diameter of the circle of the blurred point. For each of these cases, we calculate from (13) 2 or 3 filters $c(x, y)$ with the dimensions of $n = m = 2kr + 1$, where:

- for $r = 5$, was assumed $k = 5, 7$ and 8, obtaining $n = m = 51, 71$ and 81,
- for $r = 10$ was assumed $k = 3$ and 4 obtaining $n = m = 61$ and 81,
- for $r = 14$ was assumed $k = 2$ and 3 obtaining $n = m = 57$ and 85.

The calculations assume that the dimensions $N = M$ of images $h(x,y)$ are the same as the dimensions $n = m$ of the calculated filters $c(x,y)$. The resulting images $u(x,y)$ and $\underline{u}(x,y)$ according to (9) had the following dimensions: $\underline{N} = \underline{M} = 101, 141, 161$ for $r = 5$; 121, 161 for $r = 10$ and 113, 169 for $r = 14$. The resulting dimensions of the matrices are included in Table 1.

The $H^{\mathrm{T}}H$ product computation time was from 30 min for $n = 51$ to 15 h for $n = 81$ and approx. 26 h for $n = 85$. The computation time of solving Eqs. (13) was much shorter and ranged from a few minutes for 2601 equations, about 1 h and 45 min for 6561 equations to approx 2 h and 30 min for 7225 equations.

The efficiency of the equalization of the point which was blurred into a circle with a radius of $r = 5$ pixels, is shown in Fig. 1. Figure 1a shows an image of a blurred point

Table 1. The dimensions of the matrices H, $H^{T}H$, C and U

r	k	n	N	Rows and columns of H	Rows and columns of $H^{T}H$	Rows of C	Rows of U
5	5	51	101	10201, 2601	2601, 2601	2601	10201
5	7	71	141	19881, 5041	5041, 5041	5041	19881
5	8	81	161	25921, 6561	6561, 6561	6561	25921
10	3	61	121	14641, 3721	3721, 3721	3721	14641
10	4	81	161	25921, 6561	6561, 6561	6561	25921
14	2	57	113	12769, 3249	3249, 3249	3249	12769
10	3	85	169	28561, 7225	7225, 7225	7225	28561

where pixels are lightened from level 3 to 255 to make a circle visible. Figure 1b shows the image after equalization with one visible bright pixel. Dark invisible pixels appearing after their brightening are shown in Fig. 1c.

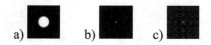

Fig. 1. Images for $r = 5$ and $k = 8$: (a) blurred point, (b) blurred point after equalization, (c) blur-red point after equalization with the artificially brightened environment around the middle point

4 Application of Filters for Image Equalization

To demonstrate the sharpness equalization, two known photos were selected from [10]: Cameraman with sizes 256×256 pixels and Lena_color with sizes 512×512 pixels. These two images were convoluted with the PSF given by formula (7) for $r = 5$, 10 and 14 to get the out of focus images for experiments, like in [4]. These images being out of focus were equalized using the filters calculated in the previous chapter.

In Fig. 2 are shown the Cameraman image blurred by PSF with $r = 5$ and the equalization results by using the filters of various sizes for $k = 5$, 7 and 8. The equalized image around the objects shows unpleasant reflections. As the filter size increases, these reflections become less visible, but the visible noise increases. However, for $r = 10$, the larger filter gave an image with less noise, as shown in Fig. 3. For $r = 14$, a better effect was obtained using a smaller filter for $k = 2$, because for $k = 3$ a very large noise was obtained, which can be seen in Fig. 4.

Fig. 2. Cameraman, $r = 5$: (a) before equalization, (b) after equalization using the filter with $k = 5$, (c) after equalization using the filter with $k = 7$, (d) after equalization using the filter with $k = 8$

Figures 5, 6 and 7 show the best results of the equalization of Lena_color images blurred successively for $r = 5$, 10 and 14. In the case of $r = 5$ and 10, a better effect was obtained by using a larger filter. For $r = 14$, the smaller filter gave the better result, because the larger one gave too much noise, similar how in case the Cameraman image equalization.

The equalizer's size k selection is the result of a compromise between the improvement of sharpness and the amount of noise in the image and depends on the equalization purpose. If the image should to be pleasant to watch and be closer to the

Fig. 3. Cameraman, $r = 10$: (a) before equalization, (b) after equalization using the filter with $k = 3$, (c) after equalization using the filter with $k = 4$

original, a smaller equalization filter will usually be better. If the image has to show more details, choose a larger equalization filter.

In the case of color images, we assume in experiments that these images are coded in the RGB system and we will use the same equalization filter three times, separately for each color.

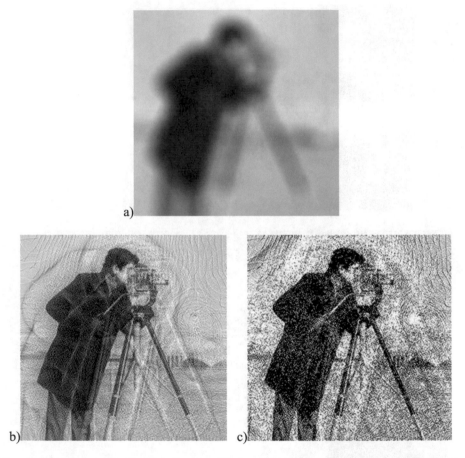

Fig. 4. Cameraman, $r = 14$: (a) before equalization, (b) after equalization using the filter with $k = 2$, (c) after equalization using the filter with $k = 3$

Fig. 5. Lena, $r = 5$: (a) before equalization, (b) after equalization using the filter with $k = 8$

Fig. 6. Lena, $r = 10$: (a) before equalization, (b) after equalization using the filter with $k = 4$

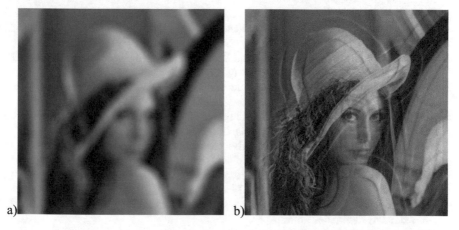

Fig. 7. Lena, $r = 14$: (a) before equalization, (b) after equalization using the filter with $k = 2$

5 Conclusion

In the equalization examples presented in this paper, the images with known point spread function (PSF) have been used. In fact, this function is not exactly known. However, you can count equalization filters for a set of typical PSFs. This requires multiple execution of several-hour computations, but such calculations will be performed once for each filter. Each filter can be used repeatedly. The equalization itself requires computations from several seconds to several minutes, depending on the size of the image. Therefore, you can equalize a blurred image many times using different filters and choose the best result.

76   J. Kisilewicz

The obtained equalized images contain noise and visible distortions. However, they are much clearer and show details that are invisible or unreadable in the blurred images. An exemplary application may be to obtain readability of blurred text. Completely invisible face of the cameraman in the images before the equalization, becomes recognizable in Figs. 2, 3c and 4b, similar to the face of Lena in Fig. 7.

Acknowledgments This work was supported by statutory funds of the Department of Systems and Computer Networks, Wroclaw University of Science and Technology, grant No. 0401/0154/17.

References

1. Chan, T.F., Wong, C.-K.: Total variation blind deconvolution. IEEE Trans. Image Process. **7**(3), 370–375 (1998)
2. Chia-Feng, C., Jiunn-Lin, W., Ting-Yu, T.: A single image deblurring algorithm for nonuniform motion blur using uniform defocus map estimation. In: Mathematical Problems in Engineering, vol. 2017 (2017)
3. Jiunn-Lin, W., Chia-Feng, C., Chun-Shih, C.: An adaptive richardson-lucy algorithm for single image deblurring using local extrema filtering. J. Appl. Sci. Eng. **16**(3), 269–276 (2013)
4. Kotera, J., Šmídl, V., Šroubek, F.: Blind deconvolution with model discrepancies. IEEE Trans. Image Process. **26**(5), 2533–2544 (2017)
5. Lucy, L.B.: An iterative technique for the rectification of observed distributions. Astron. J. **79**, 745–754 (1974)
6. Ren, D., Zuo, W., Zhang, D., Xu, J., Zhang, L.: Partial deconvolution with inaccurate blur kernel. IEEE Trans. Image Process. **27**(1), 511–524 (2018)
7. Richardson, W.H.: Bayesian-based iterative method of image restoration. J. Opt. Soc. Am. **62**, 55–59 (1972)
8. Tai, Y.W., Tan, P., Brown, M.S.: Richardson-lucy deblurring for scenes under a projective motion path. IEEE Trans. Pattern Anal. Mach. Intell. **33**(8), 1603–1618 (2011)
9. Yuzhikov, V.: Restoration of defocused and blurred images. http://yuzhikov.com/articles/BlurredImagesRestoration1.htm
10. Test images Collections: http://www.hlevkin.com/06testimages.htm

A First Summarization System of a Video in a Target Language

Kamel Smaïli[1]([⊠]), Dominique Fohr[1], Carlos-Emiliano González-Gallardo[2],
Michał Grega[3], Lucjan Janowski[3], Denis Jouvet[1], Artur Komorowski[3],
Arian Koźbiał[3], David Langlois[1], Mikołaj Leszczuk[3], Odile Mella[1],
Mohamed A. Menacer[1], Amaia Mendez[4], Elvys Linhares Pontes[2],
Eric SanJuan[2], Damian Świst[3], Juan-Manuel Torres-Moreno[2,5],
and Begona Garcia-Zapirain[4]

[1] Loria University of Lorraine, Nancy, France
kamel.smaili@loria.fr
[2] LIA Université d'Avignon et des Pays de Vaucluse, Avignon, France
[3] AGH University of Science and Technology Kraków, Kraków, Poland
[4] University of DEUSTO Bilbao, Bilbao, Spain
[5] Ecole Polytechnique de Montréal, Montreal, Canada

Abstract. In this paper, we present the first results of the project AMIS
(Access Multilingual Information opinionS) funded by Chist-Era. The
main goal of this project is to understand the content of a video in a
foreign language. In this work, we consider the understanding process,
such as the aptitude to capture the most important ideas contained in a
media expressed in a foreign language. In other words, the understanding
will be approached by the global meaning of the content of a support
and not by the meaning of each fragment of a video.

Several stumbling points remain before reaching the fixed goal. They
concern the following aspects: Video summarization, Speech recognition,
Machine translation and Speech segmentation. All these issues will be
discussed and the methods used to develop each of these components will
be presented. A first implementation is achieved and each component of
this system is evaluated on a representative test data. We propose also
a protocol for a global subjective evaluation of AMIS.

Keywords: Video summarization · Speech recognition
Machine translation · Text boundary segmentation
Text summarization · Sentence compression

1 Introduction

Nowadays, the information is widely available in different medias: TV, social networks, newspapers, etc. The main difference in comparison to what we had one
or two decades before is that people can access to the videos of social networks

Supported by Chist-Era (AMIS project).

in foreign languages. When, the video does not necessitate any understanding, there is no main problem. In the opposite, when the information necessitates to understand the language, a human being is limited in terms of mastering foreign language, even if YouTube proposes a rough translation of some contents. In AMIS, a Chist-Era project[1], the main objective is to make available a system, helping people to understand the content of a source video by presenting its main ideas in a target understandable language. The understanding process is considered here to be the comprehension of the main ideas of a video. We think that the best way to do that, is to summarize the video for having access to the essential information. Henceforth, AMIS focuses on the most relevant information by summarizing it and by translating it to the user if necessary. As a result, AMIS will permit to have another side of story of an event since we can, for instance, have the Russian version of the war in Syria. Several skills are necessary to achieve this objective: video summarization, automatic speech recognition, machine translation, text summarization, etc. Each output of a subsystem of AMIS can enrich in upstream or downstream the other modules. That makes AMIS working such as a workflow where the flow refers to the information necessary for a component. In this article, we will present the first result that works such as a pipeline system connecting the output of each component to the input of the next one.

2   Different Components of AMIS

2.1   Video Summarization

We designed and developed an operational framework for summarization of newscasts and reports [11]. The framework is designed in such a way, that it allows for easy experimentation with different approaches to video summarization. The framework hosts several high- and low-level meta-data extraction algorithms (referring to our former research conducted within the scope of e.g. IMCOP project [1]) that include detection of the anchor-person, recognition of day and night shots and extraction of low-level video quality indicators. The main summarization processes start with Shot Boundary Detection (SBD). This algorithm helps in prediction whether video is static or dynamic. Also, through SBD we can calculate and compare data per shots instead of frames which is a way more efficient while analyzing video clips over longer durations. The video quality indicators mentioned above are used for calculating coefficient of activity which is a product of two indicators – Spatial Activity and Temporal Activity. These indicators show amount of details appearing on the frame and how dynamic the frame is in comparison to the previous frame, respectively. The coefficient of activity is calculated in two steps – firstly per each frame and then as an average per each shot. The final summarization is built from shots with higher or equal value of coefficient of activity than average value for entire video.

---

[1] http://deustotechlife.deusto.es/amis/.

2.2 Speech Recognition System

Automatic speech recognition (ASR) and machine translation (MT) are among the AMIS project key technologies for understanding videos. Although these items were developed as separate modules, the goal is to include them more tightly with other modalities and other processes. AMIS project deals with videos in French, English and Arabic languages. Arabic is considered as one of the foreign input language of the videos. That is why an Arabic automatic speech recognition system has been developed in Loria. This system is based on the state-of-the-art methods and is trained on large acoustic and text corpora. For that we developed *ALASR: Arabic Loria Automatic Speech Recognition system* [16]. A speech recognition system needs at least two components: an acoustic model and a language model. In the following, each component is described by presenting the different steps to train such models. Training necessitates two kinds of data: acoustic and textual, which are presented in the following.

Acoustic Model. The development of the acoustic model is based on Kaldi [18] recipe, which is a state-of-the-art toolkit for speech recognition based on Weighted Finite State Transducers [17]. The ASR system uses 13-dimensional Mel-Frequency Cesptral Coefficients (MFCC) features with their first and second order temporal derivatives, which leads to 39-dimensional acoustic features. For Arabic, 35 acoustic models (28 consonants, 6 vowels and silence) are trained. The emission probabilities of the HMM models are estimated by DNN (namely DNN-HMM). The DNN-HMM are trained by applying sMBR criterion [24] and using 40 dimensional features vector (fMLLR) [6] for speaker adaptation. The topology of the neural network is as follows: a 440-dimensional input layer (40 × 11 fMLLR vectors), 6 hidden layers composed by 2048 nodes and a 4264-dimensional output layer, which represents the number of HMM states. And finally, the total number of weights to estimate is about 30.6 million.

Language Model. In Arabic, even in newspapers, several words could be simplified especially at the beginning or at the end by replacing a specific letter by another one or by omitting the *hamza* symbol. Unfortunately, both writing of a word may exist in a same document. This leads to share a probability over two forms that correspond to the same word. For instance, the word إِسْتِعْمَال (*uses*) is the right way to write it, but people could omit the *hamza* and write it such as: استِعْمَال. Obviously, this is not the only case, several other points have to be treated. Some of them are specific to Arabic and others are used in the majority of other natural languages. Several preprocessing tasks have been done on the corpora before calculating the language model [16]. The training corpus is composed by GigaWord and the speech transcripts. As this data set is unbalanced, a 4-gram language model, for each part of this corpus, has been developed and combined linearly. The optimal weights are determined in order to maximize the likelihood of the development corpus. Due to the memory constraints, the full 4-gram language model has been pruned by minimizing the relative entropy between the full and the pruned model [21].

2.3 Machine Translation

A statistical machine translation has been developed in the direction Arabic – English, since Arabic is considered such as the foreign language of the video to translate to a summarized video in English. For training, we use a parallel (Arabic – English) corpus of 9.7 million sentences extracted from United Nation corpus concerning the period of January 2000 – September 2009. A 4-gram language model has been trained on the target language of the mentioned parallel corpus. The vocabulary contains 224,000 words. The development and the test corpus are composed by 3,000 parallel sentences. This component is a statistical machine translation, but in the next version of AMIS, we will use a neural network machine translation [15].

2.4 Text Summarization

Automatic Text Summarization (ATS) is a Natural Language Processing (NLP) task [23]. The main objective of ATS is to find the most important information from a text source in order to produce an abridged and informative version. In the AMIS project, the ATS component aims to summarize a newscast or report video in French, English or Arabic languages based only on the textual information provided by the ASR or by the MT modules. ATS is divided in three different families depending on the followed approach to generate a summary: summarization by extraction, summarization by abstraction and summarization by sentence compression [23]. We implemented mainly the extractive summarization paradigm in the project because it is more robust to external noise like speech disfluencies and ASR errors [2,4]. Also, we performed some exploratory experiments using the sentence compression paradigm.

Extractive Text Summarization. Extractive Text Summarization (ETS) aims to select the most pertinent segments of the transcribed video based on different criteria like information content, novelty factor and relative position. ARTEX (Autre Résumer de TEXtes), originally developed for French, English and Spanish is an ETS system described in [22] by Torres-Moreno *et al.* Within the AMIS project, a Modern Standard Arabic (MSA) extension has been developed and added to ARTEX, making possible the generation of summaries in all the languages involved in AMIS.

Multi-sentence Compression. Multi-Sentence Compression (MSC) combines the information of a cluster of similar sentences to generate a new sentence, hopefully grammatically correct, which compresses the most relevant data of this cluster. Among several state-of-the-art MSC methods, Linhares Pontes *et al.* [12] used an Integer Linear Programming (ILP) formulation to guide the

MSC using a list of keywords. Our system incorporates this approach to use the keywords of a transcribed video to guide the compression of similar sentences and to improve the informativeness of summaries.

Independently of the summarization approach ATS relies on the existence of sentences either to select, reformulate or compress the source text. One main issue to take into account is that in the AMIS project, the source from which a text summary is created is the transcript of a ASR system or its translation; this transcript does not contain any punctuation mark, hence sentences are nonexistent. In order to solve this problem, a specialized module of sentence boundary detection has been developed.

Sentence Boundary Detection. Deep research has been done concerning Sentence Boundary Detection (SeBD) of the ASR transcripts covering the three languages of the project: English, French and Arabic. Written text differs from spoken language in the way the writer/speaker expresses its ideas. In spoken language, sentences are not as well defined as in written text, in this context a segment is defined by a sentence-like unit (SU). Well-formed sentences, phrases and single words can be considered SUs [13]. The developed SeBD system uses mainly textual features and convolutional neural networks (CNN) to segment the transcripts and generate SUs [7]. Arabic SeBD module was trained with a subset of 50 million words of the Arabic Gigaword Corpus. Besides ATS, SeBD has shown to be of vital importance for other NLP tasks in the AMIS project like NMT. For these reason a stand-alone version of the module has been deployed to be used independently by other AMIS modules.

2.5 Audio Summarization

Audio summarization task is a new approach to generate speech-to-speech summaries. This is a difficult task, because in the audio signal there is not any linguistic information (words, sentences, etc.). Although low level features such as signal intensity, starting and ending time could be used for extracting and selecting informative segments, reliable methods for automatic extraction of high level features (prosodic and linguistic features) from spontaneous speech have not yet been established [5,25]. We explore several neural architectures using deep learning in order to find the best features in the hidden layer. These features will allow us to capture the most important abstract structure (linguistic level) from a low level resource (signal).

3 Global Architecture

Several architectures are proposed to summarize a video in a target language. These architectures are presented in Fig. 1. The four ($SC1$, $SC2$, $SC3$ and $SC4$) architectures are different. In $SC1$, a summary video is achieved directly without using any knowledge on the audio content of the video. The content of the result

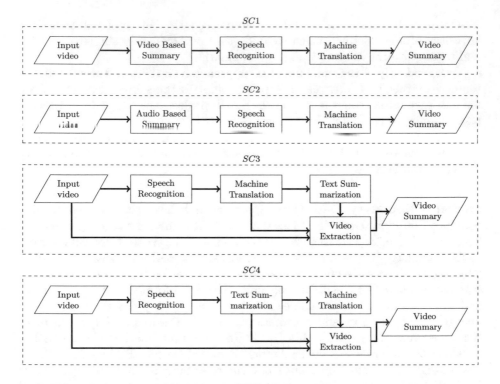

Fig. 1. Different architectures for summarizing a video to a target language

of this summary is then transcribed thanks to ALASR (our speech recognition system) then translated to English. The result of the translation is integrated such as subtitles to the summarized video.

$SC2$ is an original architecture in which an audio summary is proposed on the audio part of the original video. The result is then transcribed and translated such as in $SC1$. $SC3$ and $SC4$ are similar architectures, they take benefit from the result of ALASR. The result of this step is then a text in Arabic, then the blocks (Machine Translation + Text Summarization) and (Text Summarization + Machine Translation) are respectively performed on $SC3$ and $SC4$. In this paper, only the first architecture ($SC1$) is presented.

4 Objective Evaluation

In order to appreciate the relevance of the whole system developed, it is necessary to evaluate it globally, but also each component should be evaluated. In the following, we will evaluate each component since the consequences of the weakness of a system could be propagated to the other components.

Video Summarization. The framework is evaluated using annotated video sequences. A pool of experts decides which frames are key frames (meaning: very important frames, core of the video) and which have to be in the summary. We use VLC media player to extract frames from single shots. Obviously, this evaluation process is time-consuming and subjective when the key frames are chosen. In order to describe the obtained results, we are calculating Precision, Recall and F1 score for each sequence and algorithm. We considered video summarization based only on the visual evaluation. Both human and algorithm were provided with video without any additional audio description. Of course a person creating the evaluation could understand some written text appearing on the screen. Our goal was to validate if the summary created by human is similar to the summary created by algorithm, focusing on the visual part only. Such evaluation can be found in literature [8,9,19,20] just to name a few. Nevertheless the summary of news was not considered. In our research we noted that making a reasonable summary for human observer is very difficult. As a consequence comparing to an algorithm is not as precise as for other cases considered in literature. The first problem we found is the length of summary. In Table 1 the length of summaries provided by a human observer are presented. We can see that the shortest is just 15% of the original video while the longest is 61%. Comparing such different solutions is difficult. The difference comes from very different natures of the video news, which can span from a talking head to a report from a field where there is an action. In order to help with comparing human and automatic summarization, the automatic algorithm has the information about the length of the summary provided by humans.

Table 1. Summary length comparison. The summaries in this table are created by human.

Video ID	Summary length	Source length	Percentage
1	205	473	43%
2	86	187	46%
3	76	277	27%
4	69	200	35%
5	85	186	36%
6	119	194	**61%**
7	41	281	**15%**
8	41	233	18%

The automatic and human summaries were compared by precision and recall metrics. We calculated how many frames marked by human were also marked by an algorithm. So true positive means that both human and algorithm marked the same frame. True negative means that both human and algorithm did not mark specific frame. The results obtained for 50 evaluated sequences are presented in Table 2. The obtained results are not very good but even comparing summaries provided by two humans are not much better. The problem is the content, which is already a summary.

Table 2. Performance of video summarization

Recall	Precision	F1	Accuracy
0.13	0.36	0.19	0.36

Arabic Automatic Speech Recognition: The ASR system for Arabic has been trained on an acoustic corpus of 63 h (Nemlar [14] and NetDC [3]). The Language Model has been trained on the GigaWord. The vocabulary is composed of 95k words with an average of 5.07 pronunciations for each entry. After several tests, tuning and improvements ALASR achieves the performance presented in Table 3. This performance is achieved on a tuning and a test corpora of 31,000 sentences for each of them. This test is done on data not extracted from our video database. The issue is that we do not have any reference transcription corpus for these videos. The evaluation is then impossible. To overcome this problem, we decided to build a pseudo-reference by aligning, for each YouTube video from Euronews channel, the automatic transcription and textual data from the corresponding Youtube and Euronews webpages. The transcription is considered as a reference if the WER is under a chosen threshold [10]. Experiments have been done on a corpus of 1,300 sentences (a mixture of transcriptions from YouTube and Euronews). We have to notice, that this transcription does not correspond exactly to what has been pronounced. Consequently, the performance we provide below is under-estimated. Under these conditions, ALASR achieves a WER of 36.5.

Table 3. Performance of ALASR in terms of WER

	Dev WER	Test WER
ALASR	13.07	14.02

Machine Translation. The machine translation system has been evaluated on separated corpora: on a general one and on data extracted from our video database. In Table 4, we give the result of Arabic-English machine translation system evaluated on 3,000 sentences extracted from a corpus of the United Nations [26]. As for our ASR system, the evaluation on our database is difficult because we do not have any reference corpus for Machine Translation. To overcome this limit, we decided to create two reference corpora. The corpora are composed respectively by the translation of 197 videos of Euronews that correspond to 1,253 sentences achieved by Google and SYSTRAN. It means that we consider the result of the translation of Google respectively SYSTRAN as the

Table 4. The evaluation of the Arabic–English MT system.

	Test (3k sentences)
BLEU	39

references. The results are given in Table 5. The achieved BLEU for our system when SYSTRAN is considered such as the reference is 9.9, while the results when the translation of Google is considered as the reference is equal to 26.7. To understand the weak performance we get with SYSTRAN as reference, we used Google as a translator. The achieved BLEU for this experience is 12.8. This shows that both Google and our system fails to get good results with a translation produced by SYSTRAN.

Table 5. MT evaluation on AMIS data.

System	AMIS	Google	Systran
AMIS		26.7	9.9
Google	26.7		12.8
Systran	10	12.9	

Sentence Boundary Detection. The SeBD module was approached as a classification task, where the system should decide if a target word corresponds or not to a boundary between two SUs. Table 6 shows the results of a strict evaluation for the Arabic SeBD module over 12M samples of the Arabic Gigaword corpus. The method seems to perform really well concerning the "no boundary" class (NO_BOUND); both Precision and Recall achieve a value over 92%. By contrast the performance related to the "boundary" class (BOUND) drops almost 15% for Precision and 32.5% for Recall. The unbalanced nature of the data influences this drop in performance. The "no boundary" class represents the 84% of the samples, against the 16% from the "boundary" class. Further work is in develop to reduce the gap between both classes.

Table 6. Performance of the AMIS SeBD

Class	Precision	Recall	F1
NO_BOUND	0.928	0.963	0.945
BOUND	0.782	0.638	0.700

5   Subjective Evaluation

In the previous section, we evaluated each component independently from the others, in this section we propose a method to evaluate the system as a whole.

The method is based on a questionnaire that involves the final users, which is considered as the best indicator of the quality of this type of systems. The evaluators of the summarized videos will be 12 participants in total, 6 in Arabic (3 men and 3 women per language). The inclusion criteria are being 18 years old, with at least high school level, while the exclusion criteria is having understanding problems, reading or writing impairment. The material to be evaluated will consist of 25 videos tackling mixed topics: Politics, Soccer, War, Homosexuality. The number of summarized videos per user participating as tester is 3. The aim of the designed questionnaires is analyzing the quality of the video summarization with the best precision, taking into account the proposed resources. So, we will include 2 generic questions for all the videos, that can be evaluated from 0 ("Not done") to 4 ("Excellent"), and a set of 3/5 specific questions (depending of the length of the video) with 3 possible answers. In that way, we will know if the summary is understandable and if there is any part out of context, using the generic questions, and if the main ideas of the original video have been gathered by the summarization, using the video specific questions. The assessment data analysis will consist on statistical analysis of questionnaires and the application of some machine learning techniques, if possible for clustering and comparison purposes between genders, languages, etc.

6   Conclusion

In this article, several research aspects have been investigated through AMIS project. An understanding system of a foreign video has been developed by summarizing the source video. The objective was to capture the main idea of a video and to restore it into English. A whole system has been carried out by implementing several sub-systems, where each of them represents a real scientific challenge. The different sub-systems were assembled to give rise to the first version of AMIS. The system is operational and the results are certainly improvable, but are better than what we expected. We thought that the serialization of these systems had to produce very bad results, but this is not the case. We are currently working on developing more relevant architectures that would probably yield better results. In parallel, each sub-system is subject to regular improvements making the global system better.

Acknowledgment. We would like to acknowledge the support of Chist-Era for funding this work through the AMIS (Access Multilingual Information opinionS) project. Research work funded by the National Science Center, Poland, conferred on the basis of the decision number DEC-2015/16/Z/ST7/00559.

References

1. Baran, R., Zeja, A.: The IMCOP system for data enrichment and content discovery and delivery. In: 2015 International Conference on Computational Science and Computational Intelligence (CSCI), pp. 143–146, December 2015. https://doi.org/10.1109/CSCI.2015.137

2. Bell, P., Lai, C., Llewellyn, C., Birch, A., Sinclair, M.: A system for automatic broadcast news summarisation, geolocation and translation. In: INTERSPEECH, pp. 730–731 (2015)
3. Choukri, K., Nikkhou, M., Paulsson, N.: Network of data centres (NetDc): BNSC-an Arabic broadcast news speech corpus. In: LREC (2004)
4. Christensen, H., Kolluru, B., Gotoh, Y., Renals, S.: From text summarisation to style-specific summarisation for broadcast news. In: European Conference on Information Retrieval, pp. 223–237. Springer (2004)
5. Furui, S., Kikuchi, T., Shinnaka, Y., Hori, C.: Speech-to-text and speech-to-speech summarization of spontaneous speech. IEEE Trans. Speech Audio Process. **12**(4), 401–408 (2004)
6. Gales, M.J.: Maximum likelihood linear transformations for hmm-based speech recognition. Comput. Speech Lang. **12**(2), 75–98 (1998)
7. González-Gallardo, C.E., Torres-Moreno, J.M.: Sentence boundary detection for French with subword-level information vectors and convolutional neural networks. arXiv preprint arXiv:1802.04559 (2018)
8. Gygli, M., Grabner, H., Gool, L.V.: Video summarization by learning submodular mixtures of objectives. In: 2015 IEEE Conference on Computer Vision and Pattern Recognition (CVPR), pp. 3090–3098, June 2015. https://doi.org/10.1109/CVPR.2015.7298928
9. Huang, M., Mahajan, A.B., Dementhon, D.F.: Automatic performance evaluation for video summarization. Technical report
10. Jouvet, D., Langlois, D., Menacer, M.A., Fohr, D., Mella, O., Smaïli, K.: Adaptation of speech recognition vocabularies for improved transcription of YouTube videos. In: Proceedings of the ICNLSSP Conference (2017)
11. Leszczuk, M., Grega, M., Koźbiał, A., Gliwski, J., Wasieczko, K., Smaïli, K.: Video summarization framework for newscasts and reports - work in progress. In: Dziech, A., Czyżewski, A. (eds.) Multimedia Communications, Services and Security, pp. 86–97. Springer International Publishing, Cham (2017)
12. Linhares Pontes, E., Huet, S., Linhares, A.C., Torres-Moreno, J.M.: Multi-sentence compression with word vertex-labeled graphs and integer linear programming. In: Proceedings of TextGraphs-12: The Workshop on Graph-based Methods for Natural Language Processing. Association for Computational Linguistics (2018)
13. Liu, Y., Chawla, N.V., Harper, M.P., Shriberg, E., Stolcke, A.: A study in machine learning from imbalanced data for sentence boundary detection in speech. Comput. Speech Lang. **20**(4), 468–494 (2006)
14. Maegaard, B., Choukri, K., Jørgensen, L.D., Krauwer, S.: NEMLAR: Arabic language resources and tools. In: Arabic Language Resources and Tools Conference, pp. 42–54 (2004)
15. Menacer, M.A., Langlois, D., Mella, O., Fohr, D., Jouvet, D., Smaïli, K.: Is statistical machine translation approach dead? In: ICNLSSP 2017 - International Conference on Natural Language, Signal and Speech Processing, pp. 1–5. ISGA, Casablanca, December 2017. https://hal.inria.fr/hal-01660016
16. Menacer, M.A., Mella, O., Fohr, D., Jouvet, D., Langlois, D., Smaïli, K.: Development of the Arabic loria automatic speech recognition system (ALASR) and its evaluation for Algerian dialect. In: ACLing 2017 - 3rd International Conference on Arabic Computational Linguistics, Dubai, United Arab Emirates, pp. 1–8, November 2017. https://hal.archives-ouvertes.fr/hal-01583842
17. Mohri, M., Pereira, F., Riley, M.: Speech recognition with weighted finite-state transducers. In: Springer Handbook of Speech Processing, pp. 559–584. Springer (2008)

18. Povey, D., Ghoshal, A., Boulianne, G., Burget, L., Glembek, O., Goel, N., Hanne-mann, M., Motlicek, P., Qian, Y., Schwarz, P., Silovsky, J., Stemmer, G., Vesely, K.: The kaldi speech recognition toolkit. In: IEEE 2011 Workshop on Automatic Speech Recognition and Understanding. IEEE Signal Processing Society, December 2011. IEEE Catalog No.: CFP11SRW-USB
19. Quemy, A., Jamrog, K., Janiszewski, M.: Unsupervised video semantic partitioning using IBM watson and topic modelling. In: Proceedings of the Workshops of the EDBT/ICDT 2018 Joint Conference (EDBT/ICDT 2018), pp. 44–49, March 2018
20. Sharghi, A., Laurel, J.S., Gong, B.: Query-focused video summarization: dataset, evaluation, and a memory network based approach. In: 2017 IEEE Conference on Computer Vision and Pattern Recognition, CVPR 2017, Honolulu, HI, USA, 21–26 July 2017, pp. 2127–2136. IEEE Computer Society (2017). https://doi.org/10.1109/CVPR.2017.229
21. Stolcke, A.: Entropy-based pruning of backoff language models. arXiv preprint cs/0006025 (2000)
22. Torres-Moreno, J.M.: Artex is another text summarizer. arXiv preprint arXiv:1210.3312 (2012)
23. Torres-Moreno, J.M.: Automatic Text Summarization. Wiley, London (2014)
24. Veselỳ, K., Ghoshal, A., Burget, L., Povey, D.: Sequence-discriminative training of deep neural networks. In: Interspeech 2013 (2013)
25. Zhang, J.J., Fung, P.: Active learning with semi-automatic annotation for extrac-tive speech summarization. ACM Trans. Speech Lang. Process. (TSLP) 8(4), 6 (2012)
26. Ziemski, M., Junczys-Dowmunt, M., Pouliquen, B.: The united nations parallel corpus v1. 0. In: LREC (2016)

Comparative Analysis of Accuracy of Fuzzy Clustering Methods Applied for Image Processing

Hossein Yazdani and Kazimierz Choroś[(✉)]

Faculty of Computer Science and Management,
Wroclaw University of Science and Technology, Wroclaw, Poland
{hossein.yazdani,kazimierz.choros}@pwr.edu.pl

Abstract. Fuzzy C-Means (FCM) has been used in different aspects of machine learning, specifically in clustering and image processing. Several modified versions of FCM have been proposed to improve the accuracy. This paper also makes use of FCM in image processing using a new membership function applied in Bounded Fuzzy Possibilistic Method (BFPM) to provide a more flexible search space for data points with respect to all clusters. The method is proposed with regard to the challenges with conventional fuzzy methods. The paper evaluates the accuracy of BFPM in its partitioning strategy and membership assignments in compare with other advanced fuzzy, possibilistic, and other prototype-based partitioning methods. Promising results of tests performed on the benchmark image dataset Libras prove that BFPM performs better than other conventional methods.

Keywords: Image processing · Fuzzy C-Means
Bounded fuzzy possibilistic method · Member function · Clustering
Uncertainty

1 Introduction

Uncertainty is one of the aspects that causes learning methods to miss-classify samples in their learning procedures. Fuzzy, probability, and possibilitic methods among other methods have been introduced to overcome the issues with uncertain conditions. The proposed approaches have been applied in learning methods such as clustering methods in their similarity and membership assignments. In fuzzy and possibilistic methods, objects are capable of obtaining a degree of memberships (partial or full membership) with respect to different clusters. The degree of membership sometimes makes the learning methods uncertain in their assignments.

1.1 Clustering and Image Processing

Representing images and sub-images of an image in image processing concept is a main concern in vision systems. According to difficulty of the task of image

K. Choroś et al. (Eds.): MISSI 2018, AISC 833, pp. 89–98, 2019.
https://doi.org/10.1007/978-3-319-98678-4_11

processing for machines, the main divide and conquer algorithm has been utilized to divide an image into different non-overlapped regions. The procedure of division is called segmentation. Several approaches have been introduced in segmentation clustering with respect to memberships or belongingness of image's pixels. Each pixel (object) obtains different memberships based on the type of membership function that has been used in learning procedures. The membership assignments are applied based on the similarity between objects. In fact, similarity functions and membership assignments are two main components in clustering problems which several advanced approaches have been introduced to assign the proper memberships to objects. Clustering is a form of supervised learning that splits objects into some clusters based on the similarity between objects with respect to the selected centroids. Assume a set of n objects represented by $O = \{o_1, o_2, \ldots, o_n\}$ in which each object is typically described by numerical $feature - vector$ data that has the form $X = \{x_1, \ldots, x_d\} \subset R^d$, where d is the dimension of the search space or the number of features. A cluster or a class is a set of c values $\{u_{ij}\}$, where u represents a membership value, j is the jth object and i is the ith cluster. A partition or membership matrix is often represented as a $c \times n$ matrix $U = [u_{ij}]$ [1]. Fuzzy [2] approaches are different from crisp methods, as each object can have partial membership degrees in more than one group. This condition is stated in Eq. (1), where data objects may have partial nonzero membership degrees in several clusters, but only a full membership degree in one cluster.

$$M_{fcn} = \left\{ U \in \Re^{c \times n} |\ u_{ij} \in [0,1],\ \forall j, i; \right.$$

$$\left. 0 < \sum_{j=1}^{n} u_{ij} < n,\ \forall i;\ \sum_{i=1}^{c} u_{ij} = 1,\ \forall j \right\} \tag{1}$$

According to fuzzy condition, Eq. (1), each column of the partition matrix must sum to 1 ($\sum_{i=1}^{c} u_{ij} = 1$). Based on this property of fuzzy methods, as c becomes larger, the u_{ij} values must become smaller. This paper considers fuzzy c-means (FCM) method and studies some related works about the issues with fuzzy memberships and FCM method applied in clustering and image processing.

1.2   Fuzzy C-Means (FCM) Algorithm

Fuzzy C-Means is one of the well-known prototype-based (centroid-based) clustering method used in machine learning and image processing [1]. There are two important types of prototype-based FCM algorithms [2]. One type is based on the fuzzy partition of a sample set and the other is based on the geometric structure of a sample set in a kernel-based method [2].

Partition-Based FCM. The FCM function may be defined as [1]:

$$J_m(U, V) = \sum_{i=1}^{c} \sum_{j=1}^{n} u_{ij}^m \, ||X_j - V_i||_A^2 ; \qquad (2)$$

where U is the $(c \times n)$ partition matrix, $V = \{v_1, v_2, ..., v_c\}$ is the vector of c cluster centers (or prototypes) in $\Re^d$, $m > 1$ is the fuzzification constant, and $||.||_A$ is any inner product A-induced norm [3], i.e., $||X||_A = \sqrt{X^T A X}$ or the distance function such as Minkowski distance presented by Eq. (3). Equation (2) makes use of Euclidean distance function by assigning $(k = 2)$ in Eq. (3).

$$d_k(x, y) = \left(\sum_{j=1}^{d} |\, x_j - y_j \,|^k \right)^{(1/k)} \qquad (3)$$

Kernel FCM. In kernel FCM the dot product $\left(\phi(X).\phi(X) = k(X, X) \right)$ is used to transform feature vector X, for non-linear mapping function $\left(\phi : \; X \longrightarrow \phi(X) \; \in \; \Re^{D_k} \right)$, where D_k is the dimensionality of the feature space. Equation (4) presents a non-linear mapping function for Gaussian kernel [2].

$$J_m(U; k) = \sum_{i=1}^{c} \left(\sum_{j=1}^{n} \sum_{k=1}^{n} (u_{ij}^m u_{ik}^m \; d_k(x_j, x_k))/2 \sum_{l=1}^{n} u_{il}^m \right) \qquad (4)$$

where $U \in M_{fcn}, \quad m > 1$ is the fuzzification parameter, and $d_k(x_j, x_k)$ is the kernel base distance (replaces Euclidean distance function) between the jth and k^{th} feature vectors as:

$$d_k(x_j, x_k) = k(x_j, x_j) + k(x_k, x_k) - 2k(x_j, x_k) \qquad (5)$$

2 Related Work

Among other advanced methods, some recent methodologies in image processing have been analysed. H.H. Pie et al. proposed a new density-based technique applied in Fuzzy C-Means algorithm to obtain better results in image processing applications [4]. Authors discussed about the issues with FCM algorithm regarding the good initialization in advance and the number of clusters. The authors' proposition is based on ordering the samples based on their density which indicates the importance of each sample. Then, the centroids or prototypes will be relatively collected while the densities of the selected prototypes must be higher than the densities of the samples. The number of clusters will be chosen by the proposed technique.

T.C. Havens et al. compared partitioning-based and kernel-based FCM algorithms on very large datasets based on some parameters such as time and space complexity, speed, and quality of approximations for loadable data [5]. The authors proposed different kernel-based FCM algorithms: approximate kernel FCM (akFCM), weighted kernel FCM (wkFCM), and random sample and extend kernel FCM (rsekFCM). Authors also made use of sampling strategies: non-iterative sampling and incremental techniques for unloadable datasets.

G. Teng et al. made use of cluster ensemble techniques to propose a new method called Group Method for Data Handling (GMDH) [6], beside introducing a new cluster ensemble framework named as Cluster Ensemble Group Method of Data Handling (CE-GMDH). The authors compared their methods with other methods using different measurements: Calinsky-Harabasz index (CH), Rand Index (RI), and Normalised Mutual Information (NMI) to evaluate the separation and the compactness of the clustering methods.

Y. Zheng et al. discussed about two issues with Fuzzy C-Means algorithm, regarding similarity function and insufficient robustness to image noise [7]. The authors proposed two different algorithms named generalized FCM (GFCM) and hierarchical FCM (HFCM) to overcome the issues with conventional FCM algorithm. The authors claimed that their new strategies in similarity measurements, in compare with Euclidean distance, perform better by calculating the modified mean value which is obtained by its immediate neighbourhood. In the final stage of their experiments, both proposed methods by the authors are combined to improve robustness and accuracy of the original algorithms.

Z. Yu et al. presented a new method called Random Subspace ensemble framework on HyBrid K-Nearest Neighbour (RS-HBKNN) to deal with sparse, imbalance, and noise problems processing [8]. The proposed method is inspired from the well-known algorithm k-nearest neighbours for noisy datasets. Authors also provided a survey KNN algorithms, in addition to present three different versions of RS-HBKNN algorithms. The proposed methods have been used in supervised learning methods with the aim of overcoming the issues with conventional KNN algorithms.

Z. Yang et al. discussed about the issues with fuzzy c-means algorithms to deal with noisy data when the noisy image segmentation is required [9]. Authors presented a robust modified version of FCM algorithm to perform as good as fuzzy regression method, presented by Hoppner et al. [10]. In [9] authors claimed that some issues with Fuzzy C-Means algorithms are due to using Euclidean distance function and non-unimodal property of its membership function. The authors applied a penalty term to overcome the issue with membership function. They also utilized the less-squared distance to obtain better results in compare with Euclidean distance function. In contrast with the proposed method, Hoppner et al. [10], believed that the improper accuracy of the main FCM method is based on the local model which is induced by fuzzy clustering. The authors proposed a new regression model to reward crisp membership degrees to obtain desirable membership function in fuzzy clustering. The authors presented their new method to build a new fuzzy model of Takagi-Sugeno (TS) type from data.

Anderson et al. [2] compared the functionality of fuzzy, probability, and possibilistic membership assignments in clustering problems. Authors discussed about the validity of the best partition and the number of clusters using different validity indices. Authors also generalized several classical indices that have been applied for crisp outputs of clustering methods in order to make them applicable for both crisp and soft partitioning results. Their study was covered by some numerical examples. The main contribution of the authors was on fuzzy-Rand Index (RI) and its modifications, in addition to discussion about the complexity of the proposed validity index in compare with other validity indices.

3   BFPM and Image Processing

Bounded Fuzzy Possibilitic Method (BFPM) has been introduced to overcome the issues with conventional crisp, fuzzy, probability, and possibilistic partitioning methods in their membership assignments. The proposed method overcomes the issues with restrictions and limitations for samples in their freely participation in other clusters [11,12]. The method relaxes boundary conditions in fuzzy and possibilistic methods. BFPM differs from previous approaches by allowing the membership function to assign much higher values. Furthermore, BFPM makes it possible for a data object to be assigned full memberships in multiple clusters, if necessary, even including all clusters. BFPM membership assignment is presented by Eq. (6).

$$M_{bfpm} = \left\{ U \in \Re^{c \times n} \mid u_{ij} \in [0,1], \ \forall j, i; \right.$$

$$\left. 0 < \sum_{j=1}^{n} u_{ij} < n, \ \forall i; \ 0 < 1/c \sum_{i=1}^{c} u_{ij} \leq 1, \ \forall j \right\} \tag{6}$$

3.1   BFPM Algorithm

Bounded fuzzy possibilistic membership assignments have been applied in clustering problems, presented by Algorithm 1 [13]. The main goal of the algorithm is to provide a flexible search space for data objects in the procedure of membership assignments.

U is the $(c \times n)$ partition matrix, $V = \{v_1, v_2, ..., v_c\}$ is the vector of c cluster centers in $\Re^d$, m is the fuzzification constant, and $||.||_A$ is any inner product A-induced norm [1], and Euclidean distance function used as a similarity function by the algorithm. Equations (7) and (8) show how the algorithm calculates (u_{ij}) and how the prototypes (v_i) will be updated in each iteration. The algorithm runs until reaching the condition:

$$max_{1 \leq k \leq c}\{||V_{k,new} - V_{k,old}||^2\} < \varepsilon$$

The value assigned to ε is a predetermined constant that varies based on the type of objects and clustering problems. According to Eqs. (7) and (8),

---

Algorithm 1. BFPM Algorithm

Input: X, c, m

Output: U, V

Initialize V;

while $max_{1 \leq k \leq c}\{||V_{k,new} - V_{k,old}||^2\} > \varepsilon$ **do**

$$u_{ij} - \Big[\sum_{k=1}^{c} \big(\frac{||X_j - v_i||}{||X_j - v_k||}\big)^{\frac{2}{m-1}}\Big]^{\frac{1}{m}}, \ \forall i,j \tag{7}$$

$$V_i = \frac{\sum_{j=1}^{n}(u_{ij})^m x_j}{\sum_{j=1}^{n}(u_{ij})^m}, \ \forall i; \quad (0 < \frac{1}{c}\sum_{i=1}^{c} u_{ij} \leq 1). \tag{8}$$

end while

---

prototypes are updated in each iteration with respect to memberships assigned to data objects, which mis-assignments lead to selecting wrong prototypes for the next iteration. Consequently, the wrong selection of prototypes causes the learning methods to obtain improper accuracy. BFPM is proposed to remove restrictions on data objects to obtain proper memberships with respect to all clusters. The proposed membership assignment makes the procedure of choosing the proper prototypes more accurate.

4 Experimental Verification

The accuracy of the proposed bounded fuzzy possibilistic method is compared with other fuzzy methods in the experimental verification. An available benchmark dataset "Movement-Libras" from UCI repository is chosen with regard to compare the accuracy of the proposed method with available results published by other papers. The "Movement-libras" dataset is composed of 360 objects in 90 dimensional search space (attributes) with respect to 15 clusters. The Libras dataset contains frames from videos with hand movements of sign language gestures. In the first experiment, the accuracy of conventional fuzzy methods with recent clustering methods have been compared with the BFPM method. Table 1 presents the accuracy rates of several unsupervised and supervised fuzzy methods for Libras dataset.

The modified methods are Fuzzy Criteria Multi-Objective Feature Selection Approach (FC-MOFSA) [14], Non-Fuzzified Multi-objective Unsupervised Feature Selection (NF-MOUFS) [14], Fuzzy weighted K-Nearest Neighbors Density Peak Clustering (FKNN-DPC) [15], and fuzzy Random Subspace ensemble framework HyBrid K-Nearest Neighbor (RS-HBKNN2) [8]. According to the results, BFPM performs better than other methods while the fuzzification constant is precisely assigned, which in this experiment (m = 1.2) leads to obtaining better performance. This experiments indicates the necessity of considering the affect of different norms on final results. Table 2 presents the accuracy rates of

Table 1. Accuracy rates of several unsupervised and supervised fuzzy and Bounded Fuzzy Possibilistic Methods for Libras dataset.

Libras dataset					
Unsupervised fuzzy methods			Supervised fuzzy methods	BFPM versions	
FC-MOFSA	NF-MUFS	FKNN-DPC	RSHBKNN2	BFPM (m = 2.0)	BFPM (m = 1.2)
47.79 ± 4.44	44.44 ± 4.29	43.60	86.00	57.00	95.11

several unsupervised prototyp-based learning methods for Libras dataset in compare with BFPM for two different values of fuzzification constant. The most common prototype-based methods, rather than Fuzzy C-Means, such as k-centroid, k-modiod, and k-means approaches have been selected for this experiment. The modified methods are: Mediod [16], Dynamic Medoid (D-Medoid) [16], Centroid [16], Dataset Seeded Centroid (DS-Centroid) [16], K-Means Seeded Centroid (KMS-Centroid) [16], and k-means [16]. BFPM makes use of different fuzzification constants (or similarity norms [10]) $m = 1.2$ and $m = 2.0$ in this comparison. According to the results, BFPM performs better than other prototype-based methods with regard to fuzzification constant ($m = 1.2$).

Table 2. Accuracy rates of several unsupervised prototype-based methods in compare with bounded fuzzy possibilistic method for Libras dataset.

Libras dataset							
Unsupervised learning methods						BFPM Versions	
Mediod	D-Mediod	Centroid	DS-Centroid	KMS-Centroid	k-means	BFPM (m = 2.0)	FPM (m = 1.2)
90.31	81.43	73.38	88.96	90.12	90.44	57.00	95.11

Table 3 presents the accuracy rates of several supervised learning methods in compare with the BFPM method for Libras dataset. The modified methods are Random Subspacc ensemble framework HyBrid K-Nearest Neighbor (RS-HBKNN) [8], Radial Basis Function (RBF) [17], Dynamic time warping (DTW) [18], Auto-Regressive kernel (AR) [18], the Fisher kernel (Fisher) [18], and Reservoir kernel (RV) [18]. Promising results form BFPM prove that the proposed method is also competitive with even supervised learning methods, while the fuzzificaiton constant and other parameters are assigned properly.

Table 3. Accuracy rates of several supervised methods in compare with bounded fuzzy possibilistic method for Libras dataset.

Libras dataset							
Supervised learning methods						BFPM versions	
RS-HBKNN	RBF	DTW	AR Tree C4.5	Fisher Bayes	RV	BFPM (m = 2.0)	BFPM (m = 1.2)
88.56	95.77	94.02	91.79	94.93	93.25	57.00	95.11

Figure 1 shows the obtained memberships of the first 150 objects from Libras dataset provided by BFPM with respect to the cluster they participated in. The vertical axis presents the membership degrees and the horizontal axis illustrate objects. According to the figure, the objects' membership degrees are reasonable $(u(x_j) > 0.7)$, which indicates that objects are close to their centroids with good values of membership degree. According to the obtained accuracy and the membership degrees depicted by the figure, it can be concluded that not only BFPM performs better than other conventional fuzzy and other prototype-based partitioning methods and also competitive with supervised learning methods, but also the objects are being clustered with well defined membership assignments.

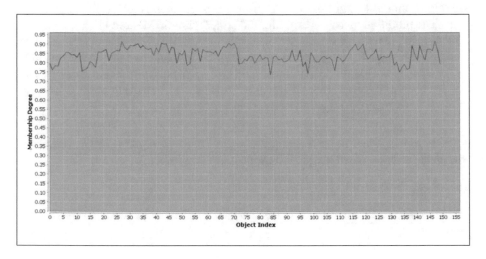

Fig. 1. Membership degrees' plot for the first 150 images from Libras dataset provided by bounded fuzzy possibilistic method.

5   Conclusion

The paper discussed about the recent Fuzzy C-Means methods on image processing with aim of comparing the accuracy of the modified fuzzy methods. The main goal of the proposed method is to introduce a flexible search space for data objects to participate in other clusters instead of restricting them to just participate in limited number of clusters. The flexible search space is provided by Bounded Fuzzy Possibilistic Method. The methodologies allow objects to show their potential abilities which can be tracked and studied in a better way. The accuracy of the proposed method has been compared with recent prototype-based methods. Promising results proved that BFPM obtained better accuracy in addition to the provided flexible search space. The proposed method also compared with some advanced supervised methods, which indicated that the unsupervised proposed method is also comparative even with supervised methods.

The results also presented how proposed method accurately assigned memberships to data objects in addition to obtaining the promising results.

References

1. Cannon, R.L., Dave, J.V., Bazdek, J.C.: Efficient implementation of the fuzzy c-means clustering algorithms. IEEE Trans. Pattern Anal. Mach. Intell. **PAMI**–**8**(2), 248–255 (1986)
2. Anderson, D.T., Bezdek, J.C., Popescu, M., Keller, J.M.: Comparing fuzzy, probabilistic, and possibilistic partitions. IEEE Trans. Fuzzy Syst. **18**(5), 906–918 (2010)
3. Hoppner, F.: Fuzzy Cluster Analysis: Methods for Classification, Data Analysis and Image Recognition. Wiley, Chichester (1999)
4. Pei, H.H., Zheng, Z.R., Wang, C., Li, C.N., Shao, Y.H.: D-FCM: density based fuzzy c-means clustering algorithm with application in medical image segmentation. Elsevier, Int. Conf. Inf. Technol. Quant. Manag. **122**, 407–414 (2017)
5. Havens, T.C., Bezdek, J.C., Leckie, C., Hall, L.O., Palaniswami, M.: Fuzzy c-means algorithms for very large data. IEEE Trans. Fuzzy Syst. **20**(6), 1130–1146 (2012)
6. Tenga, G., Heb, C., Xiaob, J., Heb, Y., Zhub, B., Jiang, X.: Cluster ensemble framework based on the group method of data handling. Elsevier, Appl. Soft Comput. **43**, 35–46 (2016)
7. Zhenga, Y., Jeond, B., Xua, D., Jonathan Wua, Q.M., Zhanga, H.: Image segmentation by generalized hierarchical fuzzy c-means algorithm. J. Intell. Fuzzy Syst. **28**, 961–973 (2015)
8. Yu, Z., Chen, H., Liu, J., You, J., Leung, H., Han, G.: Hybrid k-nearest neighbor classifier. IEEE Trans. Cybern. **46**(6), 1263–275 (2016)
9. Yang, Z., Chung, F.L., Shitong, W.: Robust fuzzy clustering-based image segmentation. Elsevier Appl. Soft Comput. **9**, 80–84 (2009)
10. Hoppner, F., Klawonn, F.: Improved fuzzy partitions for fuzzy regression models. Int. J. Approximations Reason. **32**(2), 85–102 (2003)
11. Yazdani, H.: Fuzzy possibilistic on different search spaces. In: Proceedings of the International Symposium on Computational Intelligence and Informatics, pp. 283–288. IEEE (2016)
12. Yazdani, H., Ortiz-Arroyo, D., Choroś, K., Kwaśnicka, H.: Applying bounded fuzzy possibilistic method on critical objects. In: Proceedings of the International Symposium on Computational Intelligence and Informatics, pp. 271–276. IEEE (2016)
13. Yazdani, H., Ortiz-Arroyo, D., Choroś, K., Kwaśnicka, H.: On high dimensional searching space and learning methods. In: Data Science and Big Data: An Environment of Computational Intelligence, pp. 29–48. Springer, Heidelberg (2016)
14. Caia, F., Wangb, H., Tanga, X., Emmerichb, M., Verbeeka, F.J.: Fuzzy criteria in multi-objective feature selection for unsupervised learning. In: Elsevier, International Conference on Application of Fuzzy Systems and Soft Computing, pp. 29–30 (2016)
15. Xie, J., Gao, H., Xie, W., Liu, X., Grant, P.W.: Robust clustering by detecting density peaks and assigning points based on fuzzy weighted k-nearest neighbors. Elsevier Inf. Sci. **354**, 19–40 (2016)

16. Lensen, A., Xue, B., Zhang, M.: Particle swarm optimisation representations for simultaneous clustering and feature selection. In: IEEE Symposium Series in Computational Intelligence, pp. 1–8 (2016)
17. Escobedo-Cardenas, E., Camara-Chavez, G.: A robust gesture recognition using hand local data and skeleton trajectory. In: IEEE International Conference in Image Processing, pp. 1240–1244 (2015)
18. Chen, H., Tang, F., Tino, P., Cohn, A.G., Yao, X.: Model metric co-learning for time series classification. In: Proceedings of the International Joint Conference on Artificial Intelligence, pp. 3387–3394 (2015)

An Efficient Method for Sign Language Recognition from Image Using Convolutional Neural Network

Sebastian Kotarski and Bernadetta Maleszka(⊠)

Faculty of Computer Science and Management, Wroclaw University of Science
and Technology, Wybrzeze Wyspianskiego 27, 50-370 Wroclaw, Poland
sebastian.kotarski94@gmail.com, bernadetta.maleszka@pwr.edu.pl

Abstract. Recognition of the sign language is one of the most important milestone in image recognition field. Such systems can help deaf people to communicate with the world. We feel privileged to present a new method which translates from American Sign Language (ASL) fingerspelling into a letter using Convolutional Neural Network and transfer learning. The method is using Google pre-trained model named MobileNet V1 which was trained on the ImageNet image database. Our model was trained on the dataset from Surrey University. We developed a useful model not only for desktop computers but it is also possible to apply it into mobile systems, because of low memory consumption.

Keywords: Sign language recognition
Convolutional neural network · Deep learning

1 Introduction

In the USA there are 500 000 citizens who speak American Sign Language (ASL) [11]. That language is also used abroad. The number of people communicating with sign language is very low and there is a lack of system which can translate the sign language into the readable form. Such a system can help in emergency cases and is less impersonal than paper and pencil. Most used sign languages on the world is Chinese, Brazilian and American Sign Language [20]. Most of the studies are focused on ASL analysis. There are two letters 'j' and 'z' in ASL alphabet which requires motion from the speaker.

When communicating with deaf person, there is a possibility to be taken wrongly. The problem is, that sign language cannot be translated one to one into readable form. Sign language sentences are not structured the same as in the phonic languages. For example, more important things or larger are shown at the beginning of the sentence. Sign language contains its own visual and space grammar. It also has a different syntax, which not corresponds to the phonic one. Other important topic is facial expression, head movement and location of the speaker. Although eye contact, eye direction and pauses. These physical

© Springer Nature Switzerland AG 2019
K. Choroś et al. (Eds.): MISSI 2018, AISC 833, pp. 99–108, 2019.
https://doi.org/10.1007/978-3-319-98678-4_12

gestures are not only representing emotional state, but also perform a grammar functionality. They also replace elements of phonic language such as intonation.

In this paper we present a fast and compact method for sign language recognition. This method stands alone in case of time to learn, which was reduced to achieve good results on typical machine, without high efficient GPU. This method also requires less memory than any other solution which had come into existence before. It is also possible, to use this method on the mobile systems without usage of external servers.

This paper is organised as follows. In Sect. 2 we analyse existing methods in related works. Section 3, which was called Approach and Methods, presents how our method was developed and implemented. In Sect. 4 we present how our model was tested and experiment results. Section 5 describes our summary and opinion about presented method. It also presents our thoughts about next steps and possibilities to improve our model.

2   Related Work

Systems for sign language recognition could be divided into two groups on the grounds of input signal. Part of the architectures are prepared to take as an input continuous signal, other solutions are processing only a static signal.

Sign language recognition systems can be also divided according to method for acquiring data. First group of the methods gets data from special sensor gloves [6,10,17]. The second group uses machine vision systems [14,15]. In the second case, systems can use mono cameras [7] or stereo cameras with more than one lens [4].

The growth of popularity for capturing the image increased with device called Microsoft Kinect. Microsoft company made SDK (Software Development Kit) available for game developers. Information about the depth in the image is the most important advantage. It is very significant for the gesture recognition. The additional advantage is the fact, that Microsoft's SDK delivers initial features extraction and ready model of the human body. There were many paper connected with sign language recognition which had used that technology [13,18].

Many competitions which involve creating systems for sign language recognition are organised. One of such contest was ChaLearn Looking at People Challenge 2014 [1]. For the purpose of the task, special data set was formed. The data set contains 20 different gestures in Italian Sign Language and was recorded with 27 volunteers with different light conditions, clothes and scenes. Everything was captured by Microsoft Kinect. In results, every participant had an access to example data.

It can be noticed, that except of RGB signal, also depth map and binarized view when only the speaker is selected and model of body can be used. One of the articles [13] took 5th place for all 17 participants. They achieved accuracy 95.68% on the test data set. In 2017 Microsoft decided to discontinue the production of the Kinect due to lack of interest. These several years with that technique had a large impact on the field of sign language recognition research and analysis.

The history of methods for recognising gestures started since 90s. One of the first publicised thesis was written by Murakami and Taguchi in 1991 [12]. Authors were acquiring data from the special gloves called "data gloves". The model was trained on the isolated static (without movement) gestures. Theirs system segmented gestures in time and worked on single gestures. In regards of success of Hidden Markov Models (HMM) on the field of speech recognition, it was also used to gesture recognition [3,8]. That approach has dominated the research field from mid 90s.

Nowadays, Convolutional Neural Network are used [2,13]. Nevertheless, there are works using H-CRF (Hierarchical Conditional Random Field) [18], which is an abstract model for Hidden Markov Models. Using the HMM or similar ones is reasonable because gesture processing is sequential. Convolutional Neural Network has enormous impact and success in image recognition, it is also reasonable to use it in gesture recognition [9].

In the paper we propose to use Convolutional Neural Network to recognise sign language from static images. The main advantage of our approach is lower computational complexity.

3   Approach and Methods

The method presented in this paper takes as an input an image in size 128x128px. The input image is processed by MobileNet model. In MobileNet all layers are followed by batchnorm and ReLU nonlinear activation function except the last one, the fully connected layer. The fully connected layers were removed from the MobileNet in our method. We added our custom fully connected layer and softmax layer with 24 output neurons (in actual MobileNet version there were 1000 outputs), where each neuron relates to one letter from ASL alphabet. We added Global Average Pooling (GAP) to reduce size of the last output of the convolution layer.

The output of the system is a vector of 24 rows with probability of each letter appearance. The final decision is selected as an argument of maximum probability - in other words the letter with the highest probability.

In this paper we proposed three models. For the purpose of identification, the first one is named 'fc' and only fully connected layers were learned which was shown in Fig. 1. Next model is named 'layer_1' as presented in Fig. 2 was learned not only fully connected layers, but also the last convolution layer was fine tunned. The last model, named 'layer_2' is similar to the previously described, but the last two convolution layers were fine tunned. 'layer_2' was presented in Fig. 3.

3.1   Dataset

We used data from Surrey University, which was created for the public article [16]. This data consists of 24 classes (each class is representing one letter in

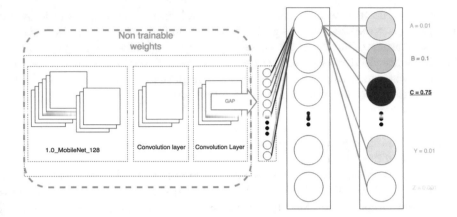

Fig. 1. Scheme for 'fc'

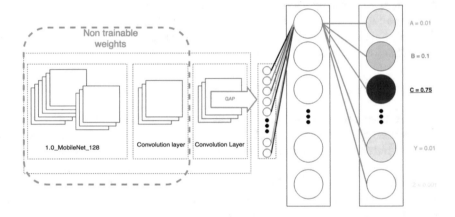

Fig. 2. Scheme for 'layer$_1$'

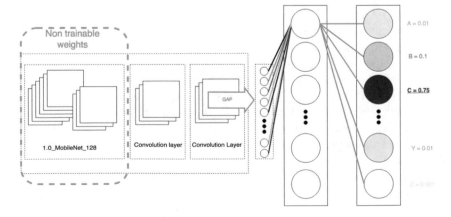

Fig. 3. Scheme for 'layer$_2$'

ASL). It was performed by 5 different volunteers. The recordings were taken with Microsoft Kinect and OpenNI+NITE framework to hand detection and track in similar light condition and surroundings. As a result we have in total more than 65 500 images. The examples from Surrey University's data set is presented in Fig. 4. The data set doesn't contain letters 'j' and 'z' because as it was mentioned before, those letters require movements.

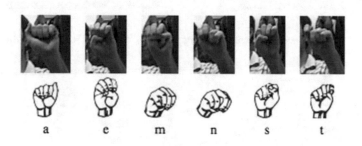

Fig. 4. Example data from Surrey University [16]

3.2 Learning

Proposed method uses pre-trained model MobileNet from Google [5]. The transfer learning uses other model (probably trained on the bigger dataset) and then it is fine tunned to the specific classification problem. In our method, we used MobileNet version 2017. It has a good accuracy on ImageNet database and better efficiency on mobile systems than Inception models. Pre-trained models could be supervised or unsupervised learned. The advantage of using such a technique is less time consuming to be tuned for a specific problem. We used mini batch gradient descent for learning, because it is quicker to learn than stochastic gradient descent (learning by one example).

For the purpose of learning, the dataset was divided into training, validation and testing subsets. We augmented the training set with randomly zoomed and rotated images to be learned more general features. Most of the images from the dataset had 100x100, but the smallest MobileNetV1 model requires 128x128, so the images were resized to that specific size.

4   Experiments and Results

We evaluated our model with the testing set, which was isolated from training and validation process of the model. The training mini batch size was 64. The validation set contained 20% of the whole dataset (13 156 examples). The testing set was also 20% of the dataset. The measures were accuracy, F-score, precision and recall. As a cost function, cross entropy was chosen. Training steps were set to 120 steps for

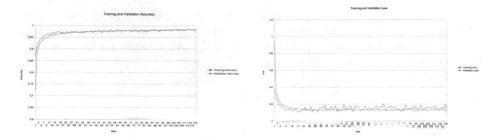

Fig. 5. 'fc' model's accuracy on training and validation set

Fig. 6. 'fc' model's loss on training and validation set

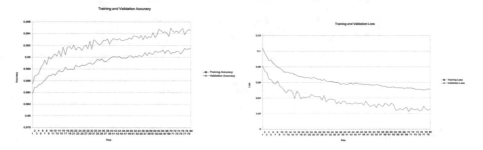

Fig. 7. 'layer_1' model's accuracy on training and validation set

Fig. 8. 'layer_1' model's loss on training and validation set

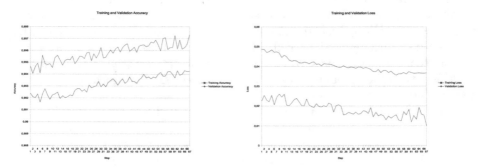

Fig. 9. 'layer_2' model's accuracy on training and validation set

Fig. 10. 'layer_2' model's loss on training and validation set

'fc' model, 80 steps for 'layer_1' and 64 steps for the 'layer_2'. Dropout was set to 0,001. We chose RMSProp as an optimiser. We have used cross entropy as a cost function. The accuracy on validation and training step in each state and the loss were presented for each model in Figs. 5, 6, 7, 8, 9 and 10. The confusion matrixes for all three models were presented in Figs. 11, 12 and 13. It is possible to see, that letters 'E' and 'S' were sometimes selected wrongly. More detailed results for each letter were presented in Table 2.

Table 1. Results for accuracy for proposed models in comparison with Garcia [2] research.

Model identifier	Accuracy [%]
B. Garcia 'full_lr'	72
'fc'	78
'layer_1'	84.6
'layer_2'	85.5

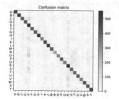 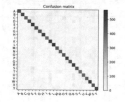 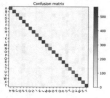

Fig. 11. Confussion matrix for model 'fc'
Fig. 12. Confussion matrix for model 'layer_1'
Fig. 13. Confussion matrix for model 'layer_2'

Obtained results have shown (Table 1) that our best model was 13.5% better than the best model in Garcia's work [2]. The F-score measure was not outstanding very much from accuracy, the model is robust and precise and doesn't miss many instances what can be seen in Figs. 11, 12 and 13.

Table 2. Results for the 'fc', 'layer_1' and 'layer_2' models for each letter

Letter	Model 'fc'				Model 'layer_1'				Model 'layer_2'			
	Precission	Recall	f1-score	Support	Precission	Recall	f1-score	Support	Precission	Recall	f1-score	Support
A	0.95	0.85	0.90	543	0.98	0.84	0.91	543	0.97	0.86	0.91	543
B	0.93	0.94	0.93	541	0.94	0.91	0.93	541	0.94	0.94	0.94	541
C	0.99	0.71	0.82	584	0.99	0.71	0.83	584	0.99	0.76	0.86	584
D	0.76	0.82	0.79	534	0.85	0.85	0.85	534	0.86	0.87	0.86	534
E	0.40	0.99	0.57	562	0.61	0.98	0.75	562	0.61	0.98	0.75	562
F	0.96	0.80	0.87	515	0.92	0.90	0.91	515	0.94	0.90	0.92	515
G	0.97	0.79	0.87	541	0.97	0.83	0.90	541	0.98	0.86	0.91	541
H	0.93	0.95	0.94	549	0.94	0.95	0.94	549	0.96	0.93	0.95	549
I	0.89	0.65	0.75	522	0.95	0.75	0.84	522	0.95	0.77	0.85	522
K	0.84	0.86	0.85	557	0.74	0.96	0.84	557	0.79	0.94	0.86	557
L	0.97	0.89	0.93	583	0.92	0.97	0.95	583	0.93	0.97	0.95	583
M	0.80	0.78	0.79	547	0.84	0.78	0.81	547	0.89	0.73	0.80	547
N	0.92	0.48	0.63	545	0.87	0.75	0.80	545	0.87	0.73	0.79	545
O	0.91	0.59	0.71	552	0.83	0.81	0.82	552	0.84	0.83	0.84	552
P	0.92	0.73	0.81	570	0.88	0.91	0.90	570	0.90	0.90	0.90	570
Q	0.95	0.65	0.78	533	0.92	0.84	0.88	533	0.92	0.82	0.87	533
R	0.53	0.95	0.68	569	0.68	0.93	0.78	569	0.70	0.93	0.80	569
S	0.60	0.72	0.66	563	0.75	0.81	0.78	563	0.73	0.86	0.79	563
T	0.67	0.74	0.71	527	0.88	0.72	0.79	527	0.88	0.72	0.79	527
U	0.86	0.82	0.84	517	0.88	0.78	0.83	517	0.93	0.80	0.86	517
V	0.84	0.63	0.72	556	0.85	0.74	0.79	556	0.82	0.83	0.83	556
W	0.94	0.73	0.82	630	0.94	0.85	0.89	630	0.95	0.85	0.89	630
X	0.59	0.84	0.69	519	0.63	0.85	0.72	519	0.63	0.86	0.73	519
Y	0.99	0.82	0.90	556	1.00	0.87	0.93	556	1.00	0.88	0.94	556
Avg/total	0.84	0.78	0.79	13215	0.87	0.85	0.85	13215	0.87	0.86	0.86	13215

5   Conclusions and Future Works

We developed a compact and quick classifier using CNN for sign language recognition. It is possible to fine tune a model in responsible time. As also was mentioned in Garcia's paper [2], the Surrey University's dataset has poor variation, all images were taken in similar light and environment conditions. If the images were more diversified, the model could have a better accuracy and would be more generalised for all 24 letters.

The difference between our model and previously created is that, we can use it offline on low efficient machines. We can create a mobile application using Apple's framework for applying machine learning models named CoreML [19] for iOS or use Tensorflow Lite Java API [21] for Android. Our model is very quick and doesn't require to be stored on high priced servers. Mobile application can store the model offline due to small size. Our model also requires less computation time, which lead to less power consumption, which is very important in nowadays mobile devices. The model could be used commercially, because these days it doesn't require any additional device. The only necessary apparatus is what majority of us have in our pocket.

The model could be improved by using images with larger resolution. Resetting lower levels of convolution layers in the pre-trained model will adapt the network to focus on features which exists in our dataset. In fact, the disadvantage is that the training process is more time consuming and it requires more data.

Acknowledgments. This research was partially supported by Polish Ministry of Science and Higher Education. Calculations have been carried out using resources provided by Wroclaw Centre for Networking and Supercomputing (http://wcss.pl), grant No. 469.

References

1. Escalera S., Baró X., Gonzàlez J., Bautista M. A., Madadi M., Reyes M., Ponce-López V., Escalante H. J., Shotton J., Guyon I.: ChaLearn looking at people challenge 2014: dataset and results. In: Proceedings of Computer Vision - ECCV 2014 Workshops, pp. 459–473 (2015)
2. Garcia, B., Viesca, S.A.: Real-time american sign language recognition with convolutional neural networks. In: CS231n: Convolutional Neural Networks for Visual Recognition (2016)
3. Grobel K., Assan M.: Isolated sign language recognition using hidden Markov models. In: Proceedings of IEEE SMC. Computational Cybernetics and Simulation, vol. 1, pp. 162–167 (1997). https://doi.org/10.1109/ICSMC.1997.625742
4. Hasanuzzaman, M., Ampornaramveth, V., Zhang, T., Bhuiyan, M. A., Shirai , Y., Ueno, H.: Real-time vision-based gesture recognition for human robot interaction. In: Proceedings of IEEE International Conference on Robotics and Biomimetics, pp. 413–418 (2004). https://doi.org/10.1109/ROBIO.2004.1521814
5. Howard, A.G., Zhu, M., Chen, B., Kalenichenko, D., Wang, W., Weyand, T., Andreetto, M., Hartwig, A.: MobileNets: Efficient Convolutional Neural Networks for Mobile Vision Applications (2017). CoRR http://arxiv.org/abs/1704.04861

6. Kadous, M.W.: Machine recognition of Auslan signs using PowerGloves: towards large-lexicon recognition of sign language computer science engineering. In: Proceedings of the Workshop on the Integration of Gesture in Language and Speech, pp. 165–174 (1996)
7. Kraiss, K.F.: Non-intrusive sign language recognition for human computer interaction. In: Proceedings of IFAC/IFIP/IFORS/IEA symposium on analysis, design and evaluation of human machine systems (2004)
8. Liang, R.H., Ouhyoung, M.: A sign language recognition system using hidden Markov model and context sensitive search. In: Proceedings of the ACM Symposium on Virtual Reality Software and Technology, pp. 59–66 (1996)
9. Liu, Q., Zhang, N., Yang, W., Wang, S., Cui, Z., Chen, X., Chen, L.: A Review of Image Recognition with Deep Convolutional Neural Network. In: Intelligent Computing Theories and Application, pp. 69–80 (2017)
10. Mehdi, S.A., Khan, Y.N.: Sign language recognition using sensor gloves. In: Proceedings of ICONIP 2002. vol. 5, pp. 2204–2206 (2002)
11. Mitchell, R.E., Young, T.A., Bachleda, B., Karchmer, M.A.: How Many People Use ASL in the United States? Sign Lang. Stud. **6**(3), 306–335 (2006)
12. Murakami, K., Taguchi, H.: Gesture recognition using recurrent neural networks. In: Proceedings of the SIGCHI, pp. 237–242 (1991)
13. Pigou, L., Dieleman, S., Kindermans, P.J., Schrauwen, B.: Sign language recognition using convolutional neural networks. In: Proceedings of Computer Vision - ECCV 2014 Workshops. Springer International Publishing, pp. 572–578 (2015)
14. Segen, J., Kumar, S.: Shadow gestures: 3D hand pose estimation using a single camera. In: Proceedings of IEEE Computer Society Conference on Computer Vision and Pattern Recognition, vol. 1, pp. 485 (1999). https://doi.org/10.1109/CVPR.1999.786981
15. Starner, T., Pentland, A.: Real-time american sign language recognition from video using hidden Markov models. In: Proceedings of ISCV, pp. 265–270 (1995). https://doi.org/10.1109/ISCV.1995.477012
16. Pugeault, N., Bowden, R.: Spelling it out: real–time ASL fingerspelling recognition. In: Proceedings ICCV 2011: 1st IEEE Workshop on Consumer Depth Cameras for Computer Vision, pp. 1114–1119 (2011). https://doi.org/10.1109/ICCVW.2011.6130290
17. Waldron, M.B., Kim, S.: Isolated ASL sign recognition system for deaf persons. In: IEEE Transactions on Rehabilitation Engineering, vol. 3, no. 3, pp. 261–271 (1995). https://doi.org/10.1109/86.413199
18. Yang, H.D.: Sign language recognition with the kinect sensor based on conditional random fields. (Sens. Basel, Switz.) **15**, 135–147 (2015). http://www.ncbi.nlm.nih.gov/pmc/articles/PMC4327011/
19. Core ML. https://developer.apple.com/documentation/coreml. Accessed 30 05 018
20. List of sign languages by number of native signers. https://en.wikipedia.org/wiki/List_of_sign_languages_by_number_of_native_signers. Accessed 30 05 2018
21. Abadi, M., et al.: TensorFlow: Large-Scale Machine Learning on Heterogeneous Systems (2015). https://www.tensorflow.org/

Audio/Speech Coding Based on the Perceptual Sparse Representation of the Signal with DAE Neural Network Quantizer and Near-End Listening Enhancement

Vadzim Herasimovich$^{(\boxtimes)}$, Alexey Petrovsky, Vladislav Avramov,
and Alexander Petrovsky

Belarusian State University of Informatics and Radioelectronics, Minsk, Belarus
vadim.gerasimovich@gmail.com,
alexey@petrovsky.eu, avramov.vladislav@gmail.com,
palex@bsuir.by

Abstract. The article presents universal sound coding framework. The encoding algorithm works at the junction of the transform and parametric approaches. The input signal goes through the decorrelation transform – wavelet packet decomposition (WPD) that is tuned to perceptual structure of the analyzed signal with the psychoacoustic modelling. The parameterization stage is the matching pursuit (MP) algorithm with the WPD based dictionaries. Selected parameters then quantized and coded for the transmission to the decoder. Quantization algorithm based on the artificial neural networks with a deep autoencoder (DAE) architecture is presented. The decoding part of the coder has the listening enhancement function. Since the decoder input is the parameters that are distributed in the subbands it is only necessary to decompose the noise signal with the corresponding filterbank and estimate the subband gain factor based on this information. The results of the conducted research like objective difference grade and performance demonstration are shown.

Keywords: Audio/speech coding · Wavelet packet · Matching pursuit
Psychoacoustics · Neural networks · Deep autoencoder
Listening enhancement

1 Introduction

Modern multimedia systems such as streaming audio, digital audio broadcasting, digital communications and so on require many constraints for the encoders. Bit stream scalability and invariance to the input sound signal become as important as the big compression ratio (CR) with high quality of the reconstructed signal. For the reason, the research and development of the new coding algorithms got a second wind.

The article introduce the universal scalable audio/speech coding algorithm based on the matching pursuit (MP) with perceptually optimized wavelet packet (WP) dictionary. A neural network (NN) based quantizer with deep autoencoder (DAE) architecture and near-end listening enhancement algorithm embedded in the decoder are presented.

© Springer Nature Switzerland AG 2019
K. Choroś et al. (Eds.): MISSI 2018, AISC 833, pp. 109–119, 2019.
https://doi.org/10.1007/978-3-319-98678-4_13

Since sparse signal representation is very convenient for the sound encoding tasks there are many works devoted to it. Classic work [1] presents greedy approach for the sparse approximation – MP with time-frequency dictionaries. The signal modelling based on it uses in such works as [2] where MP with Gabor dictionary and time-frequency masking are introduced; [3] shows the alternative dictionary formed by MDCT basis. In the algorithms like [4, 5] hybrid approach is introduced, where MP modelling for some part of the input signal is used. As for the universal audio coders, one of the examples is Opus [6, 7]. It is effective solution for both speech and sound signals but it has two separate models for them and a detector to select one.

Quantization stage of the encoding algorithm uses NN as in [8]. Classic approaches – scalar and vector quantization (VQ) [9] are time-tested algorithms but they have some cons that NN can fixes. For example, it is discarding of the internal information structure for the scalar quantization and fixed codebook size for the VQ. Comparison of the scalar and NN approaches is also shown in this article.

Originally, near-end listening enhancement for the telecommunication tasks was introduced for example in [10]. In contrast of this work, current research takes attempt to expand the idea of such a listening enhancement for the whole range of the sound signals. The perceptual characteristics of the processed signal will be also utilized.

2   Coding Algorithm Workflow

2.1   Brief Algorithm Overview

The encoding process can be divided into three main parts: input signal transform via WPD; the most perceptually relevant frame components selection via MP; compact representation of the chosen parameters via quantization and coding algorithms. The common coding scheme presented in Fig. 1.

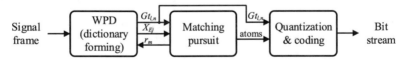

Fig. 1. Encoding algorithm main stages

WPD stage is a process that forms a dictionary of the time-frequency functions for the MP procedure. This block is consist of the WPD and the psychoacoustic modelling for the WPD tree optimization [11]. The outputs are global time-frequency masking threshold $Gt_{l,n}$ and the set of WP coefficients X_{Ej}. Details of that block are viewed in the Subsect. 2.2. The second stage of the coding algorithm is a sparse representation of the modelled signal via MP: a procedure that selects the most relevant components (atoms) from the redundant dictionary of the time-frequency functions [1]. In case of the developed encoder, MP chooses such a parameters that have the best perceptual matching of the input and modelled scalograms. This block has two outputs: a set of selected atoms in forward coding flow direction; and a residue signal r_m in backward direction since it is necessary to subtract the selected atom contribution for the signal forming. In the Subsect. 2.3 the details of this algorithm are shown. The third part of the encoder is the quantization and coding stage, which is necessary for the compact signal parameters representation. Section 3 shows the architecture of the quantizer for the developed coder. The decoder side of the coding algorithm is shown in Fig. 2.

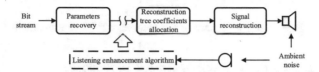

Fig. 2. Output signal decoding workflow with the listening enhancement function

The task of the decoder is to restore the parameters from the input bit stream and reconstruct the output sound signal via the inverse WPD. The decoder input information consists of the atoms and their position in the WPD tree (tree level l, node number n, coefficient number k). Thereby, there is no need to construct reconstruction tree for the each frame, as it was made in the encoder. It is available to use one limiting tree structure for the decoding and just allocate atoms in it [12]. Signal reconstruction is not the only decoder function. Since this part of the algorithm is at the sound reproducing end, there can be a noise masking situation. For the reason, there is embedded listening enhancement algorithm that provide the perceptual loudness correction of the masked signal components. Details are viewed in the Sect. 4.

2.2 Time-Frequency Functions Dictionary Collecting

WPD – is a transform that allows getting non-uniform time-frequency plan of the processed signal. WPD has a tree-like form, where one branch is a low-frequency filter and the other is a high-frequency filter [13]. Therefore, there is an ability to construct such a tree that fits best the input signal. This can be done by the cost functions that will weigh the information of the tree nodes and help to decide if there is a necessity of the node growth. As far as the input signal is a sound, the most convenient way to make tree nodes analysis is to use the psychoacoustic features of them. In this way there are two cost functions used in the coder: wavelet time entropy (*WTE*) and perceptual entropy (*PE*) [11]. The first one can show the information density of the entire WPD tree level; the second function analyzes the perceptual properties of each tree node. The algorithm of the optimal in the perceptual way WPD tree estimation is shown in Fig. 3.

As it can be seen from figure the optimization process works "on the go" from the very first root node of the WPD tree without any return to the previous levels. Both of the cost functions are estimated in wavelet domain from WP coefficients $X_{l,n,k}$. The algorithm will stop when the *WTE* value will begin to decrease. Received optimized WPD tree is a mapping of the dictionary for the MP stage.

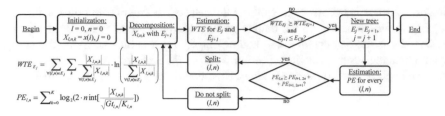

$$WTE_{E_j} = \sum_{\forall (l,n)\in E_j}\sum \frac{|X_{l,n,k}|}{\sum_{\forall (l,n)\in E_j} |X_{l,n,k}|} \cdot \ln\left(\frac{|X_{l,n,k}|}{\sum_{\forall (l,n)\in E_j} |X_{l,n,k}|} \right)$$

$$PE_{l,n} = \sum_{k=0}^{K} \log_2(2 \cdot n\,\mathrm{int}[\frac{|X_{l,n,k}|}{\sqrt{Gt_{l,n}/K_{l,n}}}])$$

Fig. 3. Block diagram of the optimal WPD tree estimation

2.3   MP Stage with Perceptually Optimized Dictionaries

MP is a greedy algorithm that maps the input data $(x(i))$ to the redundant dictionary of the time-frequency functions (g_γ). Since the presented coder needs to be invariant to the input sounds, the dictionary forming must be adaptive to them. Convenient way to achieve this is to construct it from the input with help of the adaptive WPD [12]. MP algorithm can be seen below:

```
Initialization: r₁ = x(i),  m = 0,  Gt^m_{l,n},  STOP=0;
REPEAT ∀(l,n,k):
   Select: X*_{l,n,k} ∈ X^m_{l,n,k} with max weight;
   Find: aₘ in X*_{l,n,k} with max(X*_{l,n,k}²/Gt^m_{l,n}), save current (l,n,k);
   Synthesize g_γ based on aₘ and WPD⁻¹;
   Residue estimation: r_{m+1} = rₘ - aₘg_γ;
   m = m + 1;
   Recalculate Gt^m_{l,n};
   Decomposition of rₘ;
   IF stop criterion == TRUE THEN STOP=1
WHILE STOP!=1.
```

First step, coefficients with maximum excitation weight in each frequency band are selected. Second step is devoted to the finding of such coefficient that will maximize matching between the modelled scalogram and the original one. In this case, it is equivalent to the estimation of the ratio between signal and the masking threshold. The global masking threshold $Gt_{l,n}$ that considers both simultaneous and time masking is used. After atom is selected it is necessary to subtract its contribution to the forming of the input signal, recalculate masking threshold for the residue signal and decompose it for the next iteration. Algorithm stops when the stop criterion is reached. It can be the fixed iterations (i.e. atoms) number or some energy thresholds.

Scalogram is the graphical representation of the wavelet coefficients as a 3-dimesional plot. X-axis is a duration of the frame, Y-axis – frequency bands and Z-axis shows coefficients magnitude. Examples for some signal frame are shown in Fig. 4.

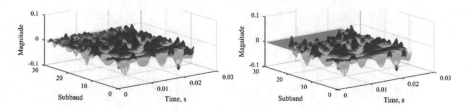

Fig. 4. Original (left) and modelled with 200 atoms (right) scalograms

3   DAE Neural Network Quantizer

The neural network quantization (NNQ) is the joint quantization of the parameter vector with real, continuous component amplitudes into some discrete set. The process of NNQ eliminates redundancy due to the effective use of the interconnected properties of the vector parameters. The definition of NNQ is formulated as follows: NNQ of dimension M is a mapping from the vector $X \in \mathbb{R}^M$ into N-dimensional code vector Y containing K discrete output values:

$$NNQ : X \in \mathbb{R}^M \rightarrow Y,$$

where $X \in [x_1, x_2, \cdots, x_M]$; $Y \in [y_1, y_2, \cdots, y_N]$; $y_i \in [y_{i1}, y_{i2}, \cdots, y_{ik}]$, $i = \overline{1, N}$.

An appropriate architecture of NN for implementing quantizer is autoencoder (AE) [8]. Conventional AE consists of two parts: an encoder and a decoder part. The encoder part (1) is a function that maps the input $X \in \mathbb{R}^{N_x}$ to hidden representation $h \in \mathbb{R}^{N_h}$ and the decoder unit (2) maps it back to a reconstruction $X' \in \mathbb{R}^{N_x}$ of the input data:

$$h = f(w_1 \cdot X + b_1), \tag{1}$$

$$X' = f(w_2 \cdot h + b_2), \tag{2}$$

where f is a nonlinear activation function; $w_1 \in \mathbb{R}^{N_h \times N_x}$, $w_2 \in \mathbb{R}^{N_x \times N_h}$ – weight matrices; $b_1 \in \mathbb{R}^{N_h}$, $b_2 \in \mathbb{R}^{N_x}$ – bias units; N_h and N_x are the number of hidden and input units respectively. AE training process consists of learning the parameters $W = \{w_1, w_2\}$ and $B = \{b_1, b_2\}$ which provide the smallest possible reconstruction error.

A single-layer AE is very limited in its computational capabilities, especially in case of solving NP-complete quantization task [14], which imposes discrete constraints on the internal representation. In this way, the choice should be made in favor of a multilayer architecture DAE [8]. Configuration of the NNQ was set as viewed in Fig. 5.

Greedy layer-wise training strategy [16] was used to solve the vanishing gradient problem [15]. The training process of DAE is divided into the following two stages: pre-training with the three single-layer AEs: $\{200 \times 100\}$, $\{100 \times 50\}$, $\{50 \times 20\}$ and fine-tuning. The first AE was trained on the prepared training set. Each subsequent AE was trained using the hidden layer outputs of each previous AE as the input training data.

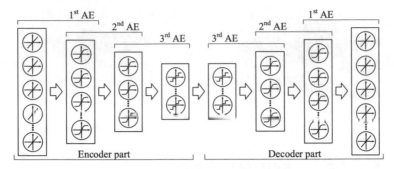

Fig. 5. DAE architecture quantizer. Bias units are not shown to simplify the structure

The activation function of the hidden layers for the first two AEs is hyperbolic tangent. To learn discrete hidden representation staircase-like activation function [17] was used for the middle layer (i.e. code layer), which formulates by the following equation:

$$f(z) = \frac{1}{M-1} \sum_{i=1}^{M-1} \left(tanh\left(\alpha(z - \frac{2 \cdot i}{M} + 1) \right) \right), \qquad (3)$$

where parameter $\alpha = 100$ is the step transition rate and $M = 32$ is a number of steps.

Each data vector was normalized to unit variance, and then sorted in ascending order. Such a data representation breaks the structure and relationships in them, and makes all the vectors highly correlated. This approach makes the proposed DAE architecture invariant to the input sound signal.

After the pre-training every single level AE is connected with each other to create a multilayer DAE. This DAE is then fine-tuned using backpropagation algorithm to make its output as close as possible to its input. To force the codes to be discrete, the outputs of the staircase units in the central code layer are rounded to the nearest flat region during the forward pass, and the straight-through estimator [18] is used to backpropagate through these discrete variables.

4   Decoder Based Near-End Listening Enhancement Algorithm Description

Noise masking is a very common case in the personal portable multimedia devices, an example of such a masking is shown in Fig. 6. As it can be seen, an ambient noise spectrum almost completely masks the input sounds. In such a surrounding listener is forced to increase the loudness of the playable audio. However, setting volume up for all components of the sound simultaneously can harm ears (quite sounds become louder but loud sound become very loud).

The idea of the listening enhancement algorithm implemented in the audio coder is based on the subband decomposition of the input signal and the ambient noise combined with the adaptive compression of the dynamic range [19]. In contrast of this work, filter bank for subband decomposition will be psychoacoustically motivated for the processed

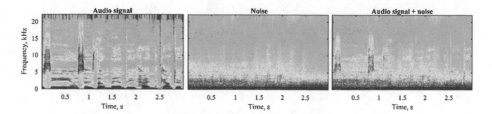

Fig. 6. An example of the ambient noise masking audio signal up to 6 kHz

frame (optimized WPD) in conducted research. The algorithm works as follows: both input signal and ambient noise are decomposed in the subbands in order to determine which frequency range of the audio can be masked. The next step is loudness correction of those sound components that have been masked. This must be done adaptively to the noise level, so gain factors are estimated as ratio between the input signal subband part and the ambient noise part levels. The decoding side of the developed audio coder with the embedded listening enhancement algorithm is shown in Fig. 7.

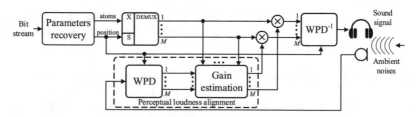

Fig. 7. Near-end listening enhancement embedded in the decoder

Listening enhancement algorithm is implemented after the decoding and dequantization (parameters recovery block in Fig. 7). Ambient noise enters the system through a microphone. Then, it is necessary to make subband decomposition of that information in the same way as the input signal. Since the bit stream contains the atoms and their position it is convenient to restore the same WPD structure for the noise decomposition, which was used for the audio signal coding (DEMUX block). Moreover, each frame will be processed with the perceptually adapted WPD tree, so the psychoacoustic properties of the signal will be preserved at the gain estimation stage. Atoms processing occurs in the perceptible loudness alignment area. This part is shown in Fig. 8.

Adaptive compressor is depicted at the right part of Fig. 8. It makes an input sound level X recalculation to Y level depending on the ambient noise level $P_{l,n}^{amb.noise}$ and two additional parameters: threshold offset from the current noise level ΔG and gain increase offset from current noise level ΔR in the following way:

$$\widehat{P_{l,n}^{rec.signal}} = \begin{cases} k \cdot P_{l,n}^{rec.signal} + b, & if \ P_{l,n}^{rec.signal} < P_{l,n}^{amb.noise} + \Delta G, \\ P_{l,n}^{rec.signal}, & otherwise \end{cases} \tag{4}$$

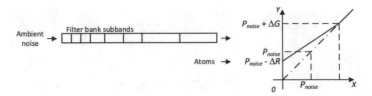

Fig. 8. Adaptive compression

where $b = P_{l,n}^{amb.noise} - \Delta R$, $k = (\Delta G + \Delta R)/\left(P_{l,n}^{amb.noise} + \Delta G\right)$. Both $P_{l,n}^{rec.signal}$ and $P_{l,n}^{amb.noise}$ are calculated as: $P = 10 \cdot \log_{10}\left(1/K \sum_{i=1}^{K_{l,n}} \left(MAG_i^2\right)\right)$, where $K_{l,n}$ is a number of the coefficients in subband (received form the encoder) with magnitude MAG. The final gain is calculated as follows:

$$gain_{l,n} = 10^{\left(\widehat{P_{l,n}^{rec.signal}} - P_{l,n}^{rec.signal}\right)/20}, \tag{5}$$

where (l, n) is a band number (in WPD tree). Algorithm dynamically estimates gain for every subband and if there is no masking of the audio signal, its value will be one.

5　Experimental Results

Test sequence used in the experiments consists of the samples with different content and shown in Table 1. All sounds are one-channel signals with 44.1 kHz sampling rate, 16-bit resolution and with 7 s minimum duration.

Table 1. Test samples description

Item	Description	Item	Description
es01	Vocal (Suzan Vega)	si01	Harpsichord
es02	German speech	si02	Castanets
es03	English speech	si03	Pitch pipe
sc01	Trumpet solo and orchestra	sm01	Bagpipes
sc02	Orchestra piece	sm02	Glockenspiel
sc03	Contemporary pop music	sm03	Plucked strings

There are two types of experiments are shown in the article: reconstructed signal quality evaluation based on ITU-R Recommendation BS.1387-1 PEAQ (Perceptual Evaluation of Audio Quality) and demonstration of the developed listening enhancement algorithm performance. PEAQ output is Objective Difference Grade (ODG – difference between tested sample and the original using perceptual criteria). Its scale is defined as follows: 0 - imperceptible impairment; –1 - perceptible, but not annoying; –2 - slightly annoying; –3 - annoying; –4 - very annoying impairment.

For quality estimation, conducted experiments were carried out on samples with using of the scalar quantization with the maximum quantization step $\Delta_{l,n} = \sqrt{12 \cdot G_{l,n}/K_{l,n}}$ and NNQ. Approximate CR can be seen in Table 2.

Table 2. Approximate CR comparison for different quantizers

Quantizer	Atoms			
	200	300	400	500
Scalar / DAE, times	19.3/70.5	13.1/56.4	9.9/47.0	8.0/40.3

The objective quality results are shown in Fig. 9. It is point to mention that CR for the scalar quantization and NNQ is very different so all the comparison must proceed from this fact. Nevertheless, it can be seen that most of ODG values for test items with scalar quantization do not exceed corresponding values for NNQ by no more than 0.5. Only two samples have no stable decrease of ODG: sc01 and sm02. This can be because of the training dataset – it may have contain small amount of such an information. In addition, for the both variations of quantization there are no ODG values below –4: all test samples have 'annoying impairment' and better.

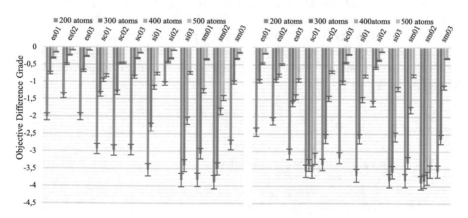

Fig. 9. ODG marks for the scalar quantization (left) and the DAE quantization (right)

Comparison of the listening enhancement function turned on and off in the decoder is shown in Fig. 10. As it seen, unprocessed signal is almost fully masked with the subway noise. Opposite of it, reconstructed signal with the listening enhancement function turned on, shows the main structure of the original signal through the noise mask.

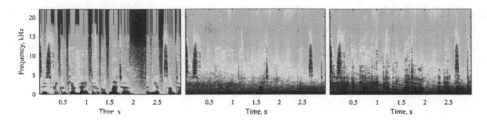

Fig. 10. Pure reconstructed with 500 atoms speech signal (left) and reconstructed speech + noise with listening enhancement turned off (center) and turned on (right)

Estimation of SII (Speech Intelligibility Index) shows that the presented approach gives improvement of speech intelligibility about 0.4 in aggressive noises (SNR between −20db and −10db) in favor of the processed one.

6 Conclusions

The article shows the universal scalable encoding algorithm using MP with the psychoacoustically motivated adaptive WPD based time-frequency dictionaries. DAE architecture quantizer is presented. Experimental results showed high quality of the approach with deep CR. Comparison with the scalar quantization tells that NNQ is very effective in the sound coding tasks. The results also show that developed coding algorithm has comparable quality with such coders as Opus and Vorbis, but NNQ can give bigger CR in perspective. A listening enhancement function was also introduced. It was shown that embedding it in the decoding part of the algorithm leads to the effective realization since it is necessary to make subband decomposition of only ambient noise signal.

References

1. Mallat, S., Zhang, Z.: Matching pursuits with time-frequency dictionaries. IEEE Trans. Sig. Process. **41**(12), 3397–3415 (1993)
2. Chardon, G., Necciari, T., Balazs, P.: Perceptually matching pursuit with Gabor dictionaries and time-frequency masking. In: ICASSP 2014, Florence, Italy, pp. 3126–3130 (2014)
3. Ravelli, E., Richard, G., Daudet, L.: Union of MDCT bases for audio coding. IEEE Trans. Audio Speech Lang. Process. **16**, 1361–1372 (2008)
4. Ruiz Reyes, N., Vera Candeas, P.: Adaptive signal modelling based on sparse approximations for scalable parametric audio coding. IEEE Trans. Audio Speech Lang. Process. **18**(3), 447–460 (2010)
5. Petrovsky, Al., Azarov, E., Petrovsky, A.: Hybrid signal decomposition based on instantaneous harmonic parameters and perceptually motivated wavelet packet for scalable audio coding. Sig. Process. **91**, 1489–1504 (2011)
6. Valin, J.-M., Maxwell, G., Terriberry, T., Vos, K.: High-quality, low-delay music coding in the Opus codec. In: AES 135th Convention, paper 8942, New York, USA (2013)

7. Vos, K., Sørensen, K. V., Jensen, S. S., Valin, J.-M.: Voice coding with Opus. In: AES 135th Convention, paper 8941, New York, USA (2013)
8. Sercov, V., Petrovsky, A.: Neural network quantizer of the parameters of the low bitrate vocoder with "speech + noise" speech formation model. In: Proceedings of the 4th International Conference "Digital signal processing and its applications" DSPA-2002, pp. 426–428 (2002). (in Russian)
9. Chu, W.C.: Speech Coding Algorithms: Foundation and Evolution of Standardized Coders. Wiley, Hoboken (2003)
10. Sauert, B., Vary, P.: Near end listening enhancement: speech intelligibility improvement in noisy environments. In: ICASSP 2006, Toulouse, France, pp. 493–496 (2006)
11. Petrovsky, A., Krahe, D., Petrovsky, A.A.: Real-time wavelet packet-based low bit rate audio coding on a dynamic reconfiguration system. In: AES 114th Convention, paper 5778, Amsterdam, The Netherlands (2003)
12. Petrovsky, Al., Herasimovich, V., Petrovsky, A.: Scalable parametric audio coder using sparse approximation with frame-to-frame perceptually optimized wavelet packet based dictionary. In: AES 138th Convention, paper 9264, Warsaw, Poland (2015)
13. Mallat, S.A.: Wavelet Tour of Signal Processing. The Sparse Way, 3rd ed. Academic Press, Burlington (2008)
14. Mumey, B., Gedeon, T.: Optimal mutual information quantization is NP-complete. In: Neural Information Coding Workshop, Snowbird, Utah, USA (2003)
15. Bengio, Y., Simard, P., Frasconi, P.: Learning long-term dependencies with gradient descent is difficult. IEEE Trans. Neural Netw. 5(2), 157–166 (1994)
16. Bengio, Y., Lamblin, P., Popovici, D., Larochelle, H.: Greedy layer-wise training of deep networks. In: Proceedings of the 19th International Conference on Neural Information Processing Systems (NIPS), pp. 153–160 (2006)
17. Hecht-Nielsen, R.: Replicator neural networks for universal optimal source coding. Science 269(5232), 1860–1863 (1995)
18. Bengio, Y., Leonard, N., Courville, A.: Estimating or propagating gradients through stochastic neurons for conditional computation. In: arXiv preprint arXiv:1308.3432 (2013)
19. Azarov, E., Vashkevich, M., Herasimovich, V., Petrovsky, A.: General-purpose listening enhancement based on subband non-linear amplification with psychoacoustic criterion. In: AES 138th Convention, paper 9265, Warsaw, Poland (2015)

The Features and Functions of an Effective Multimedia Information Retrieval System (MMIR)

Jolanta Szulc$^{(\boxtimes)}$ iD

University of Silesia in Katowice, Bankowa 12, 40-007 Katowice, Poland
jolanta.szulc@us.edu.pl

Abstract. The aim of the study is to identify the features of the effective Multimedia Information Retrieval System (MMIR). To achieve this goal, a literature analysis of the subject searched in the databases was carried out: Library and Information Science Abstracts (LISA) and Library, Information Science & Technology Abstracts (LIST). The following areas of research were discussed: criteria for information retrieval, standards for the description of the content of multimedia objects, features of an effective MMIR, implemented projects and experiments. The specified features of MMIR, such as completeness, description of behavior, including changing requirements, can be used when defining and analyzing requirements in the software development process. At the end of the article the possibilities and areas of application of MMIR in the work of libraries were indicated.

Keywords: Multimedia information retrieval system (MMIR)
Features of MMIR · Functions of MMIR · MMIR in modern libraries

1 Introduction

The principles and practice of multimedia information retrieval systems (MMIR), understood as a set of Visual Retrieval (VR), Video Retrieval (VDR), Audio Retrieval (AR) and Text Retrieval (TR), are well known to scientists and practitioners dealing with computers, computer science, information science, mathematics and many other fields. It is now necessary for librarians to become familiar with MMIR technologies. Literature registered in databases is helpful in this regard: Library and Information Science Abstracts (LISA, 1969-) and Library, Information Science & Technology Abstracts (LIST, mid-1960s-).

The analysis of the literature on the subject was carried out on the basis of searching databases: LISA and LISTA. In both databases, the same search criterion was used: "multimedia information retrieval". The bibliographic descriptions of articles from scientific journals, trade magazines and books were searched and analyzed. Search results are shown in Table 1.

© Springer Nature Switzerland AG 2019
K. Choroś et al. (Eds.): MISSI 2018, AISC 833, pp. 120–128, 2019.
https://doi.org/10.1007/978-3-319-98678-4_14

Table 1. Search results according to the search term "downloading multimedia information" in selected databases (access: 25.04.2018).

Name of the database/period	Search term	The number of records found
LISA/1990–2018	"multimedia information retrieval"	144
LISTA/1975–2018	"multimedia information retrieval"	92

The number of publications on multimedia information retrieval systems in selected databases is systematically increasing. This is illustrated in Fig. 1.

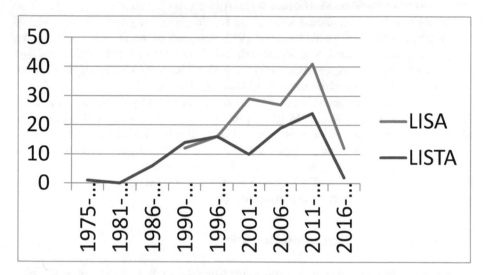

Fig. 1. Number of publications on multimedia information retrieval systems in selected databases in the period.

The first publications on multimedia information retrieval systems appeared in 1975 (in the LISA database) and in 1990 (in the LIST database). The number of publications increased and reached the highest level in both databases in the period 2011–2015. After 2015, the number of publications in both databases began to decline, which may be due to delayed indexing of documents. Visible differences may also result from different thematic areas of the literature registered in individual databases.

On the basis of the collected data, the following areas of research were identified: criteria for information retrieval, standards and standards for the description of the content of multimedia objects, features of an effective MMIR, other research problems. We begin the review of solutions in this area from the definition and division of MMIR.

2   Definition and Division of MMIR

Multimedia Information Retrieval System (MMIR or MIR) is a metasystem that is specialized in the effective processing of digital multimedia objects.

The definition of goals and assumptions of MIR is often presented in the context of differences between Information Retrieval (IR) and MMIR. IR is a search system that uses terms to retrieve text information, which is also used for audio, visual and audio-visual documents. MMIR as a methodology has evolved towards strengthening the reference document as a whole. MIR is proposed as a search system using text, images and sounds, any kind of documents or multimedia in the full sense of the word.

As part of the general MMIR methodology, we can distinguish: a system of Text Retrieval (TR), based on textual information for the processing and search of textual documents; a method of Visual Retrieval (VR), designed on visual data for the search of visual documents; a method of Video Retrieval (VDR), founded on audiovisual data for the processing of videos; and a criterion of Audio Retrieval (AR), based on sonorous data for the processing and the retrieval of audio documents [18].

In the subject literature there is also the term Multi-multiple Media Information Retrieval Systems (MMIRS). MMIRS is a multi-database system in which more than one of the component databases is managed by a Multimedia Information Retrieval System (MIRS). Application examples are many and include digital libraries, museum consortiums and cooperating news agencies. As an example, a set of Data Management System (DBMS) supported MIRS' containing thematically related data can define the scope of a multi- multimedia database system [16].

3   Criteria for Information Retrieval

The criteria for downloading multimedia information define the rules according to which each document/digital content can be analyzed and searched using appropriate language elements or metalanguage. The MIR system - as a system composed of TR, VR, VDR and AR systems - is built on the fundamental principles of analysis and search methodology based on the content of documents, referred to as Content Based Information Retrieval (CBIR). As part of the CBIR logic, the analysis and search methods are defined as content-based. They are based on the use of storage and retrieval keys of the same nature as the specific content of the resources to which they are applied. These keys are based on the language appropriate for each resource typology, able to consistently indicate specific content, as well as to the semantic aspects of a specific document.

The second way to organize the criteria of information resources is an innovative approach based on the representation of content, through visual, audio and audiovisual elements. This is a semantic and interpretative approach. In multimedia search, a more effective system is a system that can act almost directly on content, on objective data about words, forms, movements and sounds. In this structure, the formulated questions can be freely entered into the system by the user by means of words, forms, sounds and movements, and immediately analyzed and met [19].

The following conclusion can be drawn from the analysis of the literature on the subject: a high level of accuracy in downloading documents/digital content can only be achieved through a combination of different technologies. Harmonization of these techniques is assumed. A time-based inquiry can be a good preliminary method for selecting parts of documents in relation to titles or authors of works. The interaction consists in combining this form of inquiry with a query containing numbers, sounds, texts useful for searching for multimedia documents [21].

4   Standards and Regulations

The description of the content of multimedia objects is now a big challenge, which directly affects the result of searching for information. For example, an interface for describing multimedia content is a complex standard based on metadata templates that works based on optimizing the description of this type of content.

4.1   Standards for the Description of Multimedia Objects

The standards for the description of multimedia objects include:

- MPEG-7 (see: ISO/IEC TR 15938-11:2005)
- Dublin Core (see: ISO 15836-1:2017)
- Material Exchange Format (MXF) (see: ISO 26429-3:2008, ISO 26429-4:2008, ISO 26429-6:2008)
- Advanced Authoring Format (AAF) [1].

4.2   Standards Developed by the International Organization for Standardization (ISO)

In addition, standards developed by the ISO are available. These are as follows:

- ISO/IEC 15938-12: 2012 Information technology – Multimedia content description interface – Part 12: Query format – which describes the query format tools which may be used independently or in combination with other parts of ISO/IEC 15938. Each query format tool is described in two normative sections: (1) syntax - normative query specification and management format and (2) semantic - normative definition of the semantics of all the components of the corresponding query format specification. The query format provides a standardized interface for multimedia information retrieval systems (e.g. MPEG-7 databases) in three aspects, which include input query format, output query format, and query managements [9].
- ISO/IEC 13249-1:2016 Information technology – Database languages – SQL multimedia and application packages – Part 1: Framework
- ISO/IEC 13249-2:2003 Information technology – Database languages – SQL multimedia and application packages – Part 2: Full-Text
- ISO/IEC 13249-3:2016 Information technology – Database languages – SQL multimedia and application packages – Part 3: Spatial

- ISO/IEC 13249-5:2003 Information technology – Database languages – SQL multimedia and application packages – Part 5: Still image
- ISO/IEC 13249-6:2006 Information technology – Database languages – SQL multimedia and application packages – Part 6: Data mining.

Areas of application included in ISO/IEC 13249 implementations include, among others, automated mapping, desktop mapping, object management, geoengineering, graphics, location services, terrain modelling, multimedia and resource management applications.

5   Features of an Effective MMIR

The research results indicate that the development of public multimedia information services depends on four factors: (1) access to high-bandwidth networks; (2) inexpensive user devices capable of handling multimedia; (3) adoption of standards for the representation, compression, packaging and transport of multimedia information; and (4) developing a body of multimedia information and related infrastructure for its organization and search. Some researchers believe that the first three are already implemented in practice and express moderate optimism about the fourth [4].

On the basis of analyses of various MMIR descriptions, its features can be distinguished. The description of requirements for MMIR should take into account the following issues:

5.1   Completeness and Consistency

The description of requirements for MMIR should be complete and consistent. Creating a database and index should contain:

- Analysis: identification of elements in the content of the document.
- Data submission: create a generic data file for each document.
- Characteristics: extraction of characteristic data related to key aspects of content and linking to a generic data file.
- Indexing: updating the index using content properties and general data.
- Description and classification: a combination of relevant terminological information.

5.2   External Behavior of the System

The description of requirements for MMIR should describe the external behavior of the system, and not how to implement it. The search and retrieval process should include:

- Pre-search: terminological search to select parts of documents from the entire database.
- Multimedia search: use of some documents extracted as an example model to run a multimedia search.
- Match similarities: automatic capturing of documents whose similarity to the example is to a certain extent included in the set parameters.

- Development: use of additional extracted documents, modification of characteristics, selection of parts, connection of elements, terminological explanation, re-launch of search.

5.3   Future Potential Changes in System Requirements

The description of requirements for MMIR should take into account future potential changes in system requirements. Advanced functions implemented in the system should enable:

- Search analysis: carried out automatically, not only when creating or updating a database.
- External models: using samples taken from outside the database.
- Model composition: the ability to freely create a search template via system functions.

5.4   Limitations Under Which the System Will Work

The description of requirements for MMIR should include the limitations under which the system will work:

- The limitations of objective access to documents should be taken into account, as well as the gap between the gap and concept-interpretative access, referred to as semantic gap.
- Because the importance of a multimedia document is rarely unambiguous, the goal of the system must be to provide support to overcome this gap between the simplicity of content processing offered by the device and the semantic expectations of the user.

5.5   Easy Modification

The description of requirements for MMIR should be easy to modify.

5.6   Behavior of the System in Undesirable Situations

The description of requirements for MMIR should describe the behavior of the system in undesirable situations [19].

6   Projects and Experiments

Please note that the first paragraph of a section or subsection is not indented. The first paragraphs that follows a table, figure, equation etc. does not have an indent, either.

Subsequent paragraphs, however, are indented. Numerous projects and experiments are carried out, which concern, among others:

- a model for integrating indexes of textual and visual features via a multi-modality ontology and the use of DBpedia to improve the comprehensiveness of the ontology to enhance semantic retrieval [15, 20];
- the application of structured sparse representation at image annotation [14];
- the role of social Q&A in music information seeking [7];
- a graph-based approach for visual analytics of large image collections and their associated text information (e.g. iGraph system) [6];
- relevance diversity trade-off enhancement work-flows, which integrate multiple information from images, such as: visual features, textual metadata, geographic information, and user credibility descriptors [5];
- the joint usage of multiple information sources positively impacted the relevance-diversity balancing algorithms [5];
- a new generation of multimedia information retrieval systems that incorporate emotion in order to help users discover documents in meaningful ways that move beyond keyword and bibliographic searches [17];
- different tagging behaviors by analysing the book tags in different languages [11];
- the semantic gap between content and concept based multimedia retrieval, indexing vocabularies used for multimedia retrieval [12];
- the Snake Table which improves the efficiency of k-NN searches in systems, avoiding the building of a static index in the offline phase [3];
- searching of text and image repositories by keywords [13];
- downloading multimedia information from recorded presentations (e.g. through integration with the Authoring on the Fly (AOF) system [8].

7 Conclusion

Please note that the first paragraph of a section or subsection is not indented. The first paragraphs that follows a table, figure, equation etc. does not have an indent, either.

Subsequent paragraphs, however, are indented. Defining the IR methodology in accordance with traditional search parameters, basically focused on the user, concentrating on conceptual, interpretative and terminological methods, was most commonly adopted. However, the nature of multimedia documents in the contemporary information society determined the creation of multimedia databases with greater complexity than traditional ones. Searching for multimedia information has become an important issue in the work of libraries, especially since users can have access to the Internet both at home and in libraries. In addition to bibliographic information, graphic resources are now required to meet the new and demanding needs of users.

The areas of MMIR applications in modern libraries include:

- indexing of multimedia documents (which information can be effectively extracted from multimedia documents);
- searching and downloading multimedia information from images, films or words (determining how this information could be represented and organized to handle content-oriented search requests, definition of the concept of multimedia search, development of a context-orientated search rules, creation of interfaces allowing for

the formulation of searches in different dimensions, not only through words, but also through images and sounds);

- optimization of multimedia information search processes (the research was extended to data modelling; the strict categorization of documents; direct assessment; detailed rules concerning the features of browsing; the ratio of interactivity between user and machine; the processes of filtering; and the architecture of multimedia searches).
- creation and development of multimedia digital libraries (they must offer access to typological multimedia documents and be interpretable with traditional analysis, indexing and terminological scarch systems, architecture and logic of a multimedia database should be object-oriented and based on the actual content of documents);
- management of multimedia databases (the principles of such multimedia data treatment; multimedia data modelling; the analyzing, archiving or indexing best suited to the different characteristics of multimedia materials; creating indexes understood as an access link to documents; spatial and temporal relationships involving more complex objects in multimedia data processing).

The most important functions and areas of MMIR applications are shown in Fig. 2.

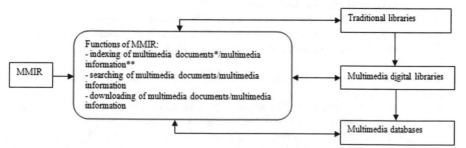

* Multimedia document – is a natural extension of a conventional textual document in the multimedia area. It is defined as a digital document that is composed of one or multiple media elements of different types (text, image, video, etc.) as a logically coherent unit [10].
** Multimedia information - is the most comprehensive method of data delivery, it reaches the recipient via image and sound (a combination of text description, spoken comment, music, photos, graphics, animation, film) [2].

Fig. 2. Selected functions and areas of MMIR applications in modern libraries.

Research on various aspects of multimedia information search can improve the quality of the practice in different ways, and the results can be beneficial for many specialists: programmers, indexers, reference librarians, web designers, major librarians and administrators.

Research results should be disseminated so that they are crucial to establishing a new approach to practice. Especially in the LIS sector, the analysis of projects through appropriate publications and implementations, as well as IT and other disciplines, will allow to track future changes and trends.

128   J. Szulc

References

1. AAF Association Specification. Advanced Authoring Format (AAF). Object Specification v 1.1. https://www.google.com/search?q=ISO&ie=utf-8&oe=utf-8&client=firefox-b. Accessed 25 Apr 2018
2. Abramowicz, W., Nowicki, A., Owoc, M. (eds.): Zarządzanie wiedzą w systemach informacyjnych, pp. 89 Wydawnictwo Akademii Ekonomicznej im. Oskara Langego, Wrocław (2004)
3. Barrios, J.M., Bustos, B., Skopal, T.: Analyzing and dynamically indexing the query set. Inf. Syst. **45**, 37–47 (2014)
4. Bulick, S.: Future prospects for network-based multimedia information retrieval. Electron. Libr. **2**(8), 88–99 (1990)
5. Calumby, R.T.: Diversity-oriented multimodal and interactive information retrieval. SIGIR Forum **1**(50), 86 (2016)
6. Gu, Y., Wang, Ch., Ma, J., Nemiroff, R.J., Kao, D.L., et al.: Visualization and recommendation of large image collections toward effective sensemaking. Inf. Vis. **1**(16), 21–47 (2017)
7. Hertzum, M., Borlund, P.: Music questions in social Q&A: an analysis of Yahoo! Answers. J. Documentation **5**(73), 992–1009 (2017)
8. Huerst, W., Mueller, R., Mayer, Ch.: Multimedia information retrieval from recorded presentations. In: SIGIR 2000 Proceedings of the 23rd Annual International ACM SIGIR Conference on Research and Development in Information Retrieval, pp. 339–341. ACM, New York (2000)
9. International Organization for Standardization. https://www.iso.org/standard/61195.html. Accessed 25 Apr 2018
10. Khosrowpour, M.: Dictionary of Information Science and Technology, vol. 2, 2nd edn, p. 617. Information Science Reference, Hershey (2013)
11. Lu, C., Zhang, Ch., He, D.: Comparative analysis of book tags: a cross-lingual perspective. Electron. Libr. **4**(34), 666–682 (2016)
12. MacFarlane, A.: Knowledge Organisation and its Role in Multimedia Information Retrieval. Knowledge Organization **3**(43), 180–183 (2016)
13. Magalhães, J., Rüger, S.: An information-theoretic framework for semantic-multimedia retrieval. ACM Trans. Inf. Syst. **4**(28), 19.1–19.32 (2010)
14. Maihami, V., Yaghmaee, F.: A review on the application of structured sparse representation at image annotation. Artif. Intell. Rev. **3**(48), 331–348 (2017)
15. Yanti Idaya Aspura, M.K., Mohd Noah, S.A.: Semantic text-based image retrieval with multi-modality ontology and DBpedia. Electron. Libr. **6**(35), 1191–1214 (2017)
16. Nordbotten, J.C.: Multimedia information retrieval systems. http://nordbotten.com/ADM/ADM_book/MIRS-frame.htm. Accessed 25 Apr 2018
17. Pennington, R.D.: The most passionate cover I've seen: emotional information in fan-created U2 music videos. J. Documentation **3**(72), 569–590 (2016)
18. Raieli, R.: Introducing multimedia information retrieval to libraries. JLIS.it **3**(7), 9–42 (2016)
19. Raieli, R.: Multimedia Information Retrieval. Theory and Techniques. Chandos Publishing, Oxford (2013)
20. Raieli, R.: The semantic hole: enthusiasm and caution around multimedia information retrieval. Knowl. Organ. **1**(39), 13–22 (2012)
21. Raieli, R., Innocenti, P.: L'innovazione possible nella prospettiva del multimedia information retrieval (MMIR) [The achievable innovation by the way of multimedia information retrieval (MMIR)]. Bollettino AIB **1**(45), 17–47 (2005)

Web Systems and Network Technologies

YouTube Timed Metadata Enrichment Using a Collaborative Approach

José Pedro Pinto[1] and Paula Viana[1,2(✉)]

[1] INESC TEC, Porto, Portugal
{jppinto,pviana}@inesctec.pt
[2] School of Engineering, Polytechnic of Porto, Porto, Portugal

Abstract. Although the growth of video content in online platforms has been happening for some time, searching and browsing these assets is still very inefficient as rich contextual data that describes the content is still not available. Furthermore, any available descriptions are, usually, not linked to timed moments of content. In this paper, we present an approach for making social web videos available on YouTube more accessible, searchable and navigable. By using the concept of crowdsourcing to collect the metadata, our proposal can contribute to easily enhance content uploaded in the YouTube platform. Metadata, collected as a collaborative annotation game, is added to the content as time-based information in the form of descriptions and captions using the YouTube API. This contributes for enriching video content and enabling navigation through temporal links.

Keywords: Video tagging · Video retrieval · Crowdsourcing
Multimedia content annotation · Gamification · Social media · YouTube

1 Introduction

Online video platforms have provided the ground for this type of content to become widely used and a source of information and entertainment for millions of users. There are numerous video sharing websites such as Vimeo, YouTube and Dailymotion. Among those, YouTube is certainly the most popular, with videos shared amongst 1.3 billion of users across the whole world. With more than 300 h of video uploaded every minute, YouTube offers a novel kind of knowledge base for multimedia data. Apart from allowing users to upload and share their videos, it also encourages them to enrich the visual content with context information that includes tags, categories, title, etc. This process results in coupling massive amounts of content with user-generated metadata that greatly facilitates video retrieval and browsing by using text-based search engines.

However, the existing metadata is usually linked to the whole video and no time-coded annotations are available. Therefore, search performance and accuracy are reduced since users must watch the entire video to find parts of their interest – a tedious and time-consuming task. This lack of timecoded annotations will make some of the users to miss the chance to watch the intended scenes at a specific time [1].

© Springer Nature Switzerland AG 2019
K. Choroś et al. (Eds.): MISSI 2018, AISC 833, pp. 131–141, 2019.
https://doi.org/10.1007/978-3-319-98678-4_15

This drawback can be overcome by generating descriptions associated to specific points in the video. However, manually annotating video content is an expensive and time-consuming process. These almost incompatible aspects are the drivers for finding new methods that enable the creation of richly described video assets. Although some video annotation systems have been proposed, no solution has been yet provided for creating metadata to improve third party platforms like YouTube.

In this paper, we propose using a collaborative video annotation game platform to extend YouTube metadata with timecoded descriptions. Crowdsourced metadata created in the game produces tags of YouTube videos which are then exported back to YouTube as description and captions files in order to be indexed by YouTube's search engine. Our work will contribute to create better video content descriptions, which will enhance video searching results, as well as help users to quickly find scenes of their interest without having to watch the full stream.

2   Related Work

Commenting on videos is an approach for users to contribute with opinions and discuss some of the content on the video. Using YouTube comments to facilitate access and retrieval of online videos has already been exploited. On their proposal [2], authors describe a set of temporal transformations for multimedia content that allows end-users to create and share personalized timed-text comments that are combined chronologically. A survey confirms a better user experience when watching videos together with these timed related comments. However, this solution uses the deprecated Flash Player.

Based on social activities, especially user comments and weblog authoring, the work presented in [3] describes a mechanism that helps users to associate video scenes with user comments, to generate entries that quote video scenes, and to extract deep-content-related information about video contents for automatic annotation. This solution is made available as a standalone Windows application, not web-based. Moreover, it is too complex for a standard user and requires the user to fill a lot of specific metadata fields to annotate a simple video.

Annotations that include sentiment analysis and emotion modelling based on YouTube channel comments has been proposed by [4]. The solution developed uses gamification approaches to help on the collection of the information that is then used to enable content recommendation.

A video scene annotation method based on tag clouds has also been proposed [5]. Comments associated to a video and collected from YouTube are processed and a tag cloud is generated based on those comments. Based on the user clicks on a tag in the cloud while watching the video, the tag gets associated with the scene in the video. However, the presented web videos don't use HTML5 technologies and the set of available tags is restricted to the ones that are extracted from the comments.

A collaborative annotation system of social media that includes temporal duration of the scenes and uses ontological themes of the selected domain has been proposed in

[6] The application allows users to annotate content using free-text or following onto-logically rules, with the objective of enhancing faster retrieval when browsing and searching for videos (specific scene, events, object, etc.).

Speech recognition has also been used to improve descriptions of online content. A framework for extracting relevant information from the audio track is exploited in [7]. Results show that the superimposition of relevant text and image-based information could be used for augmenting the viewing experience, as well as to give a full context-aware perception.

A browser extension that enables crowdsourcing of event detection in YouTube videos through a combination of textual, visual and behavior analysis techniques has also been proposed [8]. Based on the analysis of the visual content, it offers the user the choice to quickly jump to a specific shot in the video by clicking on a representative frame. The available metadata combines the one uploaded by the video owner such as title, description, tags, etc. and closed captions, which can be user-generated or auto-generated via speech-recognition. Interest-based event detection is achieved by counting clicks on shots. Although this seems promising, it requires installing a browser exten-sion, what could be an obstacle for a significant number of people. Furthermore, as in the examples above, the produced metadata is not stored on YouTube and does not then contribute to enrich the platform's video content.

Besides the limitations already identified, it is worth mentioning that the most important drawback of these proposals is the fact that metadata is stored locally, and only local users who provided the annotations have access to this information. So, there is no solution for making videos enhanced with richer content information available for the full YouTube's community. Additionally, although user comments are quite popular in YouTube, they are also extremely controversial and usually acknowledged as very noise. Given that heaps of comments are continuously posted every day, the task of filtering good video related comments is not easy.

YouTube captions mechanism has proven to be a good method to increase views. An experiment has found an overall increase of 7.32% in views for captioned videos [9]. Professional services are even available for adding captions to YouTube videos [10]. However, the service is paid and only transcribes the audio.

In our proposal, we try to overcome the identified limitations by implementing mechanisms for the validation of descriptions provided by users and by making this validated metadata available to the full YouTube's community using the captions and description fields.

3   YouTube Video Platform

Social web videos have pervaded on the web, along with contextual information that describes the content, easing video browsing and searching. However, a great part of the online platforms provides only succinct and generic information, with no temporal pointers to specific happenings. There are numerous video sharing websites, such as Vimeo, YouTube, Dailymotion, etc., where people can share their ideas and thoughts by sharing videos.

The online video sharing website YouTube was originally created in February 2005 to help people share videos of personal or well-known events. It provides a forum for people to engage with video content across the world and acts as a distribution platform for content creators. YouTube content ranges from professional to amateur and the diversity of videos goes from TV clips to short videos with a variety of content types, such as tutorials, educational videos, music clips, video blogging, etc. The popularity of the platform is driven by the easiness of sharing and reproducing video content [11].

3.1   Search Engine

YouTube doesn't yet include the tools for automatic scene description. This means that the system depends on provided metadata and relevant information to help users to find content when searching for something. Therefore, uploaders should create useful and optimized metadata to have better chances for their video to be found. Information as title, tags, description, closed captions and user's reactions, are some of the useful information that YouTube search engine will use to present more precise results on a search query.

For the purpose of this paper, we will focus on a new approach of using the captions and description features to improve search for videos and, more important, search within videos on YouTube.

3.2   Captions Mechanisms

Under the video library management page, YouTube allows owners to upload their own closed captions file as subtitles to YouTube videos. A subtitle or closed caption file contains the text transcription of the audio stream and the time codes for when each line of text should be displayed. Some files also include position and style information enabling the customization of the presentation.

YouTube supports a vast list of captions' file formats. The most usual formats are SubRip (.srt), SubViewer (.sub) and YouTube's preferred format - Scenarist Closed Caption (.scc). In 2009, a new feature that allows users to upload a simple text transcript of the spoken content and leave synchronization decision to the speech recognition mechanisms of YouTube was introduced. This feature saves users' time and trouble of transcribing the spoken video content and marking start/end times.

Figure 1 shows a YouTube video with automatic generated captions. These captions are superimposed on the video content, by clicking on the "CC button". As shown in the bottom part of the figure, close captions allow the video to be browsed by clicking on a given segment in the scrollable part of the entire caption track.

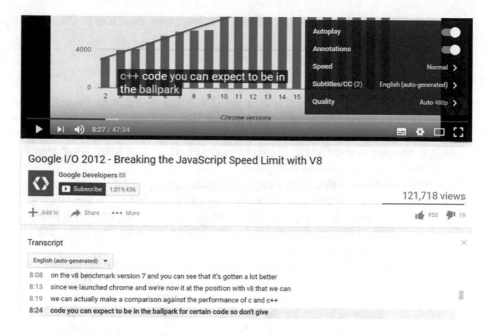

Fig. 1. YouTube automatic captions

4 Proposed Methodology

Figure 2 presents the overall architecture of the implemented platform that includes the combination of a game-based annotation system and YouTube. After making YouTube videos available for the annotation application, and after having metadata contributed

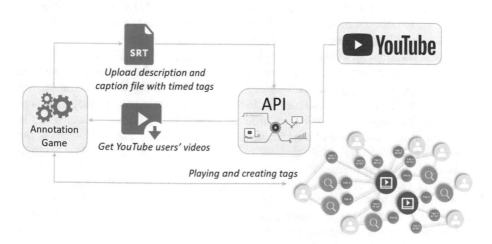

Fig. 2. Proposed system architecture

and validated by the crowd, this information is uploaded to YouTube in the form of captions and descriptions.

4.1 Annotation Game

The annotation system [12, 13] relies on a collaborative process and on gamification mechanisms to engage users on the tagging process. Tags may be freely introduced and players are rewarded if their contributions are considered valid. The created labels, or tags, are associated to specific time instants of the video, contributing to enhance the access to the exact moments of a video clip.

A scoring mechanism that takes into consideration past introduced information is used as an incentive for users to provide correct tags. Tag validation is achieved through a collaborative process, by analyzing the matchup between players' contributions. Additionally, semantically correlated tags are also considered, enabling enhancing and improving the quality of the dataset.

Three main aspects are considered on the process of tag validation: the tag itself and correlated tags from a dictionary; groups of tags organized in clusters; the number of times a tag, or correlated tag, appears in the respective cluster.

Clusters are groups of matching tags located nearby each other. They have a pre-defined length and are characterized by their centroid. Scoring is influenced by the distance of a tag to its centroid: 100, 50 and 10 points are considered. The higher the score, the closer the user is to the centroid. In this experiment, clusters were assigned windows of 12 s width, while scoring was linked to 8, 6 and 4 s time slots.

On the contribution of a player, the system verifies if the introduced tag is assigned to any of the existing clusters, or if a new one needs to be created. That assignment considers a pre-defined distance from the cluster centroid and its correlated tags. The introduced tag can result, or not, on the award of a score and on the validation of a tag if the number of agreements reaches a defined threshold.

To avoid player penalization for being the first one to introduce a specific tag, that later is validated by other players, an offline system is implemented to compensate this first effective contribution. An additional bonus is considered for having antedated useful metadata when the requirements for a score attribution are reached.

Besides contributing with metadata, players may also provide information that helps on the quality control of the tags. Moreover, different types of rewarding mechanisms that help on motivating good contributions and on maximizing the performance are considered in the game. This includes, besides scoring, prizes for completing an action, such as special badges, the definition of different game levels that the player may access and a leaderboard that shows his performance. Game levels are used on benefit of the annotation process as more difficult tasks (videos with less metadata are less likely to produce scoring) are provided to more qualified players.

The implemented mechanisms allow filtering erroneous information usually found on free tags and comments, to link metadata to timecodes and to have a collaborative effort on enhancing video information. A detailed description of the game functionalities is provided in [12, 13]. Aspects related to the performance, including usability, user engagement and tag accuracy, have been assessed in a user testbed [14].

4.2 Integrating the YouTube Platform and the Annotation Mechanism

The YouTube Data API enables developers to incorporate a variety of YouTube functionalities into their applications. The API allows the communication with YouTube by providing the developer access to the videos and user information. This can be used to personalize a web site or application with the user's existing information.

As a first step and on user's consent, the system retrieves the videos on a YouTube account, integrating them in our platform for annotation. This is the only action a user willing to use the annotation game for enhancing his YouTube video description content is requested. Future requests on user's behalf are enabled by extracting an access token that is used for all the needed communication between our platform and YouTube, providing then a transparent, non-intrusive process. Token renewing is also automatic, enabling access to the user account even when he is offline.

On a first step, the video owner selects his YouTube videos for annotation. This process enables sharing those videos with the game engine by uploading the video IDs into the annotation system, making them available for the annotation community.

Inserted tags are initially stored locally to enable comparing and matching time-related information. When conditions for a tag to become validated are reached [12], the system will automatically create caption and description files that include the tag and the associated timecode, and will upload this information to the YouTube account making this metadata fully available for everyone.

Figure 3 illustrates the use of the description field in YouTube, enhanced with timed tags imported from the annotation game. The initial description provided by the owner is kept but the field is updated with the navigable tags that, besides helping enhancing

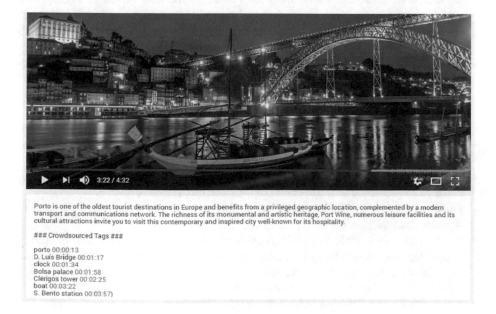

Fig. 3. Crowdsourced timed tags description

search precision, enable jumping to exact moments on the video. Figure 4 presents the use of captions to increment metadata. Tags may be overlaid in the video presentation and, additionally, they are listed to enable hyperlinking to video instants.

Fig. 4. Timed caption track

4.3 System Work Flow Process

The data flow process is depicted in Fig. 5 showing the interactions between the annotation system, the browser and the YouTube API. The process can be summarized as:

- User logs into our platform and chooses the built-in functionality for sharing his YouTube videos with our system.
- To enable retrieving his videos from the YouTube, and later on to publish captions and descriptions under his account, an authorization is required. This will allow our system to use the YouTube API methods. The user's Google Account is used to authorize the application to access the videos and to upload metadata.
- The OAuth2 authorization process is initiated on the first attempt to use the functionality in the game. On user consent, Google returns an access or/and a refresh token that is/are stored in the database for future use. This token allows uploading information from the annotation process into the user YouTube account.
- User' videos are listed and can be selected and shared with the annotation platform.
- Video IDs are extracted and added to our database to make them available for the crowd and playable along with other existing videos.
- According to the user gaming level, videos from our database will be retrieved in order to be presented to the user and played during the game.
- During the game and following the validation rules, tags can become valid due to exact or concept matching.
- A YouTube compatible.srt (SubRip) caption file and a description, both including timed tags, are automatically generated. The original description is merged with the new description so that no information is lost.
- Data is uploaded to the YouTube account by using the user ID and the token.
- Caption track and description become publicly available and ready to use/view.

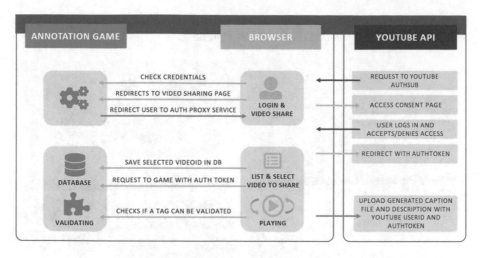

Fig. 5. Data workflow

5   System Evaluation

We have conducted a quantitative and qualitative experiment to evaluate the performance of the system. To assess the annotation approach, a set of volunteers that simulated
a crowdsourcing environment was asked to interact with the system. Different genres
of content were considered to make the annotation process not focusing in just one kind
of material. Besides analyzing the quality of the annotations, this experiment enabled
also checking the usability of the system and collecting feedback from the users by
means of a survey.

Findings show that players tend to be very accurate in time when typing some tag:
92.4% of the users were very consistent and introduced the same tag, or correlated tags,
near other players' tags. This can be explained by the fact that the gamification approach
encourages users to provide accurate information in order to score and progress in the
game. 60% of the contributed tags were validated by the system showing tag matchup.
These contributions allowed indexing 71 moments of the videos.

Motivation and engagement features included in the annotation process proved to
be effective. Not only accuracy was achieved, as well as it was evident that players
interacted actively with the game, competition within the top positions was acknowledged and answers to the questionnaire enable identifying motivation and enthusiasm
(82% of the players declared having enjoyed the game and it's features) [14]. These
findings are quite important, as productivity will depend on the motivation and enthusiasm that the gamification concepts can provide.

The metadata obtained collaboratively is expected to make searching and navigation
in video archives more efficient and to reduce the need for professional and expensive
processes of describing content. Figure 6 presents the navigation efficiency increase
achieved by uploading into YouTube the crowd contributed tags resulting from the

experiment. Direct access to specific parts of the same video was enabled by navigating on hyperlinked tags introduced in the description field.

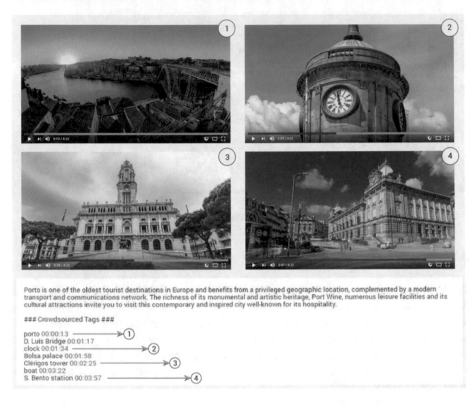

Fig. 6. Example of a YouTube video with timed tags obtained collaboratively

6 Conclusion

This paper proposes the use of a collaborative annotation game to collect metadata contributed by the crowd and enhance content in the YouTube platform using the caption feature and the description field of YouTube. By using a validation mechanism that enables cleaning erroneous information before uploading it to YouTube, the approach has benefits over the popular use of comments. To the best of our knowledge this is the first implementation of a system that uses YouTube caption and description features to improve YouTube search results using a collaborative approach. Apart from adding keywords to the video, it also associates timecodes to video descriptions, enhancing the navigation on YouTube content. Our method tries to solve two problems: the lack of useful metadata for accurate video retrieval and the difficulty on video navigation due to the lack of timed descriptions. Future work includes adding others video sharing platforms besides YouTube and creating a browser extension that allows the player to directly play the game on YouTube without the need to access another application.

Acknowledgements. The work presented was partially supported by FourEyes, a Research Line within project "TEC4Growth: Pervasive Intelligence, Enhancers and Proofs of Concept with Industrial Impact/NORTE-01- 0145-FEDER-000020" financed by the North Portugal Regional Operational Programme (NORTE 2020), under the PORTUGAL 2020 Partnership Agreement, and through the European Regional Development Fund (ERDF).

References

1. Zhou, R., Khemmarat, S., Gao, L., Wan, J., Zhang, J.: How YouTube videos are discovered and its impact on video views. Multimedia Tools Appl. **75**, 6035–6058 (2016)
2. Guimarães, R., Cesar, P., Bulterman, D.C.A.: Let me comment on your video: supporting personalized end-user comments within third-party online videos. In: Proceedings of the 18th Brazilian Symposium on Multimedia and the Web, pp. 253–260. ACM, Brazil (2012)
3. Yamamoto, D., Masuda, T., Ohira, S., Nagao, K.: Video scene annotation based on web social activities. IEEE Multimedia **15**, 22–32 (2008)
4. Mulholland, E., Kevitt, P.M., Lunney, T., Schneider, K.-M.: Analysing emotional sentiment in people's YouTube channel comments. In: Interactivity, Game Creation, Design, Learning, and Innovation, pp. 181–188. Springer, Cham (2016)
5. Yamamoto, D., Masuda, T., Ohira, S., Nagao, K.: Collaborative video scene annotation based on tag cloud. In: Huang, Y.-M.R., et al. (eds.) Advances in Multimedia Information Processing - PCM 2008, pp. 397–406. Springer, Heidelberg (2008)
6. Khusro, S., Khan, M., Ullah, I.: Collaborative video annotation based on ontological themes, temporal duration and pointing regions. In: Proceedings of the 10th International Conference on Informatics and Systems, pp. 121–126. ACM, Giza (2016)
7. Gatteschi, V., Lamberti, F., Sanna, A., Demartini, C.: An audio and image-based on-demand content annotation framework for augmenting the video viewing experience on mobile devices. In: Proceedings of 2015 IEEE International Conference on Mobile Services, pp. 468–472 (2015)
8. Steiner, T., Verborgh, R., Van de Walle, R., Hausenblas, M., Vallé: crowdsourcing event detection in YouTube video. In: Proceedings of the 1st workshop on detection, representation, and exploitation of events in the semantic web (2011)
9. 3Play Media: Adding Closed Captions to YouTube Videos Increases Views. http://www.3playmedia.com/customers/case-studies/discovery-digital-networks
10. Captions for YouTube. http://www.captionsforyoutube.com/
11. Burgess, J., Green, J.: YouTube: Online Video and Participatory Culture. Wiley, Hoboken (2013)
12. Pinto, J.P., Viana, P.: TAG4VD: a game for collaborative video annotation. In: 2013 ACM International Workshop on Immersive Media Experiences, pp. 25–28. ACM, Spain (2013)
13. Pinto, J.P., Viana, P.: Using the crowd to boost video annotation processes: a game based approach. In: Proceedings of the 12th European Conference on Visual Media Production, p. 22:1. ACM, London (2015)
14. Viana, P., Pinto, J.P.: A collaborative approach for semantic time-based video annotation using gamification. Hum. Centric Comput. Inf. Sci. **7**, 13 (2017)

Improving the Responsiveness of Geospatial Web Applications Through Client-Side Processing

Kamila Środa, Marek Łabuz, and Sebastian Ernst[(✉)]

Department of Applied Computer Science,
AGH University of Science and Technology, Kraków, Poland
`ernst@agh.edu.pl`

Abstract. Web applications which enable interactive editing of geospatial data are becoming increasingly popular, partly due to the growing interest in development of smart city concepts. However, performing calculations on geospatial objects, often expressed using a spheroidal (non-Cartesian) coordinate system, is not trivial. This paper tries to compare two approaches to developing such applications – one based on GIS backend queries, and one which performs the calculations locally on the client's machine. Prototypes implementing both approaches have been prepared, and tests based on real-life scenarios have been carried out. The results are presented along with conclusions, which are used to develop recommendations for the most efficient architectures in various situations.

Keywords: Spatial data · Web GIS · Web mapping
Geometry transformation · Smart cities

1 Introduction

Along with the growing popularity of social and mobile applications, the use of geospatial data is becoming increasingly widespread. For many purposes, it is sufficient to represent locations of individual points. If the data is used merely for visualisation purposes, the coordinates – latitude and longitude – may be represented as floating-point numbers stored in a file or a database.

However, once the need for any form of analytics arises (e.g., *how many points are there within a given distance?*), this representation is no longer suitable, due to the multidimensionality and the fact that the most common coordinate systems are not isometric, as outlined in Sect. 2.1. Hence, specialised geospatial data storage technologies have to be used; a short characteristic of such systems is presented in Sect. 2.2. This requirement becomes more pronounced for sophisticated queries, often used for operations called *Location Intelligence* in the enterprise world [6].

Of course, geospatial data is not limited to individual points. Standards define a broader set of primitives, including lines, linestrings and polygons [10].

© Springer Nature Switzerland AG 2019
K. Choroś et al. (Eds.): MISSI 2018, AISC 833, pp. 142–150, 2019.
https://doi.org/10.1007/978-3-319-98678-4_16

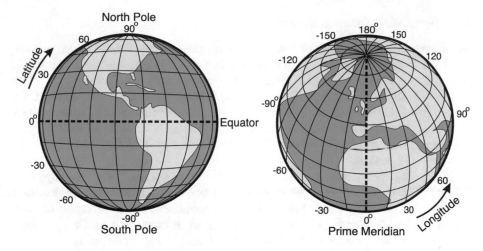

Fig. 1. Principles of a spheroidal spatial reference system (source: *Wikipedia*)

More advanced applications may replace polygonal chains with curves, including Bézier curves for a more efficient representation of complex shapes [7].

Visual representation also plays a major role when dealing with geospatial data. It is relatively easy to read a dataset of numbers, even without plotting it as a chart. Meanwhile, a dataset containing geographic coordinates is virtually incomprehensible without displaying the corresponding points on a map.

However, an even greater challenge is related to *editing* the geospatial data in a visual manner. Modern web technologies, such as HTML 5 [11], provide tools which facilitate development of interactive and responsive applications. For instance, the `canvas` element allows for flexible and graphics-rich websites. This allows, among others, for development of drawing tools with desktop-like user experience. Unfortunately, because spatial projections are rarely isometric, standard graphics libraries cannot be used. This poses a challenge for web geospatial applications, as providing the user with a fluent editing experience means the need to consult the spatially-enabled database backend constantly. This problem is discussed in more detail in Sect. 3.

Based on that, an approach to mitigate this problem is proposed in Sect. 4, and the following Sect. 5 provides the scenario for experiments and their results.

2   State of the Art

This section tries to present the important issues related to both handling geographic data as well as developing visual web applications with graphics editing functionalities.

2.1   Spatial Reference Systems

To understand the difficulties related to transforming geospatial data, a basic intuition of how points can be located on the globe is necessary. As the world is a spheroid, a natural way of defining the coordinates of a given point – one that is actually taught in schools – is by expressing them using two angles from certain defined 'zero' points. As shown in Fig. 1, the *latitude* is the angle between the *equator* and the location of the point in the *north–south* dimension, and the *longitude* is the angle between the *prime meridian* and the location in the *east–west* dimension.

This constitutes the principles of *spheroidal* spatial reference systems (SRSs in short), which have the advantage of being able to express the location of any point on the globe. One of the most commonly-used examples of such systems is WGS84, which offers uncertainty of at most several centimetres [9].

The issue with spheroidal SRSs is that there is no direct correspondence between degrees and distance on the globe. The actual distance between points with longitudes differing by $1°$ can vary greatly depending on the latitude. Also, spheroidal SRSs are inherently not Cartesian, which makes any measurements or transformations impossible without re-projection.

Local Cartesian spatial reference systems exist, but these can only be valid for smaller areas. This is a major drawbacks for web applications, which often deal with geospatial data distributed in various parts of the globe.

2.2   Spatially-Enabled Databases

Database management systems (DBMSs), regardless of the model they use – be it relational or other – usually use a basic set of common data types, such as numbers, strings, dates, etc. Some database designers resort to storing the coordinates in two separate floating-point number columns, but that approach is inefficient and requires significant effort when performing any sort of spatial analyses. While some DBMSs, such as PostgreSQL [4], support *geometric* data, that does not help with non-Cartesian coordinate systems.

This issue is solved with the help of *spatial databases*, such as the PostGIS [3] add-on to PostgreSQL. Such systems usually supplement the DBMS with the following functionalities:

1. support for various spatial reference systems, with the ability of converting spatial data between them,
2. geometric data types, providing support for primitives such as points, lines, linestrings (polygonal chains) or polygons,
3. geometric measurement and transformation functions,
4. spatial indexes to speed up querying of georeferenced data.

It must be noted that the aforementioned functions (item 3) treat the data as being plotted on a Cartesian plane, and hence yield useless results when used directly with spheroidal SRSs. An example of such mistake is shown in Fig. 2. The *Buffer* function is supposed to expand a given geometry equally in

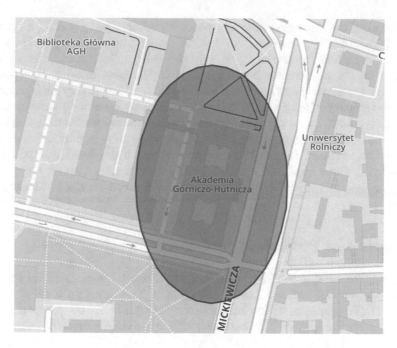

Fig. 2. Effects of direct application of spatial transformations to data in a spheroidal SRS

every direction. But, applied directly to a point in WGS84, it returned an ellipse instead of the expected circle.

This results in the need of converting coordinates, which is enabled due to multiple SRS support (item 1). Some spatial DBMSs overcome this by providing data types which express latitude/longitude coordinates in degrees, but use metres for any measurements or transformations. An example of such solution is the `geography` type, available in PostGIS since version 2.0.

3   Problem Statement

The problem described in the paper was first encountered during development of a web application which enables manipulation of geospatial elements inside the browser. A fluent editing experience for the user was a basic requirement, with editing operations including precise modifications of the elements' shapes, sizes and positions.

Such requirements are related to the increasing number of specialised geospatial web apps, aimed at developing designs and concepts for smart city infrastructures. An illustrative example for such applications are tools for interactive determination of relations between sensors (detectors) and end devices (actuators), for instance in dynamically-controlled street lighting installations [12, 13].

Because of the limitations and characteristics described in Sect. 2, a spatially-enabled database was utilised to handle these geospatial operations. The database was queried from the frontend of the web application in order to obtain new coordinates for elements being modified by the user. In order to achieve an interactive experience, such requests need to be made at a very high frequency. When the amount of information being exchanged between the frontend and the backend increases, the application's performance and responsiveness may suffer. To prevent such problems, some performance tests were conducted and a new solution that excluded the need to communicate with the database backend was implemented. Both approaches to the problem, with and without the presence of the backend in calculations, are outlined with more details in the following section.

4   Outline of Possible Approaches

In web applications, geospatial transformations can be performed by the backend, or they can be executed at the client side. The first approach consists in sending HTTP requests to the backend service where all the geospatial transformations are computed. The backend, after receiving a request, consults the spatially-enabled database. For this purpose, a PostgreSQL database was used along with a spatial extension – PostGIS. The second approach moves the calculations from the backend side to the client side and utilises a JavaScript geospatial library, Turf.js.

Turf.js [5] is an open source JavaScript library for spatial data analysis and manipulation. It includes many spatial operations for GeoJSON data and it can be used both on frontend and backend side. Turf.js also provides drivers that enable usage in Java and Swift languages.

PostgreSQL [4] is a powerful, open source object-relational database system that is known for its reliability, performance and variety of features. It can be extended with PostGIS [3], which adds support for geographic objects. It comes with a rich library of functions that allow geospatial transformations in SQL.

Both aforementioned approaches have pros and cons. On the client side, operations can affect the general performance of a website. On the other hand, when a backend is used to support geospatial transformations, the communication tunnel may become the bottleneck. In applications that use transformations rarely or in which the transformations are complex, querying the backend side seems more efficient and safe – it is better to receive the input parameters from a user and perform transformations on the backend side than to receive the result of the applied transformations from the frontend. That is because validation of input parameters is easier than the validation of a complete geospatial structure. Integrity errors in geospatial structures may result from malfunctions of the client or even deliberate attempts to hack the system.

On the other hand, if the application performs many transformations, e.g. in order to visualise animations, the responsiveness and the speed of such operations is crucial. To improve performance, developers have the possibility to increase the processing power of web servers or scale them horizontally and distribute the load. However, they do not have any influence on the users' devices and their network bandwidth, which largely determine their experience. When targeting mobile users, whose network bandwidth is usually limited, the experience may be unsatisfying, or the application may even become unusable.

Nevertheless, when performing calculations on the client side, we have better control over what transformations are executed. If we detect that a device is not capable of performing some tasks, we may limit or exclude them. It relates especially to mobile devices, where the processing power is significantly lower than that of computers. Animations aim to improve the user experience but, in some cases they may actually worsen it if they are badly designed or if their requirements overwhelm the device's capabilities. The key advantage of client side transformations over backend processing is the lack of network load, which often imposes additional costs on the users.

5   Experiments and Results

This section describes the scenario used to perform the experiments and presents their results.

5.1   Experiment Setup

The experiment consisted of two tests. The first one was aimed at demonstrating how the system behaves when many consecutive transformations are performed. Therefore, each approach has been tested for one minute of constant geospatial operations. It tested how efficient each solution is and how many operations per second can be performed. Moreover, the average execution time of a single operation was measured, along with CPU and network loads.

The second test was based on the use case of many operations being executed in parallel. It measured the time needed to accomplish the specified task.

In both tests, the transformation of a GeoJSON linestring buffered with a 5-m radius was used. This operation is supported both by PostGIS and Turf.js and consists in finding a geometry that covers all points within a given distance from the input geometry. One thousand unique linestrings downloaded from OpenStreetMap [2] representing the shapes of city streets have been used as the input geometries.

Each test has been carried out in a web browser environment (Chrome 65.0.3325.181) on an Apple MacBook Pro (2.7 GHz Intel Core i5, 8 GB RAM, Mac OS X 10.13.4). The server that has been used was Node.js (v9.6.1) [1] with PostgreSQL (v10.1) and PostGIS extension (v2.4.0).

Because of the fact that the solution using PostgreSQL/PostGIS relies heavily on HTTP communication, it was crucial to test this approach using different network profiles [8]. Table 1 presents 5 profiles with different download and upload speeds used to simulate various network scenarios.

Table 1. Network throttling profiles

Type	Download	Upload	Latency
GPRS	50 kb/s	20 kb/s	500 ms
Regular 3G	750 kb/s	250 kb/s	100 ms
Good 3G	1 Mb/s	750 kb/s	40 ms
Regular 4G	4 Mb/s	3 Mb/s	20 ms
WiFi	10 Mb/s	1 Mb/s	30 ms

5.2   Results

Test 1. After analysing 1 min of constant transformations execution, the number of operations per second, averaged time of a single execution, CPU load and network load values were observed (Table 2).

Table 2. Test 1 measurements

Network Profile	Updates/second	Single execution [ms]	CPU load	Network load
Client side processing (Turf.js)				
Wi-Fi	172.95	1.02	15–25%	NA
Backend side processing (PostgreSQL/PostGIS via HTTP)				
GPRS	0.82	1208.32	0–2%	Up to 5 KB per request
Regular 3G	6.72	146.57	0–5%	
Good 3G	12.91	75.62	3–7%	
Regular 4G	28.31	33.81	8–15%	
Wi-Fi	83.59	10.47	30–40%	

The results clearly show the advantage of the client-side solution over the backend solution. The main bottleneck is the communication between the web browser and the server. The better the network is, the difference between the two described approaches is less significant. However, even taking into account the best network profile, the Turf.js solution remains undefeated. Moreover, approach with Turf.js does not involves network communication so it does not impose additional cost on the user.

Test 2. Test 2 consisted in buffering 1,000 unique GeoJSON linestrings with a 5-m radius in a single query. After performing many repetitions of the second test scenario, the average execution speed and download time for each approach and network profile was calculated. With regard to client-side processing, the network profile did not affect the results because no HTTP communication was involved (Table 3).

Measurements for Test 2 indicate that there is a large difference in the overall waiting time for the transformation to be completed considering client side

Table 3. Test 2 measurements

Network profile	Execution time [s]	Download time [s]
Client-side processing (Turf.js)		
Wifi	1.67	NA
Backend-side processing (PostgreSQL/PostGIS via HTTP)		
GPRS	2.79	150
Regular 3G	2.44	26.63
Good 3G	2.3	19.50
Regular 4G	2.4	4.88
WiFi	2.13	0.671

processing and the PostgreSQL/PostGIS via HTTP solution even taking into account the best examined network profile.

6   Conclusions

Precise editing of geospatial data is becoming more important as applications are driven towards integration and processing of spatial data from various sources. This paper presents results obtained during experiments performed while developing an interactive application dedicated to editing of objects included in smart city concepts.

Considering the two presented approaches towards developing responsive geospatial transformations in web applications, the conclusion is that it is better to delegate some work to the client side.

In the presented case, the Turf.js JavaScript library on the frontend side was compared with a PostgreSQL/PostGIS instance accessed via HTTP on the backend side.

Based on the two tests performed, the results clearly show that the main workload is put onto the communication. Local processing-based approach gives better responsiveness, especially when network is poor. Reducing the intensity of communications pays off in reduction of the overall execution time.

References

1. Node.js Documentation. https://nodejs.org/en/docs/
2. OpenStreetMap Wiki. https://wiki.openstreetmap.org/
3. PostGIS 2.4 Documentation. https://postgis.net/docs/manual-2.4/
4. PostgreSQL 10 Documentation. https://www.postgresql.org/docs/10/static/index.html
5. Turf.js Website. http://turfjs.org/
6. ESRI: Using Location Intelligence to Maximize the Value of BI. Technical report (2011)

7. Farin, G.E.: Curves and Surfaces for Computer-Aided Geometric Design: A Practical Guide, 3rd edn. Academic Press, Boston (1992)
8. Kearney, M., Garbee, J.: Tools for web developers: optimize performance under varying network conditions, January 2018. https://developers.google.com/web/tools/chrome-devtools/network-performance/network-conditions
9. NASA: The EGM96 geoid undulation with respect to the WGS84 ellipsoid. In: EGM96, The NASA GSFC and NIMA Joint Geopotential Model, NASA/TP-1998-200801, February 2002
10. Open Geospatial Consortium: OpenGIS® Implementation Standard for Geographic information - Simple feature access - Part 1: Common architecture. OpenGIS Implementation Standard OGC 06-103r4 (2011)
11. W3C: HTML 5.2, December 2017
12. Wojnicki, I., Kotulski, L.: Empirical study of how traffic intensity detector parameters influence dynamic street lighting energy consumption: a case study in Krakow. Poland. Sustain. **10**(4), 1221 (2018)
13. Wojnicki, I., Kotulski, L.: Improving control efficiency of dynamic street lighting by utilizing the dual graph grammar concept. Energies **11**(2), 402 (2018)

Isotone Galois Connections and Generalized One-Sided Concept Lattices

Peter Butka[1](✉), Jozef Pócs[2,3], and Jana Pócsová[4]

[1] Department of Cybernetics and Artificial Intelligence, Faculty of Electrical Engineering and Informatics, Technical University of Košice,
Letná 9, 04200 Košice, Slovakia
peter.butka@tuke.sk
[2] Department of Algebra and Geometry, Faculty of Science,
Palacký University Olomouc, 17. listopadu 12, 771 46 Olomouc, Czech Republic
pocs@saske.sk
[3] Mathematical Institute, Slovak Academy of Sciences,
Grešákova 6, 040 01 Košice, Slovakia
[4] Institute of Control and Informatization of Production Processes, BERG Faculty,
Technical University of Košice, Boženy Němcovej 3, 043 84 Košice, Slovakia
jana.pocsova@tuke.sk

Abstract. We provide an approach to one-sided (crisp-fuzzy) concept lattices based on isotone Galois connections. Isotone Galois connections and concept lattices provide an alternative to the classical, antitone Galois connections based concept lattices, which are fundamental structures in formal concept analysis of object-attribute models with many-valued attributes. Our approach is suitable for analysis of data tables with different structures for truth values of particular attributes. A possible applications of this approach for approximation of object subsets is also presented.

Keywords: Isotone Galois connection · One-sided concept lattice
Closure operator · Interior operator · Formal concept analysis

1 Introduction

Formal concept analysis (FCA) is one of the mathematical tools for identification of conceptual structures among data sets. Mathematical theory of FCA is based on the notions of formal context and formal concept. This theory is well developed in the monograph of Ganter and Wille [12]. The set of all formal concepts of a formal context forms a complete lattice, called the concept lattice, which reflects the relationship of generalization and specialization among the formal concepts. Nowadays, FCA has been applied to a variety of fields such as information retrieval, machine learning and knowledge discovery. In the classical formal contexts, the relationship between the objects and the attributes is

© Springer Nature Switzerland AG 2019
K. Choroś et al. (Eds.): MISSI 2018, AISC 833, pp. 151–160, 2019.
https://doi.org/10.1007/978-3-319-98678-4_17

described by a binary form that can only specify whether or not an attribute is possessed by an object. In many real applications, however, the relationship may be many-valued (fuzzy). Therefore, many attempts have recently been devoted to introduce fuzzy concept lattice with properties similar with classical ones. We mention the general approaches dealing with fuzzy subsets of objects and fuzzy subsets of attributes, e.g., [1,3] or [15].

A special role in fuzzy FCA play one-sided concept lattices, where usually objects are considered as a crisp subsets and attributes obtain fuzzy values. In this case interpretation of object clusters is straightforward as in classical FCA. These types of models were introduced in papers of Yahia and Jaoua [5], and Krajči [13]. In [7] we have described an approach to one-sided concept lattices involving different types of truth value structures, which generalizes all recently known approaches and is more convenient for analysis of object-attribute models with different types of attributes.

Theory of classical as well as fuzzy concept lattices is based on the mathematical notion of Galois connections. Galois connections used in FCA are special pairs of order reversing (antitone) mappings between complete lattices. Antitonicity of this mappings in some sense determines the range of applicability of concept lattices. From the theoretical point of view, and also from practical reasons, there is reasonable to consider also "positive" (order preserving or isotone) situations to be analyzed by FCA-like tools. The isotone Galois connections and concept lattices were studied for the case of residuated lattices in [4].

The aim of this paper is to describe an approach to one-sided concept lattices based on isotone Galois connections. One of the advantage of an approach described in this paper is its applicability for object-attribute models with possible different truth value structures. Another one is the fact that our approach is not tied up with any fuzzy logical framework, i.e., we do not require any additional assumptions on truth value structures.

The paper is organized as follows. In the next section we start with necessary mathematical details for introduction of concept lattices based on isotone Galois connections. Next, we describe the creation of isotone generalized one-sided concept lattices from the given formal context, which represents the real object-attribute model. We also present an illustrative example. In Sect. 3 we combine the obtained results and the classical generalized one-sided lattices for description of open and closed approximation of subsets of objects, which can be applied in different analysis of object-attribute models.

2   Isotone Galois Connections and One-Sided Concept Lattices

Theory of concept lattices is built within the framework of lattice theory, hence we assume that the reader is familiar with the basic notions of lattice theory.

Let $(P, \leq)$ and $(Q, \leq)$ be an ordered sets and let

$$\varphi : P \longrightarrow Q \quad \text{and} \quad \psi : Q \longrightarrow P$$

be maps between these ordered sets. Such a pair (φ, ψ) of mappings is called an *isotone Galois connection (residuated mappings)* between the ordered sets if they satisfy the following condition:

$$p \geq \psi(q) \quad \text{if and only if} \quad \varphi(p) \geq q \tag{1}$$

In this case the mapping φ is called *residual (upper adjoint)* and ψ is said to be *residuated (lower adjoint)*. It is well known fact that for a residual mapping there is unique residuated map and vice versa, thus in the sequel for a given residual map φ we will denote the corresponding unique residuated map ψ by the symbol φ_*.

Isotone Galois connections between complete lattices are closely related to the notion of *closure operator (closure system)* and *interior operator (interior system)*, respectively. Let L be a complete lattice. Recall that by a closure operator in L we understand a mapping $c: L \to L$ satisfying:

(c1) $x \leq c(x)$ for all $x \in L$,
(c2) $c(x_1) \leq c(x_2)$ for $x_1 \leq x_2$,
(c3) $c(c(x)) = c(x)$ for all $x \in L$, (i.e. c is idempotent).

Dually, by an interior operator in L we understand a mapping $i: L \to L$ satisfying:

(i1) $i(x) \leq x$ for all $x \in L$,
(i2) $i(x_1) \leq i(x_2)$ for $x_1 \leq x_2$,
(i3) $i(i(x)) = i(x)$ for all $x \in L$.

A subset X of a complete lattice L is called a closure system in L if X is closed under arbitrary meets and it is called an interior system if it is closed with respects to arbitrary joins. Let us note that there is one-to-one correspondence between closure (interior) operators on the one side, and closure (interior) systems on the other side.

From the properties of isotone Galois connections one can deduce that if L, M are complete lattices and (φ, φ_*) is an isotone Galois connection, then the composition $\varphi \circ \varphi_* : L \to L$ induces an interior operator in L and the composition $\varphi_* \circ \varphi: M \to M$ induces a closure operator in M. Moreover the corresponding interior system and closure system forms isomorphic complete lattices.

The isotone Galois connections allow us to construct complete lattices which are in our special interest. One such possibility is to create one-sided concept lattices based on isotone Galois connections. In order to define it, first we provide formalization of object-attribute models (input data tables) in the form of so-called generalized one-sided formal context.

A 4-tuple $c = (B, A, \mathcal{L}, R)$ is said to be a *generalized one-sided formal context* if the following conditions are fulfilled:

(i) B is a non-empty set of objects and A is a non-empty set of attributes.
(ii) $\mathcal{L}: A \to \mathsf{CL}$ is a mapping from the set of attributes to the class of all complete lattices. Hence, for any attribute a, $\mathcal{L}(a)$ denotes a structure of truth values for attribute a.

(iii) R is generalized incidence relation, i.e., $R(b, a) \in \mathcal{L}(a)$ for all $b \in B$ and $a \in A$. Thus, $R(b, a)$ represents a degree from the structure $\mathcal{L}(a)$ in which the element $b \in B$ has the attribute a.

Using the notion of formal context, we are able to define a pair of mapping between the power set of objects and the "generalized fuzzy subsets" of attributes, which is represented by direct product of truth value structures $\mathcal{L}(a)$ assigned to particular attributes. Let $(B, A, \mathcal{L}, R)$ be a generalized one-sided formal context. Then we define a pair of mapping $⅃ \colon 2^B \to \prod_{a \in A} \mathcal{L}(a)$ and $⅃_* \colon \prod_{a \in A} \mathcal{L}(a) \to 2^B$ as follows:

$$⅃(X)(a) = \bigwedge_{b \notin X} R(b, a), \quad \text{for all } X \subseteq B, \tag{2}$$

$$⅃_*(g) = \{b \in B : \exists a \in A, \ g(a) \nleq R(b, a)\}, \quad \text{for all } g \in \prod_{a \in A} \mathcal{L}(a). \tag{3}$$

Similarly to classical one-sided concept lattices, this pair of mappings $(⅃, ⅃_*)$ for any generalized one-sided formal context $(B, A, \mathcal{L}, R)$ forms an isotone Galois connection between 2^B and $\prod_{a \in A} \mathcal{L}(a)$.

Now we are able to define isotone generalized one-sided concept lattice. For generalized one-sided formal context $(B, A, \mathcal{L}, R)$ denote ${}^i\mathcal{C}(B, A, \mathcal{L}, R)$ the set of all pairs (X, g), where $X \subseteq B$, $g \in \prod_{a \in A} \mathcal{L}(a)$, satisfying

$$⅃(X) = g \quad \text{and} \quad ⅃_*(g) = X.$$

In accordance with terminology commonly used in FCA we will refer to the set X as *extent* and to the element g of direct product of complete lattices $\mathcal{L}(a)$ as *intent* of the concept (X, g). Further, we define partial order on $\mathcal{C}(B, A, \mathcal{L}, R)$ as follows:

$$(X_1, g_1) \leq (X_2, g_2) \quad \text{iff} \quad X_1 \subseteq X_2 \text{ iff } g_1 \leq g_2. \tag{4}$$

According to applied pair of mapping $(⅃, ⅃_*)$, we are then finally able to characterize isotone one-sided concept lattice as a set of concepts ${}^i\mathcal{C}(B, A, \mathcal{L}, R)$ with the partial order defined above which forms a complete lattice, where

$$\bigvee_{i \in I} (X_i, g_i) = \left(\bigcup_{i \in I} X_i, ⅃(⅃_*(\bigvee_{i \in I} g_i)) \right) \tag{5}$$

$$\bigwedge_{i \in I} (X_i, g_i) = \left(⅃_*(⅃(\bigcap_{i \in I} X_i)), \bigwedge_{i \in I} g_i \right) \tag{6}$$

for each family $(X_i, g_i)_{i \in I}$ of extent-intent pairs from ${}^i\mathcal{C}(B, A, \mathcal{L}, R)$.

At the end of this section we provide an illustrative example of isotone one-sided concept lattice defined by given generalized one-sided formal context.

Example 1. As we have already mentioned, generalization of one-sided concept lattices allow to consider various types of attributes in object-attribute models. In order to demonstrate this benefit we will consider the following one-sided context $(B, A, \mathcal{L}, R)$, where $B = \{b_1, b_2, b_3, b_4\}$ is the set of objects, $A = \{a_1, a_2, a_3, a_4\}$ is the set of attributes. Further, attributes a_1 and a_2 are binary, thus $\mathcal{L}(a_1) = \mathcal{L}(a_3) = \mathbf{2}$. Attribute a_2 is quantitative with values from real unit interval, i.e. $\mathcal{L}(a_2) = [0, 1]$. Finally, the values of a_4 are represented by modular non-distributive lattice M_3 (also known as a diamond), which consists of the smallest element o, the greatest element i, and three incomparable elements a, b, c satisfying $o < a < i$, $o < b < i$, $o < c < i$. The generalized incidence relation R of input context is depicted in Table 1. Isotone one-sided concept lattice created according to provided generalized one-sided formal context $(B, A, \mathcal{L}, R)$ is then shown on Fig. 1.

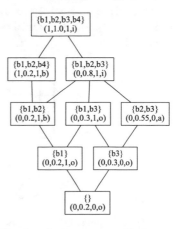

Table 1. Generalized incidence relation R.

R	a_1	a_2	a_3	a_4
b_1	0	0.55	0	a
b_2	0	0.30	1	o
b_3	1	0.20	1	b
b_4	0	0.80	1	i

Fig. 1. Isotone one-sided concept lattice for input context $(B, A, \mathcal{L}, R)$.

3 Open and Closed Approximation of the Subsets of Objects

In this section we describe combination of our isotone approach to one-sided concept lattices with classical one-sided concept lattices on possible application to approximation of subsets of objects. The process of approximation is motivated by general topology methods. In this case, to any subset of points of topological space there is corresponding the smallest closed subset containing it (closure) and the largest open subset contained in it (interior), respectively. However the systems of subsets defined via isotone Galois connections does not form a topology on the set of objects in general, there are several generalizations of topological spaces where weaker conditions on open subsets (topology) are considered. One of this concept represent generalized topologies, where the collection of open subsets forms interior systems. Since isotone Galois connections produce interior system on the side of objects, one can consider this interior system as a generalized topology and consequently the open and closed approximation of subsets is possible.

In topological space theory the closure operator is defined from given system of open sets via set-theoretical subtraction, i.e., a set $X \subseteq B$ is closed if there exists an open set $Y \subseteq B$ such that $X = B \setminus Y$. In our generalization we will go further and we will consider the closed sets as elements of an arbitrary closure system defined by generalized one-sided concept lattices [7]. This approach allows to consider different object-attribute models for approximations of object elements, which can be more convenient for various applications to approximative reasoning (rough-sets like approach to concept lattices) in information sciences. Hence our situation can be schematically described as on Fig. 2. There is given a non-empty set B of objects and two generalized one-sided formal contexts $(B, A_o, \mathcal{L}_o, R_o)$ and $(B, A_c, \mathcal{L}_c, R_c)$ for open and closed approximation respectively.

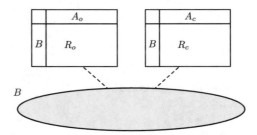

Fig. 2. Generalized one-sided context for open and closed approximation.

In order to introduce the closed approximation we recall the basic definitions concerning generalized one-sided concept lattices (see [7] for more details).

Let $(B, A, \mathcal{L}, R)$ be a generalized one-sided formal context. Using the incidence relation R there is defined a pair of mapping (forming Galois connection) $\uparrow: 2^B \to \prod_{a \in A} \mathcal{L}(a)$ and $\downarrow: \prod_{a \in A} \mathcal{L}(a) \to 2^B$ as follows:

$$\uparrow(X)(a) = \bigwedge_{b \in X} R(b, a) \tag{7}$$

$$\downarrow(g) = \{b \in B : \text{ for each } a \in A, \ g(a) \leq R(b, a)\}. \tag{8}$$

Similarly as in the case of isotone Galois connections we can construct the corresponding concept lattice $\mathcal{C}(B, A, \mathcal{L}, R) = \{(X, g) : \uparrow(X) = g, \downarrow(g) = X\}$, where

$$(X_1, g_1) \leq (X_2, g_2) \quad \text{iff} \quad X_1 \subseteq X_2 \text{ iff } g_1 \geq g_2.$$

Important fact for our further consideration is that the extents of the lattice $\mathcal{C}(B, A, \mathcal{L}, R)$ form closure system on the set of objects.

Now we define the open and closed approximation of any subset of the set of all objects. Denote by $\beth^o$ and $\beth^o_*$ the pair of mappings defined by (2) and

(3) corresponding to formal context $(B, A_o, \mathcal{L}_o, R_o)$ and similarly denote by $\uparrow^c$, $\downarrow^c$ the pair of mappings defined (7) and (8) corresponding to formal context $(B, A_c, \mathcal{L}_c, R_c)$.

Let $X \subseteq B$ be any subset. We define interior X^o and closure X^c of X as follows:

$$X^o = \beth^o\big(\beth^o_*(X)\big) \quad \text{and} \quad X^c = \downarrow^c\big(\uparrow^c(X)\big). \tag{9}$$

Since composition of operators $\beth^o$ and $\beth^o_*$ forms an interior operator and the composition of the operators $\uparrow^c$ and $\downarrow^c$ forms a closure operator on the set B, we obtain the following inequality:

$$\beth^o\big(\beth^o_*(X)\big) = X^o \subseteq X \subseteq X^c = \downarrow^c\big(\uparrow^c(X)\big).$$

Similarly as in topology, one can consider *clopen* subsets of the objects set, i.e., the subsets $X \subseteq B$ satisfying $X = X^o = X^c$. It is well known fact that clopen subsets of arbitrary topological space form boolean algebra. In our case, unlike the topological spaces, the structure of clopen subsets does not form any specific partially ordered structure.

Although, in analysis of different object attribute models these clopen sets may prove to be very useful. In classical FCA and other fuzzy modifications of FCA based on Galois connections the extents of concepts are formed by closed subset of the object set, i.e., extents are precisely members of the closure system induced by Galois connection. These closed subsets are considered to be important from the point of view of the given object attribute model, e.g., these closed subsets determine the hierarchical structure of concepts which helps to find hidden structures among data sets, inclusion relation between closed subsets describes dependencies of attributes, etc. Clopen subsets in our setting are closed (form the range of Galois connection) and moreover they are members of interior system induced by isotone Galois connection. Hence, consideration of clopen subsets instead of only closed (open) subsets can produce at least two benefits:

– Size reduction of the considered hierarchical structure.

The upper bound for number of concepts in binary object attribute models is $\min\{2^n, 2^m\}$ where n, m denotes the number of objects and attributes respectively. In practice there are large data structures, thus number of concepts rise dramatically. Clopen subsets form a subset of closed as well as open subsets, hence obviously the number of these clopen concepts will be lower than the number of closed or open one.

– Clopen subsets can provide finer information of hierarchical structure inherited from two object-attribute models.

Since clopen subsets are both closed and open they are part of hierarchical structure derived from both considered models. Thus one can see the clopen subsets "more important" than closed or open subsets considered alone.

Finally, we give an example of open and closed approximation defined by simple object attribute models.

Example 2. Consider four element set of objects $B = \{u, v, w, z\}$. Next, we have two object attribute models with two sets of attributes $A_o = \{a_1, a_2, a_3\}$ and $A_c = \{a_4, a_5, a_6, a_7\}$. In first model attributes a_1, a_2 are binary (represented by two element chain) and a_3 is real-valued attribute (represented by interval $[0, 1]$ of reals). In second model attributes a_4, a_5 are binary, a_6 is real-valued attribute and a_7 is characterized by four element chain. Generalized incidence relations R_ϱ and R_c are depicted in corresponding Tables 2 and 3, respectively.

Table 2. Generalized incidence relation R_o for isotone concept lattice.

R_o	a_1	a_2	a_3
u	0	1	0.20
v	0	1	0.40
w	1	0	0.30
z	1	1	0.55

Table 3. Generalized incidence relation R_c for antitone concept lattice.

R_c	a_4	a_5	a_6	a_7
u	1	0	0.25	2
v	1	1	0.50	3
w	0	1	0.60	0
z	0	0	0.40	1

Applying the appropriate pairs of mapping to the given one-sided formal contexts we obtain concept lattices depicted on Fig. 3.

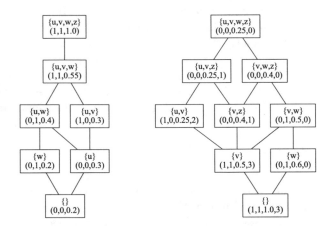

Fig. 3. Concept lattices - isotone ${}^i\mathcal{C}(B, A_o, \mathcal{L}_o, R_o)$ and antitone $\mathcal{C}(B, A_c, \mathcal{L}_c, R_c)$.

Now we can find objects which are both open and closed. The structure of $2^{\{u,v,w,z\}}$ with open and closed objects is depicted in the Fig. 4 (open objects are with gray interior, closed with dark border).

As we can see, there are four clopen subsets $\emptyset, \{u, v\}, \{w\}, \{u, v, w, z\}$.

The provided example showed some insight for both expected benefits. At first, the reduction of original FCA-based analysis can be achieved using combination of closed and open subsets of objects. The reduction of concept lattices is

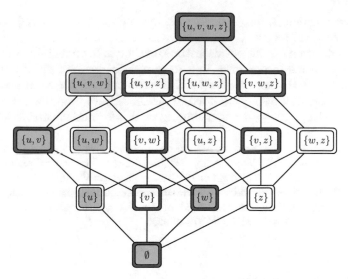

Fig. 4. Open and closed subsets in the order structure of $2^{\{u,v,w,z\}}$.

currently one of the main topics, due to fact that traditional FCA analysis can produce large amount of output concepts. Therefore, several approaches were tested in order to achieve smaller output structures and number of concepts, based on different ideas of ranking of concepts (cf. [2,8]) or usage of combined (internal and/or external) information (cf. [9,14]).

Our method is actually one of those which are able to combine different attributes on analyzed objects, so it is actually the case where we can see isotone concept lattice and its concepts as external information structure on objects originally structured by antitone concept lattice (or vice-versa). Then strong reduction can be achieved in selection of those subsets, which are important for both setups. Additionally, understanding the idea of closed and open sets and their relations leads to better (or finer) understanding of selected subsets, which can be seen as most important also from logical point of view, as fixed point for both classical and isotone interpretation of concept lattices (such concepts are of higher rank than other concepts/subsets). In the future work we would like to test our approach using real world contexts, mostly related to information retrieval and data/text mining tasks. Also, we try to extend the results to be applicable in framework of more general algebraic structures, e.g., [6,10] or [11].

4   Conclusions

In this paper we have introduced an approach for generalization of one-sided concept lattices based on isotone Galois connection. This approach is applicable in concept data analysis of various object-attribute models in similar way like ordinary concept lattices are usually used. The basic idea is the usage of isotone Galois connections, which allow the analysis of the given object-attribute model in "positive reasoning", i.e., subsets with more objects are related with higher degrees of truth values.

Main application benefit of this isotone method is the possibility of combining it with the antitone (order reversing) approach in order to obtain approximation operators on the set of objects. The presented approach is applicable in all domains where all other one-sided approaches were used, e.g., concept data analysis, knowledge discovery, text mining, and information retrieval.

Acknowledgments. The first author was supported the Slovak VEGA grant no. 1/0493/16 and Slovak APVV grant no. APVV-16-0213. The second author was supported by the project of Grant Agency of the Czech Republic (GAČR) no. 18-06915S and by the Slovak Research and Development Agency under the contract no. APVV-16-0073. The third author was supported by the Slovak Research and Development Agency under the contract no. APVV-14-0892.

References

1. Antoni, L., Krajči, S., Krídlo, O., Macek, B., Pisková, L.: On heterogeneous formal contexts. Fuzzy Set. Syst. **234**, 22–33 (2014)
2. Antoni, L., Krajči, S., Krídlo, O.: On stability of fuzzy formal concepts over randomized one-sided formal context. Fuzzy Set. Syst. **333**, 36–53 (2018)
3. Antoni, L., Krajči, S., Krídlo, O.: On fuzzy generalizations of concept lattices. Stud. Comput. Intell. **758**, 79–103 (2018)
4. Bělohlávek, R., Konecny, J.: Concept lattices of isotone vs. antitone Galois connections in graded setting: mutual reducibility revisited. Inf. Sci. **199**, 133–137 (2012)
5. Ben Yahia, S., Jaoua, A.: Discovering knowledge from fuzzy concept lattice. In: Kandel, A., Last, M., Bunke, H. (eds.) Data Mining and Computational Intelligence, Physica-Verlag, pp. 167–190 (2001)
6. Brajerčík, J., Demko, M.: On sheaf spaces of partially ordered quasigroups. Quasigroups Relat. Syst. **22**(1), 51–58 (2014)
7. Butka, P., Pócs, J.: Generalization of one-sided concept lattices. Comput. Informat. **32**(2), 355–370 (2013)
8. Butka, P., Pócs, J., Pócsová, J.: Reduction of concepts from generalized one-sided concept lattice based on subsets quality measure. Adv. Intell. Syst. Comput. **314**, 101–111 (2015)
9. Butka, P., Pócs, J., Pócsová, J., Sarnovský, M.: Multiple data tables processing via one-sided concept lattices. Adv. Intell. Syst. Comput. **183**, 89–98 (2013)
10. Demko, M.: On congruences and ideals of partially ordered quasigroups. Czechoslovak Math. J. **58**(3), 637–650 (2008)
11. Demko, M.: Lexicographic product decompositions of half linearly ordered loops. Czechoslovak Math. J. **57**(2), 607–629 (2007)
12. Ganter, B., Wille, R.: Formal Concept Analysis: Mathematical Foundations. Springer, Heidelberg (1999)
13. Krajči, S.: Cluster based efficient generation of fuzzy concepts. Neural Netw. World **13**(5), 521–530 (2003)
14. Krídlo, O., Krajči, S., Antoni, L.: Formal concept analysis of higher order. Int. J. Gener. Syst. **45**(2), 116–134 (2016)
15. Medina, J., Ojeda-Aciego, M., Ruiz-Calviño, J.: Formal concept analysis via multi-adjoint concept lattices. Fuzzy Set. Syst. **160**, 130–144 (2009)

Transfer Learning in GMDH-Type Neural Networks

Aminu Abdullahi[1] and Mukti Akter[2(✉)]

[1] Department of Computer Science, Federal University Dutse, Dutse, Nigeria
[2] School of Computer Science, University of Bedfordshire, Luton, UK
mukti.akter@study.beds.ac.uk

Abstract. Recognition of difficult patterns with the accuracy comparable to that of the human brain is a challenging problem. The ability of the human to excel at this task has motivated the use of Artificial Neural Networks (ANNs) which under certain conditions provide efficient solutions. ANNs are still unable to use the full potential of modular and holistic operations of biological neurons and their networks. The ability of neurons to transfer learned behaviour has inspired an idea to train ANN for a new task by using the behaviour patterns learnt from a related task. The useful patterns transferred from one task to another can significantly reduce the time needed to learn new patterns, and gives the neurons the ability to generalise instead of memorising patterns. In this paper we explore the ability of transfer learning for a face recognition problem by using Group Method of Data Handling (GMDH) type of Deep Neural Networks. In our experiments we show that the transfer learning of a GMDH-type neural network has reduced the training time by 31% on a face recognition benchmark.

Keywords: Transfer learning · Deep Neural Networks · GMDH
Face recognition

1 Introduction

Recognition of complex patterns which have to be learnt from observed data is a challenging task. Recognition difficulties increase when noise from the environment as well as other factors distort given image data [18,24,28]. The human brain can relatively easily learn and therefore recognise patterns of interest in a complex environment. The complexity of recognition problems has motivated the simulation of the human ability to learn difficult patterns. In the case of face recognition, various approaches described in [3,27] have discussed the following findings:

(i) Recognition of faces is not strongly dependent on image resolution. Humans can tolerate degradation of facial images proportional to their familiarity with the subject.

© Springer Nature Switzerland AG 2019
K. Choroś et al. (Eds.): MISSI 2018, AISC 833, pp. 161–169, 2019.
https://doi.org/10.1007/978-3-319-98678-4_18

(ii) Humans analyse facial images as a whole and not as distinct facial features. Certain features, such as eyebrows, are statistically more relevant for facial recognition than other features.

(iii) Facial pigmentation can be as important as facial shape. The human perception system has a bias towards face-like objects.

(iv) Facial identity recognition and facial expression recognition are performed by separate systems. Research in cortical responses has hinted at a feedforward mechanism for the human facial recognition system.

(v) Observations of a face at different orientations increase the accuracy of recognition.

According to [25], Machine Learning methods and Artificial Neural Networks (ANN) have been primarily focused on feature selection, e.g. [6], and feature extraction [8,15]. The face expression identification methods mainly extract features which represent specific emotions.

In general, facial recognition systems include the following steps: (i) detection of faces and their alignment, (ii) feature extraction, and (iii) matching of the extracted features. The detection of face location can be made with ANNs, as shown in [3]. Feature extraction can be made by using statistical approaches for extracting the meaningful patterns.

Face recognition algorithms have been developed by using Bayesian methods employing Markov chain Monte Carlo for integration over posterior parameters, as described [7,21]. The Bayesian methodology of probabilistic inference has provided highly competitive solutions, as shown in [16], whilst Markov models have been efficiently used for designing a highly competitive solution on the Cambridge face database [14].

According to [27], features extracted from images can involve: (i) structures such as edges and curves, (ii) baseline templates, and (iii) geometrically described structures. Some face recognition methods follow the brute-force technique of using pixel values to compare faces in a given data set. These methods are computationally costly and can perform poorly when facial orientation and lighting conditions change. Other methods which can be used for feature extraction include statistical methods such as principal component analysis, independent component analysis, and linear discriminant analysis.

Although face recognition is an active research area, high recognition accuracy in the presence of natural variations in face images has not been achieved yet. The analysis of the relevant literature shows that ANNs cannot outperform biological neurons and their networks. The human brain has the important ability to efficiently transfer learned behaviour from one network to another. Such an ability reduces the amount of time needed to learn new patterns, and gives neurons an ability to generalize and learn behaviour patterns instead of merely memorizing tasks.

Neurons in ANNs are defined with inputs, output, weights, and an activation function. The operations of an neuron are geared towards either learning of a particular set, operations, or parameters, or recalling the information embedded in the learned parameters. While the weights are updated during the learning of a neuron, the activation function specifies neuron behaviour outputs.

In this paper we explore the usefulness of transfer learning in Deep Neural Networks in the case of a face recognition task. We will be focused on obtaining the efficient transfer learning of Deep Neural Networks synthesised by Group Method of Data Handling (GMDH) and Convolutional Neural Networks (CNNs), which are outlined in the following sections. The efficiency of the proposed method is explored on the facial recognition benchmarks.

2   Convolutional Neural Networks for Face Recognition

According to [11], neuron structures of CNNs are designed to provide overlapping areas in the input space of individual neurons. Inspired by receptive fields in sensory neurons of animal visual cortex, CNNs use overlapping structures to smooth the effect of transitions and dimensional changes in images. A group of neurons establishing a receptive field is focused on a particular area of the image along with the overlapping regions which operate as transition zones. Subsampling or pooling layers are then used for averaging over features inputs at the upper layers. The resilience of CNNs and their tolerance for domain transitions have inspired numerous applications for image recognition.

A CNN has been used for an image recognition system proposed in [1]. Using a single batch for training and validation of a network, and using max pooling which takes the maximum of the pooled block as opposed to its average, the system has demonstrated an error rate 0.35 on the MINST database.

Another CNN (Deepface) has been designed for facial recognition in Facebook [23]. The Deepface has achieved an accuracy of 97.35%. Over 120 million parameters were integrated into a CNN to identify more than 4,000 persons. Starting with faces from the Labelled Faces in the Wild data and using a novel alignment system employing the fiducial point detectors using Support Vector Regression, the system has identified faces by translating multiple views.

The CNN system proposed in [4] has a layer for facial recognition, which uses the depth information extracted from images. Gradients represent the depth variations which feed a mesh of the system convolutional layers preceded by a convolutional subsampling layer forming 12 feature maps. Analysing the EURE-COM Kinect face data set, and using batches of 24 images over 1000 epochs, the system performance has been shown 88%.

A Deep CNN described in [10] has achieved an error rate of 0.17 on the Image Net database. The system uses rectified linear units described by function $f(x) = \max(0, x)$. Using a batch size of 128 images with a momentum of 0.9 and a weight decay of 0.0005, the training error has been decreased to the minimum rate 0.17.

Another CNN has been used for face recognition, as described in [22]. Top-level convolutional layers in the proposed system are succeeded by shared layer common shallow networks included for purposes of local feature selection. The system has a cascade structure which computes multi-level regressions in lower layers coupled with the convolutional structure. The system has been tested on the BioID and Labelled Face Parts in the Wild databases of facial images, and demonstrated a significant improvement in the accuracy.

3   Group Method of Data Handling and Transfer Learning

GMDH-type ANNs generate multiple layers using an activation function defined by a polynomial, as described in [5,19]. The neurons of the first layer are generated by the polynomials using the input variables. The neurons of the following layers are generated by using the outputs of the previous layers. A given set of best preforming neurons are then selected at each layer, so that new features are generated in each new layer.

The new layer is added to the ANN when its performance increases. The performance is evaluated with a predetermined validation function using some external data samples. A GMDH network is described by a set of polynomial functions, the complexity of which depends on the recognition task, as described e.g. in [12,17].

Activation functions in GMDH-type ANNs typically employ short-term polynomials. The activation functions can be non-linear, as shown in [5]. For linear modelling, the polynomials are described as follows:

$$y^G = a^G + b^G x_1^{(G-1)} + c^G x_2^{(G-1)} + d^G x_1^{(G-1)} x_2^{(G-1)}, \tag{1}$$

where the above polynomial activation function is represented by inputs x_1, x_2, and weights a, b, c, and d for layers G.

A GMDH network has been integrated with statistical learning for generating new features in subsequent network layers, as described in [2]. The network has been tested on a benchmark including handwritten digits, and shown to outperform the conventional Support Vector Machines as well as CNNs.

Inspired by the theory explaining the human learning processes, ANN can efficiently learn a new task by using the behaviour patterns learned from a similar task. In this context, patterns represent the subject knowledge learned from given data, as described in [9,13]. In [13], it has been shown that the useful patterns learnt from one task and transferred to an ANN for learning another task, improve the recognition accuracy and reduce the time required for training the ANN.

Figure 1 illustrates the idea of the transfer learning. The knowledge obtained from a handwritten digit recognition task is used for training a new ANN to solve a face recognition task.

In terms of learning speed and accuracy, Machine Learning methods have different performances dependent on the scale of problem, see e.g. [20]. Deep learning and CNNs have shown efficient for feature generation. Cascade structures of CNNs intent to integrate local features to enhance the performance, as shown in [22]. The research [23] shows that the accuracy of CNNs is dependent on the training time, whilst the transfer learning improves the recognition accuracy, as described in [26].

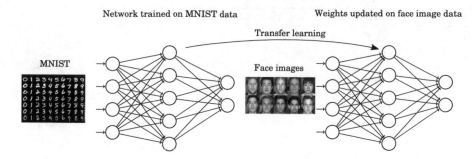

Fig. 1. Transfer learning for the face recognition problem.

The idea behind transfer learning is to use the knowledge, or behaviour patterns as well as generated useful features learnt from a domain similar to the new task for an ANN learned to solve the new task, as illustrated in Fig. 1. For example, the knowledge obtained from a handwritten image set can be embedded into an ANN for learning a face recognition task. Obviously, the use of weights defined for a similar domain instead of a random initiation of a new ANN will increase the performance.

In the case of a face recognition task, the above transfer learning can be implemented at two stages. First a GMDH-type ANN is learnt to recognise images from the MNIST dataset, as shown in Fig. 1. The network behaviour patterns learnt from the data and generated features are then embedded into a new ANN which is designed for the face recognition task on a facial image set. The embedded parameters are used for training the ANN and generating new features which can improve the network accuracy.

An algorithm for training a GMDH-type ANN is described as follows, The main steps of generating a GMDH-type ANN are shown by Algorithm 1. The network, **Net**, is initialised for training on the given data $[\mathbf{X}|\mathring{\mathbf{Y}}]$, where $\mathbf{X}$ and $\mathring{\mathbf{Y}}$ are the input data represented by m variables $\{x_i\}_{i=1}^{m}$ and labelled outcomes, respectively. GMDH algorithm requires to set F neurons which are selected in each layer in terms of neuron performance on a validation subset. Then the algorithm generates neurons of the thirst layer, $r = 1$, by using the specified activation function, Eq. 1. The new layer is added to the network while the performance increases. The algorithm stops if the number of selected neurons, F_r, becomes fewer than a given threshold F_0.

4   Experiments

We have implemented the transfer learning for CNN and GMDH-type ANNs. The first experiments have used an ANN trained on the MNIST dataset. The resultant ANN then has been used for learning to recognise human faces from the FACES 94, FACES 96, and Yale datasets.

The FACES 94 dataset comprises 3060 images of 153 individuals. Each individual is represented by 20 images. The images were taken on a plain Green

---

Algorithm 1. Training of GMDH-Type ANN

1: Inputs: $\mathbf{X}$, $\mathring{\mathbf{Y}}$, F
2: $InitiateNet()$
3: $r \leftarrow 1$ ▷ Layer index
4: $GenerateNewLayer(r)$
5: $newlayer \leftarrow SelectF(r)$ ▷ Selection of best neurons
6: **while** $newlayer$ **do**
7:     $r \leftarrow r + 1$ ▷ New Layer
8:     $GenerateNewLayer(r)$
9:     $newlayer \leftarrow SelectF(r)$
10: **end while**
11: **return Net**

Algorithm 2. $GenerateNewLayer(r)$

1: Global $\mathbf{X}, \mathbf{Net}, \mathring{\mathbf{Y}}, m$
2: $k \leftarrow 0$
3: **for** $i_1 \leftarrow 1 : m - 1$ **do**
4:     **for** $i_2 \leftarrow i_1 + 1 : m$ **do**
5:         $k \leftarrow k + 1$
6:         $A \leftarrow [i_1, i_2]$
7:         $\mathbf{U} \leftarrow \mathbf{X}(:, A)$
8:         $UpdateNet(\mathbf{U}, A, k, r)$
9:     **end for**
10: **end for**

---

background without scale variation. The illumination conditions throughout the images remain constant but there is some variation in expressions. The variation in poses and head angles are also small.

The FACES 96 dataset includes 3040 images of 152 individuals, 20 images per individual. The background of the images is relatively complex with glossy posters, but the variations in poses (such as head tilts, turns and slants) are relatively small. The critical element in this database is the translation in the position of the face in the image, this property distinguishes it from the FACES 94 dataset and makes it a good candidate to test for the robustness to translation in the proposed CNN.

The Yale database includes 2280 images of 38 individuals, 60 images for each. The pose and the position of the face in the image are kept constant. However, strong variations in illumination conditions are present. The recognition accuracy obtained on these datasets are listed in Table 1. In our first attempt, the transfer learning has not improved the accuracy. This can happen because the ANN has been trained on the MNIST data to recognise 10 classes, whilst the new recognition task is represented by 38 classes. Thus the difference of the network structures can affect the ability of ANNs to recognise the facial images. In our experiments the GMDH-type ANN and CNNs have reduced the training time. The GMDH-type ANN requires ca 9 min, whilst the CNN ca 20 min.

Table 2 shows the effect of transfer learning on training time for both ANNs.

Table 1. Accuracy of face recognition, %

ANN type	Yale	FACES 94	FACES 96
GMDH	96.2	97.1	95.2
Convolutional	97.2	99.3	98.4

Table 2. Average training times, min, before and after transfer learning

ANN type	Before	After	Reduction, %
GMDH	13	9	31
Convolutional	26	20	23

5   Conclusion

The outstanding ability of the human brain has motivated the use of Artificial Neural Networks (ANNs) which under certain conditions provide efficient solutions. However ANNs are still unable to achieve the full potential of the human brain.

An efficient approach to this problem is based on transfer learning, which assumes that a new ANN can be designed and trained by using behaviour patterns learnt from related tasks. Using this approach we can expect (i) a significant reduction of the time needed for learning a new task, and (ii) that the ANN will obtain an important ability to generalise patterns comparable with that of the human brain.

In this paper we have explored the transfer learning within two frameworks: Convolutional networks (CNN) and ANN synthesised by Group Method of Data Handling (GMDH). We have compared the performances of these networks on face recognition tasks which include patterns difficult for recognition. We have shown that in our experiments the transfer learning has reduced the training time by 31% for the GMDH-type ANN and for CNNs by 23%. The performances of these networks were competitive on the face recognition benchmarks.

Further work will be carried out on adjusting parameters of GMDH-type ANN, which will allow us to explore the ability of recognising difficult patterns in more details required in order to achieve a better recognition accuracy.

Acknowledgements. The authors would like to thank Dr Livija Jakaite, a member of the supervisory team at the School of Computer Science of University of Bedfordshire, for useful and constructive comments.

References

1. Ciresan, D.C., Meier, U., Masci, J., Gambardella, M.: Flexible, high performance convolutional neural networks for image classification. In: International Joint Conference on Artificial Intelligence (IJCAI), pp. 1237–1242 (2011)
2. Dhawan, P., Dongre, S., Tidke, D.J.: Hybrid GMDH model for handwritten character recognition. In: International Mutli-conference on Automation, Computing, Communication, Control and Compressed Sensing (iMac4s), pp. 698–703 (2013)
3. Farfade, S.S., Saberian, M.J., Li, L.J.: Multi-view face detection using deep convolutional neural networks. In: The 5th ACM on International Conference on Multimedia Retrieval (ICMR), pp. 643–650. ACM (2015)
4. Ijjina, E.P., Mohan, C.K.: Facial expression recognition using Kinect depth sensor and convolutional neural networks. In: IEEE International Conference on Machine Learning and Applications (ICMLA), vol. 2014, pp. 392–396 (2014)
5. Ivakhnenko, A.: Polynomial theory of complex systems. IEEE Trans. Syst. Man Cybern. **SMC–1**(4), 364–378 (1971)
6. Jakaite, L., Schetinin, V.: Feature selection for Bayesian evaluation of trauma death risk. In: 14th Nordic-Baltic Conference on Biomedical Engineering and Medical Physics (NBC), pp. 123–126 (2008). https://doi.org/10.1007/978-3-540-69367-3_33
7. Jakaite, L., Schetinin, V., Maple, C.: Bayesian assessment of newborn brain maturity from two-channel sleep electroencephalograms. Comput. Math. Meth. Med. **2012**, 1–7 (2012)
8. Jakaite, L., Schetinin, V., Schult, J.: Feature extraction from electroencephalograms for Bayesian assessment of newborn brain maturity. In: 24th International Symposium on Computer-Based Medical Systems (CBMS), pp. 1–6 (2011). https://doi.org/10.1109/CBMS.2011.5999109
9. Kawewong, A., Tangruamsub, S., Kankuekul, P., Hasegawa, O.: Fast online incremental transfer learning for unseen object classification using self-organizing incremental neural networks. In: IEEE International Conference on Neural Networks (IJCNN), pp. 749–756 (2011)
10. Krizhevsky, A., Sutskever, I., Hinton, G.E.: Imagenet classification with deep convolutional neural networks. In: Advances in Neural Information Processing Systems, pp. 1097–1105 (2012)
11. Nebauer, C.: Evaluation of convolutional neural networks for visual recognition. IEEE Trans. Neural Netw. **9**(4), 685–696 (1998)
12. Nyah, N., Jakaite, L., Schetinin, V., Sant, P., Aggoun, A.: Learning polynomial neural networks of a near-optimal connectivity for detecting abnormal patterns in biometric data. In: 2016 SAI Computing Conference, pp. 409–413 (2016). https://doi.org/10.1109/SAI.2016.7556014
13. Raina, R., Ng, A.Y., Koller, D.: Constructing informative priors using transfer learning. In: The 23rd International Conference on Machine Learning (ICML), pp. 713–720 (2006)
14. Samaria, F.S., Harter, A.C.: Parameterisation of a stochastic model for human face identification. In: The IEEE Workshop on Applications of Computer Vision, pp. 138–142 (1994)
15. Schetinin, V., Jakaite, L.: Extraction of features from sleep EEG for Bayesian assessment of brain development. PLOS ONE **12**(3), e0174027 (2017). https://doi.org/10.1371/journal.pone.0174027

16. Schetinin, V., Jakaite, L., Krzanowski, W.J.: Prediction of survival probabilities with Bayesian decision trees. Expert Syst. Appl. **40**(14), 5466–5476 (2013). https://doi.org/10.1016/j.eswa.2013.04.009

17. Schetinin, V., Jakaite, L., Nyah, N., Novakovic, D., Krzanowski, W.: Feature extraction with GMDH-type neural networks for EEG-based person identification. Int. J. Neural Syst. **28**(6), 1750064 (2018). https://doi.org/10.1142/S0129065717500642

18. Schetinin, V., Schult, J.: A neural-network technique to learn concepts from electroencephalograms. Theory Biosci. **124**(1), 41–53 (2005). https://doi.org/10.1016/j.thbio.2005.05.004

19. Schetinin, V., Schult, J.: Learning polynomial networks for classification of clinical electroencephalograms. Soft Comput. **10**(4), 397–403 (2006). https://doi.org/10.1007/s00500-005-0499-3

20. Schetinin, V., Schult, J., Scheidt, B., Kuriakin, V.: Learning multi-class neural-network models from electroencephalograms. In: Knowledge-Based Intelligent Information and Engineering Systems (KES), pp. 155–162 (2003)

21. Schönborn, S., Egger, B., Morel-Forster, A., Vetter, T.: Markov chain monte carlo for automated face image analysis. Int. J. Comput. Vis. **123**(2), 160–183 (2017)

22. Sun, Y., Wang, X., Tang, X.: Deep convolutional network cascade for facial point detection. In: IEEE Conference on Computer Vision and Pattern Recognition (CVPR), pp. 3476–3483 (2013)

23. Taigman, Y., Yang, M., Ranzato, M., Wolf, L.: Deepface: closing the gap to human-level performance in face verification. In: The IEEE Conference on Computer Vision and Pattern Recognition, pp. 1701–1708 (2014)

24. Uglov, J., Jakaite, L., Schetinin, V., Maple, C.: Comparing robustness of pairwise and multiclass neural-network systems for face recognition. EURASIP J. Adv. Signal Process. **2008**, 7 (2008)

25. Valenti, R., Sebe, N., Gevers, T., Cohen, I.: Machine learning techniques for face analysis. In: Anonymous Machine Learning Techniques for Multimedia, pp. 159–187 (2008)

26. Xu, Y., Du, J., Dai, L., Lee, C.: Cross-language transfer learning for deep neural network based speech enhancement. In: IEEE International Symposium on Chinese Spoken Language Processing (ISCSLP), vol. 9, pp. 336–340 (2014)

27. Zhao, W., Chellappa, R., Phillips, P.J., Rosenfeld, A.: Face recognition: a literature survey. ACM Comput. Surv. **35**(4), 399–458 (2003). https://doi.org/10.1145/954339.954342

28. Zharkova, V.V., Schetinin, V.: A neural-network technique for recognition of filaments in solar images. In: 7th International Conference on Knowledge-Based Intelligent Information and Engineering Systems, (KES), pp. 148–154 (2003). https://doi.org/10.1007/978-3-540-45224-9_22

Modelling of Taxi Dispatch Problem Using Heuristic Algorithms

Mateusz Adamczyk[1] and Dariusz Król[2(✉)]

[1] Faculty of Computer Science and Management,
Wrocław University of Science and Technology, Wrocław, Poland
[2] Department of Information Systems,
Faculty of Computer Science and Management,
Wrocław University of Science and Technology, Wrocław, Poland
Dariusz.Krol@pwr.edu.pl
http://krol.ksi.pwr.edu.pl

Abstract. Taxi routing is a complex task that involves the pickup and delivery by a fleet of vehicles in a specified time, taking into account many parameters and criteria. This paper describes issues related to this problem. It proposes a formal description of the problem and a goal function using a wide variety of criteria. Three heuristic algorithms: Ant Colony Optimization, Artificial Bee Colony and Genetic Algorithm are selected for testing. As a result three variants of these algorithms are implemented: Ant Colony System (ACS), Predict and Select - Artificial Bee Colony (PS-ABC) and Genetic Algorithm (GA) to conduct a survey. Operators used in the algorithms are adapted to a taxi dispatch problem. The analysis is performed in order to find the best parameters for the algorithms according to the data input. Finally, the efficiency of the algorithms are compared in order to determine the best algorithm. In several experiments the Ant Colony System outperforms any other algorithm presented here with respect to taxi working time currently in service.

Keywords: Evolutionary computation · Real world applications
Vehicle routing · Collective decision · Reinforcement learning
Fleet management

1 Introduction

For many years taxi business has been operating successfully as an inseparable part from public transport. Nevertheless, it has been possible to observe significant developments in taxi companies, which have slightly changed the approach to the customers. Their websites were modernized, many mobile applications became available, thanks to which the customer can order taxi, specifying all the parameters of the route and having a real time preview of the taxi heading towards the customer location. Distribution systems have been improved, in

K. Choroś et al. (Eds.): MISSI 2018, AISC 833, pp. 170–179, 2019.
https://doi.org/10.1007/978-3-319-98678-4_19

particular, the quality and speed of customer service. Their operating characteristics are very complex and sometimes they may seem chaotic. In addition, the number of taxi trips dispatched, customers and the specification of routes (number of people, type of vehicle) or the need to evenly distribute work on various taxi drivers causes another problem because it is hard to compare, analyze, plan and control driving trips related to a lot of data. Therefore, the overall taxi management is often ineffective and has been a challenging issue [2].

New technology comes to presence in the realm of IT. Along with its development, most modern devices, such as smartphones, navigation systems are equipped with GPS, which can locate their user with high accuracy. Currently, most taxi companies use GPS sensors in their cars to be able to locate their employees. By using these locations and using the road map, you can optimize travel times driven by urban congestion or alternative routes [7]. A system that solves the problem of routing with pickup and delivery has to be a dynamic system. Each time a new request is submitted or an old one is changed, it will be added to the queue of pending requests. The system taking the next elements from the queue must find the best vehicle from the currently available ones that will handle the given fare. The arrival of a new request may also change the status of the existing route if this can improve the final effect. We define this problem as a dynamic vehicle routing problem (DVRP) [4].

There are a number of previous studies on taxi dispatch problem. This problem is described and formulated as robust optimizations. Such problems are NP-hard and cannot be solved at polynomial time [1]. Therefore, the use of heuristic algorithms seems almost ideally suited to give optimal or suboptimal results in a relatively short time. There are many articles that describe the route optimisation methods from point A to point B [8,9], determining the location of taxis with the highest probability of gravity (occurrence of the order), to reduce the distance between a taxi and a potential customer [3], or the dynamic matching of taxis to orders and/or the applications of the car-pooling principle, if this improves the performance of the system [4,6].

Our work uses the third way to optimise the schedule model of taxi routing to which the heuristic algorithms will be applied. We check whether it is possible to use such algorithms and to identify the best one. The aim of the work is to: (a) formally define the problem of dynamic routing planning in a taxi company, (b) establish heuristic algorithms that will enable finding the optimal or suboptimal solution and finally (c) adjust the operation of selected algorithms. Initially we involved ourselves into the selection of values for the algorithm parameters depending on the characteristics of the input data and the tests allowing comparison of the results. Then, during the comparison the following criteria were taken into account: the number of routes, the number of taxis, operating time of the algorithm, the size of the input data, and finally total working time of taxi drivers.

2   Model Description

The problem of routing is a decisive problem consisting in minimizing transport costs needed to accomplish a specific transport task. It belongs to the basic problems of fleet management. It is an extension of the problems of the traveling salesman and the Chinese postman. In traditional terms, this problem is represented as a graph with weights, and its solution is to find the shortest path in the graph. The edges are related to weights, which most often represent distance, travel time, transport cost, etc. There are deterministic algorithms that allow to solve such problem, among others algorithm A*, Bellman-Ford or Dijkstra [1]. However, this problem is defined as an NP-hard optimizing problem. To illustrate the issue, an example describing a taxi company can be used. Assuming that n is the number of customers waiting for the fare, and m is the number of taxis ready for the route, there is $(m \cdot n) \cdot (m \cdot (n-1)) \cdot ... \cdot m = m^n \cdot n!$ practical solutions. For example, for $m = 10$ and $n = 10$, we'll get $3,6288E+16$ solutions. It is easy to imagine that using a standard computer, finding the optimal solution can take an unacceptable amount of time.

In order to find a solution, the following parameters should be taken into account:

- the number of available taxis for the rides,
- the distance in which the taxi is located from the position of the customer,
- the distance of the final location of the fare from the start positions of other fares awaiting execution and locations where new customers are likely to appear,
- time of taxi travel to the starting position of the customer,
- the duration of the fare and the maximum waiting time for a taxi,
- the parameters for a taxi met by the description given by the customer.

More formally, the problem of taxi routing is defined using a graph directed with weights $G = (V, E, \omega)$, where V is a set of vertices, E a set of edges, and ω a set of weights in this graph. Let n be the number of customers waiting for the ride, then $C = \{c_0, ..., c_n\}$ is the set of customers. Let m be the number of taxis ready for a ride, then $T = \{t_0, ..., t_m\}$ is a collection of taxis. A single vertex noted as v_{ct} denotes the fare at which the t taxi performs a course for the c customer. The vertex set is given by $V = \{v_{c_0 t_0}, ..., v_{c_n t_0}, ..., v_{c_n t_m}\}$. The edge of the graph represents a vector containing two vertices $\vec{e} = e = [v_{c_1 t_1}, v_{c_2 t_2}]$, where $v_1, v_2 \in V$, $c_1, c_2 \in C$, $t_1, t_2 \in T$ and $c_1 \neq c_2$. The cost function $w(e)$ is associated with each edge $e \in E$.

Let L_t be a collection of locations specifying the position of the taxi on the map, $L_t = \{l_{t0}, ..., l_{ti}, ..., l_{tm}\}$, where m is the number of taxis and l_{ti} is the location for the taxi $t_i \in T$. The function returning the taxi's position will be defined as $lt(t)$, where $t \in T$. Let L_c be a collection of locations specifying the starting and ending position of the fare on the map, $L_c = \{l_{c0}, ..., l_{ci}, ..., l_{cn}, l_{c(n+1)}, ..., l_{c(2n)}\}$, where n is the number of customers, l_{ci} is the starting position, and $l_{c(n+i)}$ is the final position for the customer's fare $c_i \in C$. The function that returns the starting position of the customer's c fare will be noted as $ls(c)$, and the final position as $le(c)$, where $c \in C$. Let P be a collection of taxi parameters. As a PT, a set containing sets of parameters P for taxis, $PT = \{P_0, ..., P_i, ..., P_m\}$, where $misspecified$ is the number of taxis, and P_i is the set of parameters for the taxi $t_i \in T$. The function that returns the taxi parameters will be defined as $pt(t)$, where $t \in T$. As a PC set is defined having sets of parameters \constraints of P, which customers expect from a taxi, $PC = \{P_0, ..., P_i, ..., P_n\}$, where n is the number of customers, and P_i is the set of parameters specified by the customer $c_i \in C$.

The function returning the parameters specified in the fare will be defined as $pc(c)$, where $c \in C$. The travel time between locations will be determined by the function $r(l_1, l_2)$, where $l_1, l_2 \in L_c \cup L_t$. The vector of completed fares will be denoted as $\overrightarrow{u_t} = u_t = [v_0, ..., v_k]$, where k is the number of fares provided by taxi t, $k \leq n$, where n is the number of customers. The set of fare vectors will be defined as $U = \{u_0, ..., u_i, ..., u_m\}$, where m is the number of taxis, and u_i is the position for the taxi $t_i \in T$.

The set of solutions is given by the set $S = \{s_0, ..., s_z\}$, where z is the number of possible solutions. The solution is the graph path, which is the edge vector $s = [e_0, ..., e_i, ..., e_{n'}]$, where $e_i \in E$, $e_i = (v_{i_1}, v_{i_2})$ for $i > 0 \rightarrow v_{i_1} = v_{i-1_2}$. Assuming that the function specifying the assignment of a taxi to the customer is marked by $x(c, t)$, where $c \in C$, $t \in T$, the solution must meet the following constraints:

- Up to one vehicle can be assigned to each customer:

$$\forall_{c \in C} \sum_{t \in T} x(c, t) \in \{0, 1\} \tag{1}$$

- The taxi must meet the criteria set by the customer:

$$\forall_{c \in C} \forall_{t \in T} (x(c, t) = 1 \Rightarrow pc(c) \subset pt(t)) \tag{2}$$

- A taxi can perform several fares, but it can also not do any of the following:

$$\forall_{u \in U} |u| \in \{0, .., n'\} \tag{3}$$

In order to find the best solution $s*$, the target function $W(s)$ should be minimized. It returns a reliable value, thanks to which it is possible to compare the cross results. It is necessary to optimize the solution, to find the best solution, the total time of driving all taxis will be minimized. In order to be able to calculate the $W(s)$ target function, it is necessary to specify the weights for each edge in the defined graph.

The weight $w(e)$ of edge e between two vertices $v_{c_1 t_1}$ i $v_{c_2 t_2}$, where $c_1, c_2 \in C$, $t_1, t_2 \in T$ equals:

$$w(e) = \begin{cases} r(lt(t), ls(c_2)), & t = t_1 = t_2 \text{ and } |u_{t_2}| = 0 \\ r(ls(c_1), le(c_1)) + r(le(c_1), ls(c_2)), & t = t_1 = t_2 \text{ and } |u_t| > 0 \\ r(ls(c(u_{t_2})), le(c(u_{t_2}))) + r(le(c(u_{t_2})), ls(c_2)), & t_1 \neq t_2 \text{ and } |u_{t_2}| > 0 \end{cases} \tag{4}$$

where $c(u)$ is the function that returns the customer from the top of the vector of taxi fares $u \in U$.

The minimized $W(s)$ function is the sum of the cost of the s solution's edge:

$$W(s) = \sum_{e \in s} w(e) \tag{5}$$

3   Numerical Analysis and Experiments

We have discussed the theoretical model, and in this section we perform some numerical analysis based on the model with three heuristic algorithms:

– ant algorithm,
– bee algorithm,
– genetic algorithm.

Due to the limits of space here we omit the details of the examined algorithms (Table 1).

For each of the selected algorithms, one of its variants was chosen. For the multi-criteria form-based algorithm, the standard versions: the MAX-MIN Ant System (MMAS) algorithm and the Ant Colony System (ACS) were considered. As demonstrated in the papers on the use of multi-criteria algorithms for optimizing car navigation the ACS algorithm achieves the best results and it was selected. For the bee algorithm, a choice between the standard version of the bee algorithm (ABC), the Gbest Artificial Bee Colony algorithm (GABC), the Improved Artificial Bee Colony algorithm (I-ABC) and the Predict and Select algorithm - Artificial Bee Colony (PS-ABC) was considered. The last one selected, PS-ABC algorithm was created as the combination of three algorithms (ABC, GABC and I-ABC) in order to maintain fast convergence, a large variety of solutions, and stable improvement. However, for the genetic

Table 1. The characteristics of the datasets

Test	No. of taxis	No. of fares	No. of fares/round	Max no. of vertices
1	9	16	16	144
2	16	10	10	160
3	16	16	16	256
4	26	10	1–10	260
5	17	17	5–17	289
6	15	30	14–30	450
7	37	19	1–19	703
8	32	22	2–22	704
9	73	16	16	1168

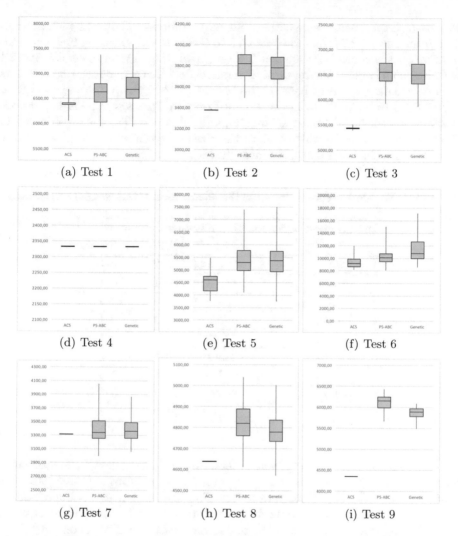

(a) Test 1   (b) Test 2   (c) Test 3

(d) Test 4   (e) Test 5   (f) Test 6

(g) Test 7   (h) Test 8   (i) Test 9

Fig. 1. Total working time for all three algorithms: ACS, PS-ABC and genetic algorithm in the box chart for the maximum runtime equal to 2 s, which is the minimum value where quantitative measurements can be conducted (tests 1–9, respectively, where the box represents the second quartile and the ends of the whiskers standard deviation above and below the mean of the data)

algorithm (GA), its standard version was used in the experiments. This will make it easier to compare results obtained from the use of other algorithms (Fig. 1).

When implementing all algorithms, the Node.js to build scalable and asynchronous test application using JavaScript. Asynchronous mode allows for parallel execution of the graph search activity, thus increasing the number of solutions that can be checked. In order to determine the time of taxi travel and the route by which taxis move between certain points, the Google Maps API library was used. Experiments were carried out on a computer equipped with a quad-core Intel Core i5 2.7 GHz processor and 16 GB of RAM operating memory, running under the control of the macOS Sierra operating system in version 10.12.6 with installed Node.js environment in version 8.9.3. To carry out the experiments a set of real data was used based on routes completed by taxis in the period from 01/07/2013 to 30/06/2014 in the city of Porto in Portugal [5]. These data are available to download from the Kaggle public data set. The data was saved in the form of an array in which each of the rows represents one completed route. The collection includes 3,381,999 fares (Table 2).

Table 2. The time needed to initialise the algorithms: ACS, PS-ABC and genetic

Test	Number of vertices	Initialisation times [msec]		
		ACS	PS-ABC	Genetc
1	144	131–144	0–2	0–1
2	160	141–149	0–1	0–1
3	256	409–447	0–2	0–1
4	26–260	0–475	0–6	0–1
5	80–289	28–639	0–5	0–1
6	182–450	215–1981	0–5	0–1
7	37–703	0–4464	0–8	0–5
8	64–704	8–5934	0–4	0–4
9	1168	8921–9338	1–8	0–4

The experimental results reveal interesting insights. Due to the stochastic nature of algorithms, the result may vary for the same input data and using the same values of the algorithm parameters. For this reason, it was important to perform not one and many tests for each data set. Each experiment consisted of simulating the actual work of the managing board at the taxi company. This simulation assumes that a new solution is searching for two minutes taking into account all requests that would appear and all available taxis and routes awaiting implementation. Each run of the algorithm to search for a solution is called a round. For each of the variants, one hundred simulations were carried out, from which the median value was calculated, the value of the first $Q1$ and the third $Q3$ quartiles. These values were used to compare the results of experiments (Fig. 2).

Once more, due to the limit we omit the details of the experiments. However, the advantage of the ACS algorithm over the others is evident. In each experiment, both the median and the difference between the first and third quartiles are the smallest. This means that the effectiveness and stability of the ACS algorithm is the largest of all three (Fig. 3).

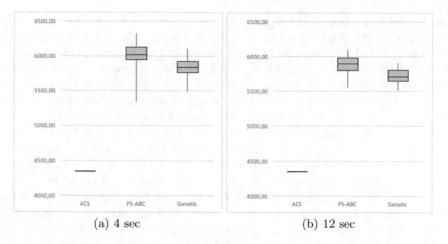

(a) 4 sec   (b) 12 sec

Fig. 2. Total working time for test 9 for all three algorithms: ACS, PS-ABC and genetic algorithm in the box chart for the maximum runtime equal to 4 s (left) and 12 s (right), which is the minimum value where quantitative measurements can be conducted (the box represents the second quartile and the ends of the whiskers standard deviation above and below the mean of the data)

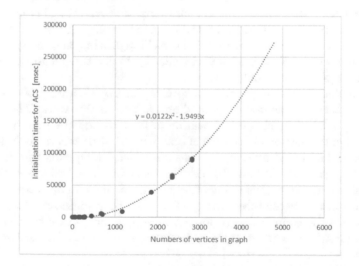

Fig. 3. Distribution of initialisation times for ACS in msec over numbers of vertices in graph

4 Concluding Remarks

The first stage of the research was to determine the best parameter values for the algorithms selected in the work. These values differed depending on the data on which the algorithm operated. The most commonly observed pattern was between the size of the input data determined by the number of vertices in the graph and the value of the algorithm parameters. The resulting choices of parameter values found to be appropriate and then tests were carried out to determine the best algorithm from among three examined. This algorithm proved to be the ACS algorithm, whose results are characterized in most cases by the smallest median and not always reaching optimal values. The difference between the first and third quartiles of obtained results is always the smallest among all compared algorithms. An obvious disadvantage of this algorithm is the long initialisation time, which includes generating the initial pheromone values at the edges of the graph. This time depends on the size of the input data. In both remaining cases, algorithm initialization time is 0–9 ms whereby being independent of the characteristics of the input data. For this reason, in the case of using the ACS algorithm, it is necessary to pay attention to the size of the input data. For example, assuming the round time of 2 min, the data size must be less than 3,600 vertices in the graph for system with a computing power of 3465 FLOPS. Then it was possible to get a result before the next round. In the case of a larger data size, it is advisable to use unit with higher computing power or to divide data between different components using, for example, parallel programming. It is also possible to choose one of the two remaining algorithms: PS-ABC, GA. Our work has shown that these algorithms are less stable and in most cases they return inferior quality solutions. However, the difference between the quality of results obtained from the use of these algorithms and the quality of solutions obtained from the use of the ACS algorithm may be so small that it will be more cost-effective for a taxi company to use these algorithms than to buy better or additional computer.

The problem of planning transport services for taxis is a very extensive problem. Only part of it has been described and solved in this work. The presented solution assumes optimising the allocation of fares to taxis by minimizing costs, determined in proportion to the time of taxi drivers' work, where work time is the time when the taxi is in motion and consumes fuel. The quality of this proposal is based only on one criterion. In fact this may be multi-criteria optimisation (MSO) assuming the maximization of profits, which depend on the revenues and costs from the completed fares, and the fuel used by vehicles. In some cases the prime rule for the company, it is not the directly profit, but customer satisfaction, which may also depend on many other parameters, such as taxi availability or utilization.

The ACS algorithm has been estimated as the best of three tested, but it still has disadvantages. The way in which it was implemented could be improved so that the initialisation time, which is the greatest weakness of this algorithm, could be significantly reduced. Thanks to this, the algorithm could be used to optimise solutions for larger input data sets. The way of problem definition allows us to extend the target function by further criteria, bringing it closer to the real problem. Other selection of algorithms should also be checked. In conclusion, the current results confirm the validity of research findings into modelling of taxi dispatch problem.

Acknowledgments. The authors acknowledge the support from the statutory funds of the Wrocław University of Science and Technology.

References

1. Fortnow, L., Homer, S.: A short history of computational complexity. Bull. EATCS **80**, 95–133 (2003)
2. Gan, J., An, B., Miao, C.: Optimizing efficiency of taxi systems: scaling-up and handling arbitrary constraints. In: Proceedings of the 2015 International Conference on Autonomous Agents and Multiagent Systems, AAMAS 2015, International Foundation for Autonomous Agents and Multiagent Systems, Richland, SC, pp. 523–531 (2015)
3. Ge, Y., Xiong, H., Tuzhilin, A., Xiao, K., Gruteser, M., Pazzani, M.: An energy-efficient mobile recommender system. In: Proceedings of the 16th ACM SIGKDD International Conference on Knowledge Discovery and Data Mining, KDD 2010, pp. 899–908. ACM, New York (2010)
4. Jung, J., Jayakrishnan, R., Park, J.Y.: Dynamic shared-taxi dispatch algorithm with hybrid-simulated annealing. Comput.-Aided Civ. Infrastruct. Eng. **31**(4), 275–291 (2016)
5. Kaggle: ECML/PKDD 15: Taxi trajectory prediction (2015). https://www.kaggle.com/c/pkdd-15-predict-taxi-service-trajectory-i
6. Kuo, M.: Taxi dispatch algorithms: why route optimization reigns, March 2016. https://blog.routific.com/taxi-dispatch-algorithms-why-route-optimization-reigns-261cc428699f
7. Xu, M., Wang, D., Li, J.: DESTPRE: a data-driven approach to destination prediction for taxi rides. In: Proceedings of the 2016 ACM International Joint Conference on Pervasive and Ubiquitous Computing, UbiComp 2016, pp. 729–739. ACM, New York (2016)
8. Yuan, J., Zheng, Y., Zhang, C., Xie, W., Xie, X., Sun, G., Huang, Y.: T-drive: driving directions based on taxi trajectories. In: Proceedings of the 18th SIGSPATIAL International Conference on Advances in Geographic Information Systems, GIS 2010, pp. 99–108. ACM, New York (2010)
9. Zheng, Y., Liu, Y., Yuan, J., Xie, X.: Urban computing with taxicabs. In: Proceedings of the 13th International Conference on Ubiquitous Computing, UbiComp 2011, pp. 89–98. ACM, New York (2011)

Malware Detection Using Black-Box Neural Method

Dominik Pieczyński and Czesław Jędrzejek[✉]

Poznań University of Technology, Poznań, Poland
{dominik.pieczynski, czeslaw.jedrzejek}@put.poznan.pl

Abstract. Because of the great loss and damage caused by malwares, malware detection has become a central issue of computer security. It has to be fast and very accurate. To develop suitable methods on needs very good quality benchmarks. One such benchmark is the Microsoft Kaggle malware challenge system run in 2015. Since then over 50 papers were published on this system. The best result were achieved with complex feature engineering. In this work we analyze the black-box neural method and what is novel analyze its results against the Microsoft Kaggle malware challenge benchmark. It is tempting to use convolution neural networks for malware analysis following the great success with analysis of images. Even the use of balanced classes and drop-out convergence does not beat XGBoost with feature engineering, although some room for improvement exists. The situation is similar to that for language analysis. The language is much more hierarchical than image, and apparently malware is too. The malware analysis still awaits optimal neural network architecture.

Keywords: Malware detection
Microsoft Kaggle malware classification challenge
Malware convolution neural networks

1 Introduction

With new threats appearing on a daily basis and attackers continuously evolving their techniques, it can be extremely difficult to keep up. 2017 was the year of the biggest ransomware outbreak in history (WannaCry) and a data breach that exposed the personal information of 150 mln United States' population (Equifax). Ponemon Institute estimates the average cost of a successful cyber attack at over $5 million in the US [2]. To counter risks, according to Gartner [1] worldwide cybersecurity spending will reach $96 billion in 2018. In less advanced countries, such as Poland, the critical infrastructure is very vulnerable to cyber attacks [3]. With the development of the Internet, the amount of malware has grown rapidly, and standard methods of its detection, such as, for example, comparing signatures of files (some current anti-virus technologies still use signature-based approach) are no longer sufficient. At the same time, methods of data processing based on so-called cloud software became popular - globally available for many users at the same time and easily scalable when increased traffic occurred. Security in the cloud can have its own problems. Poorly configured systems can allow

© Springer Nature Switzerland AG 2019
K. Choroś et al. (Eds.): MISSI 2018, AISC 833, pp. 180–189, 2019.
https://doi.org/10.1007/978-3-319-98678-4_20

hackers to exploit vulnerabilities and lead to malicious intrusion. It is worth noting, however, that the trend of transferring calculations to the cloud is no longer possible to reverse, and the increased complexity and number of viruses would require the implementation of algorithms on users' computers using most of their machines' power or often impossible to run due to too outdated hardware specifications. One of the most difficult tasks to execute by anti-malware software are the enormous amounts of data which needs to be evaluated for potential malicious intent. Software giants generate tens of millions of daily data points to be analyzed as potential malware. For example, Microsoft analyses data from 600 million computers. In order to evade detection, malware criminals introduce polymorphism to the malicious components thus obfuscating malware typology. Therefore, malware classification using machine learning algorithms is a complex task.

There are several possible methods of detecting maliciousness based on the code representation:

1. Analysis of the binary code.
2. Analysis of the decompiled code.
3. Analysis of the presence and absence of particular pairs of consecutive Application Program Interface function calls (APIs) in the API-sequence graph and comparison of those features in the executable code [24]. The methods 1th and 2nd treat two code forms separately.

In this paper we analyze only the binary code which for Microsoft Kaggle Windows systems is in the form of the Portable Executable (PE) format, a file format for executables, object code, DLLs, FON Font files, and others used in 32-bit and 64-bit versions of Windows operating systems. Contrary to text classification there is plenty of features to be considered, mostly because of in the absence of dominant features in raw executable binary files. To build a malware detection system, most works rely on feature engineering. At minimum a set of features may be derived from executable file (raw bytes) or opcode of a malicious file. Here we restrict the analysis to static analysis, where we look at information from the binary program that can be obtained without running it.

2   Related Work

The feature selection for a malware detection system can be very laborious as demonstrated during and after the Microsoft Kaggle malware classification challenge [4, 5]. For this contest Microsoft in 2015 released a malware with a huge dataset of near 0.5 terabytes of data, containing more than 20 K malware samples of 9 types (all the samples were malware). Here, we briefly summarize the effort of the winners [6] (team "Say no to overfitting"), the runner-up team [7] of the contest and the results of the team that took the 3rd place [8].

Both teams used a learning-based system which uses different malware characteristics to effectively assign malware samples to their corresponding families without doing any deobfuscation and unpacking process. The proportion of packed malware has been growing rapidly and now comprises more than 80% of all existing malware

[9]. It was shown [9] that precision of 97.7% in identifying packing algorithms was achieved. All of the samples in this study was malware.

2.1   Feature Representation

In [6] a combination of opcode and bytecode features were used. initially opcode features appearing 200 times in at least one file we selected, namely: 165 1-gram, 27225 2-gram, 22285 3-gram, 22384 4-gram – altogether 70894 opcode n-gram features. 448 segment count features were found in the "golden features" section data. The machine learning algorithm takes into account the division into segments by analyzing the size of each of them, the number of occurrences of a given operating code (opcode), the number of instances of consecutive instructions (n-grams) and image analysis of assembly language (asm) file pixel intensity [10].

In addition, 15000 feature from the categories proposed in [11] were considered. These so called other features were 1-grams, and 4-grams of the bytecode, DLL function call features and derived assembly features. All of the acquired features have been normalized and saved to feature vectors for each file. Then using RF (random forest) a set of features was reduced to 4416 n-gram opcodes, 19 segment features, 800 asm file pixel intensities and 2193 other features.

In [7] features of three types were used:

(1) on file properties: size, compressed size and ratios between them - for both bytes and asm files: ab_ratio: asm file size, divided by bytes file size, abc_ratio: asm file compression rate, divided by bytes file compression rate.

(1) ab2abc_ratio: ab_ratio, divided by abc_ratio.

(2) on contents of the asm files: sections, DLLs, opcodes and intepunction statistics by Section.

(3) on contents of the bytes files: 1, 2, and 4 grams, full lines and distribution of entropy.

2.2   Feature Representation

Random Forest is the most often used for feature reduction, and it was used in [6]. The reason is because the tree-based strategies used by random forests naturally ranks by how well they improve the purity of the node (equivalently, bring information gain). Nodes with the greatest decrease in impurity happen at the start of the trees, while nodes with the least decrease in impurity occur at the end of trees. Thus, by pruning trees below a particular node is the most effective). One can as in [8] use feature fusion. One approach is the best subset selection technique which starts with subsets containing just one feature. Then a classifier is trained, and the subsets with the highest value of the objective function used to assess the performance (e.g., accuracy, loss functions, etc.) are retained. Then, the process is repeated for any subset containing f features, where f is increased by one at each step so that, for example, all the possible subsets consist of two features.

2.3 Classification

The Microsoft Kaggle malware classification challenge problem is a large system but not extremely large. Experience with Kaggle challenges shows that boosting methods give the best results for this size of a system. The results obtained before 2016 claiming the best results with SVM and Random Forest have to be treated cautiously because XGBoost was proposed in 2016 [12] and whose parameters are completely tunable and thus easy to use. All leading Kaggle entries [6–8] combined results of different methods using the ensemble method [13].

None leading contributions in Microsoft Kaggle malware classification challenge used neural networks. Such works appeared later.

3 Convolution Neural Networks

For malware detection, all of these previous applications use a significant amount of domain knowledge for feature extraction. Recently two works [14, 15] proposed and evaluated a simple convolution deep neural network architecture detecting malicious Windows executable files by learning from their raw sequences of bytes and labels only, that is, without any domain-specific feature extraction nor preprocessing. The problem reduces to finding methods for scalable learning of neural networks for long sequences: the whole byte file. Convolution neural networks (CNN) have no competition for recognition of objects from very large annotated image databases [16].

The works concentrated on scalability. In [14] data were divided into two groups. Group B data was provided by an anti-virus industry partner, where both the benign and malicious programs were meant to be representative of files seen on real machines. The Group B training set consisted of 400,000 files split evenly between benign and malicious classes. The testing set had 77,349 files, of which 40,000 were malicious and the remainder were benign.

In [15] a deep convolutional malware classifier was designed on 20 million of Windows EXE files represented as raw sequences of bytes. The approach did not perform quite as well as hand engineered features.

4 Convolution Neural Networks for the Microsoft Kaggle System

The results of [14, 15] are not in the public domain (mainly due to the size of the systems). So we decided to repeat the calculation using the architecture of [15] and [14] (for the bytecode part for the Microsoft Kaggle system). Class membership of 9 malware types is highly disparate. Therefore, a validation set contains 1092 samples with the following weights (matching distribution in the training set) (Fig. 1).

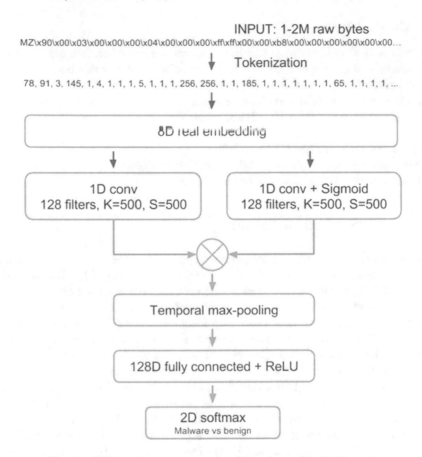

Fig. 1. CNN architecture processing full malware file (1-d image)

We present results in the following Tables. Table 1 presents weights of balanced malware classes after undersampling. Table 2 displays over-allresults for 2 versions of convolution networks performance. AUC was approximated to the case of many-label classification. Tables 3 and 4 show the performance for particular classes for the pure Malconv, and the confusion matrix, respectively. Tables 5 and 6 show analogous results for the Malconv+ Drop-out method. Over-all application of Drop-out for better convergence improves results, except for the smallest class #4, where the results are wrong (Tables 5 and 6).

Microsoft does not disclose annotated testing results. However, even after the challenge, the Kaggle site in response to a submission still returns log-loss. In our case we obtain:

```
Private Score: 0.123234446
Public Score: 0.127822114.
```

Table 1. Weights of balanced malware classes.

Classes	Weights
Class 0 (Ramnit)	0.027
Class 1 (Lollipop)	0.017
Class 2 (Kelihos_ver3)	0.014
Class 3 (Vundo)	0.088
Class 4 (Simda)	1.000
Class 5 (Tracur)	0.056
Class 6 (Kelihos_ver1)	0.106
Class 7 (Tracur)	0.034
Class 8 (Gatak)	0.041

Table 2. Networks performance

Model	Accuracy	Weighted accuracy	Log loss	Weighted log loss	AUC	Weighted AUC
MalConv	0.9725	0.8730	0.11964	0.4657	0.9981	0.9969
MalConv + Dropout	0.9826	0.8585	0.1212	0.4900	0.9998	0.9990

The results for particular classes are:

Table 3. The performance for particular classes for MalConv

Class	Precision	Recall	f1-score	Support
0	0.89	1.00	0.94	155
1	1.00	0.96	0.98	248
2	0.99	1.00	0.99	295
3	1.00	0.96	0.98	48
4	1.00	0.20	0.33	5
5	1.00	0.99	0.99	76
6	1.00	0.97	0.99	40
7	0.99	0.91	0.95	123
8	0.95	0.99	0.97	102
Avg/total	0.97	0.97	0.97	1092

Table 4. Confusion matrix for MalConv

```
        0    1    2    3  4  5  6   7    8
0 [155   0    0    0  0  0  0   0    0]
1 [  3  238   4    0  0  0  0   0    3]
2 [  0    0  295    0  0  0  0   0    0]
3 [  2    0    0   46  0  0  0   0    0]
4 [  3    0    0    0  1  0  0   0    1]
5 [  0    0    0    0  0 75  0   1    0]
6 [  0    1    0    0  0  0 39   0    0]
7 [ 10    0    0    0  0  0  0 112    1]
8 [  1    0    0    0  0  0  0   0  101]
```

Table 5. The performance for particular classes for MalConv + Drop-out

Class	Precision	Recall	f1-score	Support
0	0.94	1.00	0.97	155
1	1.00	0.98	0.99	248
2	1.00	1.00	1.00	295
3	0.96	0.96	0.96	48
4	0.00	0.00	0.00	5
5	0.99	1.00	0.99	76
6	1.00	1.00	1.00	40
7	0.98	0.96	0.97	123
8	0.97	0.99	0.98	102
Avg/total	0.97	0.97	0.97	1092

Table 6. Confusion matrix for MalConv + Drop-out

```
        0    1    2    3  4  5  6   7    8
0 [155   0    0    0  0  0  0   0    0]
1 [  0  243   0    1  0  0  0   0    1]
2 [  0    0  294    1  0  0  0   0    0]
3 [  1    1    0   46  0  0  0   0    0]
4 [  4    0    0    0  0  0  0   1    0]
5 [  0    0    0    0  0 76  0   0    0]
6 [  0    0    0    0  0  0 40   0    0]
7 [  4    0    0    0  0  1  0 118    0]
8 [  1    0    0    0  0  0  0   0  101]
```

Our results are significantly worse than the winning teams of the Microsoft Kaggle original challenge. But they are much better than presented in Tables 1 and 2 of [14] and in [20], obtained with deep belief network (Table 7).

Table 7. The original Kaggle challenge winning log-loss results

Place	Log-loss
1st	0.0028
2nd	0.0034
3rd	0.0040

5 Conclusions

Convolution neural networks in pure form for the Microsoft Kaggle malware system (only byte file) are relatively easy to implement and demand no feature engineering. Though our hyper-parameters may not be optimal it is clear that giving up on opcode analysis bears significant quality cost. Malware file can be treated as image but it is not enough. The situation is similar to that for language analysis [17, 18]. The language is much more hierarchical than image and apparently the malware is too. Recently, very deep neural network was applied to for the Microsoft Kaggle malware system [19]. The processing of the MalNet was divided into two stages. The first stage is to preprocess malware sample data, it takes a binary form of a Windows executable file, generates a grayscale image from it, and extracts opcode sequence and metadata feature using disassembler tool. The grayscale image was processed by CNN, opcode by LTSM, and the results went to stacking ensemble to integrate two networks' output and metadata. In [18] 99.36% overall accuracy was achieved still worse than in [6–8] during the challenge using XGBoost. Malware analysis still awaits optimal neural network architecture. One should try other neural network methods. In [21] it was shown that ADAM, the stochastic gradient descent (SGD) neural network [22] with suitable hyper-parameters performs almost as well as booster feature methods for the relatively small Reuters corpus. More efficient oversampling of rare type virus samples such as SMOTE should be investigated [23].

The progress in this area crucially depends on large, open and sharable annotated datasets. Some step in this direction was done by EMBER: An Open Dataset for Training Static PE Malware Machine Learning Models [25]. Though not entirely open, this dataset is more open than the Microsoft Kaggle dataset and much larger.

We acknowledge the Poznan University of Technology grant (04/45/DSPB/0185).

References

1. Gartner. https://www.gartner.com/newsroom/id/3836563. Accessed 6 June 2018
2. Ponemon Institute: COST OF CYBER CRIME STUDY 2017: https://www.accenture.com/t20171006T095146Z__w__/us-en/_acnmedia/PDF-62/Accenture-2017CostCybercrime-US-FINAL.pdf#zoom=50. Accessed 6 June 2018
3. PWC: Cyber-ruletka po polsku Dlaczego firmy w walce z cyberprzestępcami liczą na szczęście, Polish cyber-roulette Why companies count on luck in the fight against cybercriminals: https://www.pwc.pl/pl/pdf/publikacje/2018/cyber-ruletka-po-polsku-raport-pwc-gsiss-2018.pdf. Accessed 6 June 2018
4. Microsoft Kaggle challenge: https://www.kaggle.com/c/malware-classification. Accessed 6 June 2018
5. Ronen, R., Radu, M., Feuerstein, C., Yom-Tov, E., Ahmadi, M.: Microsoft Malware Classification Challenge: https://arxiv.org/abs/1802.10135. Accessed 6 June 2018
6. Microsoft malware winner 1st place: http://blog.kaggle.com/2015/05/26/microsoft-malware-winners-interview-1st-place-no-to-overfitting; https://www.youtube.com/watch?time_continue=979&v=VLQTRlLGz5Y. Accessed 6 June 2018
7. Microsoft malware winners 2nd place: https://www.kaggle.com/c/malware-classification/discussion/13863. Accessed 6 June 2018
8. Ahmadi, M., Ulyanov, D., Semenov, S., Trofimov, M., Giacinto, G.: Novel feature extraction, selection and fusion for effective malware family classification. In: Proceedings of the CODASPY 2016, pp. 183–194. ACM, New York (2016)
9. Bat-Erdene, M., Kim, T., Park, H., Lee, H.: Packer detection for multi-layer executables using entropy analysis. Entropy **19**(3), 125 (2017)
10. Masud, M.M., Khan, L., Thuraisingham, B.M.: A scalable multi-level feature extraction technique to detect malicious executables. Inf. Syst. Front. **10**(1), 33–45 (2008)
11. Nataraj, L., Karthikeyan, S., Jacob, G., Manjunath, B.S.: Malware images: visualization and automatic classification. In: Proceedings of the 8th VizSec 2011. ACM, New York (2011). Article 4
12. Chen, T., Guestrin, C.: XGBoost: a scalable tree boosting system. In: KDD, pp. 785–794 (2016)
13. Zhou, Z.-H.: Ensemble Methods: Foundations and Algorithms. CRC Press, Florida (2012)
14. Raff, E., Barker, J., Sylvester, J., Brandon, R., Catanzaro, B., Nicholas, C.K.: Malware Detection by Eating a Whole EXE. CoRR abs/1710.09435 (2017)
15. Krčál, M., Švec, O., Bálek, M., Jašek, O.: Deep convolutional malware classifiers can learn from raw executables and labels only. In: ICLR 2018 Workshop (2018)
16. http://imageimage-net.org/challenges/LSVRC/2017/results. Accessed 6 June 2018
17. Zhang, X., LeCun, Y.: Byte-Level Recursive Convolutional Auto-Encoder for Text. CoRR abs/1802.01817 (2018)
18. Schwenk, H., Barrault, L., Conneau, A., LeCun, Y.: Very deep convolutional networks for text classification. EACL **1**, 1107–1116 (2017)
19. Yan, J., Qi, Y., Rao, Q.: Detecting malware with an ensemble method based on deep neural network. Security and Communication Networks 2018:1–7247095:16 (2018)
20. Yuxin, D., Siyi, Z.: Malware detection based on deep learning algorithm: Neural Comput & Appl. (2017). https://doi.org/10.1007/s00521-017-3077-6
21. Zdrojewska, A., Dutkiewicz, J., Jędrzejek, C., Olejnik, M.: Comparison of the novel classification methods on the reuters-21578 corpus. In: Choroś, K., et al. (eds.) Proceedings of MISSI 2018 (2018)

22. Kingma, D.P., Ba, J.: Adam: a method for stochastic optimization. In: The International Conference on Learning Representations (ICLR), San Diego (2015)
23. Chawla, N.V., Bowyer, K.W., Hall, L.O., Kegelmeyer, W.P.: SMOTE: synthetic minority over-sampling technique. J. Artif. Intell. Res. **16**, 321–357 (2002)
24. Iwamoto, K., Wasaki, K.: Malware classification based on extracted API sequences using static analysis. In: Proceedings of the Asian Internet Engineeering Conference (AINTEC 2012). ACM, New York (2012)
25. Anderson, H.S., Roth, P.: EMBER: An Open Dataset for Training Static PE Malware Machine Learning Models. CoRRabs/1804.04637 (2018)

Intrusion Detection and Risk Evaluation in Online Transactions Using Partitioning Methods

Hossein Yazdani and Kazimierz Choroś[✉]

Faculty of Computer Science and Management,
Wrocław University of Science and Technology, Wrocław, Poland
{hossein.yazdani,kazimierz.choros}@pwr.edu.pl

Abstract. Security is the main issue for real time systems, specially for financial and banking systems. Some of the customers who pay much attention to confidentiality and security on their network activities and transactions prefer to use the most secure channels, and for the others speed and the ease of services are more important. An optimized method should be a solution, but both strategies follow one common idea that any anomaly, abnormality, and intrusion should be handled in advance, as the reputation of each organization is based on trust. This paper proposes a new method with the aim of considering any anomaly in advance, in addition to partitioning strategy. The BFPM method makes use of the well-known Fuzzy C-Means clustering algorithm to evaluate whether packets or transactions are risky or not, and in what extent they will be risky in the near future. The proposed method aims to provide a flexible search space to cover prevention and prediction techniques at the same time.

Keywords: Network security · Intrusion detection
Online transaction · Partitioning methods · Fuzzy C-Means algorithm
BFPM method · Clustering · Risk evaluation · Object movement

1 Introduction

Providing confidentiality (keeps customers' information in a safe box out of other's hands), integrity (guarantees the accuracy of information), and availability (makes the information accessible any time and anywhere) at the same time is a big issue in all security systems [1]. Risk managements are taken into consideration where the optimal systems have faced several risks on transactions and packets, reputation, information security, price, liquidity, foreign exchange, and compliance risks [1]. The paper aims to provide an advanced method to cover risk management among other factors without running the learning methods for several times. The proposed method was applied on some datasets in classification and clustering problems.

© Springer Nature Switzerland AG 2019
K. Choroś et al. (Eds.): MISSI 2018, AISC 833, pp. 190–200, 2019.
https://doi.org/10.1007/978-3-319-98678-4_21

1.1 Network Security

All security methods follow a common approach to detect any fault or abnormal transaction, but a comprehensive method needs to monitor and handle the behaviour of transactions/packets in advance. Sophisticated systems need to be aware of the further movements of each transaction to prevent the system form any unwanted events beforehand. According to Table 1, regarding the cyber threat taxonomy [2], this paper studies the types of threat that can be detected and prevented using machine learning approaches. Further more, the paper considers both classification and clustering approaches in its experimental verifications to evaluate and illustrate the ability of the proposed methods in intrusion detection and prevention systems. As Table 1 shows, cyber threats are mainly categorized in different categories: intrusion, fraud, malware, availability attacks, and abuse content, but some of the threats are commonly categorized into different groups. Cyber security is designed to protect computer systems, networks, programs, and data from any virus or attack using a set of technologies, processes, and components [3]. Malware or virus is designed to get an unauthorized access to computer systems, which sometime replicates or spreads itself (worm), or prevents being detected (trojan horse). Malwares also designed to collect information from the systems (spyware), to inject unsolicited advertising material (adware), to access and control the computer systems (rootkit), or to remotely take over and control computer systems (bot). Different types of malwares and viruses have been designed for different purposes, and in all these cases detecting and preventing the viruses are the main goals of anti-viruses, firewalls, intrusion detection (IDS), and intrusion prevention (IPS) systems. Three main types of cyber analytics can be categorized as: misuse-based,

Table 1. The taxonomy of cyber attacks.

Cyber threat				
Intrusion	Fraud	Malware	Availability attack	Abuse content
Login attempts	Phishing	Worm	Denial of service	Spam
Account takeover	Masquerading	Ransomware		Incendiary
Social engineering			Sabotage	Speech
Scanning			:	Inappropriate
Sniffing				Violence
Privilege escalation	Unauthorized use of resources	Rootkit		:
Application	:	Trojan		
Compromise		Advance persistent Threat		
:		:		

anomaly-based, and hybrid approaches. Misuse-based approaches detect known attacks by using their signatures, where anomaly-based techniques model the normal network's behaviour and identify anomalies from normal behaviour. The former methods are effective without generating large number of false alarm but need frequent updates, while the latter approaches have high false alarm rates, but are needed to detect new attacks which cannot be detected by misuse-based methods. The hybrid methods make use of both strategies to reduce the false alarm rates in addition to raise the detection rates. Different learning methods have been used in anomaly detection techniques which this paper considers just classification and clustering methods by applying Bounded Fuzzy Possibilistic Method in their membership assignments.

1.2   Partitioning Methods

Classification or supervised learning methods work by splitting the data sets into two groups as training and testing data sets. The training data set is used to build a classier based on the class label, and the classifier will be used to estimate the class label for the objects in the testing data set [4]. The accuracy of the method can be improved by getting feedback from the testing data set and updating the classifier. Clustering or unsupervised learning is a method for categorizing data objects into different clusters with respect to their similarity [5]. To classify or cluster data objects into some clusters based on the existing similarity between data objects, mostly several common membership and distance functions are used in both supervised and unsupervised learning methods. In general, a cluster or a class is a set of cn values $\{u_{ij}\}$, where j represents the jth object and i implies the ith cluster. The partition matrix is often represented as a $c \times n$ matrix $U = [u_{ij}]$ [6]. Assume a set of n objects represented by $O = \{o_1, o_2, \ldots, o_n\}$ in which each object is typically introduced by numerical $feature - vector$ data that has the form $X = \{x_1, \ldots, x_d\} \subset R^d$, where d is the dimension of the search space or the number of features. Regardless of the type of clustering methods, membership assignments can be categorized as crisp, fuzzy, probabilistic, possibilistic, and BFPM methods [7].

2   Related Work

Edge et al. [1] presented strategies to protect financial systems from attacks using attack trees and protection trees in a cost effective way. Authors discussed the importance of considering security in financial systems to maintain the customers' trust. They made use of attack trees to highlight the weaknesses of the financial systems, which the protection trees were used to cover the weaknesses. The authors followed the idea that high level threats on online banking should be analysed and decomposed into different levels or actions, and the probability of occurrence of each attack should be calculated. Based on the estimated probabilities, the protection strategy will be provided. The authors did not pay much attention to any specific prevention method, but instead they aimed to show a perspective on implementation of a secure system.

Cao et al. [8] discussed about the infrastructure, security threats, risk management, and security strategies of mobile banking systems. The authors briefly discussed the main topics in security and threats by presenting an architecture of mobile banking systems into three layers: connector, service, and channel layers. The authors also categorized the security threats into two main categories: page security threats: (1. cross frame scripting 2. cross site request forgery) and service security threats: (1. reflection injection 2. denial of service attack [9]). Finally, the authors discussed about the security requirements: authentication, single sign-on, out-of-band authentication, second-factor authentication, inline password protection, and authorization. The paper did not discuss more about each individual solution.

Shon et al. [10] presented a hybrid machine learning method on network anomaly detection using the well-known method – Support Vector Machine (SVM) in classification strategy. The authors discussed how zero-day cyber attack can be detected through different stages of the proposed method by dividing and separately labelling the normal and attack traffic. The authors also presented another methodology to obtain better accuracy by evaluating packets according to their behaviour. In the first stage, the profile of normal packets using Self-Organized Feature Map (SOFM), without having any knowledge in advance, will be created. Then, the packets filtering and genetic algorithms have been used to evaluate the network traffic for the packets that violate TCP/IP standard or other abnormality. Meanwhile, some data preprocessing and knowledge gathering have been used to increase the accuracy of the proposed method.

Tsai et al. [11] reviewed some machine learning approaches in anomaly detection for intrusion detection systems by discussing about the pattern classification, single classifier (including: k-nearest neighbour, support vector machine, naive bayes networks, decision trees, self-organization maps, artificial neural network, genetic algorithms, and fuzzy logic), hybrid classifiers, and ensemble classifiers. In the next part of their discussion, different classifiers have been compared on their functionalities. The authors also briefly discussed about feature selection techniques. Ahmed et al. also reviewed some advanced machine learning approaches by providing a survey on classification based network anomaly detection (support vector machine, Bayesian network, and neural network rule-based), statistical anomaly detection (mixture model, signal processing technique, and Principal Component Analysis (PCA)), and clustering-based (regular clustering and co-clustering) [12]. The authors also discussed about some types of abnormal activities, e.g., credit card fraud, mobile phone fraud, cyber attacks, etc., by categorizing anomaly into: point anomaly (deviating from the normal pattern of the dataset), contextual anomaly (anomalously behaviour of a particular context), collective anomaly (anomalously behaviour of collection of similar data instances with respect to the entire dataset).

Asmir et al. [13] proposed a new method on fuzzy based semi-supervised learning for intrusion detection systems. The authors made use of a Single hidden Layer Feed-forward Neural network (SLFN) to train the output of fuzzy membership vector by assigning random weights. They discussed the issues with supervised learning methods in intrusion detection systems that make only use of known labelled samples to train the classifiers. They also briefly studied about the semi-supervised methods and strategies: self-training, graph-based models, generative models, and Transductive Support Vector Machines (TSVMs). The authors presented an algorithm to evaluate their proposed method using different preprocessing techniques to convert symbolic and discrete data into continuous data which can be handled by their presented neural network method. There are also several studies on different aspects of security: using different fuzzy learning methods [14], using other machine learning methods [15], $R2L$ and $U2R$ attacks [16], and other advanced methods [17].

3   Supervised and Unsupervised BFPM

Bounded Fuzzy Possibilitic Method (BFPM) [18] has been introduced to overcome the issues with the conventional crisp, fuzzy, probability, and possibilistic partitioning methods in their membership assignments. The method overcomes the issues with restrictions and limitations for samples in their freely participation in other clusters and classes. BFPM have been applied in both classification and clustering problems to provide the flexible environment to evaluate the potential ability of each object individually with respect to all classes or clusters. Two different algorithms introduced for classification and clustering concepts presented by Algorithms 1 and 2 respectively. The proposed algorithms also cover the evaluation of both types of cyber attacks detection categories: misused-based and anomaly-based detection techniques, where the classification approach is processed by the BFPCM algorithm, and the clustering approach will be handled by the BFPM algorithm.

3.1   BFPCM Algorithm

This algorithm is designed for supervised learning concept, which performs on selecting the training and the testing datasets using random sampling, where training and testing data sets can be chosen using different methods such as holdout, random sampling, k-fold cross validation, and bootstrap. The main idea of the algorithm is to cover both feature and vector spaces by assigning membership degrees and weights to features and objects during the learning procedures, presented by Algorithm 1 [19].

Algorithm 1. BFPCM Algorithm

Input: X, c,
Output: U
for $(\forall\ X_j\ \text{in}\ D_T)$ do

$$u'_{ij} = \min_{i=1}^{c} \left[\sum_{l=1}^{n'} \frac{\sum_{k=1}^{d}(||X_{jd} - X_{ld}||^2 * w_d)}{d} \right], \quad \forall i,j \tag{1}$$

$$u''_{ij} = \min_{i=1}^{c} \left[\sum_{l=1}^{n'} \sum_{k=1}^{d}(||X_j - X_l||^2) \right], \quad \forall i,j \tag{2}$$

$$u_{ij} = \frac{u'_{ij} + u''_{ij}}{2}, \quad \forall i,j \tag{3}$$

end for

where c is the number of classes or class labels, D_T is the testing dataset, U is the membership matrix, u_{ij} is the membership degrees for the jth object for the ith class, n' is the number of objects in the training dataset, and $||.||$ is the L_2 norm distance function. The algorithm assigns a membership degree u'_{ij} to each object in the testing dataset based on Eq. (1). This stage assigns a weight to each feature of each object with regard to the distance between the features of the object and the feature of other objects in training dataset. Equation (2) shows how u''_{ij} will be generated for objects in the test dataset with respect to classes based on the L_2 norm of distance function. Consequently, the final membership matrix for objects will be calculated using Eq. (3).

3.2   BFPM Algorithm

Algorithm 2 [20] is designed for unsupervised learning concept, using BFPM membership assignments for anomaly detection, while there is no label for packets or objects.

Algorithm 2. BFPM Algorithm

Input: X, c, m
Output: U, V
Initialize V;
while $max_{1\leq k\leq c}\{||V_{k,new} - V_{k,old}||^2\} > \varepsilon$ **do**

$$u_{ij} = \Big[\sum_{k=1}^{c}\big(\frac{||X_j - v_i||}{||X_j - v_k||}\big)^{\frac{2}{m-1}}\Big]^{\frac{1}{m}}, \quad \forall i,j \tag{4}$$

$$V_i = \frac{\sum_{j=1}^{n}(u_{ij})^m x_j}{\sum_{j=1}^{n}(u_{ij})^m}, \quad \forall i \; ; \quad (0 < \frac{1}{c}\sum_{i=1}^{c} u_{ij} \leq 1). \tag{5}$$

end while

$V = \{v_1, v_2, ..., v_c\}$ is the vector of c cluster centroids in $\Re^d$, m is the fuzzification constant, U is the $(c \times n)$ partition matrix, and $||.||_A$ is any inner product A-induced norm. Equations (4) and (5) show how the algorithm calculates (u_{ij}) and how the prototypes (v_i) will be updated in each iteration. The algorithm runs until reaching the condition:

$$max_{1\leq k\leq c}\{||V_{k,new} - V_{k,old}||^2\} < \varepsilon \tag{6}$$

The value assigned to ε is a predetermined constant that varies based on the type of objects and clustering problems. The proposed algorithms are designed to detect zero-day cyber attacks (new attacks) using partitioning methods to evaluate whether the new packet or transaction is normal or not.

4   Experimental Verification

In the first scenario for the risk management's analysis, a dataset has been chosen with regard to accuracy's assessments of the proposed method in classification problems. The proposed algorithm was designed to cover both feature and vector spaces to evaluate objects and their behaviour with respect to different dimensions. According to the better results obtained by BFPCM in compare with other learning methods, we can conclude that the proposed algorithm can satisfactorily classify the labelled data objects. This stage of experiment was designed with respect to the nature of misused-based analysis which performs on known threat and attacks. Then, for the next experiment, an arbitrary benchmark dataset has been chosen to evaluate the functionality of the second algorithm (BFPM) which is designed for unsupervised learning methods from two different perspectives. In the first step the accuracy of the proposed algorithm is considered, and in the second step, the ability of the proposed method in prediction strategy has been considered. According to the very good results obtained by BFPM, we concluded that the proposed method is capable of providing a flexible search space for both supervised and unsupervised learning strategies with promising results. For the final stage, the proposed method evaluated the potential ability of each data object on their movement from one cluster to another. In other words, an arbitrary dataset has been chosen to evaluate the functionality of the proposed method in terms of prevention and prediction strategies.

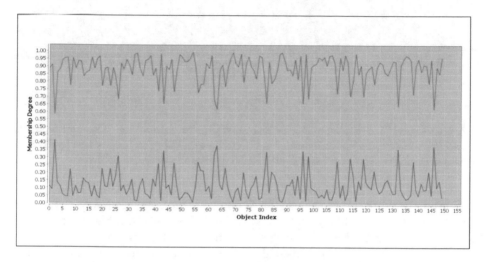

Fig. 1. Objects' movements analysis from one cluster to another. The plot presents how likely objects are about to move to other cluster. The plot is obtained by fuzzy methods, which most of the objects are completely separated.

Figures 1 and 2 demonstrate how likely objects are willing to move from one cluster to another by small changes in their feature spaces by fuzzy and BFPM methods respectively. Horizontal axes depict data objects and vertical axes present membership values obtained by data objects. The upper points (red points) present the memberships of objects with respect to the cluster that data objects are clustered in, and the lower points (blue points) are the objects' memberships with regard to the closest clusters that objects might move to by small changes in their feature spaces. Figure 1, which is obtained from fuzzy method, shows fuzzy methods assign the proper membership values to data objects in order to cluster them in precise clusters. There is no concern about the closest clusters for data objects by these types of methods, although data objects are clustered with desirable memberships.

Figure 2 presents data objects and their memberships obtained by BFPM with respect to two clusters: the cluster that data objects are clustered in and the closest cluster. The proposed method allows objects to show their potential ability to participate in other clusters instead of putting constraints to cluster them in normal or abnormal clusters. This flexible search space allows to track

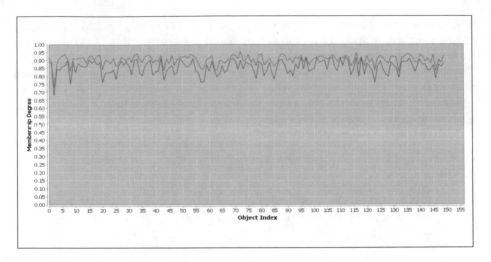

Fig. 2. Objects' movements analysis from one cluster to another. The plot presents how likely objects are about to move to other cluster. The plot is obtained by BFPM, presenting which objects are considered in high risks for the near future.

and study the objects' movement for the near future. As the figure shows, the method presents how likely each object is about to move to other cluster which is very important for intrusion detection and prevention systems.

As Fig. 1 shows, the memberships obtained by each data object are totally separated which is the property of fuzzy methods that separate objects in order to cluster them after some iterations. Contrary to Fig. 1, Fig. 2 depicts that membership degrees from different clusters can be close to each other and small changes in even one feature value may result in objects movements.

5   Conclusion

The paper discussed the importance of using sophisticated methods on analysing packets and online transactions to protect systems from any fraud or attack, where the security systems on the server side cannot easily detect such abnormalities. The paper proposed two different strategies in supervised and unsupervised concepts by offering two different algorithms to handle misused-based and anomaly-based detection techniques. The proposed method aims to detect and prevent attacks according to both categories, in addition to analyse the behaviour of each packet or transaction for the near future. Security maps and plans are capable of securing the systems from attacks when the format of the transaction are mostly changed, but for analysing the correct transactions, systems need to make use of learning methods to evaluate the behaviour of the packets or transactions. The proposed method presented how learning methods can detect and prevent anomalies, abnormalities, intrusions, and attacks by studying the object movements in advance.

References

1. Edge, K., Raines, R., Grimaila, M., Baldwin, R., Bennington, R., Reuter, C.: The use of attack and protection trees to analyze security for an online banking system. In: Proceedings of the Annual Hawaii International Conference on System Sciences, p. 144b. IEEE (2007)
2. Chio, C., Freeman, D.: Machine Learning and Security. O'Reilly (2017)
3. Buczak, A.L., Guven, E.: A survey of data mining and machine learning methods for cyber security intrusion detection. IEEE Commun. Surv. Tutorials $18(2)$, 1153–1176 (2016)
4. Hoppner, F.: Fuzzy Cluster Analysis: Methods for Classification, Data Analysis and Image Recognition. Wiley (1999)
5. Cannon, R.L., Dave, J.V., Bazdek, J.C.: Efficient implementation of the fuzzy c-means clustering algorithms. IEEE Trans. Patt. Anal. Mach. Intell. **PAMI–8**(2), 248–255 (1986)
6. Anderson, D.T., Bezdek, J.C., Popescu, M., Keller, J.M.: Comparing fuzzy, probabilistic, and possibilistic partitions. IEEE Trans. Fuzzy Syst. $18(5)$, 906–918 (2010)
7. Yazdani, H.: Fuzzy possibilistic on different search spaces. In: Proceedings of the International Symposium on Computational Intelligence and Informatics, pp. 283–288. IEEE (2016)
8. Cao, B., Fan, Q.: The infrastructure and security management of mobile banking system. In: IEEE International Conference on E-Service and E-Entertainment, pp. 1–3 (2010)
9. Paliwal, S., Gupta, R.: Denial-of-Service, probing and remote to user (R2L) attack detection using genetic algorithm. Int. J. Comput. Appl. $60(19)$, 57–62 (2012)
10. Shon, T., Moon, J.: A hybrid machine learning approach to network anomaly detection. J. Inf. Sci. $177(18)$, 3799–3821 (2007)
11. Tsai, C.F., Hsu, Y.F., Lin, C.Y., Lin, W.Y.: Intrusion detection by machine learning: a review. Expert Syst. Appl. $36(10)$, 11994–12000 (2009)
12. Ahmed, M., Mahmood, A., Hu, J.: A survey of network anomaly detection techniques. J. Netw. Comput. Appl. **60**, 19–31 (2016)
13. Aamir, R., Ashfaq, R., Wang, X.Z., Huang, J.Z., Abbas, H., He, Y.L.: Fuzziness based semi-supervised learning approach for intrusion detection system. J. Inf. Sci. **378**, 484–497 (2017)
14. Zhou, J., Chen, C.L.P., Chen, L., Li, H.X.: A collaborative fuzzy clustering algorithm in distributed network environments. IEEE Trans. Fuzzy Syst. $22(6)$, 1443–1456 (2014)
15. Masduki, B.W., Ramli, K., Saputra, F.A., Sugiarto, D.: Study on implementation of machine learning methods combination for improving attacks detection accuracy on intrusion detection systems (IDS). In: International Conference on Quality in Research, pp. 56–64. IEEE (2015)
16. Jeya, P.G., Ravichandran, M., Ravichandran, C.S.: Efficient classifier for R2L and U2R attacks. Int. J. Comput. Appl. $45(21)$, 28–32 (2012)
17. Kiljan, S., Eekelen, M.V., Vranken, H.: Towards a virtual bank for evaluating security aspects with focus on user behavior. In: SAI Computing Conference, pp. 1068–1075. IEEE (2016)
18. Yazdani, H., Ortiz-Arroyo, D., Choroś, K., Kwaśnicka, H.: Applying bounded fuzzy possibilistic method on critical objects. In: Proceedings of the International Symposium on Computational Intelligence and Informatics, pp. 271–276. IEEE (2016)

19. Yazdani, H., Kwaśnicka, H.: Fuzzy classification method in credit risk. In: Proceedings of the International Conference on Computational Collective Intelligence. Lecture Notes in Computer Science, vol. 7653, pp. 495–505. Springer (2012)
20. Yazdani, H., Ortiz-Arroyo, D., Choroś, K., Kwaśnicka, H.: On high dimensional searching space and learning methods. In: Data Science and Big Data: An Environment of Computational Intelligence, pp. 29–48. Springer (2016)

Prediction of Autism Severity Level in Bangladesh Using Fuzzy Logic: FIS and ANFIS

Rahbar Ahsan, Tauseef Tasin Chowdhury, Wasit Ahmed, Mahrin Alam Mahia, Tahmin Mishma, Mahbubur Rahman Mishal, and Rashedur M. Rahman[✉]

Department of Electrical and Computer Engineering, North South University, Plot-15, Block-B, Bashundhara Residential Area, Dhaka, Bangladesh
{rahbar.ahsan,tauseef.tasin,ahmed.wasit,mahrin.mahia, tahmin.mishma,mishal.rahman,rashedur.rahman}@northsouth.edu

Abstract. A type of neurodevelopment disorder also known as autism is currently more visible than before among the people of Bangladesh. Some research works could be found on autism but very few papers are guided to measure the severity level. Hence, this research focuses on attaining the severity level of autism using fuzzy methods like Mamdani Fuzzy Inference System (MAMFIS) and Adaptive Neuro-Fuzzy Inference System (ANFIS). A survey has been conducted on autistic children to find the severity level. The levels used in this research are low, medium, high. A comparative study of those two methods has been reported in this paper. By using ANFIS we get better accuracy compared to the FIS model.

Keywords: Autism; ASD · Fuzzy · Severity · Fuzzification · Rule generation
Defuzzification · FIS · ANFIS · Triangular function

1 Introduction

Autism Spectrum Disorder (ASD) is one kind of disorder which causes problems like social interaction, communication, learning disabilities etc. According to [1], autism may affect people of all races, ethnic and socio-economic conditions. This is a life-long brain disorder that is usually detected in early childhood. People with autism have problems in interaction, dealings with others and have difficulties in comprehending the world around them. Autism is a syndrome varying in severity and influence from person to person. There may be people with no speech, learning disabilities, having unusual patterns of language, behaviors etc. Also, it is very important to detect autism spectrum disorder at younger ages because young toddlers tend to harm themselves in many ways. About 28% of the affected children unintentionally/unknowingly harm themselves by head banging, hair pulling, arm biting, eye poking, skin scratching etc. [2].

From [3], social interaction problems can be diagnosed preliminary at the age of six to ten months old. For example, prefer to be alone, not responding to his/her name and avoiding physical contacts etc. are the social interaction problems. Also, about 40% children with ASD do not like to talk and some of them like 25% to 30% learn something but they forget as days go by. There are some patterns in behavior which are found in

© Springer Nature Switzerland AG 2019
K. Choroś et al. (Eds.): MISSI 2018, AISC 833, pp. 201–210, 2019.
https://doi.org/10.1007/978-3-319-98678-4_22

kids with ASD. Eating quickly, very sensitive to noise, problems with coordination are some parameters of patterns of behavior.

With the help of core symptoms of ASD and [4], in this research a similar questionnaire has been made to conduct a survey among the teachers of children with ASD who are the current students of four institutions in Bangladesh. The institutions are 'Proyash', 'Institute of Special Education', 'Seher Autism Center', and 'Alokito Shishu'.

According to [5], fuzzy logic is a standard logic which is used to determine states between true and false. The performance of fuzzy logic is significant where more degree of membership is possible. Based on this characteristic of fuzzy logic, the severity level of autism in children can be predicted. This paper focuses on 2 fuzzy methods: Fuzzy Inference System (FIS) and Adaptive Neuro Fuzzy Inference System (ANFIS) which are applied on 98 children with ASD.

2   Background

It is a matter of sorrow that Bangladesh do not have any nationwide statistics of autistic children, but compared to our population, almost 550000 children with ASD could be found [6]. For their education at least 5000 to 10000 special schools are required but only a few private sectors are working in this domain. According to [1], in those existing schools, parents do not want to admit their autistic child because most of the teachers are not properly trained on dealing autism. As, there is no professional training for teachers.

In addition, doctors were not conscious of the clinical features of autism and its management in the past years. This creates misdiagnosis and mistreatment. Also, according to [7], one mistreatment of autism can cause severe diseases which may lead to death. Initial diagnosis of children with autism and appropriate treatment thereof is important to grow their full potential.

From 1990, the treatment of children with ASD was started to reflect in Dhaka Shishu Hospital [1]. Some private sectors like Society for the Welfare of Autistic Children (SWAC), Autistic Welfare Foundation (AWF), PROYASH and others started their service for children with ASD in the year 2000. Consciousness for autism was heightened in Bangladesh when the Centre for Neurodevelopment & Autism in Children was established in the Bangabandhu Sheikh Mujib Medical University. The National Advisory Committee on autism has been formed to provide awareness and services to autistic children. Four task forces are also working under the guidance of the committee. Now, activities on autism has gained momentum in Bangladesh. But still, some people specially people leaving in rural area feel humiliated about autism which causes mental barrier to the promotion of scientific management of ASD [8]. However, the Govt. of Bangladesh has made an incredible achievement in reducing stigma related to autism in last four years.

3   Related Works

The authors in [4] proposed a neuro-fuzzy model by applying artificial neural network (ANN) technique. This model helps to diagnosis autism. They focused on 11 core symptoms of autism. To fuzzify the inputs, they used triangular membership functions. After fuzzification, they trained their network through back propagation training algorithm and the performance of the model was in the range of 85%–90% which can be increased by more record sets for training. They also proposed that using classification of spectrum disorder, the work can be enhanced.

In [9], the authors used fuzzy cognitive maps (FCMS) for processing the diagnosis of Autism Spectrum Disorder (ASD). According to the authors, the process of M-CHAT was very complex. M-CHAT only deals with crisp value that is why the process of M-CHAT can be hard to detect autism state in children. As, FCM can handle high levels of uncertainties, deliver accurate results and compute human behavior, the authors used it for reducing the complexity of M-CHAT. They compared their results with some previous works and increased the accuracy from 79% to 85.04%.

The authors in [10] used fuzzy c-means clustering (FCM) for getting the structural information of sub cortical regions of brain such as Corpus Callosum (CC) and Brain Stem (BS) from magnetic resonance imaging for creating a framework which will be able to detect autism automatically.

Similar to our work in another paper [11] researchers detected severity level of autism by using fuzzy expert system. Their work was divided into two parts: data acquisition and fuzzy system architecture. In data acquisition part, they made a questionnaire on the basis of interview sessions with psychologists and distributed it among the parents and teachers of the autistic children. The fuzzy system architecture includes fuzzification, rule evaluation, aggregation of rule output and defuzzification. They categorized the severity level of autism into mild, moderate and severe. The accuracy of their result was more than 60%. But their dataset size was only 36 and they gained accuracy of about 60%. In this paper we work with a larger dataset of autistic children of Bangladesh and the accuracy we have got is much higher compared to 60%. Besides, we vary with different membership functions in FIS model as well use ANFIS in our current attempt.

4   Methodology

4.1   Data Acquisition

The severity level of autism can be identified in many ways. Since autism is a spectrum disorder, a person can be mildly, moderately or severely autistic. There are some core problems which have a dominant factor in the detection of severity level. Key features which are taken into consideration are given in Table 1.

Table 1. Key features/symptoms used in the questionnaire.

#	Key features/Symptoms
1	Understanding capability
2	Eye contact
3	Interacting capability with other children
4	Play kit recognition
5	Interested in gadgets
6	Responsiveness
7	Oversensitive to noise
8	Play pretend or make-believe
9	Meaningless finger movements
10	Attention seeking capability

A questionnaire is made focusing on these core problems and distributed among the teachers of special child school "Proyash", "Seher Autism Center" and "Alokito Shishu". Parents are the perfect candidate to fill the questionnaire for their children but the parents of autistic children in Bangladesh are very conservative about sharing any information about their child. Apart from parents, teachers are the best candidate for filling up questionnaire because they also have the first-hand experience with ASD child. Total 98 questionnaires were distributed among the teachers of both school and they had filled each questionnaire based on responses of individual ASD student.

4.2 Mamdani Fuzzy Inference System

Fuzzy logic can solve decision-making problem by using human critical reasoning capability. In real life, the decision-making process is not precise and it can be expressed as numeric and mathematical terms using a fuzzy linguistic variable. For designing such a complex system, fuzzy logic is a better choice [12].

Mamdani fuzzy inference system is one of the most used inference systems since it furnishes sensible outcomes with a moderately straightforward structure, and furthermore due to the instinctive and interpretable nature [13].

Fuzzification. Firstly, the user will fill the set of questions and for choosing each option they need to rate that option between 1 to 3. Here 3 means extreme and 1 means least. After that the system will automatically fuzzify the input alongside its accurate membership value. Here Triangular membership function is being used for both input and output.

$$F(x,p,q,r) = \max\left(\min\left(\frac{x-p}{q-p}, \frac{r-p}{r-q} \right), 0 \right) \tag{1}$$

Here p, q, r are the parameters for adjusting the membership function of x.

Figure 1 depicts some of the input membership function where there is a certain amount of overlapping occur due to no fine cut off point for an option.

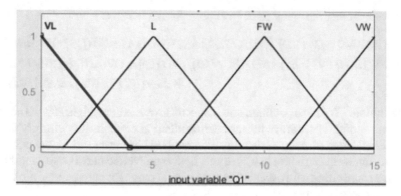

Fig. 1. One of the input membership functions using triangular function.

The linguistic variable has been used for representing output as Low, Medium, High. Further data analysis and Table 2 shows that medium autistic level is very co-related to Low and High level due to that reason medium level membership function has a substantial amount of overlap with other two regions Fig. 2.

Table 2. Range for output linguistic variables.

Fuzzy variable	Range
Low	0–0.25
Medium	0.25–0.75
High	0.75–1.0

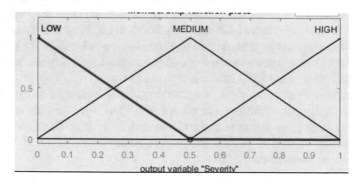

Fig. 2. Output membership function using triangular function.

Rule Generation. Rules are generated based on focusing major 10 problematic areas of detecting autism severity. A rule R_α can be expressed as

$$R_\alpha = if\ (OP_1\ is\ A_1)\ AND\ (OP_2\ is\ A_2)\ AND\ \dots\ (OP_n\ is\ A_n)\ then\ (SV\ is\ B) \qquad (2)$$

Where Rα is implication relation and AND is the operation of corresponding ELSE from fuzzy algorithm that take the min value among all options.

One rule among the many rules have been shown in Eq. (3)

$$if (Q1 \; is \; FW) \; AND \; (Q2 \; is \; UD) \; AND \; (Q3 \; is \; L) \; AND \; (Q4 \; is \; A) \; AND \; (Q5 \; is \; VL) \; AND$$
$$(Q6 \; is \; FW) \; AND \; (Q7 \; is \; LS) \; AND \; (Q8 \; is \; UD) \; AND \; (Q9 \; is \; VO) \; AND \; (Q10 \; is \; UD) \qquad (3)$$
$$then \; (SEVERITY \; is \; MEDIUM)$$

Defuzzifiction. We have got linguistic output as low, medium, high. But we need crisp value as an output. There are number of defuzzification methods for defuzzify the fuzzy value. Here 'Center of Area' Defuzzification method has been used for the defuzzification. Here, linguistic variables low, medium, high is converted into a crisp value between 0 to 1. After consulting with psychiatrist, we have set range for output linguistic variables which are shown in Table 2.

4.3 ANFIS (Adaptive Network-Based Fuzzy Inference System)

One of the benefits of using neuro-fuzzy systems is that they are more systematic. Additionally, they are less dependent on expert knowledge which contributes to this system being capable of handling complex and nonlinear problems. There has been a significant reduction in the quantity of if-then rules by introducing weighting factors for inputs.

The membership function parameters that best allow the associated Fuzzy Inference System (FIS) to track the given input and output data, in order to compute these function parameters, ANFIS provides a method for the fuzzy modeling procedure to learn information about a data set.

For this system, ANFIS model is constructed in a MATLAB environment. The input/output data is used to generate the rules and the membership values. After analyzing input dataset, membership values are assigned to each input data using triangular membership function. Since we have many attributes in our input set, subtractive clustering algorithm was used to generate the FIS as it is more efficient than the grid partition algorithm when generating rules and membership functions for a large number of attributes. The structure of the ANFIS model is showed in Fig. 3.

Hybrid method is chosen, which is a combination of least squares estimation and backward propagation to optimize and train our ANFIS as it gives us better results than using only the back-propagation method. Gaussian membership functions are used in the generated FIS. A total of 65 rules are used to evaluate the output. Figure 4 shows the rule viewer of the generated FIS.

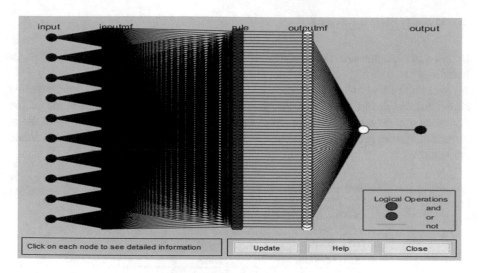

Fig. 3. ANFIS structure.

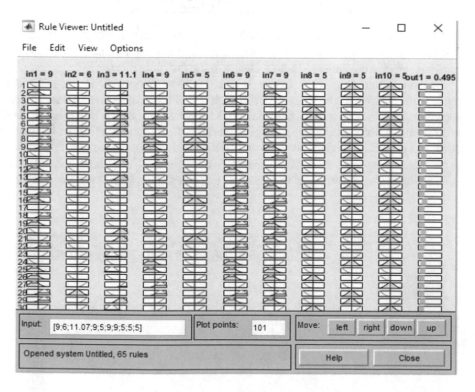

Fig. 4. Rule viewer from the model.

5   Result and Analysis

In this model, input parameters which are mentioned in Table 1 are the main 10 problematic areas. The output will be low, medium and high.

Mamdani fuzzy inference system is being validated with the data taken from the questionnaire. It has been seen that predicted result is quite accurate. Some of the predicted and actual output is shown in Table 3. It has been seen that Mamdani fuzzy inference system predicted result has an accuracy of 80%. All 98 data used to validate the result.

Table 3. Comparison of predicted output and actual output.

No	Actual output	Predicted output
1.	Medium	Medium
2.	High	Medium
3.	Medium	Low
4.	High	High
5.	Medium	Medium

When all data has been used to train ANFIS low, medium and high considered as a fixed point 0.3, 0.5 and 0.8 respectively and accuracy have been found from the differences of those fixed points and finding mean among the differences.

For ANFIS, out of the whole dataset, 60% are used to train the ANFIS and 40% are used to validate it. We run our system for 100 epochs. Testing error remained constant after epoch changed. The accuracy gained from the ANFIS model is 82% which is better than the 80% we got from FIS. In Fig. 5, the plotting of predicted and actual output is shown. RMSE has been calculated for both FIS and ANFIS with the remaining 40% data which has been used as a testing set in ANFIS model. In both cases, RMSE has been found to be 0.200. By observing the predicted results for 40 test cases, it has been found that there are

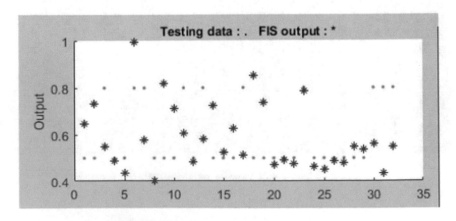

Fig. 5. Predicting and actual output for testing data in ANFIS.

5 test cases where a comparitive large difference exists between predicted and actual output in ANFIS. Due to this reason ANFIS has samilar RMSE as FIS.

6 Discussion and Future Work

Autistic severity level detection is a complex process which requires series of assessment and close observation. Early detection of autism severity will be very helpful for the parents to treat their children with more intensive care. Adaptive Fuzzy Neuro Infcrence System and Mamdani Inference System are used to model Autism severity detection system by focusing on 10 major core problems. In Mamdani FIS Inference System we need to generate rules but in ANFIS system generate rules by analyzing the dataset. ANFIS has been seen gaining better result than Mamdani FIS Inference System. Better accuracy of ANFIS validates that analyzing the dataset creates more precise rules.

By this model, it has been seen that how human intuition can be modeled mathematically using fuzzy logic which can validate real-world data. This system will also help the practitioners and the parents detecting autism severity level early. Early detection is very useful to decide required therapy for the ASD children. It also has gained better result than the previous study in the context of accuracy in both FIS and ANFIS. The comparison between two inference system also shows that using ANFIS will be a better approach than Mamdani FIS. In future, we plan to increase our dataset which will eventually help ANFIS to generate more accurate rules which will lead to achieve better accuracy in prediction.

Acknowledgments. First, we would like to thank Proyash, Institute of Special Education, Alokito Shishu, Seher Autism Center for helping us on data collection. We want to thank psychiatrist from Proyash for guiding us while making the rule set. Finally, we want to thank all the teachers who had participated in the survey.

References

1. Autism awareness in Bangladesh and its challenges: The Independent, 11 April 2016. http://www.theindependentbd.com. Accessed 1 May 2018
2. Soke, G.N., Rosenberg, S.A., Hamman, R.F., Fingerlin, T., Robinson, C., Carpenter, L., Giarelli, E., Lee, L.-C., Wiggins, L.D., Durkin, M.S., Diguiseppi, C.: Brief report: prevalence of self-injurious behaviors among children with autism spectrum disorder—a population-based study. J. Autism Dev. Disord. **46**(11), 3607–3614 (2016)
3. What Are the Symptoms of Autism? WebMD. https://www.webmd.com/brain/autism/symptoms-of-autism. Accessed 06 May 2018
4. Arthi, K., Tamilarasi, A.: Prediction of autistic disorder using neuro fuzzy system by applying ANN technique. Int. J. Dev. Neurosci. **26**(7), 699–704 (2008)
5. Tsoukalas, L.H., Uhrig, R.E.: Fuzzy and neural approaches in engineering. Wiley, New York (1997)
6. S. U. Ahmed: Situation analysis of autistic children. The Independent, 17 April 2016. http://www.theindependentbd.com. Accessed 1 May 2018

7. What Are the Treatments for Autism? WebMD. https://www.webmd.com/brain/autism/autism-treatment-overview. Accessed: 08 May 2018
8. Pervin, M.M.: Autism: The role of the family. The Daily Star, 5 April 2016. http://www.thedailystar.net. Accessed 4 May 2018
9. Farsi, A.A., Doctor, F., Petrovic, D., Chandran, S., Karyotis, C.: Interval valued data enhanced fuzzy cognitive maps: torwards an appraoch for Autism deduction in Toddlers. In: 2017 IEEE International Conference on Fuzzy Systems (FUZZ-IEEE) (2017)
10. Fredo, A.R.J., Kavitha, G., Ramakrishnan, S.: Analysis of sub-cortical regions in cognitive processing using fuzzy c-means clustering and geometrical measure in autistic MR images. In: 2014 40th Annual Northeast Bioengineering Conference (NEBEC) (2014)
11. Isa, N.R.M., Yusoff, M., Khalid, N.E., Tahir, N., Nikmat, A.W.B.: Autism severity level detection using fuzzy expert system. In: 2014 IEEE International Symposium on Robotics and Manufacturing Automation (ROMA) (2014)
12. Yilmaz, A., Ayan, K., Adak, E.: Risk analysis in cancer disease by using fuzzy logic. In: 2011 Annual Meeting of the North American Fuzzy Information Processing Society (2011)
13. Jassbi, J.J., Serra, P.J.A., Ribeiro, R.A., Donati, A.: A comparison of mandani and sugeno inference systems for a space fault detection application. In: 2006 World Automation Congress (2006)

Natural Language Processing and Information Retrieval

A Framework for Analyzing Academic Data

Dinh Tuyen Hoang[1], Trong Hai Duong[2], Ngoc Thanh Nguyen[3],
and Dosam Hwang[1(✉)]

[1] Department of Computer Engineering, Yeungnam University,
Gyeongsan, South Korea
hoangdinhtuyen@gmail.com, dosamhwang@gmail.com
[2] Institute of Science and Technology of Industry 4.0, Nguyen Tat Thanh University,
Ho Chi Minh City, Vietnam
haiduongtrong@gmail.com
[3] Faculty of Computer Science and Management,
Wroclaw University of Science and Technology, Wrocław, Poland
Ngoc-Thanh.Nguyen@pwr.edu.pl

Abstract. In recent years, academic data is growing rapidly in terms of volume, variety, velocity, value, and reliability. Analysing and managing it is therefore more difficult and challenging. Discovery of academic data can yield great benefits to the scientific community. In addition, academic data analysis helps to plan and orient development for research and industry. Fortunately, there are many scholarly data resources available such as the *DBLP Computer Science Bibliography*, *ResearchGate*, *CiteSeer*, and *Google Scholar*, which enable users easy access and analysis. In this paper, we propose a framework for analysing representative research issues in an academic context. We are currently in the process of building a system for collaborator recommendation, academic venue recommendation, expert finding, and group expert prediction, which are the primary issues in an academic context.

Keywords: Academic framework · Analyzing academic data
Data analysis

1 Introduction

Academic data has been growing very rapidly in recent years. Working out methods for managing and analysing such data is becoming increasingly challenging and complicated. Academic data includes information such as authors, co-authors, papers, citations, and academic networks. There are various academic data resources such as the *DBLP Computer Science Bibliography*, *Cite-Seer*, *ResearchGate*, and *Google Scholar*, which allow users easy access and analysis. The exploration of academic data has several important advantages. For instance, one can measure the influence of a given paper by examining the citation links collected from a large set of papers. It helps in determining the

© Springer Nature Switzerland AG 2019
K. Choroś et al. (Eds.): MISSI 2018, AISC 833, pp. 213–223, 2019.
https://doi.org/10.1007/978-3-319-98678-4_23

credibility of researchers. Thus, academic data analysis is not only crucial for educational purposes but also helps to direct future scientific development. Obtaining relevant and valuable data has become more complicated due to information overload.

Fortunately, due to the growing volume of academic data available for collection, it is possible to build frameworks for resolving different issues such as collaborator recommendation, expert finding, and academic venue recommendation. Several previous works proffered different solutions to the problem of sourcing useful academic data [1,3,4,10].

Some systems have also been built for sourcing academic data such as *ResearchGate*, *CiteSeer*, and *Google Scholar*. However, these systems try to provide statistical information regarding the publications rather than analyse their content. For example, *DBLP* built co-author and co-venue networks while *ResearchGate* built an academic social network. However, these systems do not provide an application programming interface (API) for developers and it is difficult to understand the mechanism of the systems.

In this paper, we propose a framework for analysing representative research issues in an academic context. We present concrete methods for collaborator recommendation, academic venue recommendation, expert finding, and group expert prediction, which are the primary issues in an academic context. We are developing a framework based on the proposed methods. The rest of this paper is structured as follows. In Sect. 2, related works are summarised to provide general knowledge about academic networks. In Sect. 3, the framework for analysing academic data is described including data collection, collaborator recommendation, academic venue recommendation, expert finding, and expert group prediction. Lastly, conclusions and future work are presented in Sect. 4.

2   Related Works

With the rapid developments in academic data sources, many researchers have recognised the need for educational data analysis. The subjects and connections in the network of scientific subjects are as shown in Fig. 1.

Klamma et al. built an academic event recommendation system based on shared interests [6]. They check authors' event participation to measure similarities in their research interests. The recommendation system considers similar interests and generates a set of events that authors might wish to attend.

Newman [9] conducted statistics on the features of scientific collaboration to demonstrate that the number of collaborators, on average, is different if the researchers have different research topics.

Li et al. [8] examined how to find cooperation for scholars in a network of co-authors. Some indicators were reviewed through paper co-authoring information, which helped determine the importance of the links. The authors analysed the DBLP data to measure the propensity for cooperation.

Hoang et al. [5] created a graph of researchers by considering co-author networks and their publications. The weight of the graph represented the relationship between the researchers.

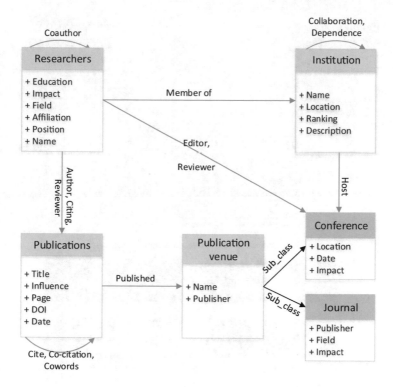

Fig. 1. A relational database among scientific objects

Some academic systems have also been developed for discovering academic data such as *DBLP*, *CiteSeer*, and *Google Scholar*. However, the system descriptions as well as the algorithms used have been kept secret by most of them. Moreover, some systems do not provide an API for researchers to collect these relevant data.

3 The Framework for Analysing Academic Data

3.1 Data Collection

Academic data has increased rapidly in recent years. A variety of academic data resources such as DBLP Computer Science Bibliography, CiteSeer, Google Scholar, and ResearchGate are available which allow users easy access and crawl. The data collection module we propose generates a database of researchers' profiles by collecting and importing data from different data resources related to academic endeavours. Exploring academic data offers significant advantages. The information about authors, co-authors, papers, citations, academic networks, etc. can be extracted from the academic data. Generally, there exist two major types of data sources for academic data collection, which are open databases and the web.

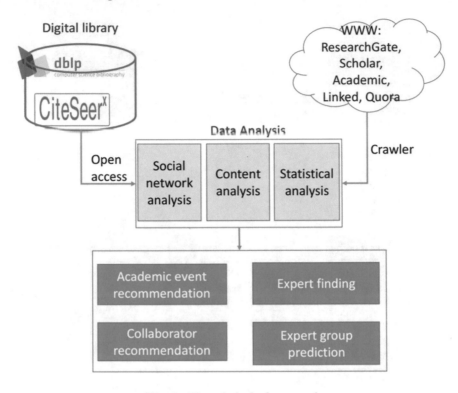

Fig. 2. The scholarly framework

Open databases, such as the DBLP Computer Science Bibliography, and Cite-Seer are well defined and structured, making it easy for the extractor process to obtain the required information. Moreover, a rich source of unstructured data can be collected from the web using crawler libraries. For example, Python provides a library named Scrapy[1] for crawling data from the internet. This data is put into an extractor for processing and then imported into the proposed database system. One major problem with academic data is the author name ambiguity. There may exist more than one researcher sharing the same distinct name, or one researcher with several names. This situation confuses the system and reduces the accuracy of results. Therefore, reviewing ambiguity in authors' names is a crucial require-ment. The name ambiguity module considers the features that can be extracted from publications such as affiliations, co-authors' names, research topics, and key-words. To improve the efficiency of the name ambiguity module, we set matching rules for each researcher's name retrieved from the database. If any rule adequately distinguishes the authors of a paper, the rest of the rules are not executed. Once all the rules have been executed, the next publication is considered for all authors independent of any previously published papers. We set the order for matching

---

[1] https://scrapy.org/.

the rules as follows: the first name and last name are priority-checked, and then the affiliation is tested. Thereafter, the co-authors are examined and the scientific fields and keywords considered (Fig. 2).

3.2   Collaborator Recommendation

In this subsection, we present a collaborator recommendation method by considering the previous research cooperation network and the similarity between these research topics. We create a weighted directed graph $G(V, E, W)$, where V is a set of nodes (a set of researchers), E is a set of directed edges, and W represents the set of values for the collaboration strength between researchers. We attempt to discover new collaborators by considering direct and indirect connections between the researchers. The details of this proposed method are as shown in Fig. 3. Let $R = \{r_1, r_2, ..., r_N\}$ be a set of researchers, and $V = \{v_1, v_2, ..., v_M\}$ be a set of academic events. We need to find a set R_K of researchers where a target researcher, r_i should explore the collaboration, so that $R_K \subset R$.

Definition 1. *A research collaboration can be understood as the working together of researchers to reach the common goal of creating new scientific knowledge.*

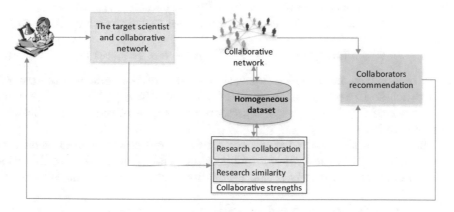

Fig. 3. The workflow of the collaborator recommender system

Definition 2. *The research cooperation $C(r_i, r_j)$ between researchers r_i and r_j is computed as follows:*

$$C(r_i, r_j) = \frac{1}{2} \times \left(\frac{|r_i^P \cap r_j^P|}{|r_i^P|} + \frac{|r_i^V \cap r_j^V|}{|r_i^V|} \right) \tag{1}$$

where $|r_i^P|$ and $|r_j^P|$ are the numbers of published papers for researchers r_i and r_j, and $|r_i^V|$ and $|r_j^V|$ are the number of attended academic events for researchers r_i and r_j, respectively. The value of $C(r_i, r_j)$ is 0 if researchers r_i and r_j had never written the same papers or attended the same academic events.

Conversely, if a researcher r_j had written all papers and attended all events with target researcher r_i, then the value of $C(r_i, r_j)$ is 1. The higher the value of $C(r_i, r_j)$, the higher the probability that researchers r_i and r_j will collaborate in the near future.

Definition 3. *The research similarity $R(r_i, r_j)$ between researchers r_i and r_j is calculate as follows:*

$$R(r_i, r_j) = \sum_{p_i \in r_i{}^P, p_j \in r_j{}^P} \frac{sim(p_i, p_j)}{|r_i{}^P| \times |r_j{}^P|} \times \sum_{v_i \in r_i{}^V, v_j \in r_j{}^V} \frac{sim(v_i, v_j)}{|r_i{}^V| \times |r_j{}^V|} \qquad (2)$$

where $sim(p_i, p_j)$ returns the content similarity between publications p_i and p_j, and $sim(v_i, v_j)$ returns the value of the content similarity between conferences v_i and v_j. In order to compute the content similarity between between publications p_i and p_j, we use Doc2Vec [7] to convert them to vectors. Then, the cosine similarity between the two vectors is used to calculate the value of the function $sim(p_i, p_j)$. From the content similarity between two events, we extract textual descriptions and topics of interest from events in the *WikiCFP* data to compute the value of the function $sim(v_i, v_j)$. The higher the value of $R(r_i, r_j)$, the closer the topic research between those researchers.

Definition 4. *The collaboration strength $T(r_i, r_j)$ between researchers r_i and r_j is computed by combining the research cooperation and the research similarity as follows:*

$$T(r_i, r_j) = (1 - \alpha) \times C(r_i, r_j) + \alpha \times R(r_i, r_j) \qquad (3)$$

where α is a parameter which controls the rates of the influence of research cooperation and research similarity on $T(r_i, r_j)$; and ($\alpha \in [0, 1]$). $T(r_i, r_j)$ which has a range of $0 \leq T(r_i, r_j) \leq 1$ represents the degree of relationship between those researchers.

The edges of the directed graph between researchers are created when they have collaborated on at least one publication. The value of the collaboration strength between researchers r_i and r_j for direct connections is set as follows:

$$D_1(r_i, r_j) = \begin{cases} T(r_i, r_j), & \text{if } r_i \text{ has collaborated on at least one publication with } r_j. \\ 0, & \text{otherwise.} \end{cases}$$

$$(4)$$

However, many researchers do not collaborate as direct connections, even though they may connect through mediate researchers. In that case, those researchers are deemed to have indirect connections. Let $R_m = \{r_1, r_2 ... r_m\}$ be a set of mediate researchers that are considered as intermediate nodes of the path from researcher r_i to researcher r_j. In order to reduce time complexity, the number of intermediate nodes should be smaller than a threshold. The collaboration strength between indirect connection researchers r_i and r_j is computed as follows:

$$D_2(r_i, r_j) = \begin{cases} Max_{k=1..p}\left(\prod_{k=1}^{m-1} T(r_k, r_{k+1})\right), & \text{if } r_i \text{ has indirect connection to } r_j. \\ 0, & \text{otherwise.} \end{cases} \tag{5}$$

When the number of mediate researchers between the two researchers with indirect connection is large, the value of $D_2(r_i, r_j)$ will be too small and time-consuming to calculate. Therefore, we set a threshold ($m \leq threshold, h$) for the number of mediate researchers when implementing the proposed method.

The recommendation score between researchers r_i and r_j can be calculated by combining Eqs. (4) and (5) as follows:

$$D(r_i, r_j) = (1 - \beta) \times D_1(r_i, r_j) + \beta \times D_2(r_i, r_j) \tag{6}$$

where the parameter β is used to adjust the weight between $D_1(r_i, r_j)$ and $D_2(r_i, r_j)$ based on the collaboration strength of those researchers and $\beta \in [0, 1]$. By ranking the recommendation scores $D(r_i, r_j)$ in descending order, we have a list of collaborators to recommend to the target researcher r_i.

3.3   Academic Venue Recommendation

In this subsection we present an academic event recommendation method based on the research similarity and interaction of researchers. Let $R = \{r_1, r_2, ..., r_N\}$ be a set of researchers, and $V = \{v_1, v_2, ..., v_M\}$ be a set of academic events. Let $V_F = \{v_{f1}, v_{f2}, ..., v_{fM}\}$ be a set of future academic events, $(V_F \subset V)$. Let $r_i(V)$ be a set of academic events that researcher r_i has attended, $(r_i(V) \subseteq V)$. This method focuses on finding a set of future academic events V_F that the target researcher r_i should attend or submit his/her publications.

Collaborative filtering methods have proved that people have a tendency for behaviours similar to their past actions, such as purchasing similar items and liking or disliking a similar set of products [2]. We use this result for our model by considering the content of events attended by target researcher, r_i. We collect the description of all the past events attended by target researcher, r_i and compute the similarity between them and future events v_f, $(v_f \in V_F)$ as follows:

$$F_1 = \frac{\sum_{v_i \in r_i(V)} sim(v_i, v_f)}{|r_i(V)|} \tag{7}$$

where the function $sim(v_i, v_f)$ returns the value of the content similarity between events v_i and v_f. One of the important features of the scientific community is that researchers tend to attend events also attended by their collaborators. Let $r_i(R_K)$ be a set of K researchers that have the largest collaboration strength with target researcher r_i, $r_i(R_K) \in R$. We calculate the similarity between all the conferences attended by $r'_i s$ collaborators (in the past) and future events v_f as follows:

$$F_2 = \frac{\sum_{r_j \in r_i(R_K)} \left(\frac{\sum_{v_j \in r_j(V)} sim(v_j, v_f)}{|r_j(V)|} \times T(r_i, r_j) \right)}{\sum_{r_j \in r_i(R_K)} T(r_i, r_j)} \tag{8}$$

where $|r_j(V)|$ is the number of events attended by collaborator r_j.

The recommendation score is determined by combining Eqs. (7) and (8), as follows:

$$F(r_i, v_f) = (1 - \gamma)F_1 + \gamma F_2 \tag{9}$$

where $\gamma \in [0, 1]$, identifies the impact of F_1 and F_2 on the recommendation score values. A set of recommended conferences for target researcher r_i is created based on the ranking of the recommendation score returned by Eq. (9).

We rank the recommendation score $F(r_i, v_f)$ in descending order to achieve a list of academic events to recommend for the target researcher, r_i.

Table 1. Features used in detecting experts

Feature name	Description
r_hindex	H-index
r_citation_count	Cumulative citation count
r_mean_citations_per_paper	Mean number of citations per paper
r_mean_citations_per_year	Mean number of citations per year
r_papers	Number of papers published
r_coauthor	Number of coauthors collaborated with
r_age	Career length (years since the first paper was published)
r_max_single_paper_citations	Max number of citations for any of the author's papers

3.4 Expert Finding

In this subsection, we present a method for finding experts to solve or answer a given query, Q. The query can be a question, a topic, or a paper. We propose a method which takes into account the researchers' scientific achievements and their behaviour to identify whether the researcher is an expert or a non-expert. We perform the expert finding using two approaches, where the first approach is a topic model. Let $T = t_1, t_2, ..., t_n$ be the set of all predefined topics. Then, we select topics similar to the given query Q and compute the weight of each researcher r_i on each topic t_i. The K researchers with the largest sum of weight will be selected. The second approach is known as the machine learning method. We extract features from the researchers' profile, as shown in Table 1 and convert these features into a form suitable for machine learning using the feature engineering technique. A number of machine learning algorithms are tried, and the most suitable algorithm selected through the ensemble method.

3.5 Expert Group Prediction

In this subsection, we propose an expert group prediction method by taking into account three criteria, including diversity and competence. When a request (paper, topic, or question) is submitted to the system, the relevant information about potential experts is discovered to find the best match. The request can be supplied to the system as a text query. An indexing method is needed to transform it into a vector that can be assigned directly to the system. We equally index the profile of the expert as a vector by applying the deep learning method proposed by Le and Mikolov [7]. Let V_Q be a vector that is converted from the text query and V_{r_i} be a vector that represents the profile of researcher, r_i. The content similarity between V_Q and V_{r_i} can be computed as follows:

$$Similarity(V_Q, V_{r_i}) = \frac{|V_Q \times V_{r_i}|}{|V_Q| \times |V_{r_i}|} \qquad (10)$$

The expert candidates list is determined through sorting the value of the similarities between the query, V_Q and the experts' profiles, V_{r_i} in descending order.

A list of N researchers may be selected to review a particular problem, Q. We consider three critical components for choosing a group of researchers: diversity, competence, and suitability.

The diversity criteria is preferred when choosing an expert group. To fit the diversity criterion, we require that the experts in the same group do not have the same affiliations or co-authors. After eliminating groups that do not satisfy the rule, the diversity value of each group (D_g) is computed as follows:

$$D_g = \frac{\sum_{i=1}^{K} \sum_{j=1}^{K} (1 - Similarity(V_{r_i}, V_{r_j}))}{\frac{1}{2} \times K \times (K - 1)} \qquad (11)$$

where $Similarity(V_{r_i}, V_{r_j})$ is the content similarity of experts r_i and r_j, and K is the number of members in the group.

The competence of an expert in each group is an influential factor affecting the quality of the review process. In order to compute the competence of an expert, we consider three main features: the number of publications, the number of citations, and the h-index. Usually, the number of citations for each author is higher than the number of papers. According to Thomson Reuters[2], the number of citations is four times the number of papers. Therefore, we put $\frac{1}{4}$ as a constant to normalise the competence function. The competence of a group (C_g) can be calculated as follows:

$$C_g = \frac{1}{1 + e^{-\sum_{i=1}^{K} \frac{1}{3}(|r_i{}^P| + \frac{1}{4}|r_i{}^C| + |r_i{}^H|)}} \qquad (12)$$

where $|r_i{}^P|$, $|r_i{}^C|$, and $|r_i{}^H|$ are the number of publications, the number of citations, and the h-index of researcher, r_i, respectively. We use the sigmoid

---

[2] https://www.timeshighereducation.com/news/citation-averages-2000-2010-by-fields-and-years/415643.article.

function to convert the value of the C_g range from 0 to 1, which is the domain range of the diversity value.

The suitability of an expert group for a given query Q is computed by summing the similarity values between the vector profiles of each member in the group with the vector, V_Q.

$$Sui\_g = \frac{1}{1 + e^{\sum_{i=1}^{K} Similarity(V_Q, V_i)}} \tag{13}$$

The recommendation score for the group of experts for given query Q is measured by combining the three component values as follows:

$$R\_score = \alpha' \times D_g + \beta' \times C_g + \gamma' \times Sui\_g \tag{14}$$

where $\alpha' + \beta' + \gamma' = 1$ is used to adjust the weight of D_g, C_g and Sui_g in the recommendation score, and R_score has a range $[0, 1]$. By ranking the R_score in descending order, we obtain a list of groups for answering or reviewing the given query, Q.

4   Conclusion and Future Work

Research on academic data has grown rapidly in recent years. Many researchers have understood the importance of analysing educational data. Inspired by real-world academic contexts, we propose a framework for analysing academic data. We present detailed methods for solving representative research issues such as collaborator recommendation, academic venue recommendation, expert finding, and expert group prediction.

In future studies, we intend to complete the framework presented here and compare our system with other systems. In addition, we will consider other issues such as scientific impact evaluation and research trend prediction.

Acknowledgment. This research was funded by Basic Science Research Program through the National Research Foundation of Korea (NRF) funded by the Ministry of Science, ICT & Future Planning (2017R1A2B4009410).

References

1. Achakulvisut, T., Acuna, D.E., Ruangrong, T., Kording, K.: Science concierge: a fast content-based recommendation system for scientific publications. arXiv preprint arXiv:1604.01070 (2016)
2. Burke, R., Felfernig, A., Göker, M.H.: Recommender systems: an overview. AI Mag. **32**(3), 13–18 (2011)
3. Chen, H.H., Gou, L., Zhang, X., Giles, C.L.: CollabSeer: a search engine for collaboration discovery. In: Proceedings of the 11th Annual International ACM/IEEE Joint Conference on Digital Libraries, pp. 231–240. ACM (2011)
4. Dong, Y., Ma, H., Shen, Z., Wang, K.: A century of science: Globalization of scientific collaborations, citations, and innovations. arXiv preprint arXiv:1704.05150 (2017)

5. Hoang, D.T., Tran, V.C., Nguyen, T.T., Nguyen, N.T., Hwang, D.: A consensus-based method to enhance a recommendation system for research collaboration. In: Asian Conference on Intelligent Information and Database Systems, pp. 170–180. Springer (2017)
6. Klamma, R., Cuong, P.M., Cao, Y.: You never walk alone: recommending academic events based on social network analysis. In: Proceeding of Complex Sciences, pp. 657–670. Springer (2009)
7. Le, Q.V., Mikolov, T.: Distributed representations of sentences and documents. arXiv preprint arXiv:1405.4053 (2014)
8. Li, J., Xia, F., Wang, W., Chen, Z., Asabere, N.Y., Jiang, H.: ACRec: a co-authorship based random walk model for academic collaboration recommendation. In: Proceedings of the Companion Publication of the 23rd International Conference on World Wide Web Companion, pp. 1209–1214. ACM (2014)
9. Newman, M.E.: Scientific collaboration networks. I. network construction and fundamental results. Phys. Rev. E **64**(1), 016131 (2001)
10. Pham, M.C., Kovachev, D., Cao, Y., Mbogos, G.M., Klamma, R.: Enhancing academic event participation with context-aware and social recommendations. In: 2012 IEEE/ACM International Conference on Advances in Social Networks Analysis and Mining (ASONAM), pp. 464–471. IEEE (2012)

Verifying Usefulness of Algorithms for WordNet Based Similarity Sense Disambiguation

Elżbieta Kukla and Andrzej Siemiński[✉]

Faculty of Computer Science and Management,
Wrocław University of Science and Technology, Wrocław, Poland
{Elzbieta.Kukla,Andrzej.Sieminski}@pwr.edu.pl

Abstract. The prospective readers of this paper belong the community of researchers and practitioners working in the area of natural language processing but do not necessarily specialize in the word sense disambiguation (WSD). It starts with a brief introduction into WSD and gives an overview of the classical approaches to the problem. The aim of the paper is to evaluate the accuracy of several already known algorithms for calculating synset similarity. These are later used to select word senses by a proposed algorithm. It uses the weighing of synset similarities. The validity of the whole process was verified using a large corpus of linguistic data that was tagged by a professional linguist. The senses were represented by the WordNet 2.1 synsets. The experiments clearly indicate that the weighing does increase the precision disambiguation process.

Keywords: WSD · WordNet · SemCore corpus
Synset similarity algorithm evaluation · Synset similarity weighing
Word sense disambiguation

1 Introduction

There are very few unambiguous words. Being able to identify the proper meaning of a word is crucial for a whole range of tasks ranging from machine translation to information retrieval. Although easy for native language speakers, the computers find it very hard to properly disambiguate words. This was the reason why the disambiguation is often referred to as an AI-complete problem [1]. A fairly exhaustive survey of the research work related to disambiguation could be found in [2].

Developing new disambiguation methods requires profound linguistic knowledge. Fortunately, over the years a great number of algorithms we proposed. The paper aims at researchers that do not specialize in WSD but still need to include it in their projects. It presents the main encountered problems with their solutions and evaluates their efficiency. It also introduces a similarity weighing algorithm that is capable of significantly increasing the precision of sense identification.

© Springer Nature Switzerland AG 2019
K. Choroś et al. (Eds.): MISSI 2018, AISC 833, pp. 224–235, 2019.
https://doi.org/10.1007/978-3-319-98678-4_24

The paper consists of 6 Sections. The introduction is followed by the presentation of the related work on the area. Any work on NLP depends heavily on the used data. In the paper, we use extensive and popular natural Language resources: WordNet for the description of English and SemCore data as a source of expert annotated texts. These datasets are introduced in Sect. 3. The next Section starts with the presentation of popular WordNet-based approaches to the estimation of synset similarity and then discusses the variety of problems of selecting proper sensets. The conducted experiments are introduced in Sect. 5. It starts with the description of preprocessing of data. After that, we describe the used algorithm for selecting synsest. In particular, we study the effect of weighing synset similarity values. The section ends with the presentation and analysis of the obtained results. The paper ends with conclusions and some suggestions for the direction of future research work.

2   Related Work

2.1   Basic Problems of Disambiguation

Generally formulated word sense disambiguation problem in fact requires the answering of four subsequent questions:

- How to specify word sense representation? There are two optional possibilities. First of them relies on choice of a representation from a finite set of meanings. In this case the set should be earlier constructed. Moreover it has to be constantly updated. Second solution tries to determine word sense representation by generating it rather than selecting. A predefined set of rules is necessary but the generation of new meanings of words is possible [3].
- Is the searched meaning of a word limited to one domain or does it refer to unrestricted language? For domain-dependent approach we have many efficient methods that could be used directly or in combination with e.g. Named Entities Recognition.
- Should we use supervised or unsupervised approach? The supervised methods require large collections of manually annotated words. Although such sets of data exist but they are limited and do not cover the whole heterogeneity of natural language texts that could be found on Internet. The SemCor set is a representative example of such source of data. It contains more than 200 000 words annotated by their meanings [4]. The unsupervised approach applies structured machine readable resources as for example thesauri [5] and dictionaries [6] or ontologies [7]. Thesauri and dictionaries are used in the context of unrestricted language while ontologies – in domain limited WSD.
- How granular should be the sense of a word? Theoretical deliberations indicate that fine granularity should ensure precise word sense identification. However in some cases this approach may give superfluous results e.g. while translating from one natural language to another. It should be noticed that many of the senses proposed by linguists are rarely (if ever) used in common language. This may negatively affect the recall of information retrieval.

2.2   Taxonomy of WSD

Methods applied to WSD could be classified according to different criteria. The most important are supervision and knowledge [2]. For supervision we have two classes:

- Supervised WSD is category of the methods that use machine learning techniques and sense-labelled training sets to learn a classifier
- Unsupervised WSD contains the methods that apply unlabeled corpuses namely none of the corpus elements are manually attributed their senses

  Similarly, knowledge criterion splits methods into two categories:

- Knowledge-based (or knowledge rich) approaches that use external resources like machine readable dictionaries, thesauri, and ontologies
- Corpus-based (or knowledge poor) methods that do not use any external lexical resources

  Different approaches applied to WSD are also distinguished with respect to the context. In this case two categories are considered:

- Token-based approaches combine a specific meaning of a word with a context it appears
- Type-based methods assume that a particular sense assigned to a word remains unchanged within a text.

  It is worth to notice that token-based approach can be easily transformed to type-based by attributing a word its most frequently used meaning in the text.

3   Language Resources for Disambiguation

3.1   WordNet

WordNet – a lexical data base of English – is widely used in WSD research. It contains English lemmas (basic form of a word) for nouns, verbs, adjectives, and adverbs [8]. Every word in WordNet is part-of-speech tagged and sense-labelled. The words are organized in sets of synonyms (synsets). A single synset may be interpreted as the set of words that have the nearly the same meaning. One word is usually contained in many synsets. For each synset WordNet defines:

- A gloss that is a textual description of the meaning together with the eventual usage examples
- Lexical and semantic relations that connect it with other synsets. While semantic relations apply to synsets as a whole (i.e. to all their elements) lexical relation may refer to specific word senses concluded in respective synsets.

  Antonymy, pertainymy and nominalization are the examples of lexical relations. Semantic relations in turn refers to particular parts-of-speech (POS). Since our experiments are limited to nouns only the most important relations that refer to nouns will be quoted:

- Hypernymy (called also kind-of or is-a), for example oak is a tree so tree is a hypernym of oak
- Hyponymy is inverse relation to hypernymy, for example iron is a hyponym of metal
- Meronymy (called part-of), for example branch is a meronym of tree
- Holonymy is inverse relation to meronymy, for example wheel is a meronym of bicycle

Semantic relations mentioned above constitute over 80% of all synset relations, and 70% of it are hypernymy – hyponymy relations. The structure of WordNet database is rather complex. A recent overview of Java libraries that facilitate its processing is described in [9].

3.2   Brown Corpus

During our experiments, we have used an extract of the Brown University Standard Corpus of Present-Day American English. The corpus was a carefully compiled selection of current American English. Although it is relatively old (it was compiled in the 1960's) and is not by far not the largest NL corpora it is still widely used in the field of NLP.

In the paper, we have used the Part 1of the SemCor corpus 2.1 which consists of semantically annotated texts from the Brown Corpus. All the words in SemCor were tagged by linguists for POS (Part of Speech). Additionally, the content words were lemmatized and sense-tagged. The used version of SemCor (2.1) has senses taken from WordNet 2.1.

The analyzed extract of SemCor 2.1 consisted of 103 texts with 47776 lemmas (canonical forms of a word). The number of different lemmas was much less – there were only 8177 of them.

A lemma in WordNet could belong to several synsets. The synsets are listed in descending order of occurrences. In what follows the position of a synset on such a list is called its index number. Almost 80% of lemmas have only one meaning (synset) within the data set. The number of lemmas with one meaning within a single text is higher but not as much as could be expected. Their number is 6944 what makes up 85% of all lemmas. The number is high but not high enough to support the "one word one sense (in a file)" assumption shared by some linguists.

It turns out that the first sense is by far the most common one. For some 32% of all words, only the first sysnet is present in the whole data set. Having only one sense within a single text is by far more popular. There we 5038 such words. This constitutes almost 62% of all words.

In general, the frequency of occurrence of word synsets declines very steeply with the sysnet index number. Such a sense distribution has to be taken into account in the process of disambiguation. It reflects the way the English language is used. That phenomenon is not reflected in algorithms of estimating similarity of senses as they take into account the structure of the WordNet or the textual properties of lemmas' descriptions, see the next Section.

4   WordNet Based Word Disambiguation

4.1   Approaches to Sysnset Similarity Estimation

WordNet as it was mentioned in Sect. 3.1 contains information about nouns, verbs, adjectives, and adverbs. In what follows algorithms were limited to nouns. Moreover, they take into account mainly ISA relation (the most frequent relation in WordNet) that forms hierarchical structure of synsets.

One of the most important problem in WordNet based WSD is determination of algorithms that compute similarity between synsets. These algorithms apply three different approaches: path based, info content-based, and gloss based. They also introduce a concept of lowest subsumer as the least common synset that surmises both considered synsets.

In path-based approach, the degree of similarity of two concepts is defined as a shortest path between them in WordNet structure. However, this simple definition makes some trouble because path length in network or taxonomy is not univocally determined. An algorithm that applies this approach is the Wu and Palmer [10] algorithm originally developed for verbs but now mostly used for nouns. The similarity SIM (C_1, C_2) of two synsets C_1 and C_2 is defined by the formula:

$$Sim(C_1, C_2) = \frac{2Len(C_{lcs}, C_{root})}{Len(C_1, C_x) + Len(C_2, C_x) + 2Len(C_{lcs}, C_{root})} \tag{1}$$

Where $Len(C_x, C_y)$ is the number of nodes that separate synsets C_x and C_y. The C_{root} is the most upper synset.

Info content-based approach uses amount of information contained in a concept or synset $IC(cpt)$ that can be calculated according to the formula:

$$IC(cpt) = -\log(p(cpt)) \tag{2}$$

Where $P(cpt)$ is the frequency of the concept (or synset) cpt and all its subordinate concepts (or synsets) that is determined basing on a large corpus. It is worth to notice that the most general concept that is the root of the hierarchy of senses has information content $IC(cpt)$ equal to zero.

Information content is then applied in calculating similarity of two concepts c_1 and c_2. One of the commonly used measures was proposed by Jiang and Conrath [11]. It assumes that the difference between the information content of the individual concepts and that of their lowest common subsumer is related to their similarity. Thus the similarity $sim(c_1, c_2)$ of two concepts c_1 and c_2 is determined by the following formula:

$$sim(c_1, c_2) = \frac{1}{IC(c_1) + IC(c_2) - 2IC(lcs(c_1, c_2))} \tag{3}$$

Where $lcs(c_1, c_2)$ is the lowest common subsumer of the concepts c_1 and c_2.

Gloss-based algorithms focus on glosses of the synsets and their intersections. They are used to measure semantic similarity of the synsets involved. Lesk [12] proposed an algorithm that uses formula (4) to calculate similarity of two synsets S_1 and S_2:

$$sim(S_1, S_2) = |gloss(S_1) \cap gloss(S_2)| \tag{4}$$

Where $gloss(S_x)$ is a textual definition of synset S_x regarded as a bag of words that constitute the definition.

Lesk algorithm in its original form is very sensible to the wording of the definition so some extended gloss overlap algorithms were proposed [13] to diminish this defect. The alternative solutions broaden the gloss by the glosses of concepts that are in semantic relations to the concept under consideration. These relations are explicitly presented in the WordNet, e.g. hypernymy or meronymy.

4.2   Approaches to the Selection of Senses

As it was mentioned in Sect. 2.2 to select the sense of a word three categories of approaches are used: supervised methods, unsupervised methods and knowledge-based methods. However, some authors distinguish additionally a group of semi-supervised methods that use relatively small amount of annotated data [2].

Supervised approaches assume that context itself contains enough amount of information to disambiguate a word occurring within it. The context is represented by a set of features which could be grouped to: local features representing local context of a word usage, topical features determining a general topic of a text (more general context), syntactic features describing syntactic relations between considered word and other words in the same sentence, and semantic features like senses of words in context. These features form a vector that represents word occurrence within a sentence.

Supervised WSD uses machine-learning techniques and manually sense-labelled data sets to train a classifier. Generally, the classifier deals with a single word (represented by its feature vector) and classifies it by assigning the appropriate sense to each its occurrence. Up to now most of the machine-learning algorithms have been used to WSD. The best results were achieved by support vector machine (SVM) and memory-based learning (e.g. k-Nearest Neighbors) methods for they can cope with the high dimension of feature vector.

Semi-supervised approach include methods that train a classifier with limited supervision. Bootstrapping is one of the main approaches that belongs to this category. It aims at building a sense classifier based on only few annotated data. Besides, at starting point the algorithm needs a large set of unannotated data, and one or more basic classifiers. First a basic classifier is trained with few annotated data then it is used to classify a part of large unlabeled corpus. The most credible classification results are included to the training set. The procedure repeats until the whole corpus is processed or the algorithm reaches predefined maximum number of iterations.

Unsupervised approach assumes that a word that occurs in the same context has always the same meaning. The methods belonging to this category can infer word senses from a text by grouping word occurrences, and subsequently assigning new word occurrences to the detected clusters. Although unsupervised methods do need large sets of manually sense-labelled data their results are sometimes inaccurate because they may not discover all clusters that refer to the dictionary senses of the word. Also their evaluation is much more difficult by comparison with supervised methods.

Knowledge based approaches use knowledge resources like dictionaries, thesauri, ontology etc. to deduce the meanings of words in context. Three main knowledge-based algorithms namely path-based, info content-based and gloss-based were presented in Sect. 4.1.

5   Experiment Setup

During the experiment, the SemCore 2.1 corpus was used. The texts are annotated by linguists. The annotation is very extensive and the data need to be specified to be efficiently processed by disambiguation algorithms.

5.1   Data Preprocessing

The following example shows an extract of the input data:

<p pnum=2>
<s snum=1>
<wf cmd=done pos=NN lemma=street wnsn=3 lexsn=1:06:00::>streets</wf>
<wf cmd=ignore pos=IN>of</wf>
<wf cmd=ignore pos=DT>the</wf>
<wf cmd=done pos=NN lemma=old_man wnsn=1 lexsn=1:18:02::>Old_Man</wf>

Each word is described by one line. The part of speech codes were adopted from the Penn Treebank Project [14]. The experiment was limited to nouns. The only used words have tags: NN (Noun, singular or mass), NNS (Noun, plural), NNP (Proper noun, singular), and NNPS (proper noun, plural). We have decided so because the nouns are most informative part of speech, their WordNet definitions are very precise and they are connected by many relations. A line with a noun description contains among others: its POS Tag, lemma, original word form and synset number are given. In addition to that special lines mark the sentence and paragraph number, e.g. in word streets if the first word of the first sentence in the second paragraph.

In the data preprocessing phase the lemmas of all nouns in with their synset numbers from one sentence are extracted and then collected to form a separate line. All other elements are rejected. The data from the above example would be reduced to the following line:

street [3] old_man [1]

The inclusion of expert disambiguation makes it possible to easily evaluate the efficiency of respective algorithms. Sentences with only one noun were eliminated.

5.2   Word Sense Selection

The word similarity measures described in Sect. 4.1 estimate the similarity between two words senses. The implementation of the algorithms was developed at the University of Sussex and is now in the public domain [15].

As described before in order to identify the meaningful word sense we have first to extract a part of text called a text window. In the next step, we have to take into account the similarities between the word senses within the window and pick a set of senses that are most like to each other.

Let us use the following notation:

- K_{max}: the number of words (nouns) in a text window (sentence).
- $w_{k,s}$: the WordNet sense number s for the k-th noun in the text window.
- $w_{k,\,Smax}$: the number of senses of the k-th word of the text window.
- $w_{k,sel}$: the selected sense of k-th word of the text window.
- $sim_x(w_{k,u}, w_{m,n})$: the value of similarity measure x between the u-th sense of word k and the n-th sense of word m from the text window.

Let $sensMaxNum(w_1, .. w_{Kmax})$ denote the number of all possible text window senses. It is equal to:

$$sensMaxNum(w_1, .. w_m) = \sum_{n=1}^{n=m} w_{n,Smax} \qquad (5)$$

For a text window with 3 words and: $w_{1,Smax} = 2$, $w_{2,Smax} = 1$, and $w_{3,Smax} = 2$ the value of $sensMaxNum(w_1, w_2, w_3) = 4$. The individual text window senses, called CD Candidate Disambiguation's, are listed below:

$$CD(w_{1,1}, w_{2,1}, w_{3,1}), CD(w_{1,2}, w_{2,1}, w_{3,1}), CD(w_{1,1}, w_{2,1}, w_{3,2}), CD(w_{1,2}, w_{2,1}, w_{3,2}).$$

The calculate the value of a $CD(w_1, .., w_{Kmax})$ we sum the similarities of all its word senses pairs. The number of such pairs (NoP) is equal to the number of all pair combinations without repetitions of its elements:

$$NoP(w_1, \ldots, w_{Kamx}) = \binom{2}{K_{max}} = \frac{2!}{K_{max}!(2! - K_{max})} \qquad (6)$$

The value of a CD is the sum of the similarities of all sense pairs in that sequence:

$$Val(w_{1,1}, w_{2,1}, w_{3,1}) = sim_x(w_{1,1}, w_{2,1}) + sim_x(w_{1,1}, w_{3,1}) + sim_x(w_{1,1}, w_{3,2})$$
$$+ sim_x(w_{2,1}, w_{3,1}) + sim_x(w_{2,1}, w_{3,2}).$$

The algorithm selects the CD with the highest value of the function Val.

The disadvantage of such an approach is that the number of possible senses could well exceed 100 even for sentences of a moderate length. This is easily acceptable for the evaluation of word similarity methods. We acknowledge that in real life applications some sort of optimization of necessary in order to reduce the complexity of sense

finding. To speed up the process of calculating the value of *Val* the individual similarities $sim_x(w_{x,y}, w_{z,v})$ are stored in a serialized Java Hash Map.

5.3   Experiment Results and Their Evaluation

The synset similarity algorithms used during the experiment belonged all three major
groups that were introduced in Sect. 4.1:

- Lin and Resnik algorithms represent the info content approach,
- Lesk algorithm is gloss based,
- Wup & Palmer and Path algorithms are path based.

To adjust the value of similarity to the disproportional high occurrence rate of the
synset with low index value we have applied the following weighing algorithm:

$$sim_x W(w_{x,y}, z_{m,n}) = W\left[\left\lfloor \frac{y+n}{2} \right\rfloor\right] * sim_x(w_{x,y}, z_{m,n}) \qquad (7)$$

where:
$\left\lfloor \frac{m}{n} \right\rfloor$ is the value of integer division of two integers m and n.
$W[x]$ is the xth position in the weighting vector.
The weighing vectors used during the experiments are listed in Table 1. In order to
simplify the entering of parameters tf the value of $\left\lfloor \frac{m+n}{2} \right\rfloor$ exceeds the size of the vector
then the last position is used, e.g. for the vector W_1 we have:

$$W_1\left[\left\lfloor \frac{1+1}{2} \right\rfloor\right] = 5; W\left[\left\lfloor \frac{3+3}{3} \right\rfloor\right] = 2 \text{ and } W_1\left[\left\lfloor \frac{10+7}{2} \right\rfloor\right] = 1;$$

The W_0 vector has a referential purpose. It represents the case when no weighing
takes place. The values stored in the last vector correspond to the distribution of synset
index values that were discussed in Sect. 3.2.

Table 1. Weighing vectors used in the experiments

Weighing vector	Synset index number				
	1	2	3	4	5
W_0	1				
W_1	10	5	3	1	
W_2	15	5	2	1	
W_3	80	12	4	2	1

The precision obtained by the 6 selected algorithms is shown in Table 2. It is clear
that the weighing has a substantial influence on the achieved precision. Without
weighing the procession achieved by the algorithms differs considerably. Introducing

weighing improves precision in all cases and the increase is more visible for the originally least efficient algorithms. As a result, the values shown in the last column are very much alike. The type of weighting vector used does not have much influence on the precision.

The paper [16] reports the results of disambiguation for the same input data and the same set of algorithms. They are lower than the values reported here. This is probably due to the differences in the used synset selection methods. That paper used methods that are available on the website prepared by T. Pedersen [17].

Table 2. The quota of properly disambiguated words

Algorithm	Weighing vector			
	W_0	W_1	W_2	W_3
LIN	0.65	0.71	0.72	0.74
Resnik	0.56	0.66	0.69	0.72
Wup & Palmer	0.61	0.71	0.73	0.73
Path	0.56	0.74	0.75	0.75
Lesk	0.65	0.70	0.71	0.72

Table 3 shows the binary precision. The binary precision is introduced to overcome the problem of excessive granularity of WordNet synsets. It is reported that in many cases even proficient English speakers have difficulties in comprehending the differences in meaning of synsets. The WordNet includes the frequencies of occurrence of all synsets in the extensive data used while creating it. It is not uncommon that the number is 0. It means there was no such synset in input data. Its introduction to the WordNet was based on the expertise in English of the linguists participating in the development of WordNet.

The binary precision accepts as proper all selections of synsets if:

- the expert and identified synsets are identical or
- both expert and identified synsets have index number different from 1.

As a result, only errors accounted stem from not proper identification of synsets with index number 1 which is most confusing for a user.

Table 3. The quota of properly disambiguated words (binary precision)

Algorithm	Weighing vector			
	W_0	W_1	W_2	W_3
LIN	0.75	0.77	0.77	0.78
Resnik	0.66	0.72	0.73	0.75
Wup & Palmer	0.67	0.74	0.76	0.76
Path	0.71	0.77	0.78	0.77
Lesk	0.77	0.76	0.75	0.76

As could be expected the binary precision has values considerably surpassing the values reported in Table 3. They are also higher than the results reported in [16] although there the ensemble methods were applied. The weighing factor has much lower influence on the efficiency than in the case of basic precision. In the cases of the Lesk and Lin algorithm, the influence is small or even negligible. It should be also noted, that when weighing is applied to both versions of precision we have measured almost the same values.

6 Conclusions

The paper compares the precision of disambiguation that is achieved by various algorithms for measuring the synset similarity. In order to compare them fairly the same synset selection method was used. The experiments clearly show that weighing of synset similarities significantly increases the basic precision. The obtained results compare favorably with the results reported in [16] where the same set of algorithms and input data were used. At the same time, it is worth noting that the form of the weighing vector does not have much influence on the resulting precision. On the other hand, the synset similarity weighing for the binary precision seems not to be justified.

The method is computationally expensive but enables to obtain decent disambiguation precision. Optimizing the synset selection process is definitely worth investigating.

References

1. Mallerey, J., C.: Thinking about foreign policy: finding an appropriate role for artificial intelligence computers, Ph.D. dissertation, MIT Cambridge, MA (1988)
2. Navigli, R.: Word sense disambiguation: a survey. ACM Comput. Surv. (CSUR) 41(2), 1–69 (2009). https://doi.org/10.1145/1459352.1459355
3. Yarowsky, D.: Hierarchical decision lists for word sense disambiguation. Comput. Human. 34(1–2), 179–186 (2000)
4. Miller, G.A., Leacock, C., Tengi, R., Bunker, R. T.: A semantic concordance. In Proceedings of the ARPA Workshop on Human Language Technology, pp. 303–308, 1993
5. Roget, P.M.: Roget's International Thesaurus, 1st edn. Cromwell, New York (1911)
6. Soanes, C., Stevenson, A. (eds.): Oxford Dictionary of English. Oxford University Press, Oxford (2003)
7. Gruber, T.R.: Toward principles for the design of ontologies used for knowledge sharing. In: Proceedings of the International Workshop on Formal Ontology (Padova, Italy) (1993). https://doi.org/10.1006/ijhc.1995.1081
8. Fellbaum, C. (ed.): WordNet: An Electronic Database. MIT Press, Cambridge (1998)
9. Finlayson M.A.: Java libraries for accessing the princeton WordNet: comparison and evaluation. In: Proceedings of the 7th International Global WordNet Conference (GWC 2014), Tartu, Estonia, pp. 78–85 (2014)
10. Wu, Z., Palmer, M.: Proceeding ACL 1994 Proceedings of the 32nd Annual Meeting on Association for Computational Linguistics, pp. 133–138 (1994)

11. Jiang J., Conrath, D.W.: Semantic similarity based on corpus statistics and lexical taxonomy. In: Proceedings of the 10th International Conference on Research in Computational Linguistics. arXiv:cmp-lg/9709008 (1997

12. Lesk, M.: Automatic sense disambiguation using machine readable dictionaries: how to tell a pine cone from an ice cream cone. In: Proceedings of the 5th SIGDOC, New York, NY, pp. 24–26 (1986)

13. Banerjee, S. Pedersen, T.: Extended gloss overlaps as a measure of semantic relatedness. In: Proceedings of the 18th International Joint Conference on Artificial Intelligence (IJCAI, Acapulco, Mexico), pp. 805–810 (2003)

14. Taylor A.: The Penn treebank: an overview. https://pdfs.semanticscholar.org/182c/4a4074e8577c7ba5cbbc52249e41270c8d64.pdf. Accessed 10 Mar 2018

15. https://github.com/greenmoon55/textclustering/tree/master/lib/edu/sussex/nlp/jws. Accessed 10 Mar 2018

16. Siemiński A.: Practice of word sense disambiguation. In: Intelligent Information and Database Systems: 10th Asian Conference, ACIIDS 2018, Dong Hoi City, Vietnam, March 19–21, 2018, pp. 159–169 (2018)

17. http://maraca.d.umn.edu/allwords/allwords.html. Accessed 10 Mar 2018

Comparison of Paragram and Glove Results for Similarity Benchmarks

Jakub Dutkiewicz and Czesław Jędrzejek[✉]

Poznan University of Technology,
Plac Marii Skłodowskiej-Curie 5, 60-965 Poznań, Poland
czeslaw.jedrzejek@put.poznan.pl

Abstract. Distributional Semantics Models (DSM) derive word space from linguistic items in context. In this paper we provide comparison between two methods for post process improvements to the baseline DSM vectors. The counter-fitting method which enforces antonymy and synonymy constraints into the Paragram vector space representations recently showed improvement in the vectors' capability for judging semantic similarity. The second method is our novel RESM method applied to GloVe baseline vectors. By applying the hubness reduction method, implementing relational knowledge into the model by retrofitting synonyms and providing a new ranking similarity definition RESM that gives maximum weight to the top vector component values we equal the results for the ESL and TOEFL sets in comparison with our calculations using the Paragram and Paragram + Counter-fitting methods. The Paragram or our cosine retrofitting method are state-of-the-art results for the SIMLEX-999 gold standard. Apparently relational knowledge and counter-fitting is more important for judging semantic similarity than sense determination for words.

Keywords: Language models · Vector spaces · Word embedding
Similarity

1 Introduction

Distributional language models are frequently used to measure word similarity in natural language. This is a basis for many semantic tasks. The DSM often consists of a set of vectors; each vector corresponds to a character string, which represents a word. Mikolov et al. [16] and Pennington et al. [21] implemented the most commonly used word embedding (WE) algorithms. Vector components in language models created by these algorithms are latent. Similarity between words is defined as a function of vectors corresponding to given words. The cosine measure is the most frequently used similarity function, although many other functional forms were attempted. Santus et al. [27] highlights the fact that the cosine can be outperformed by ranking based functions. As pointed out by many works [10], evidence suggests that distributional models are far from perfect. Vector space word representations obtained from purely distributional information of words in large unlabeled corpora are not enough to best the state-of-the-art

© Springer Nature Switzerland AG 2019
K. Choroś et al. (Eds.): MISSI 2018, AISC 833, pp. 236–248, 2019.
https://doi.org/10.1007/978-3-319-98678-4_25

results in query answering benchmarks, because they suffer from several types of weaknesses:

1. Inadequate definition of similarity,
2. Inability of accounting of senses of words,
3. Appearance of hubness that distorts distances between vectors,
4. Inability of distinguishing from antonyms,
5. In case of retrofitting distortion vector space - loss of information contained in the original vectors.

In this paper we use the existing word embedding model but with several post process enhancement techniques. We address three of these problems. In particular, we define a novel similarity measure, dedicated for language models. Similarity is a function, which is monotonically opposite to distance. As the distance between two given entities gets shorter, entities are more similar. This holds for language models. Similarity between words is equal to similarity between their corresponding vectors. There are various definitions of distance. The most common Euclidean distance is defined as follows:

$$d_e(p_1, p_2) = \sqrt{\sum_{c \in p} \left(c_{p_1} - c_{p_2} \right)^2}$$

Similarity based on the Euclidean definition is inverse to the distance:

$$sim_e(p_1, p_2) = \frac{1}{1 + d(p_1, p_2)}$$

Angular definition of distance is defined with cosine function:

$$d_a(p_1, p_2) = 1 - \cos(p_1, p_2)$$

We define angular similarity as:

$$sim_a(p_1, p_2) = \cos(p_1, p_2)$$

Both Euclidean and Cosine definitions of distance could be looked at as the analysis of vector components. Simple operations, like addition and multiplication work really well in low dimensional spaces. We believe, that applying those metrics in spaces of higher order is not ideal, hence we compare cosine similarity to a measure of distance dedicated for high dimensional spaces. In this paper we restrict ourselves to three gold standards: TOEFL, ESL and SIMLEX-999. The first two are small but reliably annotated (and therefore confidence in their performance can be assumed 100%). Other used benchmarks suffer from several drawbacks. Both WS-353 [7] and MEN [2] do not measure the ability of models to reflect similarity. Moreover, as pointed out by [10], for WS-353, RG [26] and MEN, state-of-the-art models have reached the average performance of a human annotator on these evaluations.

2 Related Work

In information retrieval (IR) document similarity to a question phrase can be expressed using ranking. The objective and repeatable comparison of rankings requires a rank similarity measure. Santus et al. [27] introduced another ranking based similarity function called APSyn. In their experiment APSyn outperformed the cosine similarity, reaching 73% of accuracy in the best cases (an improvement of 27% over cosine) on the ESL dataset, and 70% accuracy (an improvement of 10% over cosine) on the TOEFL dataset. In contrast to our work, they use the Positive Pointwise Mutual Information algorithm to create their language model. Despite such large improvement the absolute results were quite poor compared to the state-of-theart. Our motivation for the RESM method was the NDCG measure that is considered the most appropriate measure for IR evaluation. Levy et al. [14] investigated GloVe [21] proposition of using the context vectors in addition to the word vectors as GloVe's output. This led a different interpretation of its effect on the cosine similarity function. The similarity terms which can be divided into two groups: second-order includes a symmetric combination of the first-order and second order similarities of x and y, normalized by a function of their reflective first-order similarities. The best [14] result for SIMLEX-999 is 0.438; for WordSim-353 similarity and WordSim-353 relatedness their result is a few hundredths worse than that of Paragram [29].

A successful avenue to enhance word embeddings was pointed out by Faruqui et al. [5], using WordNet [17], and the Paraphrase Database 1.0 [8] to provide synonymy relation information to vector optimization equations. They call this process retrofitting, a pattern we adapt to the angular definition of distance, which is more suitable to our case. We also address hubness reduction. Hubness is related to the phenomenon of concentration of distances - the fact that points get closer at large vector dimensionalities. Hubness is very pronounced for vector dimensions of the order of thousands. To for reduce hubness Dinu and Baroni [4] used ranking instead of similarity. The method requires the availability of more, unlabeled source space data, in addition to the test instances, and is natural in application to cross-language or cross-modal applications We apply this method of localized centering for hubness reduction [6] for the language models. The Paragram vectors were further improved by Mrksic et al. [18]. This was achieved by the joint retro -fitting and counter-fitting optimization of the following objective function:

$$AP(V') = \sum_{(u,w) \in A} \tau\left(\delta - d\left(v'_u, v'_w\right)\right)$$

$$SA(V') = \sum_{(u,w) \in S} \tau\left(d\left(v'_u, v'_w\right) - \gamma\right)$$

$$VSP(V, V') = \sum_{i=1}^{N} \sum_{j \in N(i)} \tau\left(d\left(v'_i, v'_j\right) - d\left(v_i, v_j\right)\right)$$

$$C(V, V') = k_1 AP(V') + k_2 SA(V') + k_3 VSP(V, V')$$

The values of hyperparameters are the following $\delta = 1$, $\gamma = 0$ and k1 = k2 = k3. In the original Paragram + Counter-fitting work $\delta = 1$ corresponds to zero angle between a pair of vectors for synonyms, and $\gamma = 0$ corresponds to 90° angle between a pair of vectors for antonyms, which is not quite consistent with SIMLEX-999 [10] average value of cosines for synonym pairs, 0.77, and average value of cosines for antonym pairs, 0.17. However, as we verified, using these theoretically more acceptable values docs not improve SIMLEX-999 results.

3 Method

In our work we define the language model as a set of word representations. Each word is represented by its vector. We refer to a vector corresponding to a word w_i as v_i. A complete set of words for a given language is referred to as a vector space model (VSM). We define similarity between words w_i and w_j as a function of vectors v_i and v_j.

$$sim(w_i, w_j) = f(v_i, v_j)$$

We present an algorithm for obtaining optimized similarity measures given a vector space model for word embedding. The algorithm consists of 6 steps:

1. Refine the vector space using the cosine retrofit algorithm,
2. Obtain vector space of centroids,
3. Obtain vectors for a given pair of words and optionally for given context words,
4. Recalculate the vectors using the localized centering method,
5. Calculate ranking of vector components for a given pair of words,
6. Use the ranking based similarity function to obtain the similarity between a given pair of words.

The cosine retrofit algorithm follows the original [5] approach except that we use the cosine distance and rotations of retrofitted vectors in space. The formulae look different compared to [18] but give equivalent results. We use all of the methods in together to achieve significant improvement over the baseline method. We present details of the algorithm in the following sections.

3.1 Baseline

The cosine function provides the baseline similarity measure:

$$sim(w_1, w_2) = \cos(v_1, v_2) = \frac{v_1 \cdot v_2}{||v_1|| \cdot ||v_2||}$$

The cosine function achieves a reasonable baseline. It is superior to the Euclidean similarity measure and is used in various works related to word similarity. In our work we use several post-process modifications to the vector space model. We also redefine the similarity measure.

3.2   Implementing Relational Knowledge into the Vector Space Model

Let us define a lexicon of relations (rel) L. Each row in the lexicon consists of a word and a set of its related words (synonyms and antonyms).

$$L(w_i) = \{w_j, \beta_j : rel(w_i, w_j), \beta_j = w(rel)\}$$

A basic method of implementing synonym knowledge into the vector space model was previously described in [8]. We refer to that method as retrofit. It uses the iterational algorithm of moving the vector towards an average vector of its weighted related words according to the following formula.

$$v_i' = \frac{\alpha_i v_i + \dfrac{\sum_{w_j \in L(w_j)} \beta_j v_j}{||L(w_j)||}}{2}$$

In the original formula [5], variables α and β allow us to weigh the importance of certain relations. The basic retrofit method moves the vector towards its destination (e.g. shortens the distance between the average synonym to a given vector) using the Euclidean definition of distance. This is not consistent with the cosine distance used for evaluation. Instead we improve [5] idea by performing operations in spherical space, thus preserving the angular definition of distance. Given a basic rotation matrix R for a plane (i, j).

$$R_{i,j}(\theta) = \begin{bmatrix} cos\theta & -sin\theta \\ sin\theta & cos\theta \end{bmatrix}$$

For each plain (p, q) we calculate angle $\theta_{(p,q)}(vi, v_j)$ between word and its related word. We apply a rotation $R_{p,q}(\theta_{(p,q)}(v_i, v_j)\beta_j)$ to that word. Finally, we take an average of a given word vector and average of rotated, related vector.

$$v_i' = \frac{v_i + \dfrac{\sum_{w_j \in L(w_j)} (\prod_{(p,q)} R_{p,q}(\theta_{p,q}(vi, vj)\beta_j)) v_i}{||L(w_j)||}}{2}$$

We refer to this method as to generalized retrofitting. The original formula from [5] was obtained by minimizing the following objective function.

$$\Psi(\text{VSM}) = \sum_{i=1}^{n} [\alpha_i ||v_i - \widehat{v}_i||^2 + \sum_{w_j \in L(w_i)}^{n} \beta_{ij} ||v_i - v_j||^2]$$

We change the Euclidean definition of distance to cosine in the formula to obtain a new objective function.

$$\Psi'(\text{VSM}) = \sum_{i=1}^{n} [(1 - \cos(v_i, \widehat{v}_i))^2 + \sum_{w_j \in L(w_i)}^{n} \beta_{ij}(1 - \cos(v_i, v_j))^2]$$

We take first derivative of Ψ' to define a single step change of a vector v_i for each vector component $v_{(i,k)}$.

$$\frac{\delta\Psi'}{\delta v_{i,k}} = -\frac{2\delta \cos(v_i, \widehat{v}_i)}{\delta v_{i,k}} + 2\cos(v_i, \widehat{v}_i)\frac{\delta \cos(v_i, \widehat{v}_i)}{\delta v_{i,k}}$$

$$+ 2 \sum_{w_j \in L(w_i)}^{n} -\beta\frac{\delta \cos(v_i, v_j)}{\delta v_{i,k}} + \cos(v_i, v_j)\frac{\delta \cos(v_i, v_j)}{\delta v_{i,k}}$$

$$\frac{\delta \cos(v_i, v_j)}{\delta v_{i,k}} = \frac{v_{j,k}}{|v_i||v_j|} - \cos(v_i, v_j)\frac{v_{i,k}}{|v_i|^2}$$

This method and obtained results are equivalent to the work of [18]. We refer to this method as cosine retrofitting.

3.3 Localized Centering

We address the problem of hubness in high dimensional spaces with the localized centering approach applied to every vector in the space. The centered values of vectors, centroids, are the average vectors of k nearest neighbors of the given vector v_i. We apply a cosine distance measure to calculate the nearest neighbors.

$$c_i = \frac{\sum_{v_j \in k - NN(v_i)} v_j}{N}$$

In [6], the authors pointed out that skewness of a space has a direct connection to the hubness of vectors. We follow the pattern presented in the [9] and recalculate the vectors using the following formula.

$$sim(v_i, v_j) = sim'(v_i, v_j) - \cos(v_i, c_i)^{\gamma}$$

We use empirical method to find values of the hyperparameter gamma.

Table 1. a. Example of question with wrong answer in TOEFL (the correct answer is cushion, our answer is scrape). b. Set of possible questions with the same question word and different correct answers.

a	Q. word	P1	P2	P3	P4
	Lean	Cushion	Scrape	Grate	Refer
b	Q. word	P1	P2	P3	P4
	Iron	Wood	Metal	Plastic	Timber
	Iron	Wood	Crop	Grass	Grain

3.4   Ranking Based Similarity Function

We propose a component ranking function as the similarity measure. This idea was originally introduced in [27] who proposed the APSyn ranking function. Let us define the vector v_i as a list of its components.

$$v_i = [f_1, f_2, \ldots, f_n]$$

We then obtain the ranking r_i by sorting the list in descending order (d in the equation denotes type of ordering), denoting each of the components with its rank on the list.

$$r_i^d = \{f_1 : rank_i^d(f_1), f_2 : rank_i^d(f_2), \ldots, f_n : rank_i^d(f_n)\}$$

APSyn is calculated on the intersection of the N components with the highest score.

$$APSyn(w_i, w_j) = \sum_{f_k \in top(r_i^d) \cap top(r_j^d)} \frac{2}{rank_i(f_k) + rank_j(f_k)}$$

APSyn was originally computed on the PPMI language model, which has unique feature of nonnegative vector components. As this feature is not given for every language model, we take into account negative values of the components. We define the negative ranking by sorting the components in ascending order ('a' in the equation denotes type of ordering).

$$r_i^a = \{f_1 : rank_i^a(f_1), f_2 : rank_i^a(f_2), \ldots, f_n : rank_i^a(f_n)\}$$

As we want our ranking based similarity function to preserve some of the cosine properties, we define score values for each of the components and similarly to the cosine function, multiply the scores for each component. As the distribution of component values is Gaussian, we use the exponential function.

$$s_{i,f_k} = e^{-rank_i(f_k) \cdot \frac{k}{d}}$$

Parameters k and d correspond respectively to weighting of the score function and the dimensionality of the space. With high k values, the highest ranked component will be the most influential one. The rationale is maximizing information gain. Our measure is similar to infAP and infNDCG measures used in information retrieval [25], that give maximum weight to the several top results. Lower k values increase the impact of lower ranked components at the expense of 'long tail' of ranked components. We use the default k value of 10. The score function is identical for both ascending and descending rankings. We address the problem of polysemy with a differential analysis process. Similarity between pair of words is captured by discovering the sense of each word and then comparing two given senses of words. The sense of words is discovered by analysis of their contexts. We define the differential analysis of a component as the sum of all scores for that exact component in each of the context vectors.

$$h_{i,f_k} = \sum_{w_j \in context(w_i)} s_{j,f_k}$$

Finally we define the Ranking based Exponential Similarity Measure (RESM) as follows.

$$RESM^a(w_i, w_j) = \sum_{f_k \in top(r_i^a) \cap top(r_j^a)} \frac{s_{i,f_k}^a \cdot s_{j,f_k}^a}{h_{i,f_k}^a}$$

The equation is similar to the cosine function. Both cosine and RESM measures multiply values of each component and sum the obtained results. Contrary to the cosine function, RESM scales with a given context. It should be noted, that we apply differential analysis with a context function h. An equation in this form is dedicated for the ESL and TOEFL test sets. The final value is calculated as a sum of the RESM for both types of ordering.

$$RESM(w_i, w_j) = RESM^a(w_i, w_j) + RESM^d(w_i, w_j)$$

Table 2. State of the art results for TOEFL and ESL test sets

Authors	TOEFL	ESL
Bullinaria and Levy [3]	100.00%	66.00%
Osterlund et al. [19]		
Jarmasz and Szpakowicz [11]	79.70%	82.00%
Lu et al. [15]	97.50%	86.00%

4 Evaluation

This work compares the state-of-the-art word embedding methods for three most reliable gold standards: TOEFL, ESL and SIMLEX-999. TOEFL consists of 80 questions, ESL consists of 50 questions. The questions in ESL are significantly harder. Each question in these tests consists of a question word with a set of four candidate answers. It is worth pointing out, that the context given by a set of possible answers often defines the question (a selection of a sense of a word appearing in the question). Answering these tests does not require finding the most similar word out of the whole dictionary but only from multiple choice candidates; therefore TOEFL and ESL are less demanding than SIMLEX-999. A question example in Table 1 highlights the problem. In the first question, all of possible answers are building materials. Wood should be rejected as there is more appropriate answer. In second question, out of possible answers, only wood is a building material which makes it a good candidate for the correct answer. This is a basis for applying a differential analysis in the similarity measure. Table 2 illustrates state of the art results for both test sets. The TOEFL test set was introduced in [13]; the ESL test set was introduced in [28].

4.1 Experimental Setup

We use the unmodified vector space model trained on 840 billion words from Common Crawl data with the GloVe algorithm introduced in [21]. The model consists of 2.2

Table 3. Accuracy of various methods on TOEFL and ESL test sets.

Method	TOEFL	ESL	SimLex-999
HR + Cosine	91.25%	66.00%	0.438
RETRO + Cosine	95.00%	62.00%	0.435
HR + RETRO + Cosine	96.25%	74.00%	0.438
APSyn	80.00%	60.00%	0.338
RETRO + APSyn	97.50%	70%	Not applicable
RESM	90.00%	76.00%	Not applicable
RETRO + RESM	96.25%	80.00%	Not applicable
HR + RETRO + RESM	97.50%	80.00%	Not applicable
RETRO + RESM + heuristic	97.50%	84.00%	Not applicable
Paragram	97.50%	84.00%	0.688
Paragram + HR	97.50%	84.00%	0.692
Paragram + Counter-fitting	97.50%	82.00%	0.736
Paragram + cosRETRO[a]	97.50%	82.00%	0.724
Paragram + cosRETRO+[b]	97.50%	84.00%	0.716
Pilehvar and Navigli			0.436
DECONF			0.517

[a]cosRETRO is trained on Equivalence relation from PPDB for synonymy and Exclusion relation from PPDB for antonymy concatenated with WordNet antonyms.
[b]cosRETRO+ is trained on all relations from PPDB.

million unique vectors; Each vector consists of 300 components. The model can be obtained via the GloVe authors website. We also use the GloVe refinement enriched with PPDB 1.0 relations called Paragram [29] (there is an extension of this paper at https://arxiv.org/pdf/1506.03487.pdf with more details included). We run several experiments, for which settings are as follows: In the evaluation skewness $\gamma = 9$ and $k = 10$. All of the possible answers are taken as context words for the differential analysis. In our runs we use all of the described methods separately and conjuncttively. We refer to the methods in the following way. We denote the localized centering method for hubness reduction as HR. We use a Paraphrase Database lexicon introduced in [8] for the retrofitting. We denote the generalized retrofitting as RETRO and the cosine retrofitting as cosRETRO.

4.2   Heuristic Improvement

Although the hubness reduction method does not increase the number of correct answers for the ESL test set, we noticed that the average rank of the correct answer goes down from 1.32 to 1.24. That is a significant improvement. To obtain better results we combined the results with and without localized centering. The heuristic method chooses the answer with the best average rank for both sets. By applying that method we obtained two additional correct answers.

4.3   Sense Recognition

Semantic distributional methods operate on words or characters. A vector position for a word is an average of positions of the word senses. The is a major deficiency, particularly if a word with a given sense is statistically rare. In Table 4 of [18] a pair: [dumb, dense] appears amount the highest-error SIMLEX-999 word pairs using Paragram vectors (before counter-fitting). Counter-fitting does not improve this result. In Table 3 we also show the results sense recognition systems. The approach of [22] is purely based on the semantic network of WordNet and does not use any pre-trained word embeddings. In their approach, similarly to the VSM representation of a linguistic item, the weight associated with a dimension in a semantic signature denotes the relevance or importance of that dimension for the linguistic item. In [22] this method was improved using the [5] like formula, with senses in place of synonyms or antonyms. In this DECONF method the word embeddings were used following [16]. They obtained state-of-the-art results for MEN-3 K, and RG-65 (here Spearman correlation 0.896 by far exceeds the inter-annotator confidence). However, for SIMLEX-999 the multi-sense results are 0.3 worse [23] and 0.2 worse in [22] compared to the Paragram based results. We improved the accuracy results by 8.75% and 24% for TOEFL and ESL test sets respectively. We observe the largest improvement of accuracy by applying the localized centering method for TOEFL test set. Testing on the ESL question set seems to give the best results by changing the similarity measure from cosine to RESM. Thus each step of the algorithm improves the results. The significance of the improvement varies, depending on the test set. A complete set of measured accuracies is presented in Table 3. The results for the APSyn ranking method are obtained using the Glove vector, not using PPMI as in [27]. The performance of

Paragram with counter-fitting for SIMLEX-999 is better than inter-annotator agreement, measured by average pairwise Spearman ρ correlation, which varies between 0.614 to 0.792 depending on concept types in SIMLEX-999. There are many results that are significantly worse than these in Table 3. For example, in [12] the accuracy for the TOEFL standard is 88.75% and for SIMLEX-999 the Spearman ρ is 0.53.

5 Conclusions

This work compares the state-of-the-art word embedding methods for three most reliable gold standards: TOEFL, ESL and SIMLEX-999. For TOEFL and ESL the GloVe, PPDB baseline with retrofitting, our novel RESM similarity measure and hubness reduction we are able to equal the Paragram results. For SIMLEX-999 Paragram with Counter-fitting results are clearly better than the Glove based methods using the PPDB 1.0. However, we propose the cosine retrofitting that basically achieves the Paragram with Counter-fitting results. The Paragram with Counter-fitting method contains several hyperparameters which is one source of its success. Its effects can be seen in Table 10 at https://arxiv.org/pdf/1506.03487.pdf. The Spearman ρ values for SIMLEX-999 are 0.667 for Paragram300 fitted to WS353, and 0.685 for Paragram300 fitted to SIMLEX-999. The difference is even larger for WS353. Then the Spearman ρ values for WS-353 are 0.769 for Paragram300 fitted toWS353, and 0.720 for Paragram300 fitted to SIMLEX-999. Still the best word embedding based methods are not able to achieve the performance of other dedicated methods for TOEFL and ESL. The work of [15] employed 2 fitting constants (and it is not clear that they were the same for all questions) for answering the TOEFL test where only 50 questions are used. Techniques introduced in the paper are lightweight and easy to implement, yet they provide a significant performance boost to the language model. Since the single word embedding is a basic element of any semantic task one can expect a significant improvement of results for these tasks. In particular, SemEval-2017 International Workshop on Semantic Evaluation run (among others) the following tasks [1]:

1. Task 1: Semantic Textual Similarity
2. Task 2: Multilingual and Cross-lingual Semantic Word Similarity
3. Task 3: Community Question Answering

In the category Semantic comparison for words and texts. Another immediate application would be information retrieval (IR). Expanding queries by adding potentially relevant terms is a common practice in improving relevance in IR systems. There are many methods of query expansion. Relevance feedback takes the documents on top of a ranking list and adds terms appearing in these document to a new query. In this work we use the idea to add synonyms and other similar terms to query terms before the pseudo- relevance feedback. This type of expansion can be divided into two categories. The first category involves the use of ontologies or lexicons (relational knowledge). The second category is word embedding (WE). Here closed words for expansion have to be very precise, otherwise a query drift may occur, and precision and accuracy of retrieval may deteriorate. There are several avenues to further improve the similarity results.

1. Use the multi-language version of the methods [24]
2. Use PPDB 2.0 to design the Paragram vectors [20]
3. Apply the multi-sense methods (knowledge graps) with state-of-the-art relational enriched vectors
4. Recalibrate annotation results using state-of-the-art results.

References

1. Semeval-2017 tasks. http://alt.qcri.org/semeval2017/index.php?id=tasks
2. Bruni, E., Boleda, G., Baroni, M., Tran, N.-K.: Distributional semantics in technicolor. In: The 50th Annual Meeting of the Association for Computational Linguistics, Proceedings of the Conference, Jeju Island, Korea, 8–14 July 2012, Volume 1: Long Papers, pp. 136–145 (2012). http://www.aclweb.org/anthology/P12-1015
3. Bullinaria, J.A., Levy, J.P.: Extracting semantic representations from word co-occurrence statistics: stop-lists, stemming, and SVD. Behav. Res. Methods **44**(3), 890–907 (2012)
4. Dinu, G., Baroni, M.: Improving zero-shot learning by mitigating the hubness problem. CoRR, abs/1412.6568 (2014)
5. Faruqui, M., Dodge, J., Jauhar, S.K., Dyer, C., Hovy, E.H., Smith, N.A.: Retrofitting word vectors to semantic lexicons. CoRR (2014)
6. Feldbauer, R., Flexer, A.: Centering Versus Scaling for Hubness Reduction, pp. 175–183. Springer, Cham (2016)
7. Finkelstein, L., Gabrilovich, E., Matias, Y., Rivlin, E., Solan, Z., Wolfman, G., Ruppin, E.: Placing search in context: the concept revisited. In: Proceedings of the 10th International Conference on World Wide Web, WWW 2001, pp. 406–414. ACM, New York (2001). ISBN 1-58113-348-0. https://doi.acm.org/10.1145/371920.372094
8. Ganitkevitch, J., Van Durme, B., Callison-burch, C.: PPDB: the paraphrase database. In: HLT-NAACL 2013 (2013)
9. Hara, K., Suzuki, I., Shimbo, M., Kobayashi, K., Fukumizu, K., Radovanovic, M.: Localized centering: reducing hubness in large-sample data. In: AAAI, pp. 2645– 2651. AAAI Press (2015)
10. Hill, F., Reichart, R., Korhonen, A.: Simlex-999: evaluating semantic models with (genuine) similarity estimation. Comput. Linguist. **41**(4), 665–695 (2015). https://doi.org/10.1162/COLI_a_00237
11. Jarmasz, M., Szpakowicz, S.: Roget's thesaurus and semantic similarity. CoRR (2012)
12. Kiela, D., Hill, F., Clark, S.: Specializing word embeddings for similarity or relatedness. In: EMNLP, pp. 2044–2048. The Association for Computational Linguistics (2015)
13. Landauer, T.K., Dumais, S.T.: A solution to plato's problem: the latent semantic analysis theory of acquisition, induction, and representation of knowledge. Psychol. Rev. **104**(2), 211–240 (1997)
14. Levy, O., Goldberg, Y., Dagan, I.: Improving distributional similarity with lessons learned from word embeddings. TACL **3**, 211–225 (2015). https://tacl2013.cs.columbia.edu/ojs/index.php/tacl/article/view/570
15. Lu, C.-H., Ong, C.-S., Hsu, W.-L., Lee, H.-K.: Using filtered second order co-occurrence matrix to improve the traditional co-occurrence model (2011)
16. Mikolov, T., Chen, K., Corrado, G., Dean, J.: Efficient estimation of word representations in vector space. CoRR (2013)

17. Miller, G.A., Fellbaum, C.: Wordnet then and now. Lang. Resour. Eval. **41**(2), 209–214 (2007)
18. Mrksic, N., Séaghdha, D.O., Thomson, B., Gasic, M., Rojas-Barahona, L.M., Su, P.-H., Vandyke, D., Wen, T.-H., Young, S.J.: Counter-fitting word vectors to linguistic constraints. In: HLT-NAACL, pp. 142–148. The Association for Computational Linguistics (2016)
19. Osterlund, A., Odling, D., Sahlgren, M.: Factorization of latent variables in distributional semantic models. In: EMNLP (2015)
20. Pavlick, E., Rastogi, P., Ganitkevitch, J., Van Durme, B., CallisonBurch, C.: PPDB 2.0. better paraphrase ranking, fine-grained entailment relations, word embeddings, and style classification. In: ACL (2), pp. 425–430. The Association for Computer Linguistics (2015)
21. Pennington, J., Socher, R., Manning, C.: Glove: global vectors for word representation. In: Proceedings of the 2014 Conference on Empirical Methods in Natural Language Processing (EMNLP), Doha, Qatar, October 2014, pp. 1532–1543. Association for Computational Linguistics (2014)
22. Pilehvar, M.T, Collier, N.: De-conflated semantic representations. In: EMNLP, pp. 1680–1690. The Association for Computational Linguistics
23. Pilehvar, M.T., Navigli, R.: From senses to texts: an all-in-one graph-based approach for measuring semantic similarity. Artif. Intell. **228**, 95–128 (2015). https://doi.org/10.1016/j.artint.2015.07.005
24. Recski, G., Iklódi, E., Pajkossy, K., Kornai, A.: Measuring semantic similarity of words using concept networks. In: Proceedings of the 1st Workshop on Representation Learning for NLP, Berlin, Germany, pp. 193–200. Association for Computational Linguistics (2016)
25. Roberts, K., Gururaj, A.E., Chen, X., Pournejati, S., Hersh, W.R., Demner-Fushman, D., Ohno-Machado, L., Cohen, T., Xu, H.: Information retrieval for biomedical datasets: the 2016 biocaddie dataset retrieval challenge. Database, 2017:bax068 (2017)
26. Rubenstein, H., Goodenough, J.B.: Contextual correlates of synonymy. Commun. ACM **8** (10), 627–633 (1965). https://doi.acm.org/10.1145/365628.365657
27. Santus, E., Chiu, T.-S., Lu, Q., Lenci, A., Huang, C.-R.: What a nerd! beating students and vector cosine in the ESL and TOEFL datasets. CoRR, abs/1603.08701 (2016)
28. Turney, P.D.: Mining the Web for Synonyms: PMI-IR versus LSA on TOEFL, pp. 491–502. Springer, Heidelberg (2001)
29. Wieting, J., Bansal, M., Gimpel, K., Livescu, K.: From paraphrase database to compositional paraphrase model and back. TACL **3**, 345–358 (2015). https://tacl2013.cs.columbia.edu/ojs/index.php/tacl/article/view/571

Calculating Optimal Queries
from the Query Relevance File

Jakub Dutkiewicz and Czesław Jędrzejek[(⊠)]

Poznań University of Technology, Poznań, Poland
{jakub.dutkiewicz, czeslaw.jedrzejek}@put.poznan.pl

Abstract. Query Expansion could bring very significant improvement of baseline results of Information Retrieval process. This has been known for many years, but very detailed results on annotated sets provide richer insight on the preferred added word space. In this work we introduce novel expanded term adequacy measures related to term frequency and inverse document frequency in relevant and non-relevant groups of documents. Term evaluation scores are derived using two term characteristics: inverse term representativeness and term usability. We generate the Optimal Queries based on the documents contents and the Qrels files of data used in the Text Retrieval Conference 2016 – Clinical Decision Support track (TREC-CDS 2016). The improvement can be up to a factor of 2 depending on the evaluation measure. Potentially, the method can be improved by increasing the learning set and applied to retrieval of documents in biomedical contests.

Keywords: Information retrieval · Query expansion
Term representativeness keyword

1 Introduction

The theory of information retrieval (IR) is quite mature so that in terms of evaluation measures the progress is saturating. To get further progress we need to go beyond models that refer only to query terms in a particular question. One could try to use knowledge contained in average relevance of a set of questions concerning given topic over an appropriate cluster of documents. The system, within which the IR Process runs, retains knowledge about a set of documents if they belong to the same topic. IR Process starts with a Query – a text message, which generalizes information about a problem given by a user of the system. IR process implements an IR Model, which uses the Query to retrieve a subset of documents relevant to the given problem from a set of all known documents. In our work, we extend the standard Information Retrieval protocol with a process known as Query Expansion (QE). QE process occurs after a Query is submitted and before the IR Model is applied. QE process broadens or reformulates the Query. Here we propose an algorithm for calculating an Optimal Query. Our goal is to create a training set of Queries and optimal Query Expansions, which would be later used to create a Supervised Learning based Query Expansion Model.

© Springer Nature Switzerland AG 2019
K. Choroś et al. (Eds.): MISSI 2018, AISC 833, pp. 249–259, 2019.
https://doi.org/10.1007/978-3-319-98678-4_26

Query Expansion could bring very significant improvement to retrieval process. For example, in the recent bioCADDIE Retrieval Challenge [1] the winning UCSD submission used the Snippet-based query expansion model [2]. The authors point out, that removing query expansion process from the information retrieval pipeline results in larger decrease of evaluation measure infNDCG (inferred Normalized Discounted Cumulative Gain) compared to removing other process enhancements. In the same Challenge [3] claims that in some cases adding a manually selected terms to queries significantly improves the information retrieval process. For example, adding the term 'CheY' to the first query increased infNDCG for this query to 0.585 from the baseline (very low) value of 0.209.

This work presents the method of acquiring the Optimal Queries for data used in the Text Retrieval Conference 2016 – Clinical Decision Support track (TREC-CDS 2016) [4]. The track data consists of 30 queries, a set of documents, which is a subset of PubMed collection and a file called Qrels. Data in the Qrels file determines whether given document is relevant to a query or not. It should be noted that the document set is partially annotated, so a relevance of a query to a document could be denoted as 'unknown' in the Qrels file. We generate the Optimal Queries based on the documents contents and the Qrels file. The acquisition method is an extension of the commonly known TF-IDF framework. There are two main contributions of this work – a set of Optimal Queries for the TREC data with its evaluation and a method of obtaining the Optimal Queries.

The paper is organized as follows. In the second section we refer to the similar works in this area. In the third section we briefly explain the TF-IDF framework and introduce measures used in the Optimal Query generation, we also describe the Optimal Query generation method. We go over the performance measures and present the results of using the Optimal Queries on the TREC data in the fourth section. The last section is dedicated towards the conclusions and the intended use of Optimal Queries set.

2   Related Works

Probabilities of a term appearing in a group relevant and nonrelevant documents appear in existing Information Retrieval (IR) Models. A scoring function (Chap. 7.2.1 of [5]) that is used as a basis of BM25

$$scoring\ function = \sum\nolimits_{i:d_i=1} \log \frac{p_i(1-s_i)}{s_i(1-p_i)}$$

uses said probabilities. In the equation p_i refers to the probability of term i occurring in relevant documents, and s_i equals the probability that term i occurs in non-relevant documents, d_i iterates over all terms in document d. In our work, we estimate and use the probabilities of term appearing in the relevant documents.

Another application of the relevance evaluation of terms can be found Divergence From Randomness (DFR) models and its extensions. Adaptive Distributional Extensions to DFR Ranking (ADR) was proposed in [6]. DFR ranking models [7] assume that informative terms are distributed in a corpus differently than non-informative terms. Using SVM [6] first detects the best-fitting distribution of non-informative terms in a collection, and then adapt the ranking computation to this best-fitting distribution. ADR method calculates the term information distribution to enhance the DFR method. In our work we are using the term information distribution to create a set of potential query expansions. In this paper, as well as in [6], the term evaluation is used for indirect enhancement of the Information Retrieval process. While [6] focuses on the method of obtaining the best distribution from a set of existing models, we assume a distribution and create a dedicated model for that distribution.

Claveau [8] re-examined the properties of representation spaces for documents or words in Information Retrieval. He showed that so called intrinsic dimensionality is chiefly tied with the notion of indiscriminateness among neighbors of a query point in the vector space. Using word embedding he developed a guidance for selection proper terms for expansion.

3   Method

In this section we briefly go over the TF-IDF framework, and we introduce the *term representativeness* and *term usability* measures. We will also describe the aggregation methods used to generate the Optimal Queries. Before all that, we briefly present the structure and contents of data used in the process.

The input data consists of a set Q of queries Q_i, which consist of query terms q_i; a query-document relevance file – Qrels; a set of documents D, which is split with use of the Qrels file into two subsets for each query D_Q^{POS} and D_Q^{NEG}. Subsets contain respectively all documents which are annotated as relevant to a given query and all documents which are annotated as non-relevant to a given query. Each document d within the set D consists of terms t.

3.1   TF-IDF Framework

TF-IDF framework is a short for Term Frequency – Inverse Document Frequency framework. Accordingly to its name, framework uses two separate metrics - a term frequency and a document frequency. The term frequency (*tf*) measure denotes the number of occurrences of a given term in a document, we define *tf* of a term t_i in document d as

$$tf_d(t_i) = \frac{|\{t : t = t_i \, and \, t \in d\}|}{|\{t : t \in d\}|}$$

The inverse document frequency (*idf*) refer to the number of documents in which the term was used. We define *idf* of a term t_i in a document set D as:

$$idf_D(t_i) = \log\left(\frac{|D|}{|\{d : t_i \in d \, and \, d \in D\}|}\right)$$

Table 1. Intended behavior of the *ir* measure given the *idf* for two separate sets.

idf_{D1}	idf_{D2}	ir
High	Low	High
Low	High	Low
High	High	High
Low	Low	High

The *tf* measure indicates the connection between a given term and a given document. The *idf* measure identifies the amount of information carried by a given term. In principle, the TF-IDF framework aggregates these two values to obtain a scoring function for the IR Model. Score of the term t for the document d within a document set D is given by

$$tfidf_D(t, d) = tf_d(t) \times idf_D(t).$$

The aggregation, calculated as a sum of scores for all of the terms within a query Q is as follows

$$tfidf_D^Q(d) = \sum_{t \in Q} tfidf_D(t, d).$$

In this work, we are using only the raw measures – *tf* and *idf*.

3.2   Term Representativeness

Let us define a representative term as a term, which has a high probability of appearing in a document from a given set, while maintaining a lower probability of appearing in a document outside of this set. We want our Optimal Query to contain terms, which appear in the D^{POS} subset and do not appear in the D^{NEG}. We use the *inverse term representativeness (ir)* to measure the representativeness property. *Inverse representativeness* is calculated upon the *inverse document frequency*. Contrary to the *idf*, low values of this measure indicate a good term - a term which is a good representation of D^{POS}. Despite similar calculation, the nature of *idf* and *ir* is reverse, *idf* denotes specificity of a term, while *ir* denotes the representativeness of a term. To calculate the *ir* for a set D^{POS} we need a set of documents D^{NEG} (referred to as negative samples), which do not belong to D^{POS}, *inverse representativeness* is intended to discriminate which terms are representative for the set D^{POS} and are not representative for the negative samples. If the probability of a term appearing in a document is similar in both sets, its representativeness should be lower (hence, value of the measure should be higher). Table 1 presents the intended behavior of the measure. If the term represents one of the sets and does not represent the other, we just need to take ratio of *df* measures for both sets, which is denoted as follows

$$ir'(t_i)_{D^{POS},D^{NEG}} = \frac{idf_{D^{POS}}(t_i)}{idf_{D^{NEG}}(t_i)}$$

If however the term is equally represented in both sets, we should increase the value of *ir* according to the formula:

$$ir''(t_i)_{D^{POS},D^{NEG}} = \frac{1}{\left|idf_{D^{POS}}(t_i) - idf_{D^{NEG}}(t_i)\right|}$$

We denote the above part of calculation as the normalization term. We use it to avoid anomalies. Let us describe an example of such anomaly. Imagine a pair of document sets for which the *idf* value of a term t_i is equal to respectively 0.001 and 0.1 – we believe this term to be representative for both sets, so we expect the *inverse representativeness* value for this term to be high. However, due to the ratio of *idf* the *ir'* value for this term is low. As the raw difference between *idf* values is relatively low, we manage to increase the value of *ir* with use of the normalization term. Finally we calculate the aggregated value of *ir* as:

$$ir(t_i)_{D^{POS},D^{NEG}} = \frac{idf_{D^{POS}}(t_i)}{idf_{D^{NEG}}(t_i)} \cdot \frac{1}{\left|idf_{D^{POS}}(t_i) - idf_{D^{NEG}}(t_i)\right|}$$

3.3   Term Usability

Let us define term usability as a value which indicates how often do we use a term within a given set D^{POS}. Similarly to the term representativeness, we don't want to assign high value of usability of a term if it is equally represented in a set of documents, which do not belong to D^{POS}. We calculate the *term usability (s)* upon the *tf* value. We also need a set of negative samples D^{NEG} to calculate the measure. We define *tf* measure for a set of documents as an average of *tf* values for each document within this set. We calculate that value with the following formula

$$tf_D^{mean}(t_i) = \frac{\sum_{d_i \in D} \frac{|\{t : t = t_i \, and \, t \in d_i\}|}{|\{t : t \in d_i\}|}}{|D|}$$

We define *term usability* as a ratio of *tf* for the positive document set and a *tf* for the negative samples. We apply analogical normalization term as we applied for the *inverse representativeness* for similar reasons.

$$s(t_i) = \frac{tf_{D^{POS}}^{mean}(t_i)}{tf_{D^{NEG}}^{mean}(t_i)} \left| tf_{D^{POS}}^{mean}(t_i) - tf_{D^{NEG}}^{mean}(t_i) \right|$$

We want the Optimal Query to contain terms which are used frequently within the positive document set and infrequently within the set of negative documents. The higher value of *usability* of the term, the more likely it is to be included in the Optimal Query. Term evaluation measures s and ir are query dependent. Sets D^{POS} and D^{NEG} vary for different queries. We denote query Q_i for the s and ir scores as respectively $s_{Qj}(t_i)$ and $ir(t_i)_{Q_j}$.

3.4   Term Evaluation Score

As it was already mention, we want Optimal Query to contain terms with as low *inverse term representativeness* and as high *term usability* as possible. The most natural scoring function derives directly from the TF-IDF framework as a multiplication of *usability* and *representativeness*. As the *inverse representativeness* is a contradiction of *inverse document frequency*, the final score comes as a fraction:

$$score_1\left(t_i, Q_j\right) = \frac{s_{Qj}(t_i)}{ir(t_i)_{Q_j}}.$$

We apply the L2 normalization to the s and ir scores over the entire set of terms to reduce dependence on Query specific properties, such as sizes of D^{POS} and D^{NEG}. We also want to separately measure contribution of the *usability* and *representativeness*. We use another score value to perform that task

$$score_2\left(t_i, Q_j\right) = \frac{k_1}{ir(t_i)_{Q_j}} + k_2 \cdot s_{Qj}(t_i)$$

In that equation, k_1 and k_2 elements are responsible for weighting the contribution of term uniqueness and term representativeness. Finally, we assemble a set of N terms for each query with the highest scores. This set is denoted as the Optimal Query.

4   Evaluation

We evaluate the designed measures on the TREC-CDS 2016 data. Document corpus is a subset of PubMed collection and it consists of 1240582 medical publications. Query collection consists of 30 queries. There are 5339 documents annotated as relevant to any query. We analyze the term evaluation properties on a case study of 3 first queries. We measure the performance of the term evaluation properties on an entire collection of queries using infNDCG, infAP, P@10 and R-prec metrics. This set of evaluation measures was proposed for the TREC competition. We compare our Optimal Queries to the best TREC-2016 results. It should be kept in mind, that we expect significantly better performance, as we used Query Relevance file to generate the queries.

Table 2. Summaries and a number of relevant documents for three chosen topics.

Topic	Summary	Relevant
1	78 M transferred to nursing home for rehab after CABG. Reportedly readmitted with a small NQWMI. Yesterday, he was noted to have a melanotic stool and then today he had approximately 9 loose BM w/some melena and some frank blood just prior to transfer, unclear quantity	128
2	An elderly female with past medical history of hypertension, severe aortic stenosis, hyperlipidemia, and right hip arthroplasty. Presents after feeling a snap of her right leg and falling to the ground. No head trauma or loss of consciousness	34
3	A 75F found to be hypoglycemic with hypotension and bradycardia. She had UA positive for klebsiella. She had a leukocytosis to 18 and a creatinine of 6. Pt has blood cultures positive for group A streptococcus. On the day of transfer her blood pressure dropped to the 60 s. She was anuric throughout the day, awake but drowsy. This morning she had temp 96.3, respiratory rate 22, BP 102/26	143

Table 3. *Inverse representativeness* for chosen terms. TOP 5 section contains five terms with the lowest *inverse representativeness* score for the D^{POS} set for a given query. FROM QUERY section contains evaluation of some terms, which appear in the query.

	Query 1		Query 2		Query 3	
	Term	*ir*	Term	*ir*	Term	*ir*
TOP 5	endoscopy	.048	arthroplasty	.001	dysgalactiae	.029
	melena	.049	fracture	.006	hypermucoviscous	.040
	gastrointestinal	.054	loosening	.007	aerobactin	.044
	angiolipoma	.056	Cemented	.009	pyogenes	.060
	bleeding	.062	hemiarthroplasty	.007	pyrogenic	.070
FROM QUERY	melena	.049	arthroplasty	.001	klebsiella	.132
	stool	.240	Leg	.083	drowsy	3.340
	blood	1.583	anesthesia	.444	anuric	.299
	edema	1.815	Edema	1.53	hypoglycaemia	.572
	chf	.871	hypertension	1.88	bradycardia	2.035
	unresponsiveness	.654	Snap	1.21	hypotension	.289

4.1 Analysis of the Term Representativeness Property

The TREC data includes various levels of description for a given problem. Each level of description is formulated as a query. The problem itself is referred to as a topic. Three queries are assigned for each topic. First one, denoted as Description, is a vast chunk of text, which describes a given problem with details, it includes specialized medical language. The second one, denoted as Summary is a few statements long, deprived of details description of the topic. The last one, denoted as Note, is a short message, which gives a brief overview of the problem. To analyze the term properties, we use three random topics. Summary queries for the chosen topics are presented in Table 2. As we can see, each topic is provided with a set of relevant documents. Those documents are the contents of the D^{POS} set. All other documents are the contents of D^{NEG} set. We calculate *inverse term representativeness* for all queries and present results for the three topics in Table 3.

Table 4. *Usability* for chosen terms. TOP 5 section contains five terms with the highest *usability* score for the D^{POS} set for a given query. FROM QUERY section contains evaluation of some terms, which appear in the query.

	Query		Query		Query	
	Term	s	Term	s	Term	S
TOP 5	amebic	.432	Cemented	47.3	chlorite	.532
	hookworm	.172	arthroplasty	9.44	nephrol	.453
	tyrosin	.161	periprosthetic	4.07	mutator	.284
	lipomatosis	.129	acetabular	3.87	ceftaroline	.078
	ugb	.111	Hip	3.44	pyogenes	.069
FROM QUERY	melena	.021	arthroplasty	9.44	klebsiella	.007
	stool	.005	Leg	.080	drowsy	.001
	blood	.001	anesthesia	.001	anuric	.004
	edema	.001	Edema	.001	hypoglycaemia	.001
	chf	.001	hypertension	.001	bradycardia	.001
	unresponsiveness	.001	Snap	.001	hypotension	.002

Table 5. Optimal Queries for the first three Queries with a length of 10.

Query 1	Query 2	Query 3
amebic	arthroplasty	nephrol
bleeding	Cemented	ceftaroline
hookworm	Tha	meningococcemia
tyrosin	Thas	chlorite
endoscopy	acetabular	endotoxaemia
gastrointestinal	periprosthetic	sepsis
lipomatosis	arthroplasties	fasciitis
gcts	osteotomy	stss
qft	Loosening	pyogenes
melena	Varus	mutator

As we can see, the TOP 5 terms accordingly correspond to given queries. TOP 5 terms for first and second query are directly connected to the problem given in the query. TOP 5 terms for the third query are less obvious, but still could be connected to the query. The FROM QUERY section includes some terms, which appear in the query, and are also more likely to appear outside of the positive document set, as their *ir* value is relatively high. It basically means, that not having those terms in the query would most likely improve the results of an IR process.

4.2 Analysis of the Term Usability Property

We performed similar analysis for the *term usability* property. Results are presented in Table 4. Here we can see that TOP 5 terms are much more specific, but still somewhat connected to a given problem. We find it plausible to have Optimal Queries, which include terms from the TOP 5 section. *Usability* value for the most terms taken from queries is set below .001. We expected the value to be higher, hence we carry out quantitative tests for the impact of the *usability*.

4.3 Evaluation of the Optimal Queries

We create the Optimal Queries using both *score₁* and *score₂* functions. Table 5. Contains 10 random terms from each query.

Table 6. Evaluation of significance of the normalization term in the score measures. The 'Normalization' row contains the measures with the normalization term applied.

Query model	N	P@10		infAP		infNDCG	
Normalization		*ir*	*ir, s*	*ir*	*ir, s*	*ir*	*ir, s*
Score₁	25	0.4067	0.4167 (+0.100)	0.0837	0.0829 (−0.001)	0.3477	0.3769 (+0.029)
Score₁	35	0.3967	0.4567 (+0.600)	0.0763	0.0867 (+0.010)	0.2768	0.4107 (+0.134)
Score₁	45	0.4533	0.4767 (+0.023)	0.0920	0.0973 (+0.005)	0.3060	0.4127 (+0.106)
Score₂	30	0.4667	0.4500 (−0.017)	0.0883	0.0904 (+0.002)	0.3648	0.3697 (+0.005)

To perform a quantitative analysis, we use Terrier software to create an index of the TREC collection and perform IR process using generated Optimal Queries. Retrieval evaluation is compared to TREC best results with evaluation metrics used in the TREC conference. We are using Query Relevance file to obtain Optimal Queries, so we expect results to be significantly better than TREC best results. We present full set of results in Table 7. Additionally, we perform some tests against the normalization term. We carry out the same retrieval tests on Queries generated with the normalization term for the *uniqueness* property and without it. Comparison is illustrated by Table 6. Our investigation suggests, that the highest baseline is given by the DPH retrieval model. We use that model in all tests for the evaluation.

Table 7. Evaluation scores for the Optimal Queries compared to TREC results

Query Creation model	N	P@10	R-prec	infAP	infNDCG
TREC best results	-	0.4033	0.1744	0.0454	0.2815
$Score_1$	25	0.4167	0.2016	0.0829	0.3769
$Score_1$	30	0.4400	0.2024	0.0806	0.3853
$Score_1$	35	0.4567	0.2006	0.0867	0.4107
$Score_1$	40	0.5000	0.2161	0.0945	0.4060
$Score_1$	45	0.4767	0.2187	0.0973	0.4127
$Score_1$	50	0.5067	0.2296	0.0997	0.4089
$Score_1$	55	0.4867	0.2257	0.0999	0.4007
$Score_1$	60	0.5133	0.2287	0.1002	0.4057
$Score_2$ ($k_1 = 1$; $k_2 = 0$)	30	0.4500	0.2011	0.0910	0.3728
$Score_2$ ($k_1 = 10$; $k_2 = 1$)	30	0.4500	0.2006	0.0907	0.3718
$Score_2$ ($k_1 = 1$; $k_2 = 1$)	30	0.4500	0.2007	0.0904	0.3697
$Score_2$ ($k_1 = 1$; $k_2 = 10$)	30	0.4667	0.2013	0.0932	0.3809
$Score_2$ ($k_1 = 1$; $k_2 = 0$)	30	0.4667	0.2013	0.0895	0.3783
$Score_2$ ($k_1 = 1$; $k_2 = 1$)	40	0.4600	0.2038	0.0886	0.3684
$Score_2$ ($k_1 = 1$; $k_2 = 1$)	50	0.4800	0.2060	0.0950	0.3942
$Score_2$ ($k_1 = 1$; $k_2 = 1$)	60	0.4700	0.2063	0.0953	0.3887

5   Conclusions

We create a set of queries which returns desirable evaluation values for the TREC-CDS 2016 dataset. The intention is to create a Supervised Learning based Query Expansion model. The training set for this model consists of Queries and Expansion Terms with their evaluation. Query Expansion model is expected to tell us whether a given term could be a Query Expansion term or not. We believe that a dataset provided by TREC-CDS 2016 is large enough to create such a model. Even though there are only 30 queries, a number of possible expansion terms is nearly unlimited. According to our evaluation, we can assign at least 60 terms for each query as possible Query Expansions. This gives us 1800 positive samples. It is possible to create a good binary classifier, with 1800 positive samples, as it has been done multiple times [9, 10]. Even though the share number of positive samples is acceptable, the samples are similar to each other and the learning algorithm might not be able to grasp entire domain of the Expansion process. The presented method is easily replicable for any dataset, which contains the query relevance file. If the samples created with TREC-CDS 2016 won't be enough to train the model, we plan to expand our research to other TREC datasets. As for the quality of samples, the p@10 and infNDCG vary depending on the used dataset, hence we compare the results with TREC best results. Generated queries perform, as expected, better than the Queries provided with the TREC dataset. As long as the performance is better than the baseline, and the Query Expansion classifier should grasp a piece of knowledge contained in Optimal Queries. We intend to use the TREC learning data to optimize query terms for the bioCADDIE Retrieval Challenge.

Acknowledgments. We acknowledge the Poznan University of Technology grant (04/45/DSPB/0185).

References

1. Roberts, K., Gururaj, A.E., Chen, X., Pournejati, S., Hersh, W.R., Demner-Fushman, D., Ohno-Machado, L., Cohen, T., Xu, H.: Information retrieval for biomedical datasets the 2016 bioCADDIE dataset retrieval challenge. Database 2017, bax068 (2017)
2. Wei, W., Ji, Z., He, Y., Zhang, K., Ha, Y., Li, Q., Ohno-Machado, L.: Finding relevant biomedical datasets: the UC San Diego solution for the bioCADDIE retrieval challenge. Database 2018, bay017 (2018)
3. Bouadjenek, M.R., Verspoor, K.: Multi-field query expansion is effective for biomedical dataset retrieval. Database 2017, bax062 (2017)
4. Roberts, K., Demner-Fushman, D., Voorhees, E.M., Hersh, W.R.: Overview of the TREC clinical decision support track. In: TREC 2016 (2016). https://trec.nist.gov/pubs/trec25/papers/Overview-CL.pdf. Accessed 2 May 2018
5. Croft, W.B., Metzler, D., Strohman, T.: Search Engines - Information Retrieval in Practice. Pearson Education, pp. I–XXV, 1–524 (2009). ISBN 978-0-13-136489-9
6. Petersen, C., Simonsen, J.G., Järvelin, K., Lioma, C.: Adaptive distributional extensions to DFR ranking. In: CIKM 2016, pp. 2005–2008 (2016)
7. Amati, G., van Rijsbergen, C.J.: Probabilistic models of information retrieval based on measuring the divergence from randomness. ACM Trans. Inf. Syst. **20**(4), 357–389 (2002)
8. Claveau, V.: Indiscriminateness in representation spaces of terms and documents. In: ECIR 2018, pp. 251–262 (2018)
9. Cho, J., Lee, K., Shin, E., Choy, G., Do, S.: How much data is needed to train a medical image deep learning system to achieve necessary high accuracy? arXiv:1511.06348.pdf (2016)
10. Figueroa, R.L., Zeng-Treitler, Q., Kandula, S., Ngo, L.H.: Predicting sample size required for classification performance. BMC Med. Inf. Decis. Mak. **12**, 8 (2012)

Survey on Neural Machine Translation into Polish

Krzysztof Wolk[✉] and Krzysztof Marasek

Polish-Japanese Academy of Information Technology, Warsaw, Poland
kwolk@pja.edu.pl

Abstract. In this article we try to survey most modern approaches to machine translation. To be more precise we apply state of the art statistical machine translation and neural machine translation using recurrent and convolutional neural networks on Polish data set. We survey current toolkits that can be used for such purpose like Tensorflow, ModernMT, OpenNMT, MarianMT and FairSeq by doing experiments on Polish to English and English to Polish translation task. We do proper hyperparameter search for Polish language as well as we facilitate in our experiments sub-word units like syllables and stemming. We also augment our data with POS tags and polish grammatical groups. The results are being compared to SMT as well as to Google Translate engine. In both cases we success in reaching higher BLEU score.

Keywords: NMT · CNN in translation · RNN in translation
Machine translation into Polish

1 Introduction

Machine translation (MT) started ca. 50 years ago with some rule-based systems. In 90's statistical MT (SMT) systems were invented. They create statistical models by analyzing aligned source-target language data (training set) and use them to generate the translation. This has been extended to phrases [1] and additional linguistic features (part-of-speech, and such a method dominates over last decades. During training phase, SMT creates a translation model and a language model. The first one stores the different translations of the phrases while the later model stores the probability of the sequence of phrases on the target side. During the translation phase, the decoder chooses the best translation based on the combination of these two models. This of course needs huge training sets (for proper estimation of statistical models) what limits translation quality especially for grammatically rich languages. Recent achievement, Neural Machine Translation uses vector-space word representation and deep learning techniques to learn best weights for neural network to transform segments from source to the target language. This is achieved using different recurrent network architectures: recurrent networks [2], networks with attention mechanism [3] or convolutional networks [4]. After initial enthusiasm gained by better NMT results on shared tasks [5], some observations has been made that NMT not guarantees better MT performance. Koehn and Knowles [6] found that for English-Spanish and German-English pairs NMT systems (compared to SMT) have: worse out-of-domain performance, worse

© Springer Nature Switzerland AG 2019
K. Choroś et al. (Eds.): MISSI 2018, AISC 833, pp. 260–272, 2019.
https://doi.org/10.1007/978-3-319-98678-4_27

performance in low resource conditions, worse translation of long sequences, sometimes weird word alignments produced by attention mechanism, some problems with large beam decoding, but better translation of unfrequent words (perhaps because use of subword units).

This study surveys SMT and NMT toolkits for Polish-English translations.

General quality of MT systems hardly depends on language pairs, training data amount and quality and domain's match. Particularly challenging is translation to/from low resourced language with different syntax and morphology. Polish as a Slavic language, have quite free word order and is highly inflected. The inflectional morphology is very rich for all word classes, seven distinct cases affect not only common nouns, but also proper nouns as well as pronouns, adjectives and numbers, complex orthography. This, in case of Polish-English translation, forms tasks which are hard to solve by statistical systems: unbalanced dictionaries (Polish usually 4–5 times bigger than English), segments sequences (free-word order) probabilities estimation, frequent use of foreign words but with Polish inflection, limited sizes of parallel corpora.

2 Toolkits Used in the Research

The baseline system testing was done using the Moses open source SMT toolkit [7] with its Experiment Management System (EMS) [8]. The SRI Language Modeling Toolkit (SRILM) [9] with an interpolated version of the Kneser-Ney discounting (interpolate –unk –kndiscount) was used for 5-gram language model training. We used the MGIZA++ [10] tool for word and phrase alignment. KenLM [11] was used to binarize the language model, with a lexical reordering set to use the msd-bidirectional-fe model.

As a second SMT toolkit we used state of the art ModernMT (MMT) system [12]. It was created in cooperation of Translated, FBK, UEDIN and TAUS. ModernMT also has secondary neural translation engine. Form the code on Github we know that it is based on PyTorch [13] and OpenNMT [14] it also uses BPE for sub-word units [15] generation as default. More detailed information are unknown and not stated on the project manual pages. Project probably has many more default optimizations.

The third toolkit we used is based on Google's TensorFlow. TensorFlow is an open source software library for high performance numerical computation. One of the modules within the TensorFlow is seq2seq [2, 16]. Sequence-to-sequence (seq2seq) models have enjoyed great success in a variety of tasks such as machine translation, speech recognition, and text summarization. This work uses seq2seq in the task of Neural Machine Translation (NMT) which was the very first testbed for seq2seq models with wild success.

The next system we experimented on was MarianNMT [17] (formerly known as AmuNMT) that is an efficient Neural Machine Translation framework written in pure C ++ with minimal dependencies. It has mainly been developed at the Adam Mickiewicz University in Poznań (AMU) and at the University of Edinburgh. It advantages are up to 15x faster translation than Nematus and similar toolkits on a single GPU, up to 2x faster training than toolkits based on Theano, TensorFlow, Torch on a single GPU, multi-GPU training and translation, usage of different types of models, including deep

RNNs, transformer and language model Binary/model-compatible with Nematus [18] models for certain model types, adjustation for Polish language.

Though RNNs have historically outperformed CNNs [19] at language translation tasks, their design has an inherent limitation, which can be understood by looking at how they process information. Computers translate text by reading a sentence in one language and predicting a sequence of words in another language with the same meaning. RNNs [16] operate in a strict left-to-right or right-to-left order, one word at a time. This is a less natural fit to the highly parallel GPU hardware that powers modern machine learning. In comparison, CNNs can compute all elements simultaneously, taking full advantage of GPU parallelism. They therefore are computationally more efficient. Another advantage of CNNs is that information is processed hierarchically, which makes it easier to capture complex relationships in the data.

3   Data Preparation

The experiments described in this article were conducted using official test and training sets from IWSTL'13 [20, 21] conference as well as WMT'17 conference. From IWSLT we borrowed the TED Lectures corpora and from WMT we used the Europarl v7 corpus [22]. Whereas TED Lectures were ready to be used the Europarl had to be pre-processed. To be more precise it was necessary to deduplicate the dataset and assure that none of development and test data was present in the training data set. The specification of both corpora is presented in the Table 1.

Table 1. Corpora specification

	Number of sentences	Unique Polish tokens	Unique English tokens
TED	134,678	92,135	58,393
Europarl	619,858	164,140	50,474

For data sub-word division and augmentation, we used our author tool. It was implemented as part of the study the Polish language and is able to segment Polish texts into the suffix prefix core and for syllables. Its additional advantage is the possibility of dividing the text into grammatical groups and tagging texts with the POS tags. This type of tool will not only have considerable significance for scientists, but also for business. In the currently rapidly evolving machine translation based on neural networks, the morphological segmentation is necessary to reduce the size of dictionaries consisting of full word forms (the so-called open dictionary) which are used, among others, in the training of the translation system and in language modelling.

Sample Stemmed Polish Sentence:

```
dział++ --@@pos_noun++ --@@b_38++ --@@sb_1++ --ania pod++
--@@pos_past_participle++ --@@b_46++ --@@sb_2++ --j te
w++ --@@pos_other_x++ --@@b_0++ --@@sb_0 wy++ --ni++ --
@@pos_noun++ --@@infl_M3++ --@@b_0++ --@@sb_4++ --ku
```

For English side text tagging we utilized spaCy POS Tagger for which we coded python script to unify the format of both tools. The tagging speed was about 70 sentences per second. A sample result:

```
we --%%pos_pronoun++ can --%%pos_verb++ not --
%%pos_adverb++ run --%%pos_verb++ the --
%%pos_determiner++ risk --%%pos_noun++ of --
%%pos_adposition++ creating --%%pos_verb++ a --
%%pos_determiner++ regulatory
```

Whenever in experiments section "stemmed" is used it means that the data was processed to such data format.

In addition, we used byte pair encoding (BPE) [15], a compression algorithm, to the task of word segmentation. BPE allows for the representation of an open vocabulary through a fixed-size vocabulary of variable-length character sequences, making it a very suitable word segmentation strategy for neural network models. We try this method independently as well as in conjunction with our "stemmer". After tokenizing and applying BPE to a dataset, the original sentences may look like the following. Note that the name "Nikitin" is a rare word that has been split up into subword units delimited by @@.

```
Madam President , I should like to draw your attention to
a case in which this Parliament has consistently shown an
interest . It is the case of Alexander Ni@@ ki@@ tin .
```

4   Experiments

All of the experiments were conducted on the same machine. We had to our disposal 32 core CPU, 256 GB of RAM and 4 x nVidia Tesla K80 GPUs. We used only CUDA-enabled toolkits that were able facilitate cuDNN library for faster neural computing. This allowed us not only to measure the quality but also time cost needed for similar operations on different toolkits and settings [23].

The experiments on neural machine translation were started rather in most casual option which is usage of Recurrent Neural Networks (RNN). Firstly, we focused on TensorFlow toolkit provided by the Google corporation. Using its default settings and official IWSTL 2013 PL-EN test sets we tried to compare it with SMT (Moses) quality. The results of such comparison are showed in Table 2.

Table 2. TensorFlow and Moses baseline results on TED corpus.

Corpus	Direction	NMT (BLEU)	SMT (BLEU)	NMT (Training time)	SMT (Training time)
TED	PL->EN	4.46	16.02	4 days	1.5 h
TED	EN->PL	5.87	8.49	4 days	1.5 h

The quality of NMT was not only much lower but also much slower. We conducted that it might be because of the fact that TED has a very wide domain, diverse dictionary and dictionary size disproportion between PL and EN. On the other hand, most successful research on NMT were done on narrow domains on texts that had similar vocabulary on both sides. Decision was made to switch to European Parliament Proceedings (EUP) parallel corpus. Those results are showed in the Table 3. As we can see even that adagrad optimization made positive impact on training outcome still the time cost and lower quality were not satisfying [24].

Table 3. TensorFlow and Moses baseline results on EuroParl corpus.

Corpus	Direction	Iteration	NMT (BLEU)	SMT (BLEU)	NMT (Training time)	SMT (Training time)
EUP	PL->EN	5000	6.13	37,91	1 day	2 h
EUP	PL->EN	10000	9.38	37,91	2 days	2 h
EUP	PL->EN	20000	13.02	37,91	4 days	2 h
EUP	PL->EN	50000	15.69	37,91	4.5 days	2 h
EUP	PL->EN	70000	15.44	37,91	4.5 days	2 h
EUP	PL->EN	100000	14.52	37,91	5 days	2 h
EUP	PL->EN	150000	13.67	37,91	7 days	2 h
EUP	EN->PL	30000	8.17	27.11	6 days	2 h
EUP	EN->PL	60000	9.78	27.11	7 days	2 h
EUP	EN->PL	100000	10.10	27.11	8 days	2 h
EUP	EN->PL	130000	10.21	27.11	9.5 days	2 h
EUP	EN->PL	180000	9.85	27.11	12 days	2 h
EUP	PL->EN	20000	9.78	37,91	2 days	2 h
EUP - Adagrad	PL->EN	40000	11.34	37,91	4 days	2 h
EUP - Adagrad	PL->EN	55000	12.43	37,91	5 days	2 h
EUP - Adagrad	PL->EN	90000	19.43	37,91	7 days	2 h
EUP - Adagrad	PL->EN	120000	19.63	37,91	9 days	2 h

In conclusion we assumed that current baseline RNN topology and training parameters in the TensorFlow were not properly optimized for the polish language. That is why we put our attention on Polish NMT system called Marian NMT. The Table 4 shows our experiments in PL-EN direction.

Table 4. MarianNMT Polish to English results with sub-word units.

Options	Iter(k)	Non-stemmed, no-bpe, baseline	Stemmed, no-bpe	Non-stemmed, bpe	Stemmed, no-bpe
	Steps (k)	BLEU	BLEU	BLEU	BLEU
	2	5.29	-	-	4.83
	4	21.44	16.02	-	17.03
	6	30.89	-	-	22.1
	8	34.79	25.7		24.09
	10	36.74	27.11	35.54	25.18
	12	37.99	27.82	-	25.64
	14	38.57	28.08	-	25.95
	16	39.15	28.54	-	26
	18	39.54	28.42	-	25.98
	20	39.51	28.68	37.98	25.91
	22	39.83	28.79	-	25.62
	24	39.98	28.65	-	25.58
	26	40.08	28.69	-	25.02
	28	40.18	28.53	-	25.1
	30	40.45	-	38.76	24.63
	32	40.36	-	-	24.57
	34	40.37	-	-	24.26

In the Table 5 we present similar experiments on opposite direction (EN->PL).

Even that we finally obtained satisfactory results those were still too similar to SMT method. Another problem was training performance. We required one week to compute a proper experiment. We also found out that using sub-words units (stemming) improves the translation into Polish by visible factor, whereas in opposite direction the results were negative.

All this was motivation for using CNN instead of RNN. It should provide similar results in much faster time. Decision was made to use FairSeq toolkit developed in the Facebook laboratories. We show results of translation into Polish in Table 6. We were able to obtain results very similar to MarianNMT but in much less time. It required only 1.5 day in average to conduct a full training whereas in Marian NMT it was about 6 days. We also decided to translate only into Polish direction because from Marian NMT experiments we concluded that sub-word units are only usable when translating into Polish. In opposite direction they generated too many wrong hypotheses. The same conclusions could be drawn from FairSeq experiments.

266   K. Wolk and K. Marasek

Table 5. MarianNMT English to Polish results with sub-word units.

Options	Iter (k)	Stemmed, no-bpe	Stemmed, no-bpe	Non-stemmed, no-bpe, baseline	Stemmed-korrida, no-bpe	dim-rnn-2048, stemmed-korrida, no-bpe	max-length-200, dim-rnn-2048, stemmed-korrida, no-bpe
		BLEU	BLEU	BLEU	BLEU	BLEU	BLEU
	2	1.36	1.17	0.81	1.26		1.25
	4	6.58	7.98	6.8	5.96	8.32	4.75
	6	10.1	13.35	15	10.56	12.86	9.69
	8	12.31	15.9	19.56	12.98	15.15	14.41
	10	13.56	17.52	22.32	14.31	16.31	17.51
	12	14.54	18.6	23.96	15.06	17.09	19.93
	14	14.83	19.08	25.17	15.6	17.23	21.33
	16	15.15	19.63	26.25	15.95	17.31	22.35
	18	15.41	20.01	27.11	16.1	17.77	23.04
	20	15.67	20.29	27.76	16.4	17.74	23.61
	22	15.59	20.68	28.3	16.21	17.73	24.46
	24	15.77	20.83	28.29	16.75		24.89
	26	15.95	20.7	28.43	16.58		25.19
	28	15.91	21.13	28.81	16.63		25.55
	30	15.82	21.04	29.06	17.04		26.07
	32	15.86	21.07	29.14	16.92		26.17
	34	15.62	20.87	29.35	16.87		26.37
	36	15.43	21.09	29.68	16.74		26.78
	38	15.46	21.23	29.75	16.57		26.9
	40	15.27		29.85	16.51		27.04
	42	15.34		29.97	16.51		27.31
	44			30.14	16.59		27.61
	46			30.15	16.38		27.63
	48			30.25	16.47		27.8
	50			30.34			27.86
	52			30.48			28.06
	54			30.41			28.18
	56			30.63			28.37
	58			30.66			28.29
	60			30.82			28.43
	62			30.72			28.28
	64			30.28			28.59
	66						28.52
	204						31.35

Table 6. Initial experiments on CNN Translation into Polish using FairSeq.

#	Experiment	Data type	Settings	BLEU	Sentence length limit	Dict en	Dict pl
5	4_eup-n-max-175-enpl	Stemmed with grammatical group codes	With group and suffix numbers	26.77	175	28.364	37,328
7	5_eup-n-max-59-enpl-not-stemmed	Not-stemmed	Baseline	29.59	59	28,210	80,437
8	6_eup-n-max-175-enpl-stemmed-no-codes	Stemmed	-bptt 0	28.85	175	28,364	37,008
9	6_eup-n-max-175-enpl-stemmed-no-codes	Stemmed	-bptt 25 test, worse bleu, slower training	28.48	175	28,364	37,008
10	6_eup-n-max-175-enpl-stemmed-no-codes	Stemmed	5 attention modules - noutembed 768	31.29	175	28,364	37,008
11	4_eup-n-max-175-enpl	Stemmed with grammatical group codes	Do 5 attention modules work better than 2? - noutembed 768	29.74	175	28,364	37,328
12	7_eup-n-stemmer-3-enpl	Stemmed	Stemmer stemmed-3 from destemmed stemmed-5; -nenclayer 15 -nlayer 10 -nembed 256 -noutembed 256 - nhid512	31.33	175	28,364	31,655
13	7_eup-n-stemmer-3-enpl	Stemmed	Stemmer-r -neclayer 15 - nlayer 5 -nembed 512 - noutembed 768 -nhid 512	31.02	175	28,364	31,655

Our next step in research was training hyperparameter search. For this purpose, we used only 25% of dataset in order to improve the performance.

Our baseline translation system score was equal to 24.95 whereas finally we obtained score of 26.18 in BLEU metric [25]. Most importantly we proved that our "stemmer" works well and that it really improves the translation quality. Highest score was obtained using stemmed data augmented with grammatical group and base suffix code. Some minor improvement was observed while only using stemmed word forms without extra codes. What is interesting, most advanced tagging and stemming with group codes and POS tags did not work as anticipated. Most likely reason to this is that we simply added to many information which artificially extended the number of tokens in sentences. This could possibly reduce the training accuracy especially that our word embeddings encoding vector was set only to have size of 256 (–dim-embeddings).

Table 7. CNN Experiments with sub-word units (English to Polish).

#	Experiment	Data type	Parameters	BLEU
23	23_st5_15at_100pr	Stemmed with grammatical group codes	-model fconv -nenclayer 15 - nlayer 15 -dropout 0.25 - nembed 256 -noutembed 256 - nhid 512	29.91
24	24_st5_15at_100pr_512emb	Stemmed with grammatical group codes	-model fconv -nenclayer 15 - nlayer 15 -dropout 0.25 - nembed 512 -noutembed 512 - nhid 512	30.67
25	25_st5_15at_100pr_768emb	Stemmed with grammatical group codes	-model fconv -nenclayer 15 - nlayer 15 -dropout 0.25 - nembed 768 -noutembed 768 - nhid 512	30.71
29	29_stem26	Stemmed with grammatical group codes and POS tags	–dropout 0.25 –optim nag –lr 0.25 –clip-norm 0.1 – momentum 0.99 –max-tokens 5000; arch='fconv', decoder_embed_dim=512, decoder_out_embed_dim=256, encoder_embed_dim=512, max_source_positions=1024, max_target_positions=1024	25.97
30	30_stemmed_15_bpe	Stemmed with grammatical group codes and BPE	Like 29, –max-tokens 6000	31.53
31	31_clean_bpe	BPE only	Same as 30	30.23
33	33_stemmed_with_pos_191_en_pos_bpe	Stemmed with grammatical group codes, POS tags and BPE	Same as 30	31.28

Next, we choose best settings from hyperparameter search part and applied it to 100% of our data set. Because of the POS issue we also decided to increase maximum number of accepted tokens per sentence and the size of embedding vector to 512 and 768 respectively. This made the BLEU score improve event more to 31.53 without POS tags (but with stemming and BPE) and with POS tags and BPE to 31.28. By doing so we managed to obtain better results on CNN then on MarianNMT(RNN) reducing training time by factor of 4 (Table 7).

The problem of little lower score while using POS tags which we did not anticipate remained. That is why did manual system evaluation by analysing translation results. That is how we discovered a looping issue. For instance, for the following sentence the generated hypothesis was partially correct but repeated many times. This was most likely reason for BLEU disproportion.

We judged that the attention window on English side was too small and made the system go into a loop. Unfortunately, we did not succeed in eliminating this issue yet.

The final step of our research was comparison of our systems to commercial Google Translate engine and context aware ModernMT system (Table 8). Both of those systems are state of the art tools but (especially MMT) their close architecture makes it impossible to directly compare the results. Nonetheless, they put some light at the outcomes of this research.

Table 8. Translation of EUP using Google Translate

Engine	Translation direction	BLEU
Google	PL->EN	31,74
Google	EN->PL	27,61
ModernMT (SMT)	PL-EN	39,17
ModernMT (SMT)	EN->PL	28,93
ModernMT (NMT)	PL->EN	41,47
ModernMT (NMT)	EN->PL	33,01

5   Conclusions

To conclude our work, we put everything in one big table for easier comparison. We only put the best scores for every toolkit that we used in the research and we skipped the information if the data was augmented with tags, BPE or stemmed. Only best scores remained. We compare our work to Google Translate API [26] and ModernMT (Table 9).

Summing up we successfully surveyed main translation systems that are present on the market in context of polish language. We trained baseline statistical system and successfully improved its quality using neural networks. What is more we were able to achieve better system score than Google Translate Engine within the test domain. We also proved that using sub-words units in translation into polish make positive impact on translation quality. To be more precise our sub-division tool performed better than the widely used BPE method, especially when also annotating data with grammatical groups or POS tags.

Table 9. Summary of translation experiments.

System	Type	Direction	BLEU	TRANING TIME
Moses	SMT	PL->EN	37.91	2 h
TensorFlow	RNN	PL->EN	19.63	9 days
MarianNMT	RNN	PL->EN	40.45	2 days
MarianNMT	RNN (Stemmed)	PL->EN	28.74	3 days
FairSeq	CNN	PL->EN	-	-
Google	Hybrid	PL->EN	31.74	-
ModernMT	SMT	PL->EN	39.17	1 h
ModernMT	RNN	PL->EN	41.47	15 h
Moses	SMT	EN->PL	27.11	2 h
TensorFlow	RNN	EN->PL	10.21	9.5 days
MarianNMT	RNN	EN->PL	30.28	6 days
MarianNMT	RNN (Stemmed)	EN->PL	31.35	6 days
FairSeq	CNN	EN->PL	29.59	1.5 days
FairSeq	CNN (Stemmed)	EN->PL	31.53	1.5 days
Google	Hybrid	EN->PL	27.61	-
ModernMT	SMT	EN->PL	28.93	1 h
ModernMT	RNN	EN->PL	33.01	15 h

Nonetheless it must be noted that much better engines that we trained are most likely to be possible to be prepared. First of all, we used small amount of training data. Secondly, it would be a good idea to incorporate language model into NMT trained from bigdata amounts of texts. Another idea for future experiments is adding lemmatization into our data. We also plan use transfer learning methods in order to further improve quality of NMT and adapt it to other text domains.

We believe that currently CNN is best translation path to follow. In our opinion by optimizing data training parameters and CNN topology it would be easy to overscore in BLEU even the ModernMT system.

References

1. Koehn, P., Och, F.J., Marcu, D.: Statistical phrase-based translation. In: Proceedings of the 2003 Conference of the North American Chapter of the Association for Computational Linguistics on Human Language Technology, vol. 1, pp. 48–54. Association for Computational Linguistics, May 2003
2. Sutskever, I., Vinyals, O., Le, Q.V.: Sequence to sequence learning with neural networks. In: Advances in Neural Information Processing Systems, pp. 3104–3112 (2014)
3. Bahdanau, D., Cho, K., Bengio, Y.: Neural machine translation by jointly learning to align and translate. arXiv preprint arXiv:1409.0473 (2014)
4. Gehring, J., Auli, M., Grangier, D., Yarats, D., Dauphin, Y.N.: Convolutional sequence to sequence learning. arXiv preprint arXiv:1705.03122 (2017)

5. Luong, M.T., Manning, C.D.: Stanford neural machine translation systems for spoken language domains. In: Proceedings of the International Workshop on Spoken Language Translation, pp. 76–79 (2015)
6. Koehn, P., Knowles, R.: Six challenges for neural machine translation. arXiv preprint arXiv: 1706.03872 (2017)
7. Koehn, P., Hoang, H., Birch, A., Callison-Burch, C., Federico, M., Bertoldi, N., Cowan, B., Shen, W., Moran, C., Zens, R., Dyer, C.: Moses: open source toolkit for statistical machine translation. In: Proceedings of the 45th Annual Meeting of the ACL on Interactive Poster and Demonstration Sessions, pp. 177–180. Association for Computational Linguistics, June 2007
8. Vasiļjevs, A., Skadiņš, R., Tiedemann, J.: LetsMT!: a cloud-based platform for do-it-yourself machine translation. In: Proceedings of the ACL 2012 System Demonstrations, pp. 43–48. Association for Computational Linguistics, July 2012
9. Stolcke, A.: SRILM-an extensible language modeling toolkit. In: Seventh International Conference on Spoken Language Processing (2002)
10. Junczys-Dowmunt, M., Szał, A.: Symgiza++: symmetrized word alignment models for statistical machine translation. In: Security and Intelligent Information Systems, pp. 379–390. Springer, Heidelberg (2012)
11. Heafield, K.: KenLM: faster and smaller language model queries. In: Proceedings of the Sixth Workshop on Statistical Machine Translation, pp. 187–197. Association for Computational Linguistics, July 2011
12. Jelinek, R.: Modern MT systems and the myth of human translation: Real World Status Quo. In: Proceedings of the International Conference Translating and the Computer, November 2004
13. PyTorch Core Team: Pytorch: Tensors and dynamic neural networks in python with strong GPU acceleration (2017)
14. Klein, G., Kim, Y., Deng, Y., Senellart, J., Rush, A.M.: Opennmt: open-source toolkit for neural machine translation. arXiv preprint arXiv:1701.02810 (2017)
15. Sennrich, R., Haddow, B., Birch, A.: Neural machine translation of rare words with subword units. arXiv preprint arXiv:1508.07909 (2015)
16. Cho, K., Van Merriënboer, B., Gulcehre, C., Bahdanau, D., Bougares, F., Schwenk, H., Bengio, Y.: Learning phrase representations using RNN encoder-decoder for statistical machine translation. arXiv preprint arXiv:1406.1078 (2014)
17. Junczys-Dowmunt, M., Grundkiewicz, R., Grundkiewicz, T., Hoang, H., Heafield, K., Neckermann, T., Seide, F., Germann, U., Aji, A.F., Bogoychev, N., Martins, A.: Marian: Fast Neural Machine Translation in C++. arXiv preprint arXiv:1804.00344 (2018)
18. Sennrich, R., Firat, O., Cho, K., Birch, A., Haddow, B., Hitschler, J., Junczys-Dowmunt, M., Läubli, S., Barone, A.V.M., Mokry, J., Nădejde, M.: Nematus: a toolkit for neural machine translation. arXiv preprint arXiv:1703.04357 (2017)
19. LeCun, Y., Bottou, L., Bengio, Y., Haffner, P.: Gradient-based learning applied to document recognition. Proc. IEEE 86(11), 2278–2324 (1998)
20. Wołk, A., Wołk, K., Marasek, K.: Analysis of complexity between spoken and written language for statistical machine translation in West-Slavic group. In: Multimedia and Network Information Systems, pp. 251–260. Springer, Cham (2017)
21. Wołk, K., Marasek, K.: Polish-English speech statistical machine translation systems for the IWSLT 2013. arXiv preprint arXiv:1509.09097 (2013)
22. Koehn, P.: Europarl: a parallel corpus for statistical machine translation. In: MT Summit, vol. 5, pp. 79–86, September 2005

23. Wu, Y., Schuster, M., Chen, Z., Le, Q.V., Norouzi, M., Macherey, W., Krikun, M., Cao, Y., Gao, Q., Macherey, K., Klingner, J.: Google's neural machine translation system: bridging the gap between human and machine translation. arXiv preprint arXiv:1609.08144 (2016)
24. Kingma, D.P., Ba, J.: Adam: a method for stochastic optimization. arXiv preprint arXiv: 1412.6980 (2014)
25. Papineni, K., Roukos, S., Ward, T., Zhu, W.J.: BLEU: a method for automatic evaluation of machine translation. In: Proceedings of the 40th Annual Meeting on Association for Computational Linguistics, pp. 311–318. Association for Computational Linguistics, July 2002
26. Groves, M., Mundt, K.: Friend or foe? Google Translate in language for academic purposes. Engl. Specif. Purp. 37, 112–121 (2015)

Multilingual and Intercultural Competence for ICT: Accessing and Assessing Electronic Information in the Global World

Marcel Pikhart[(✉)]

University of Hradec Králové, Hradec Králové, Czech Republic
marcel.pikhart@gmail.com

Abstract. We are experiencing unprecedented changes in the way people communicate in the recent decade due to the massive introduction of network communication tools, platforms and devices in the global world. We use the Internet as an everyday tool for communicating information, however, we do not realise sufficiently that the information is coded into a language. Despite the fact that the majority of the shared information is in English as a lingua franca, i.e. English as a shared tool for communication, we have to be aware of the fact that the important part of the message is lost in translation because we still use original languages for websites. The websites are almost always created in the original language (such as Chinese) and then there is usually an option to switch to English. This English translation is usually literal and ignores the intercultural differences of the given language and the target language. The paper attempts to show potential pitfalls of this aspect of the global Internet, i.e. its interculturality, which is almost always neglected. The IT specialists must be aware of these potential intercultural misunderstandings, particularly when they create websites. This is also the reason why the author introduced adequate course for IT and business students (intercultural business communication) because of the urgent need of the market to be able to understand interculturality deeply so that the electronic communication is managed properly and without significant information loss. The conducted research focused on the awareness of the issue in ICT staff in several Czech small and medium enterprises and corporations with the result of almost zero awareness of the need to understand the intercultural situation in the global Internet communication. Therefore, this paper attempts to summarize the key aspects of intercultural communication for ICT and highlights the need of the awareness implemented into the website creation itself.

Keywords: Intercultural communication · Languages in ICT
Communication in ICT · Multilingual aspects of ICT
Multicultural aspects of ICT

1 Global Intercultural Environment in ICT: English as a Lingua Franca

In the past decade we are experiencing unprecedented trends in electronic communication perpetrated by ubiquitous Internet in global communication.

© Springer Nature Switzerland AG 2019
K. Choroś et al. (Eds.): MISSI 2018, AISC 833, pp. 273–278, 2019.
https://doi.org/10.1007/978-3-319-98678-4_28

Currently, in basically all western companies, English has become a lingua franca, i.e. a shared communicative tool used throughout the company, both for its internal and external communication. This fact must be reflected by the ICT staff inasmuch they are responsible for company bilingual website creation, PowerPoint templates, etc. Many other company tools are bilingual too, such as CRM software, etc. The reason for this approach is convenience - English as a lingua franca is present in all company communication platforms, not connected merely to business but also to ICT. The lack if intercultural skills will necessarily lead to business failures [14] and ICT departments are the key segment in creating electronic communication standards.

We can observe that English has become a global language and it means that it obtains a very special position and function alongside with the mother tongues. English as a lingua franca is a truism both in business and in ICT. We could even conclude that English has become a lingua franca in these two areas. From a linguistics point of view English as a lingua franca is characterised by a very high level of mutual understanding and cooperative principles [12]. However, we have to realise that there are not only many regional versions of English but also many versions of English used by non-native speakers, and these discrepancies lead to many misunderstandings and language conflicts [5].

The Internet has created a platform of cross-border market penetration which necessarily means that business ought to adopt universal standards. This leads to a question whether the global market which is now based on electronic transfer of information (knowledge society, ICT) should change its communication protocol. The answer is that organisational communication practices have to be altered so that they respect global rules of electronic communication which has become more uniform and neutral [13].

2 Current Situation in the ICT Regarding Interculturality

Basically all websites are bilingual, i.e. both in the mother tongue of the company, but also in other international languages - almost 100% of the websites use English version, some also provide Spanish, French and Russian options (other languages may be available too). The biggest danger comes at the phase of the translation of the documents, but also with the contents and design of the website. Moreover, it is crucial to realise that in the global intercultural world it is not only the contents of the information which is important and crucial to understand the message but more and more it is also the context we connect the information with.

Nations, companies and people are interconnected, they use languages to share the information and this happens in a cultural context which must be taken into account. Despite the fact that linguists conclude that, surprisingly, when there are two users of a shared communication which is used by the two people for whom it is not their mother tongue, this discourse is usually without any communication problems as both parties accept certain limits and their communication becomes efficient and authentic and both parties create individually supportive environment to understand each other [4]. The hegemony of the given language as a phenomenon which is fixed and invariable is therefore dubious, but it doesn't mean that we can ignore cultural differences and encoding which influence the context of the message transferred from the sender to the receiver.

Email messages are ubiquitous these days throughout companies, however, in some cultures (high-context vs. low-context cultures) this form of communication is considered as inappropriate whereas in others totally adequate. The introduction of directness through email communication may be considered rude and inappropriate in some Asian countries, however, very appropriate in European context [12]. The translation of websites into English which is a relatively direct language can be considered inappropriate in the users whose cultural background is based in more indirect communication (Europe vs. Asia).

Intercultural aspects are very important in website creation which is a form of marketing for the company. For Germanic western European countries (Germany, Austria, the UK, Switzerland) the websites should reflect high performance and achiever themes as these countries have small power-distance index, medium uncertainty avoidance, high individuality and high masculinity, to use traditional Hofstede's terms [7]. Countries of Latin origin (France, Portugal, and Spain) have medium power distance, strong uncertainty avoidance, varied individualism and low masculinity. Therefore, the information presented should appeal to consumers' status with an emphasis on functionality and stress reducing features (guarantees, warranties). However, Scandinavian countries (Denmark, Sweden, Finland, and Norway) have small power distance index, low uncertainty avoidance, high individualism and low masculinity. The creators of the websites and marketers should focus on novelty, environmental issues and social consciousness [6].

When we plan to use the website globally, we have to take into account possible intercultural implications. It is not sufficient to produce something in a source language and then merely translate it into the target language. The layout of the website will also be influenced by the translation process, as the English language text translated to other languages will vary in length (25% longer in French, 30% longer when translated into German).

Various symbols used in websites and other IT communication means will bare various meanings, usually interpreted incorrectly, and are far from universal understanding. The same rule applies in colours. White colour in Europe means purity, however, this colour is connected to death in Chinese culture.

There is a cultural checklist which may be taken into consideration:

- the brand name and the product is acceptable in target markets
- slogans might not work in other languages
- respect taboo topics, words and sex taboos
- avoid referenced not understood by people from other cultures
- be careful with symbols and icons, in many cultures they bear a different meaning
- check the balance between text and visuals, the ratio may need to be changed as they move from one culture to another
- the text length will vary when translated from one language into another [1].

To sum it up, we are facing new electronic communication challenges which must be solved urgently not to lose business opportunities [2, 9]. Therefore, the ICT departments have a great opportunity to show their importance in companies when proving their pragmatic practical approach which may lead to company improved competitiveness.

3   Research

There has been a lot of research into internal communication in companies and also its connection to ICT [8, 10, 11]. However, the reason to conduct this research was the fact that there is no research into the importance of intercultural issues in ICT despite the fact that we should pay our undivided attention to the topic due to its increased importance in the past few years. ICT staff should know that the cultural context of the information is necessary to decode the information.

3.1   Research Question

How much are the ICT specialists in the Czech Republic aware of the interculturality when creating bilingual websites?

3.2   Research Methodology

Several (27 respondents) ICT specialists (top management and middle management) from a few small and medium enterprises (total number 16) and corporations (total number 3) doing business both in the Czech Republic and also involved in international business were interviewed on their awareness of the current situation of interculturality in the ICT issues, such as web creation, PowerPoint template creation and creation of other company templates. The research was qualitative with the aim to find out the awareness of interculturality of the staff responsible for creation or maintenance of the websites. The research was conducted in January 2018 directly in these companies through guided interviews.

3.3   Research Results and Discussion

Nearly all the respondents expressed their lack of information connected to interculturality in the creation of websites of these companies. Just two out of 27 knew the potential risks and were aware of the situation. The knowledge of these was connected to their business experience from other, usually very different, culturally distant countries (e.g. China). These respondents knew the dangers when culturally different information and experience is transferred insensitively from one language into another one. These respondents are also aware that the mere literal transfer from one language into another one used in company websites is not sufficient, however, the vast majority of the respondents do not consider this issue as crucial.

The research clearly shows the lack of intercultural competence of ICT staff responsible for the creation and maintenance of company communication tools which are used in the global environment. Therefore, there are many dangers arising from this fact, of which the most important is misunderstanding. The technical aspects are considered more important, however, a lot of research mentioned above clearly shows that there are many important intercultural aspects which are still neglected.

4 Conclusion

David Crystal expressed his opinion on improper use of English in business communication already in 1997 [3] when he observed that a third of British exporters lose their opportunity to export for language reasons. Nowadays, when all companies use English for business communication (both oral and electronic), the threat is much higher. The information which is transferred electronically is vulnerable to be distorted and this threat is present mostly when the transfer crosses cultural borders, i.e. moves from one culture to a culture which is very different, e.g. Europe vs. Asia.

The intended information transfer can be easily distorted, however, when we realise the potential pitfalls we have an opportunity to minimize these situations. To stay competitive, organisations and their departments of ICT will have to adapt to the current changes and dynamics in the global market which is significantly intercultural, otherwise they will necessarily fall behind. This is the task for ICT staff who is responsible for a substantial part of electronic communication facilitation.

Acknowledgment. This paper is supported by the SPEV 2018 project of the Faculty of Informatics and Management, University of Hradec Kralove, Czech Republic. The author thanks Jan Špriňar for his cooperation.

References

1. Bassnett, S.: Studying British Cultures. London (1997)
2. Cerna, M., Svobodova, L.: Internet and social networks as the support for communication in the business environment - pilot study. In: Jedlicka, P., Maresova, P., Soukal, I. (eds.) Hradec Economic Days, vol. 7, no. 1 (2017)
3. Crystal, D.: English as a Global Language. Cambridge University Press, Cambridge (1997)
4. Firth, A.: The lingua franca factor. Intercult. Pragmat. **6**(2), 147–170 (2009)
5. Gerritsen, M., Nickerson, C.: BELF: business english as a lingua franca. In: Bargiela-Chiappini, F. (ed.) The Handbook of Business Discourse. Edinburgh University Press, Edinburgh (2009)
6. Gibson, R.: Intercultural Business Communication. Oxford University Press, Oxford (2002)
7. Hofstede, G., et al.: Cultures and Organisations. Software of the Mind (2011). http://testrain.info/download/Software%20of%20mind.pdf
8. Hola, J.: Internal communication in small and medium sized enterprises. Ekonomie Manag. **XV**(3), 32–45 (2012)
9. Maresova, P., Klimova, B.: Economic and technological aspects of business intelligence in European business sector. In: 11th International Scientific Conference on Future Information Technology (FutureTech)/10th International Conference on Multimedia and Ubiquitous Engineering (MUE), Beijing, Peoples Republic China, 20–22 April 2016. Advanced Multimedia and Ubiquitous Engineering: FutureTech & MUE, Lecture Notes in Electrical Engineering, vol. 393, pp. 79–84 (2016)
10. Pikhart, M.: Current managerial communication based on modern technology and cross-cultural transfer. In: Advances in Computer Science, pp. 495–498. World Scientific and Engineering Academy and Society, Athens (2012)

11. Pikhart, M.: The impact of advanced technologies on communication and the company efficiency in intercultural management. In: Advances in Finance and Accounting, pp. 275–280. Worlds Scientific and Engineering Academy and Society, Athens (2012a)
12. Pikhart, M., Koblizkova, A.: The central role of politeness in business communication - the appropriateness principle as the way to enhance business communication. J. Intercult. Commun. **45** (2017)
13. Samovar, L., Porter, R., McDaniel, E.R.: Communication Between Cultures. Thomson, Belmont (2012)
14. Washington, M.C., et al.: Intercultural communication in global business: an analysis of benefits and challenges. Int. Bus. Econ. Res. J. **11**(2), 217 (2012)

Selection of Most Suitable Secondary School Alternative by Multi-Criteria Fuzzy Analytic Hierarchy Process

N. S. M. Rezaur Rahman, Md. Abdul Ahad Chowdhury,
Abdullah-Al Nahian Siraj, Rashedur M. Rahman$^{(\boxtimes)}$,
Rezaul Karim, and K. M. Arafat Alam

Department of Electrical and Computer Engineering, North South University,
Plot-15, Block-B, Bashundhara Residential Area, Dhaka, Bangladesh
shaon13nov@gmail.com, arafatalam42@gmail.com,
{ahad.chowdhury, nahian.siraj, rashedur.rahman,
karim.rezaul}@northsouth.edu

Abstract. Selecting the most suitable secondary school for a student is a multi-criteria decision problem. In this paper we propose a system that uses fuzzy analytic hierarchy process (AHP) to define and evaluate linguistic variables that represent these criteria. This system takes some of them as crisp inputs and a set of fuzzy linguistic values from the user that denote relative importance measures, as deemed by the user, for every possible pair of these factors. Then it uses Fuzzy AHP to calculate absolute weights for each factor and comparative scores for a few user-selected schools for each criterion. These values are used to assign a final score to each school, which is the measurement of its suitability for the given input.

Keywords: Fuzzy logic · Analytic hierarchy process
Facility location selection · Multi-attribute decision making

1 Introduction

A student's future prospects for professional success can highly depend on the selection of educational institute for him. In Bangladesh, such decisions for millions of students are made every year by their parents and guardians. Unfortunately, the outcome of this decision-making in the conventional way often does not happen to be very suitable. In this paper, we propose fuzzy analytic hierarchy process (Fuzzy AHP) for helping the parents or the guardians to select the most suitable school for his child. Fuzzy AHP was originally introduced by Saaty [1], which combined analytic hierarchy process with fuzzy logic for solving multi-attribute decision making problems. In our method, we take fuzzy linguistic inputs for relative importance of each pair of criteria and various information of a student. We then use these values to compute suitability scores of the schools and suggest the best school(s).

© Springer Nature Switzerland AG 2019
K. Choroś et al. (Eds.): MISSI 2018, AISC 833, pp. 279–289, 2019.
https://doi.org/10.1007/978-3-319-98678-4_29

2   Related Works

Application of Fuzzy AHP for multi-attribute decision making is very common for facility location selection problems. Boltürk et al. [2] proposed a new Hesitant Fuzzy AHP where both triangular and trapezoidal fuzzy numbers were used as input through Hesitant Fuzzy AHP and comparative scores for the alternatives were calculated.

Singh [3] proposed Extent Fuzzy AHP to determine facility location, where a square matrix of triangular fuzzy numbers was used to determine relative importance of the criteria through Fuzzy AHP. Another similar approach was proposed by Chou et al. [4] that integrated Fuzzy Set Theory with Factor Rating System and Simple Additive Weighting to evaluate facility location alternatives for a supply chain management problem.

3   Methodology

We present the description of our methodology in the following five sections.

3.1   Data Collection and Pre-processing

We obtained all of our data from Bangladesh Open Data [5], the government-maintained portal for datasets on national statistics. We combined multiple datasets into a single dataset for our purpose. Table 1 depicts our dataset and its different attributes.

Table 1. Dataset and selected attributes

Datasets	Selected fields
Institutional Information of Bangladesh	EIIN, Name of institution, Division, District, Thana, Mauza, Post, Institute type
Institute-wise Teacher Student Info	EIIN, Number of total teachers, Number of female teachers, Number of total students, Number of female students
Students Information of Bangladesh	EIIN, Number of students by age in classes 6–10
Information Related Socio-Economic Background of Parents/Guardian of Secondary Level Students	EIIN, Number of students categorized by parent's or guardian's profession in classes 6–10
Grade Wise Student in School	EIIN, Number of male and female students in classes 6–10

3.2   User Inputs

We want to ensure that our system always gives the user a personalized output. For this purpose, we take student's age, sex, intended class (grade) for admission, guardian's living division, district, thana and union (smaller sub regions of a district) as input from the user. We also take linguistic input denoting relative importance for each possible pair of the attributes of a school: student-teacher ratio (TSR), socio-economic status (SES), ratio of male or female students in the school (MFR), age of school (AS), distance from the user's home (DIST) and age gap of the student from the average age of students in the specific class (ADS).

The relative importance of these attributes, which are comparative to each other, is taken by giving the user a set of linguistic values as options to choose from. These linguistic values and their corresponding triangular fuzzy number (TFN) values are given in Table 2 in (l m u) format.

Table 2. Linguistic values and corresponding triangular fuzzy numbers

Linguistic values	Triangular fuzzy numbers
Equally important	(1 1 1)
Moderately important	(2 3 4)
Fairly important	(4 5 6)
Very important	(6 7 8)
Absolutely important	(9 9 9)

3.3   Criteria Weight Calculation

First, we convert the linguistic values into TFNs according to the scale in Table 2 and construct a comparison matrix with them. For example, if criterion A is "Moderately Important" compared to criterion B, then the comparative importance value of criterion A with respect to criterion B in the matrix will be (2   3   4), and the value of criterion B against criterion A will be (1/4   1/3   1/2) in the matrix. The relation can be shown in mathematical terms like Eq. (1):

$$C = \begin{bmatrix} c_{11}c_{12} & \cdots & c_{1n} \\ \vdots & \ddots & \vdots \\ c_{n1}c_{n2} & \cdots & c_{nn} \end{bmatrix} \tag{1}$$

Here, C is the comparison matrix and c_{ij} is the comparative importance value of i-th criterion with respect to j-th criterion. We take geometric mean of TFNs for each row of this matrix. According to Buckley [6], this is given by Eq. (2):

$$r_i = \left(\prod_{j=1}^{n} c_{ij} \right)^{\frac{1}{n}} \tag{2}$$

Here, each r_i value is a TFN, denoted by TFN (l_i m_i u_i). Let us assume the new set of values for each row form a vector R is given in Eq. (3)

$$R = \begin{bmatrix} r_1 \\ \vdots \\ r_n \end{bmatrix} \tag{3}$$

Then we calculate a vector V by taking the vector summation of TFNs of R, taking the inverse of each member of the sum and sorting the members in ascending order. Since $l \leq m \leq u$ for any TFN (l m u), this process can be described in mathematical terms like Eq. (4):

$$V = \begin{bmatrix} \dfrac{1}{U} \dfrac{1}{M} \dfrac{1}{L} \end{bmatrix} \tag{4}$$

Here $L = \sum_{i=1}^{n} l_i$, $M = \sum_{i=1}^{n} m_i$ and $U = \sum_{i=1}^{n} u_i$ for each r_i from R.

Now we multiply every member of matrix R individually with V. This gives us the fuzzy weight matrix W as given in Eq. (5)

$$W = \begin{bmatrix} \dfrac{l_1}{U} \dfrac{m_1}{M} \dfrac{u_1}{L} \\ \vdots \\ \dfrac{l_n}{U} \dfrac{m_n}{M} \dfrac{u_n}{L} \end{bmatrix} \tag{5}$$

These fuzzy weight values for each criterion is defuzzified using the Center of Area (COA) method. Since these are all triangular fuzzy numbers, this can be calculated by taking the arithmetic mean of l, m, u as given in Eq. (6)

$$v_i = \frac{l_i + m_i + u_i}{3} \tag{6}$$

These crisp values are normalized by their sum as given by Eq. (7):

$$w_i = \frac{v_i}{\sum_{p=1}^{n} v_p} \tag{7}$$

Therefore, by applying this process on matrix C, we get $w_1, w_2, \ldots w_n$ for n criteria.

After this, the user selects m alternatives from the list of schools. These schools are compared against each other based on each criterion, which produces an m-by-m comparison matrix for each criterion. For this, we calculate a "score" for each criterion.

3.4   Criteria-Wise Score Calculation for Alternatives

To calculate score for some of the attributes, we apply cumulative distribution function (CDF) of standard normal distribution, in order to obtain a value between 0 and 1. The formula we use for some of these values is shown in Eq. (8). Here, μ is the mean and σ

is the standard deviation. For calculating mean and standard deviation, we consider the data of all the schools available in our dataset.

$$cdf(x, \mu, \sigma) = \frac{1}{2}\left[1 + \mathrm{erf}\left(\frac{x - \mu}{\sigma\sqrt{2}}\right)\right]$$ (8)

In this equation, $\mathrm{erf}(x)$ means the error function, denoted by Eq. (9):

$$\mathrm{erf}(x) = \frac{2}{\sqrt{\pi}}\int_0^x e^{-t^2} dt$$ (9)

To calculate the score for "student sex ratio", we apply the following equation:

$$score_{mfr} = \begin{cases} \frac{\#ms}{\#st} & \textit{if sex} = 'male' \\ \frac{\#fs}{\#st} & \textit{if sex} = 'female' \end{cases}$$ (10)

Here, $score_{mfr}$ means number of male or number of female in ratio depending on the user's input sex, $\#ms$ means the number of male students, $\#fs$ means the number of female students and $\#st$ number of total students.

For criterion "teacher-student ratio" we apply formula described by Eq. (11).

$$score_{tsr} = 1 - cdf(tsr, \mu_{TSR}, \sigma_{TSR})$$ (11)

Here, tsr means ratio of number of students and number of teachers in the school, $score_{tsr}$ means teacher-student ratio score, μ_{TSR} denotes the mean of all the TSR values and σ_{TSR} denotes the standard deviation in TSR values.

For criterion "average age of students" we take the absolute value of the difference between the user's given age and the average age of students in the class, ads. We normalize it using Eq. (8) and obtain the score for this criterion, $score_{ads}$ by Eq. (12). Here, μ_{ADS} and σ_{ADS} denote the mean and the standard deviation of all the ADS values.

$$score_{ads} = 1 - cdf(ads, \mu_{ADS}, \sigma_{ADS})$$ (12)

For criterion "age of school" we take the age of school, as, from the year of establishment of the school given in the dataset. We normalize it using Eq. (8) and calculate the score for this criterion, $score_{as}$.

$$score_{as} = cdf(as, \mu_{AS}, \sigma_{AS})$$ (13)

For the criterion "socio-economic status" we have conducted a survey on 15 participants. They have assigned 15 professions and 7 administrative levels a number between 1 and 10 (inclusive), based on their perception of social status. The data for the profession score obtained from the survey is rounded to the nearest multiple of 2.5, to represent 4 distinct social classes. The data from the survey is shown in Tables 3 and 4.

Table 3. Professions and their status

Professions	Profession status, p_j
Lawyer, Doctor, Engineer, Businessman	10.0
Government job, Private job, Teacher	7.5
Expatriate, Cultivation, Small business owner	5.0
Worker, Fisherman, Weaver, Potter, Blacksmith	2.5

Table 4. Administrative levels and their status

Administrative level	Area status, A
Metropolitan	10.0
City Corporation	9.0
District Sadar Municipality	7.0
Upazilla Sadar Municipality	5.0
Other Municipality Area	4.0
Upazilla Sadar but not Municipality	2.0
Rural Area	1.0

In our dataset the number of students whose guardians are working in these 15 professions and the level of the areas in which the schools are located are given. For each selected school, we calculate the socio-economic score by the Eq. (14):

$$S = \left(\frac{1}{N}\sum_{j=1}^{15} p_j n_j\right) \times W_P + A \times W_A \tag{14}$$

Here N is the total number of students in the school, p_j is the socio-economic status score of the j-th profession, n_j is the number of students in the school whose parents or guardians work in the j-th profession. A is the area-based status of the school's location. $W_P = 10$ and $W_A = 5$, which are the weights that guardian's occupation and area status carry. These values were also obtained from the survey. We normalize this value according to Eq. (8), shown in Eq. (15).

$$score_{ses} = cdf(S, \mu_S, \sigma_S) \tag{15}$$

Finally, for the criterion "distance", at first we assign a crisp value $d = 10$ for each school. If the school is in the same division as the user's inputted division, we subtract 4 from d. Similarly, if the school is in the same district, thana or union/ward, we subtract 3, 2 and 1 respectively. This value is used to calculate the score for this criterion, as shown in Eq. (16).

$$score_{dist} = 1 - \frac{d}{10} \tag{16}$$

3.5 Criteria-Wise Comparison Matrix Formation for Alternatives and Final Calculation

After gathering all these scores, we construct a comparison matrix for each criterion. At first, for comparing i-th alternative to j-th alternative on n-th criterion, we assign an integer, $cs_{n_{ij}}$, between -10 and 10 (inclusive) according to Eq. (17), given that $score_{n_i}$ and $score_{n_j}$ are the score values of i-th alternative and j-th alternative on n-th criterion.

$$cs_{n_{ij}} = \begin{cases} \lfloor |score_{n_i} - score_{n_j}| \rfloor & if score_{n_i} \geq score_{n_j} \\ \lceil |score_{n_i} - score_{n_j}| \rceil & else \end{cases} \tag{17}$$

Finally, for each $cs_{n_{ij}}$ value we take a TFN from Table 5 and assign this value in the i-th row and j-th column in the comparison matrix for n-th criterion. Thus, for n criteria we obtain n matrices $C_1, C_2, \dots C_n$ that are m-by-m in dimension, in which a member a_{ijk} represents the comparison TFN of j-th school with the k-th school based on the i-th criterion, shown in Eq. (18).

$$C_n = \begin{bmatrix} a_{n11} & a_{n12} & \cdots & a_{n1m} \\ & \vdots & \ddots & \vdots \\ a_{nm1} & a_{nm2} & \cdots & a_{nmm} \end{bmatrix} \tag{18}$$

Table 5. Possible crisp scores ($cs_{n_{ij}}$) and corresponding TFNs

Negative crisp scores	Corresponding TFNs	Positive crisp scores	Corresponding TFNs
−10	(1/9 1/9 1/9)	0	(1 1 1)
−9	(1/9 1/9 1/9)	1	(1 1 1)
−8	(1/9 1/8 1/7)	2	(1 2 3)
−7	(1/8 1/7 1/6)	3	(2 3 4)
−6	(1/7 1/6 1/5)	4	(3 4 5)
−5	(1/6 1/5 1/4)	5	(4 5 6)
−4	(1/5 1/4 1/3)	6	(5 6 7)
−3	(1/4 1/3 1/2)	7	(6 7 8)
−2	(1/3 1/2 1)	8	(7 8 9)
−1	(1 1 1)	9	(9 9 9)
0	(1 1 1)	10	(9 9 9)

For each of these comparison matrices, Fuzzy AHP is applied using Eqs. (2) through (7). Applying this process of matrices $C_1, C_2, \ldots C_n$ yields sets of alternative-wise scores for all the criteria, $\{v_{11}, v_{21}, \ldots v_{m1}\}$, $\{v_{12}, v_{22}, \ldots v_{m2}\}, \ldots \{v_{1n}, v_{2n}, \ldots v_{mn}\}$. Now we calculate the final scores for each of these m schools, based on the n criteria. We obtain this final score Z_i for the i-th school by

$$Z_i = \sum_{p=1}^{n} w_p v_{ip} \qquad (19)$$

Here, w_p is the absolute weight of p-th criterion that had been calculated earlier.

4   Demonstration and Result Analysis

4.1   Demonstration of Calculation

Let us go through an example to demonstrate how this process works. We assume that a user has given the crisp and fuzzy inputs shown in Tables 6 and 7. The fuzzy inputs are mapped with TFNs from Table 2 to generate a comparison matrix, shown in Table 8.

Table 6. Crisp inputs

Fields	Inputs	Fields	Inputs
Class	7	Division	RAJSHAHI
Student's age	15	District	NAWABGANJ
Student's sex	Female	Thana	CHAPAI NAWABGANJ SADAR
		Union/Ward	WARD NO-03

We apply AHP on this comparison matrix, which is described in Eqs. (2) to (7). This gives us crisp weights of each criterion.

Now we assume that the user selects 3 schools. The details of these schools are shown in Table 8. For each school, its values are used to obtain scores for each criterion using Eqs. (10) to (17). These scores are shown in Table 9, which are used to construct comparison matrices for each criterion. Each matrix is inputted through Fuzzy AHP, as described in Eqs. (2) to (7). They generate crisp values for each criterion and for each alternative. These values, along with crisp weights of each criterion, are used as described in Eq. (19) to calculate final score for each alternative. This has been shown in Table 11.

Therefore, the first alternative (A1) is the best choice for the student.

Table 7. Fuzzy inputs

Ab. Imp	Ve. Imp	Fa. Imp	Mo. Imp	Crit.	Eq. Imp	Crit.	Mo. Imp	Fa. Imp	Ve. Imp	Ab. Imp
				TSR		SES	√			
	√			TSR		MFR				
		√		TSR		AS				
				TSR		DIST			√	
		√		TSR		ADS				
	√			SES		MFR				
		√		SES		AS				
				SES	√	DIST				
	√			SES		ADS				
				MFR	√	AS				
				MFR		DIST		√		
				MFR	√	ADS				
				AS		DIST		√		
				AS	√	ADS				
	√			DIST		ADS				

Table 8. Comparison matrix

	TSR	SES	MFR	AS	DIST	ADS
TSR	(1 1 1)	(1/2 1/3 1/4)	(4 5 6)	(2 3 4)	(1/6 1/7 1/8)	(2 3 4)
SES	(4 3 2)	(1 1 1)	(6 7 8)	(2 3 4)	(1 1 1)	(6 7 8)
MFR	(1/6 1/5 1/4)	(1/8 1/7 1/6)	(1 1 1)	(1 1 1)	(1/4 1/5 1/6)	(1 1 1)
AS	(1/4 1/3 1/2)	(1/4 1/3 1/2)	(1 1 1)	(1 1 1)	(1/4 1/5 1/6)	(1 1 1)
DIST	(8 7 6)	(1 1 1)	(6 5 4)	(6 5 4)	(1 1 1)	(6 7 8)
ADS	(1/4 1/3 1/2)	(1/8 1/7 1/6)	(1 1 1)	(1 1 1)	(1/8 1/7 1/6)	(1 1 1)

Table 9. Details of user-selected school alternatives

	TSR	SES	MFR	AS	DIST	ADS
A1	128.58	82.99	50.94	91.00	1.00	2.53
A2	38.83	133.23	35.61	171.00	6.00	3.20
A3	29.75	119.87	22.69	27.00	6.00	2.97

Table 10. Scores of user-selected school alternatives

	TSR	SES	MFR	AS	DIST	ADS
A1	0.00	0.98	0.51	0.46	0.90	0.73
A2	0.46	1.00	0.36	0.46	0.40	0.39
A3	0.62	1.00	0.23	0.45	0.40	0.51

Table 11. Alternative-based scores for each criterion and final score of each alternative

Criteria	Weights	A1	A2	A3
TSR	0.13	0.09	0.42	0.48
SES	0.32	0.33	0.33	0.33
MFR	0.05	0.40	0.32	0.27
AS	0.06	0.33	0.33	0.33
DIST	0.38	0.71	0.14	0.14
ADS	0.05	0.54	0.21	0.25
TOTAL		0.46	0.27	0.27

4.2   Analysis

From Table 7 we see that the user has thought "distance" is one of the two highest-priority criteria along with "socio-economic status". For this, in Table 10, we see that criterion "distance" has the highest weight, and the alternative that has the highest score in "distance" criterion ultimately comes to be the best alternative. As "socio-economic status" is also among the top priority of the user, so it is chosen as the more important one in every comparison, except when comparing with "distance". This is why it comes a close-second in terms of criterion-weight. Then the user ranks the criteria "teacher-student ratio". So, this criterion comes third in terms of criterion-weight. But here we see the effect of linguistic terms and corresponding fuzzy numbers. The user has thought "teacher-student ratio" is important, but only "moderately" or "fairly" important, not "very" or "absolutely" important. For this reason, this criterion comes in third, but in a distant third. This affects the result, as we see that even though A1 has the lowest score in this criterion, it still has gained the highest final score. The remaining three criteria have very poor relative importance and therefore have less impact on the final score.

5   Conclusion

In context of Bangladesh, selection of secondary school is an important decision to make for any guardian. In this paper we propose to use Fuzzy AHP to assist the decision-maker. In our system we take into account most of the common criteria usually considered for choosing a school. Our target is to provide the user with various information about the schools, but at the same time give enough options to provide his or her personal preferences. This method can be used to solve many other problems of facility selection and multi-attribute decision-making as well.

References

1. Saaty, T.L.: The Analytic Hierarchy Process. McGraw-Hill, New York (1980)
2. Boltürk, E., Onar, S.C., Öztayşi, B., Kaharman, C., Göztepe, K.: Multi-attribute warehouse location selection in humanitarian logistics using hesitant fuzzy AHP. Int. Adv. Res. J. Sci. Eng. Technol. 3(2) (2016)
3. Singh, R.K.: Facility location selection using extent fuzzy AHP. Int. Adv. Res. J. Sci. Eng. Technol. 3(2) (2016)
4. Chou, S., Chang, Y., Shen, C.: A fuzzy simple additive weighting system under group decision-making for facility location selection with objective/subjective attributes. Eur. J. Oper. Res. 189, 132–145 (2008)
5. Bangladesh Open Data, (n. d.). http://data.gov.bd/dataset
6. Buckley, J.J.: Fuzzy hierarchical analysis. Fuzzy Sets Syst. 17(1), 233–247 (1985)

Comparison of the Novel Classification Methods on the Reuters-21578 Corpus

Anna Zdrojewska, Jakub Dutkiewicz, Czesław Jędrzejek$^{(\boxtimes)}$, and Maciej Olejnik

Poznań University of Technology, Poznań, Poland
anna.zdrojewska@student.put.poznan.pl,
{jakub.dutkiewicz,czeslaw.jedrzejek}@put.poznan.pl

Abstract. The paper describes an evaluation of novel boosting methods of the commonly used Multinomial Naïve Bayes classifier. Evaluation is made upon the Reuters corpus, which consists of 10788 documents and 90 categories. All experiments use the tf-idf weighting model and the one versus the rest strategy. AdaBoost, XGBoost and Gradient Boost algorithms are tested. Additionally the impact of feature selection is tested. The evaluation is carried out with use of commonly used metrics – precision, recall, F1 and Precision-Recall breakeven points. The novel aspect of this work is that all considered boosted methods are compared to each other and several classical methods (Support Vector Machine methods and a Random Forests classifier). The results are much better than in the classic Joachims paper and slightly better than obtained with maximum discrimination method for feature selection. This is important because for the past 20 years most works were concerned with a change of results upon modification of parameters. Surprisingly, the result obtained with the use of feedforward neural network is comparable to the Bayesian optimization over boosted Naïve Bayes (despite the medium size of the corpus). We plan to extend these results by using word embedding methods.

Keywords: Classification · Reuters · Boosting · Naïve Bayes

1 Introduction

The modern classification methods employ boosting algorithms. Main goal of these algorithms is to create a final classifier upon a number of classifiers created with a well known method. The boosting strategies use the initial algorithm (referred to as weak learner) to iteratively improve various classification parameters. The parameters vary from the intuitive impact of a document on a classifier, denoted as a weight in AdaBoost algorithm to technically sound implementation of adding a pseudo residuals based corrections to the initial classifier in gradient boosting. The boosting techniques outperform the baseline methods [1, 2]. In this paper we compare three boosting methods on the commonly used Reuters corpus [8]. We use the Multinomial Naïve Bayes algorithm as the baseline so that a comparison with the Joachim work [6] can be directly made. We train the classifiers with use of the AdaBoost [1], XGBoost [2] and Gradient Boost [3] boosting strategies [4]. The aim of this work is systematic comparison of all leading classification methods.

© Springer Nature Switzerland AG 2019
K. Choroś et al. (Eds.): MISSI 2018, AISC 833, pp. 290–299, 2019.
https://doi.org/10.1007/978-3-319-98678-4_30

Table 1. Distribution of the most populated categories in the Reuters corpus.

Category	Number of documents in training set
earn	1103
acq	723
money-fx	189
crude	188
grain	141
trade	121
interest	113
ship	77
wheat	59
corn	50

2 Dataset

We access the corpus trough the Python Natural Language Toolkit (NLTK) API. The corpus is divided into two parts, the training sub corpus and the testing sub corpus. The Reuters corpus consists of 21578 document and 90 categories. The Reuters-21578, Distribution 1.0 test collection is available from http://www.daviddlewis.com/resources/testcollections/reuters21578. The documents without topics and typographical errors are removed from the corpus. The categories, which contain only one document are also removed. The final, accessible through the NLTK download interface version of the corpus contains 10788 documents split into train and test sub-corpora. It is often referred to as a codename Reuters-21578 ApteMod and has been used in multiple applications [6, 7]. The training set contains 7769 documents and the test set contains 3019 documents. The distribution of the most populated categories is presented in Table 1. We believe, that a number of documents and categories is suitable for a sufficient evaluation.

3 Classification

All experiments were implemented in Python with the Scikit-learn library. For every experiment, we remove all stopwords and numbers from the text. We use the TF-IDF term weighting to create a classification space model. Each vector in the model represents one document. Components of the vector represent features. We take all words (except stopwords) as a feature set. With a number of documents $f_{t,c}$ in which the term t appears, size of a corpus C, number D of all terms in the document d and a number $f_{t,d}$ of appearances of term t in the document d we calculate the component values of vector as follows

$$v_{d,t} = \frac{f_{t,d}}{D} \times \log\left(\frac{C}{f_{t,c}} + 1\right)$$

We use the classification model to train and test various classifiers. All classifiers are created with *One-Vs-Rest* strategy. This strategy implements one binary classifier for each class. We use this strategy to achieve classification transparency and interpretability. A total of four methods are evaluated. We use a Multinomial Naïve Bayes classification method as our baseline classifier. We extend the work with Adaboost, XGBoost and Gradient Boost algorithms; some of these methods we not known at the time of publication of the original Joachims work [6].

3.1 Multinomial Naïve Bayes

Multinomial Naive Bayes classification method is based on the probabilistic approach, which defines the probability of a document d belong to the class c as

$$P(c|d) \sim P(c) \cdot \prod_{1 \leq k \leq D} P(t_k|c)$$

where the $P(t_k|c)$ is the probability that a term t_k belongs to the class c and $P(c)$ is the probability of any document occurring in the class c. To find whether a document belongs to a class or not, we compare probabilities among all classes. In this particular case, there are only two classes to be considered in each category. Either document belongs to a category (first class) or not (second class). We estimate the $P(c)$ and $P(t_k|c)$ values upon the training set. With C_c documents in the class c and a total of C documents we define $P(c)$ as

$$\hat{P}(c) = \frac{C_c}{C}$$

Given the vocabulary V and a set of all documents, which belong to the class c the $P(t_k|c)$ is estimated as follows

$$\hat{P}(t_k|c) = \frac{\sum_{d \in c} (v_{t_k,d}) + \alpha}{\sum_{d \in c} \sum_{t' \in V} (v_{t',d}) + |V| \cdot \alpha}$$

$|V|$ denotes a total number of words in the vocabulary. It should be noted that in classical definition of Naïve Bayes method the feature values v are equal to the number of occurrences of a given term. We use the TF-IDF weighting as described in Sect. 3. To determine whether the document belongs to the category, we calculate a class score

$$score(d) = \log(\hat{P}(c)) + \sum_{t_k \in d} \log(P(t_k|c))$$

If the $score(d)$ is greater for the first class than for second class, the document is assumed to belong to a given category.

$$h(d) = \begin{cases} 1 & score_1(d) > score_2(d) \\ 0 & otherwise \end{cases}$$

3.2   AdaBoost

In the classical definition of the Naïve Bayes, all of the documents are treated equally. AdaBoost algorithm uses a distribution of weights for documents. Sum of all components in the distribution must be equal to one. The initial distribution of weights can be chosen arbitrarily. We use a following distribution

$$\forall d \quad w_d^1 = \frac{1}{C}$$

so that all documents are initially treated equally, as it is assumed in the Multinomial Naïve Bayes algorithm. The algorithm boosts initial algorithm – in our case it is Multinomial Naïve Bayes by researching various distributions of weights for documents and applying them into the initial classifier. Weights distributions are computed iteratively with use of error given by current hypothesis. Given the proper, training labels $y(d)$ for each document, error for the t iteration is defined as follows

$$\varepsilon_t = \sum_d w_d^t |h(d) - y(d)|$$

In each iteration, the distribution of weights is updated with the following formula

$$\beta_t = \frac{\varepsilon_t}{1 - \varepsilon_t}$$

$$w_d^{t+1} = w_d^t \beta_t^{1 - |h(d) - y(d)|}$$

We can see, that weights for the documents which are annotated mistakenly are increased. The weights are again normalized, so that the sum of all weights is equal to one and provided to the Multinomial Naïve Bayes classifier. The weights are used in the training algorithm to calculate the conditional probabilities

$$\hat{P}(t_k|c) = \frac{\sum_{d \in c} (w_d^t v_{t_k,d}) + 1}{\sum_{d \in c} \sum_{t' \in V} (w_d^t v_{t',d}) + |V|}$$

Aside from that, Naïve Bayes algorithm works as it is described in Sect. 3.1. The algorithm is repeated T times and it produces T different hypothesis. The final hypothesis is calculated as follows.

$$h_f(d) = \begin{cases} 1 & \sum \log \frac{1}{\beta_t} h_t(x) \geq \frac{1}{2} \log \sum \frac{1}{\beta_t} \\ 0 & otherwise \end{cases}$$

An instance of a base classification algorithm is referred to as weak learner. Boosting algorithm combines a number of weak learners to create a stronger classification algorithm.

3.3 Gradient Boost

Similarly to the AdaBoost, the Gradient Boost algorithm uses weak learners to create a final classifier. Gradient Boost uses an additive approach to combine multiple classifiers. The initial classifier is trained with the base algorithm – in Gradient Bosst algorithm it is a decision tree. We denote the output of the initial classifier as $F_0(x)$. Similarly to the AdaBoost the next step is to find the error of the classifier. Error is defined as a loss function. In our case it is a logistic regression loss function given by

$$J(Y, F(x)) = \sum_d y(d) \log h(d) + (1 - y(d)) \log(1 - h(d))$$

Our goal is to minimize the loss function. We can do that by applying the Gradient Descent idea to the classifier. We calculate the gradient of the loss function and fit the initial classifier to the gradient, so we can correct the initial classifier. To do that, for each document we calculate a first derivative

$$r_i = -\frac{\partial J(Y, F(x))}{\partial F(x)}$$

Then, we fit the document vectors to its error derivatives with the base algorithm. The new predictions are denoted as $h_0(x)$. Calculated predictions could be viewed as the corrections to the initial predictions. We need to keep in mind that we are looking for the hypothesis, a classification model. We need a way to calculate the corrections from the input data as opposed to taking the corrections from the data outputs. The $h_0(x)$ is a model, which predicts the corrections from the input data. We add the corrections to the initial classifier with a weight γ, the weight is chosen so the classifier gives us the lowest value of the loss function.

$$F_1(x) = F_0(x) + \gamma h_0(x)$$

Table 2. Category (Reuters-10) specific evaluation of the Multinomial Naïve Bayes method.

Category	Precision	Recall	F1	Breakeven point
earn	0.964	0.942	0.953	0.950
acq	0.966	0.890	0.926	0.952
money-fx	0.701	0.893	0.786	0.754
grain	0.794	0.832	0.813	0.805
crude	0.867	0.867	0.867	0.867
trade	0.532	0.632	0.578	0.628
interest	0.792	0.671	0.727	0.725
ship	0.813	0.685	0.743	0.809
wheat	0.714	0.704	0.709	0.718
corn	0.707	0.517	0.597	0.607

We repeat that process iteratively to obtain the final classifier. In the next iteration, current F_1 becomes F_0. It should be noted, that the Gradient Boost algorithm can be used to minimize any differentiable loss function.

3.4   XGBoost

XGBoost [5] is an algorithm is an implementation of gradient boosted decision trees designed for speed and performance. It has recently been dominating applied machine learning and Kaggle competitions for structured or tabular data. For each split made, the split gain is calculated for every possible split for every possible feature and the one with the maximal gain is taken. Experience with Kaggle challenges shows that boosting methods give the best results for medium size of a system. For, example all leading Microsoft Kaggle malware classification challenge problem entries [9] used XGBoost for classification.

4   Evaluation

We conduct a series of experiments to evaluate the chosen methods. As the categories are unevenly populated we focus on the category specific measures. We also provide the micro-averaged evaluation over the tested algorithms. We calculate the differences between several commonly used classifiers and the boosted version of the Multinomial Naïve Bayes. First we calculate the Multinomial Naïve Bayes evaluation. We use four evaluation metrics – precision, recall, F1 measure and Precision-Recall breakeven point. We observe major differences between categories. The classifiers for the more populated categories are consistently more often correct, than the classifiers for the categories with a lesser number of samples. Results are presented in Table 2.

Table 3. Category specific evaluation of the AdaBoost method.

Category	Precision	Recall	F1	Breakeven point	Best result of Joachims [6] breakeven point
earn	0.984	0.975	0.979	0.977	0.985
acq	0.981	0.947	0.963	0.963	0.950
money-fx	0.835	0.709	0.767	0.798	0.740
grain	0.968	0.825	0.891	0.906	0.931
crude	0.920	0.793	0.852	0.888	0.889
trade	0.830	0.709	0.764	0.777	0.769
interest	0.884	0.641	0.743	0.791	0.744
ship	0.912	0.584	0.712	0.838	0.854
wheat	0.913	0.746	0.821	0.847	0.852
corn	0.936	0.785	0.854	0.880	0.851

We train the Multinomial Naïve Bayes classifier with the AdaBoost strategy. We observe equal increase in the performance across all categories. The major increase in the lower populated categories is contradictory to the remarks made by authors of AdaBoost. Classifiers with a performance worse than 0.5 accuracy are supposed to be the weakness of the algorithm. We observe the highest improvement for the classifiers with lower performance. To calculate the evaluation presented in Table 3 we used 25 iterations of AdaBoost over the Multinomial Naïve Bayes. We run the same experiments for the XGBoost and Gradient Boost algorithms.

Table 4. Category specific evaluation of the XGBoost method.

Category	Precision	Recall	F1	Breakeven point
earn	0.971	0.979	0.975	0.975
acq	0.953	0.847	0.896	0.917
money-fx	0.778	0.687	0.729	0.765
grain	0.969	0.852	0.907	0.939
crude	0.897	0.835	0.865	0.867
trade	0.735	0.760	0.747	0.735
interest	0.846	0.587	0.693	0.755
ship	0.926	0.707	0.802	0.831
wheat	0.851	0.887	0.868	0.859
corn	0.898	0.946	0.921	0.928

Table 5. Category Reuters-10 specific evaluation of the Gradient Boost method.

Category	Precision	Recall	F1	Breakeven point
earn	0.971	0.976	0.973	0.974
acq	0.965	0.845	0.901	0.931
money-fx	0.765	0.603	0.675	0.748
grain	0.938	0.919	0.928	0.932
crude	0.890	0.820	0.853	0.862
trade	0.696	0.726	0.711	0.692
interest	0.865	0.587	0.700	0.763
ship	0.838	0.584	0.688	0.696
wheat	0.840	0.887	0.863	0.845
corn	0.868	0.946	0.905	0.910

We notice a consistent improvement over the baseline methods. Out of the three AdaBoost performs best. Results for XGBoost and Gradient Boost are presented in respectively Tables 4 and 5. We provide the cross classifier comparison based on the micro-averaged Precision, Recall and F1 scores in Table 6.

Table 6. The Bayesian optimization over boosted Naïve Bayes for Reuters-10.

Category	Precision	Recall	F1	Breakeven point
earn	0.987	0.982	0.984	0.983
acq	0.987	0.950	0.968	0.971
money-fx	0.820	0.737	0.776	0.788
grain	0.976	0.826	0.895	0.906
crude	0.917	0.820	0.866	0.899
trade	0.828	0.701	0.759	0.786
interest	0.819	0.725	0.769	0.771
ship	0.905	0.640	0.750	0.854
wheat	0.914	0.746	0.822	0.789
corn	0.936	0.786	0.854	0.857

Table 7. The results for feed-forward neural network with ADAM for Reuters-10.

Category	Precision	Recall	F1
earn	0.987	0.982	0.984
acq	0.987	0.950	0.968
money-fx	0.820	0.737	0.776
grain	0.976	0.826	0.895
crude	0.917	0.820	0.866
trade	0.828	0.701	0.759
interest	0.819	0.725	0.769
ship	0.905	0.640	0.750
wheat	0.914	0.746	0.822
corn	0.936	0.786	0.854

Table 8. Micro-average evaluation of various methods.

Method	Precision	Recall	F1
Multinomial Naïve Bayes	0.883	0.866	0.874
Adaboost	0.956	0.881	0.917
XGBoost	0.928	0.871	0.899
Gradient Boost	0.926	0.861	0.891
Adaboost with Bayesian optimization of hyperparameters	0.954	0.894	0.923
Feed-forward neural network with ADAM	0.908	0.939	0.924
Best result of Joachims [6] Precision-Recall breakeven	0.864		

Table 9. Comparison of the precision-recall breakeven points for commonly used classifiers and boosting methods of a default classifier – in that case Multinomial Naïve Bayes and Gradient Trees.

Method	earn	acq	money-fx	grain	crude	trade	interest
Linear SVM	0.944	0.925	0.754	0.891	0.811	0.741	0.781
Random Forests	0.935	0.916	0.761	0.825	0.852	0.721	0.775
SVM (poly)	0.925	0.912	0.762	0.821	0.821	0.754	0.781
SVM (rbf)	0.924	0.984	0.711	0.711	0.867	0.722	0.730
Gradient Boost	0.974	0.931	0.748	0.932	0.862	0.692	0.763
XGBoost	0.975	0.917	0.765	**0.939**	0.867	0.735	0.755
AdaBoost	**0.977**	**0.963**	**0.798**	0.906	**0.888**	**0.777**	**0.791**

The best results are achieved with The Bayesian optimization [10] (Table 6) over boosted Naïve Bayes and feed-forward neural network with ADAM [11] (Table 7). This is surprising because it was believed that the medium size of the corpus causes the results obtained with neural networks (particularly, convolution neural networks) are not competitive with the boosted classification [12, 13]. The collective comparison of methods is presented in Table 8.

We compare the boosting methods with several commonly used algorithms. Namely Linear SVM, polynomial and radial based function kernel SVMs, and Random Forests classifier. The boosting algorithms achieve superior results to the baseline classifiers. The results are listed in Table 9.

5   Conclusions

We find the boosting algorithms are a solid improvement over baseline traditional methods. This is important because for the past 20 years most works were concerned with a change of results upon modification of parameters and not with the direct comparison with [6]. Improvement can be seen across classifiers with various number of training samples. On top of that, the performance of novel implementations of well-known classifiers seem to be better than the performance described in the classic Joachims original work [6], mainly due to the novel feature selection methods. The results shown are slightly better or equal compared with these obtained with maximum discrimination method for feature selection [14] (we are able to make comparison only to data presented in Fig. 5(a2)). In this work the best F1 results for Reuters-10 were achieved including 100 features, whereas the best accuracy result was achieved using 1000 features. The medium size of the corpus causes the results obtained with help of convolution neural networks are not competitive with the boosted classification [12, 13].

Recently, very good results for the sentiment Stanford corpus were obtained using Bayesian optimization and word embedding methods compared to classical methods considered here [10].

Statistical tests should be performed for the methods we used to justify the results (analogous to assigning p-value for hypothesis significance). Here we are dealing with a multidimensional problem (precision, recall, F1, break-even) and we defer these calculations to future works. We plan to extend the Reuters corpus results by larger version of the Reuters corpus [7], word embedding methods and for larger systems [15].

Acknowledgements. We acknowledge the Poznan University of Technology grant (04/45/DSPB/0185).

References

1. Banerjee, S., Majumder, P., Mitra, M.: Re-evaluating the need for modelling term-dependence in text classification problems. CoRR abs/1710.09085 (2017)
2. Chen, T., Guestrin, C.: XGBoost: a scalable tree boosting system
3. Freund, Y., Schapire, R.: A decision theoretic generalization of on-line learning and an application to boosting. J. Comput. Syst. Sci. **55**, 119–139 (1997)
4. Friedman, J.H.: Greedy function approximation: a gradient boosting machine. Ann. Stat. **29**, 1189–1232 (2000)
5. Ji, Y., Noah, A., Smith, N.A.: Neural discourse structure for text categorization. In: ACL (1), pp. 996–1005 (2017)
6. Joachims, T.: Text categorization with support vector machines: learning with many relevant features. In: ECML, pp. 137–142 (1998)
7. Lewis, D.D., Yang, Y., Rose, T., Li, F.: RCV1: a new benchmark collection for text categorization research. J. Mach. Learn. Res. **5**, 361–397 (2004)
8. Liang, H., Sun, X., Sun, Y., Gao, Y.: Text feature extraction based on deep learning: a review. EURASIP J. Wirel. Commun. Netw. **2017**(1), 211 (2017)
9. Manning, C.D., Raghavan, P., Schütze, H.: Introduction to Information Retrieval. Cambridge University Press, New York (2008)
10. Yogatama, D., Kong, L., Smith, N.A.: Bayesian optimization of text representations. In: EMNLP, pp. 2100–2105 (2015)
11. Kingma, D.P., Ba, J.: Adam: a method for stochastic optimization. In: The International Conference on Learning Representations (ICLR), San Diego (2015)
12. Salakhutdinov, R., Hinton, G.E.: Semantic hashing. Int. J. Approx. Reason. **50**(7), 969–978 (2009)
13. Yang, Y., Liu, X.: A re-examination of text categorization methods. In: Proceedings of 22nd Annual International SIGIR (1999)
14. Tang, B., Kay, S., He, H.: Toward optimal feature selection in Naive Bayes for text categorization. IEEE Trans. Knowl. Data Eng. **28**(9), 2508–2521 (2016)
15. Ji, Y., Smith, N.A.: Neural discourse structure for text categorization. In: ACL 2017, Vancouver, Canada, pp. 996–1005 (2017)

Computational Intelligence for Multimedia Understanding

Deep Learning Techniques for Visual Food Recognition on a Mobile App

Michele De Bonis[✉], Giuseppe Amato, Fabrizio Falchi, Claudio Gennaro, and Paolo Manghi

Institute of Information Science and Technologies of the National Research Council of Italy (ISTI-CNR), Pisa, Italy
{michele.debonis,giuseppe.amato,fabrizio.falchi,
claudio.gennaro,paolo.manghi}@isti.cnr.it
http://www.isti.cnr.it

Abstract. The paper provides an efficient solution to implement a mobile application for food recognition using Convolutional Neural Networks (CNNs). Different CNNs architectures have been trained and tested on two datasets available in literature and the best one in terms of accuracy has been chosen. Since our CNN runs on a mobile phone, efficiency measurements have also taken into account both in terms of memory and computational requirements. The mobile application has been implemented relying on RenderScript and the weights of every layer have been serialized in different files stored in the mobile phone memory. Extensive experiments have been carried out to choose the optimal configuration and tuning parameters.

Keywords: Convolutional Neural Network · Android App
Food recognition

1 Introduction

The research in this paper exploits the use of Convolutional Neural Networks (CNN) for food recognition. A CNN is a type of artificial neural network which is based on a large collection of neural units (artificial neurons), loosely mimicking the way a biological brain solves problems with large clusters of biological neurons connected by axons.

Since CNNs are very complex and they need powerful hardware in order to get results in a reasonable time, they usually run on a device with high computational power in terms of CPU and GPU. The goal of our work was to design a food recognition system to be used by a mobile application, providing good recognition performance even when reducing the size of the neural network model to run on the limited computational and memory resources of a smartphone.

The idea of developing algorithms for food recognition is not new. Bossard et al. [2] present a method for food recognition based on a dataset of 101 categories of food. They used Random Forests to mine discriminant components of

© Springer Nature Switzerland AG 2019
K. Choroś et al. (Eds.): MISSI 2018, AISC 833, pp. 303–312, 2019.
https://doi.org/10.1007/978-3-319-98678-4_31

each class of food. Image classification is performed relying on Support Vector Machines (SVM). Wang et al. [12] present a method for food recognition based on the same classes of food in [2]. Exhaustive experiments of recipe recognition using visual, textual information and fusion are carried out. Visual features are extracted using different methods, one of which is the OverFeat Convolutional Neural Network. Chen et al. [3] propose deep architectures for simultaneous learning of ingredient recognition and food categorization, by exploiting the mutual but also fuzzy relationship between them.

Amato et al. [1] propose a system called WorldFoodMap, which captures the stream of food photos from social media and, thanks to a CNN food image classifier, identifies the categories of food that people are sharing.

In our work, we developed an effective CNN for food recognition following these three directions:

– training from scratch various CNN architectures in order to select the most promising ones in terms of offered accuracy and network model size
– fine tuning, using food images, the most promising CNN architectures after a pre-training with the ILSVRC dataset [9]
– selecting the CNN with best accuracy/model size ratio in order to port it on the mobile phone and still guarantee high accuracy

In addition, we also performed a threshold analysis on the score of the prediction. This analysis helps to assess the confidence of the prediction made by the food recognition system.

2   Datasets

The experiments were executed on two datasets with 101 classes of very popular food types (i.e. 101 dishes). In literature, there are two versions of datasets with these 101 dishes using different sets of images: ETHZ Food101 [2] and UPMC Food101 [12] (see Fig. 1).

The way in which the images are collected to populate the two datasets are different, and this determines a difference in the number of images per class and in the type of the images.

We used two different datasets in order to have the possibility to perform cross-testing (e.g. train a CNN on a dataset and test on the other) to evaluate the so-called transfer learning (i.e. the ability to recognize images of another dataset).

Table 1. Comparison of datasets

Dataset	Number of classes	Images per class	Source
UPMC	101	790–956	Various
ETHZ	101	1000	Specific

True Pizza   Noised Pizza   True Hamburger   Noised Hamburger

Fig. 1. Example images of ETHZ dataset (pizza) and UPMC dataset (hamburger).

The datasets present some noise images (i.e. images not representative of the class). This is due to the protocol used to collect them. In fact, as will be shown in next sections, images are collected by querying two different search engines and it might happen that the result of a query leads to wrong images (Table 1).

We decided to format the images in both datasets in the same way. Moreover, we divided the datasets into three sets (training, validation and test set). The **testing set** of each dataset is exactly the same used in literature. This allows one to compare the results in terms of accuracy. UPMC dataset has a test set composed of 22,716 images, while the ETHZ dataset has a test set of 25,250 images. Remaining images have been divided into the **training** and the **validation set**. 25% of remaining images have been used for the validation, while 75% of them have been used for the training. Every single image has been formatted following the same specifications. Every image has been squashed to 256×256 pixels and encoded in PNG format in order to have no losses in the compression. During the database creation, the mean image of the dataset is computed. The mean image is subtracted from every input image in order to point out the significant regions of it.

3   Comparison of Various CNN Architectures

As we stated before, we used an approach based on Convolutional Neural Networks. A CNN is composed of a possibly large number of hidden layers, each of which performs mathematical computations on the input provided by the previous layer and produces an output that is given in input to the following layer. A CNN differs from classical neural networks for the presence of convolutional layers, which can better model and discern the spatial correlation of neighboring pixels than normal fully connected layers.

For a classification problem, the output of the CNN are the classes which the network has been trained on. The output layer is processed by a softmax function which "squashes" a K-dimensional vector of arbitrary real values to a K-dimensional vector of real values in the range (0, 1) that add up to 1.

One of the objectives of our proposal is to run the CNN entirely on an Android mobile phone. To this purpose, the CNN has to be small and at the same time accurate and fast.

We tested the following CNN architectures: the AlexNet [6], the VGG-Net with 19 layers [10], the GoogLeNet [11], the SqueezeNet [5], the ResNet50 [4], the Binary Weighted Network and the XNOR-Net [8].

As a first step, we trained these CNNs from scratch using the ETHZ and UPMC datasets. This step aims to identify the most promising CNN for our purposes. We evaluated the CNNs relying on accuracy on the validation set (e.g. percentage of images of the validation set correctly classified).

We trained every CNN architecture by setting the same parameters. Generally, we set the number of training epochs to 60 but, in some cases, we used a different value depending on the accuracy trend of the CNN. We set the Batch Size to 128 for every scenario. Moreover, we chose a Learning Rate function with a step–down policy. This kind of policy consists in a step–down of 90% every 33% of epochs. If the number of epochs is 60, Learning Rate is fixed to 0.01 in first 20 epochs, to 0.001 in second 20 epochs and to 0.0001 in last 20 epochs. The Solver Type we choose is the Stochastic Gradient Descent (SGD).

Table 2. Comparisons of various CNN architectures trained from scratch

CNN	Framework	Model size	Accuracy	
			ETHZ	UPMC
ResNet-50	Caffe	~80 Mb	16%	20%
XNOR-Net	Torch	~410 Mb	25%	24%
BWN	Torch	~380 Mb	38%	34%
AlexNet	Caffe	~230 Mb	**40%**	**37%**
SqueezeNet	Caffe	~3 Mb	**41%**	**31%**
VGG-16	Caffe	~590 Mb	44%	37%
GoogLeNet	Caffe	~40 Mb	**56%**	**43%**

Table 2 shows the result we obtained. As we can see, the most promising CNNs are the GoogLeNet (highest accuracy on both datasets and very small model size), the SqueezeNet (smallest model size and high accuracy on ETHZ dataset), and the AlexNet (good accuracy on both datasets and widely supported since it has many available implementations). Also the VGG-Net has a good accuracy on both datasets, but its model size is too big to fit in mobile memory, therefore it has been discarded.

In many cases, the training sets do not have a size sufficient to train a CNN with the required accuracy and generalization performance. This is also the case of the two training sets that we are using. In these cases, much higher performance can be obtained by pre-training the network (or use an already pre-trained network) on a different scenario, for which a very large training set is available, and then fine-tune the CNN on the scenario at hand with the small training set available.

In this respect, we used available pre-trained models of the most promising CNN architectures[1]. The models are those coming from training on the ILSCVRC [9] dataset (1,000 classes of images). For the fine-tuning process, we used the same learning parameters used for training the networks from scratch. The weights of the last FC layer (which is mostly affected by the fine-tuning process) were initialized randomly, and the output was set to 101 classes (the ILSRVC dataset, used for pre-training has 1,000 classes).

An important parameter we set in the fine tuning is the Learning Rate Multiplier. It is multiplied by the Learning Rate and is set differently in each layer of the network to allow having different Learning Rates where needed. We set this parameter to 0.1 for the intermediate layers and to 1 in the last layer. This way the Learning Rate of the intermediate layers is 10 times smaller than the Learning Rate of the last layer.

Table 3. Comparisons of various CNN architectures fine tuned

CNN	Framework	Model size	Accuracy	
			ETHZ	UPMC
AlexNet	Caffe	~230 Mb	59%	55%
SqueezeNet	Caffe	~3 Mb	**64%**	**55%**
GoogLeNet	Caffe	~40 Mb	**73%**	**63%**

Table 3 shows the result we obtained after fine-tuning. As we can see, the best CNN architectures in terms of model size and accuracy on the validation set are the SqueezeNet and the GoogLeNet.

4   Experiments

After the previous preliminary experiments, intended to identify the most promising CNN, we performed more detailed experiments focusing on the two selected CNN: the SqueezeNet and the GoogLeNet. We queried both CNNs with all the images in the testing set of both datasets and we logged the scores assigned by the CNN to each class. The most promising classes were assigned higher scores. We performed this process for both CNNs using both datasets performing cross-testing: we trained on one dataset and tested on the other in order to evaluate the transfer learning property.

We considered two metrics to evaluate:

– ACCURACY TOP-K: the percentage of images in the testing set classified in Top-K by the CNN;

---

[1] Pre-trained models can be found here: https://github.com/BVLC/caffe/wiki/Model-Zoo.

– DISTRIBUTION OF SCORES OF CORRECT CLASS (SCC ICDF): the probability
  that the Scc of the images in the testing set will take a value greater than or
  equal to x (i.e. the Inverse Cumulative Distribution Function, considering the
  Scc values as a random variable);

Table 4 shows the Top1 Accuracy of the SqueezeNet. The best scenario is the one
in which the CNN is trained and tested over ETHZ dataset (e.g. about 60%).

Table 4. SqueezeNet Cross Test: Top1 Accuracy

Train/Test	UPMC	ETHZ
UPMC	50.11%	37.51%
ETHZ	37.06%	59.71%

Table 5 shows the Top1 Accuracy of the GoogLeNet. Also in this case, the
best scenario is the one in which the CNN is trained and tested over ETHZ
dataset (e.g. about 60%).

Table 5. GoogLeNet Cross Test: Top1 Accuracy

Train/Test	UPMC	ETHZ
UPMC	60.95%	50.63%
ETHZ	46.17%	72.18%

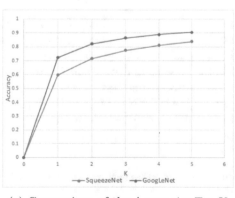

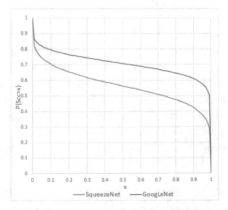

(a) Comparison of the Accuracies Top-K   (b) Comparison of the Scc ICDFs

Fig. 2. Accuracy and Scc ICDF comparisons

Since the images in ETHZ dataset have strong selfie-style (images come from social networks and can contain also people eating food), we chose models coming from training on ETHZ dataset. We referred to ETHZ dataset also for the test for the same reason. This is because the CNN is going to be used by a mobile application, so the input image will be taken by the smartphone camera and will have strong selfie-style as well.

Figure 2a shows the comparison of the Top-K accuracies of both CNN architectures. The GoogLeNet has the highest accuracy in general, but the difference between the two accuracies is decreasing over K. This means that, when K is large, there is no significant difference between the SqueezeNet and the GoogLeNet. Figure 2b depicts the comparison of the Scc ICDF of both CNN architectures. The GoogLeNet is far better than the SqueezeNet because it assigns higher scores to the correct class. Moreover, the GoogLeNet has a lower slope in the central part of the graph. This means that Scc assigned have small variation. The SqueezeNet instead assigns Scc with higher variation.

We chose the GoogLeNet as the CNN architecture to port on the mobile phone because this architecture is the best approach in term of accuracy on ETHZ dataset (i.e. 73%). Moreover, it is second only to the SqueezeNet in terms of model size (about 40 Mb vs 3 Mb). Since the mobile memory can easily store 40 Mb, we preferred the best accuracy. The SqueezeNet is slightly faster than the GoogLeNet due to its layers, but the speed of GoogLeNet is expected to be higher as soon as the mobile computational power increases.

We made a further analysis on the chosen CNN architecture: the Threshold evaluation. The purpose of this analysis is to establish a threshold on the score of the Top1 prediction in order to find the minimum score needed by the Top1 prediction to be considered correct.

The scheme in Fig. 3 depicts the usage of the threshold. If the CNN assigns a score greater than the threshold to a class, it is confident with that prediction and therefore it can be directly shown to the user. If the CNN is not confident with the prediction, the list of Top5 predictions is shown to the user.

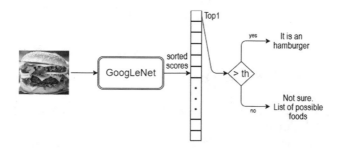

Fig. 3. Threshold usage

The graph in Fig. 4 has been obtained by varying the threshold and by counting the number of False Positives (FP) and False Negatives (FN). The result of the test is classified as Positive if the Top1 prediction score is higher than the threshold, whereas it is classified as Negative if the Top1 prediction's score is lower than the threshold. We zoomed the graph in order to better show the cross-point which is the most important part of it. The lower is the cross-point on y-axis, the higher is the accuracy. On the y-axis, the lower is the crossing point of the two functions, the better is the CNN. If the crossing point of the FP and the FN function is 0, by setting the threshold at that point both zero FP and FN are obtained. Moreover, also the maximum accuracy is reached. Table 6 shows the results obtained by choosing different thresholds.

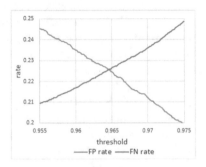

Fig. 4. Zoom on the cross-point of the threshold analysis

Table 6. Results with different thresholds

Threshold	Result	Scc ICDF (threshold)
0.7838	Less than 10% of FN	65%
0.9785	Minimum FP+FN	54%
0.9924	Less than 10% of FP	49%

5   Mobile Application Development

Our proposal consists of an Android mobile application with two different modalities: on-line and offline.

In the online mode, the CNN is deployed on a server and therefore it is implemented in Caffe, which is imported in a Java Web Application. In this modality, the smartphone acts as a client and queries the CNN on the server.

In the off-line mode, the CNN is implemented relying on RenderScript. RenderScript is a framework for utilizing heterogeneous computing on Android phones. During execution, the runtime engine distributes the computation on available processing elements such as CPU cores and GPU. The used implementation of the GoogLeNet has been presented in [7]. In our work, we modified this CNN so that it can be used in Android phones. Figure 5 shows the architecture of the Android Application. The whole code of both the application and the Web Application is available on GitHub[2].

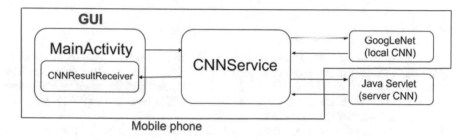

Fig. 5. Application architecture in blocks

6   Conclusions and Future Work

In this work, a mobile application for food recognition has been designed, implemented, and tested. The application recognizes images representing food by predicting the dish. The application uses a Convolutional Neural Network which has been deployed on a mobile phone after a training on two state-of-art datasets for food recognition.

In order to identify the most promising approach in terms of accuracy on the validation set, eight different CNN architectures have been trained and tested. Two scenarios have been identified: the SqueezeNet, which has a good level of accuracy and a model size extremely limited (i.e. 3 Mb of model size and about 60% of accuracy), and the GoogLeNet, which has the best accuracy and a model size second only to the SqueezeNet (i.e. 40 Mb of model size and about 70% of accuracy).

The mobile application has two versions: server-based version (i.e. CNN running on a Web Application) and local version (i.e. CNN developed by taking advantage of RenderScript framework). Hence, tests have been done in order to asses:

– The ability of the CNN to correctly predict the class of food represented in the image.
– The ability of the CNN to distinguish the correct prediction from the wrong prediction by establishing a threshold on the score assigned to the Top1 prediction.

The best scenario has been obtained by training and testing on ETHZ dataset. This is also the result that fits better with the application purpose because of the nature of the dataset (i.e. images with strong selfie style).

The GoogLeNet CNN is the best approach in terms of accuracy, but the SqueezeNet CNN is the best in terms of model size. The CNN used in both application modalities is the GoogLeNet.

References

1. Amato, G., Bolettieri, P., Monteiro de Lira, V., Muntean, C.I., Perego, R., Renso, C.: Social media image recognition for food trend analysis. In: Proceedings of the 40th International ACM SIGIR Conference on Research and Development in Information Retrieval, SIGIR 2017, pp. 1333–1336. ACM, New York (2017). https://doi.org/10.1145/3077136.3084142

2. Bossard, L., Guillaumin, M., Van Gool, L.: Food-101 – mining discriminative components with random forests, pp. 446–461. Springer International Publishing, Cham (2014). http://doi.org/10.1007/978-3-319-10599-4_29

3. Chen, J., Ngo, C.W.: Deep-based ingredient recognition for cooking recipe retrieval. In: Proceedings of the 2016 ACM on Multimedia Conference, MM 2016, pp. 32–41. ACM, New York (2016). https://doi.org/10.1145/2964284.2964315

4. He, K., Zhang, X., Ren, S., Sun, J.: Deep residual learning for image recognition. In: The IEEE Conference on Computer Vision and Pattern Recognition (CVPR), June 2016

5. Iandola, F.N., Moskewicz, M.W., Ashraf, K., Han, S., Dally, W.J., Keutzer, K.: Squeezenet: Alexnet-level accuracy with 50x fewer parameters and <1mb model size. CoRR abs/1602.07360 (2016). http://arxiv.org/abs/1602.07360

6. Krizhevsky, A., Sutskever, I., Hinton, G.E.: Imagenet classification with deep convolutional neural networks. In: Pereira, F., Burges, C.J.C., Bottou, L., Weinberger, K.Q. (eds.) Advances in Neural Information Processing Systems, vol. 25, pp. 1097–1105. Curran Associates, Inc. (2012). http://papers.nips.cc/paper/4824-imagenet-classification-with-deep-convolutional-neural-networks.pdf

7. Motamedi, M., Fong, D., Ghiasi, S.: Fast and energy-efficient CNN inference on IoT devices. CoRR abs/1611.07151 (2016). http://arxiv.org/abs/1611.07151

8. Rastegari, M., Ordonez, V., Redmon, J., Farhadi, A.: Xnor-net: Imagenet classification using binary convolutional neural networks. CoRR abs/1603.05279 (2016). http://arxiv.org/abs/1603.05279

9. Russakovsky, O., Deng, J., Su, H., Krause, J., Satheesh, S., Ma, S., Huang, Z., Karpathy, A., Khosla, A., Bernstein, M.S., Berg, A.C., Li, F.: Imagenet large scale visual recognition challenge. CoRR abs/1409.0575 (2014). http://arxiv.org/abs/1409.0575

10. Simonyan, K., Zisserman, A.: Very deep convolutional networks for large-scale image recognition. CoRR abs/1409.1556 (2014). http://arxiv.org/abs/1409.1556

11. Szegedy, C., Liu, W., Jia, Y., Sermanet, P., Reed, S., Anguelov, D., Erhan, D., Vanhoucke, V., Rabinovich, A.: Going deeper with convolutions. In: The IEEE Conference on Computer Vision and Pattern Recognition (CVPR), June 2015

12. Wang, X., Kumar, D., Thome, N., Cord, M., Precioso, F.: Recipe recognition with large multimodal food dataset. In: 2015 IEEE International Conference on Multimedia Expo Workshops (ICMEW), pp. 1–6, June 2015. https://doi.org/10.1109/ICMEW.2015.7169757

To Identify Hot Spots in Power Lines Using Infrared and Visible Sensors

Bushra Jalil[✉], Maria Antonietta Pascali, Giuseppe Riccardo Leone,
Massimo Martinelli, Davide Moroni, and Ovidio Salvetti

Institute of Information Science and Technologies,
National Research Council of Italy, Via Moruzzi, 1, 56124 Pisa, Italy
{bushra.jalil,maria.antonietta.pascali,giuseppe.leone,
massimo.martinelli,davide.moroni,ovidio.salvetti}@isti.cnr.it

Abstract. The detection of power transmission lines is highly important
for threat avoidance, especially when aerial vehicle fly at low altitude. At
the same time, the demand for fast and robust algorithms for the analysis of data acquired by drones during inspections has also increased. In
this paper, different methods to obtain these objectives are presented,
which include three parts: sensor fusion, power line extraction and fault
detection. At first, fusion algorithm for visible and infrared power line
images is presented. Manual control points describe as feature points
from both images were selected and then, applied geometric transformation model to register visible and infrared thermal images. For the
extraction of power lines, we applied Canny edge detection to identify
significant transition followed by Hough transform to highlight power
lines. The method significantly identify edges from the set of frames with
good accuracy. After the detection of lines, we applied histogram based
thresholding to identify hot spots in power lines. The paper concludes
with the description of the current work, which has been carried out in
a research project, namely *SCIADRO*.

Keywords: Image analysis · Visible images · Infrared images
Image registration · Segmentation · Unmanned Aerial Vehicles
Hot spots

1 Introduction

Defect detection in power lines at an early stage can save the life and ultimately cost of the power system maintenance. During the past, the inspection
of power lines were mainly carried out by foot patrolling which is time consuming, expensive and unsafe method. Unmanned Aerial Vehicles (UAVs) provides
an alternate solution which reduces both the risk and the cost. More recently,
thermal camera mounted on UAVs emerges as the most smart and convenient
solution of power line inspection. Thermal and infrared (IR) imaging has gained
recognitions in the field of power systems during the last two decade. It has been

© Springer Nature Switzerland AG 2019
K. Choroś et al. (Eds.): MISSI 2018, AISC 833, pp. 313–321, 2019.
https://doi.org/10.1007/978-3-319-98678-4_32

314   B. Jalil et al.

used for testing and inspection of different electric parts and also for preventive maintenance work. Infrared imaging uses infrared sensors to capture images of thermal objects based on temperature variations. The camera stores the infrared pictures of the objects as thermal and infrared images that the human can see in order to understand the conditions of the objects. In general, thermal imaging is considered as a robust, non-destructive and contact less methodology to inspect power lines as the inspection can be performed by keeping some distance and hence there is no need to halt or cut down electric supply during the inspection. More recently, UAV based power line inspection utilizes modern flight control techniques and image processing to carry out fast inspection from some distance. Compared with conventional inspection methods, UAV based inspection is more advanced and low-cost. Based on GPS data of both the UAV and towers, the embedded algorithm can also perform automatic tracking of power lines [1]. Later, these thermal images undergo different statistical and morphological processing to highlight hot spots in cables and insulators.

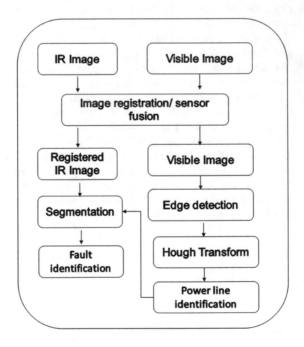

Fig. 1. Block Diagram of complete processing.

In this paper, we use different morphological methods to identify these hot spots in power lines. Several methods have been proposed in the past to identify hot spots in electrical equipments e.g. Hongwei et al. had presented a fusion algorithm for the infrared and visible power lines image [2]. Vega et al. presented a system based on a quadrotor helicopter for monitoring the power lines [3]. Lages et al.[id=R1] captured video streams from both an infrared and visible cameras, simultaneously [4]. They used both statistical and morphological methods to highlight the hot spots in the lines. Oliveira et al. discussed in detail the generation of hot spots in the transmission lines and later they had also used the same thresholding based segmentation to highlight hot spots [5]. Similarly, Wronkowicz had proposed an automated method for the hot spot detection from IR images of power transmission lines, without any reference temperature value [6]. A brief survey of some of these methods are given in [7,8] id=R1.

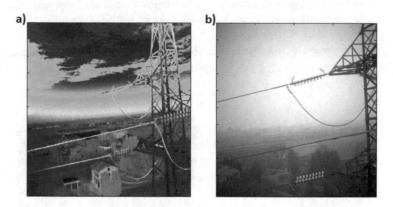

Fig. 2. (a) Infrared image of the tower and power cables. (b) Visible image of the tower and power cables.

2   The SCIADRO Project

This paper deals with the ongoing work being done in SCIADRO [1], a research project. The main idea of our work in the SCIADRO project is to provide a tool to support simultaneously the detection of the infrastructure components. Also, such algorithms should be specifically designed for the collaborative setting of an UAV swarm. The image processing aims at detection and analysis of the main components of the electrical infrastructure: electric towers, insulators, and conductors. Thermal data and images in the visible spectrum have been acquired by a drone flying at a distance of approximately 10 mt from the power lines, with different cameras, near Parma (Italy) in December, 2017. Data include also a small number of images containing common defects. These data have been used to test our methods for the detection of the infrastructure and the diagnosis of its status. At this stage, two tasks have been implemented and partially tested on real data: *i* Sensor fusion: we had applied control point selection based image registration to use both thermal and visible images to identify hot spots; *ii* detection of power lines by image processing applied to RGB images.

3   Methods

In this work, we used infrared and visible images to identify hot spots in power lines. The block diagram of processing is shown in Fig. 1. Images acquired by visible and infrared sensor as shown in Fig. 2 have different field of View, therefore, in order to identify faults from thermal images it is important to do sensor fusion first. We used manual control point selection followed by affine transformation to register both images. At second stage, we identified faults from thermal images by first segmenting power lines from the background using visible images. We applied Canny edge detection on visible images to identify significant transition and keeping in mind the linear characteristics of power lines we used Hough transform to segment power lines. At the last stage, we used histogram based thresholding to identify hot spots in thermal images. In the following section we provide a description of each step.

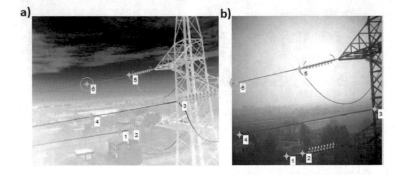

Fig. 3. Control point selection from IR and Visible image for image registration. IR image has larger FOV (field of view), so we selected IR image to be registered.

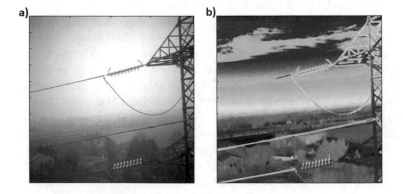

Fig. 4. Registered visible and IR image.

3.1   Image Registration

Image fusion or registration is the process of extracting and integrating the information from two or more images of the same scene taken at different times, from different viewpoints, or by different sensors. This process can provide more accurate, comprehensive and reliable image information of the same scene or object, making fusion images more suitable for human eye perception or computer understanding [2,9]. The infrared and visible light image have different imaging principles, which can generate some problems such as infrared image has a low contrast and high noise. Therefore, in order to take an advantage of both visible and infrared image properties, it is important to register them first [10]. A brief survey of some of these techniques is given in [11].

Image registration is defined as a mapping between two or more images both spatially (geometrically) and with respect to intensity. By mathematically expressing [12]:

$$I_2 = gI_1(f(x_1; x_2)); \tag{1}$$

where I_1 and I_2 are two-dimensional images (indexed by $x_1; x_2$), $f : (x_1; x_2) \rightarrow (x'_1; x'_2)$ maps the indices of the distorted frame to match those of the reference frame, and g is a one-dimensional intensity or radiometric transform. We use fusion algorithm based on manual control point selection of the infrared and visible image shown in Fig. 3. We selected 6 points on the power lines, control tower and oscillators in both images. Also, we assume that we do not need to make any radiometric adjustments, so $g = \mathbb{1}^{\text{id}=\text{R1}}$, the identity transform. $\text{S}^{\text{id}=\text{R1}}$ince there is no radiometric distortion therefore we performed affine transformation. An affine transform can perform rotation, translation, scaling and shearing operations and are linear in the sense that they map straight lines into straight lines. Affine transformation is sufficient to match two images of a scene taken from the same viewing angle but from a different position as in the present case. The detail explanation of affine transformation is given by [12]. The registerd images are shown in Fig. 4.

4   Power Line Detection

In this work, keeping in mind the typical linear characteristics of power lines, we applied Hough transform on visible images to identify power lines. Several methods based on Hough transform had been proposed in the past to identify power lines as in [13,14]. The images were processed following the steps listed here:

1. Preprocessing to improve contrast in the image.
2. Detection of edges by using Canny edge detector.
3. Hough transform to detect all lines in the images.
4. Extraction of power lines.

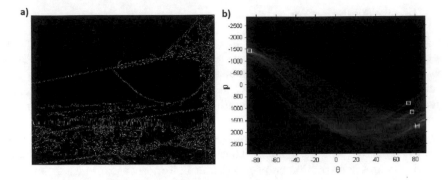

Fig. 5. (a) Edge detection by canny edge detector, (b) Detected peaks with Hough transform, where peaks correspond to the length of the line. ρ is the perpendicular distance of the peak to the origin and θ correspond to the angle. Occurrence of all positive angled peaks correspond to power lines.

As explained before, we had analyzed images acquired by a camera mounted on the drone flying close to the electric power lines. By way of example, an image is shown in Fig. 2. After image registration, we applied Canny edge detector to identify edges and remove unwanted objects from the background of interest area. Canny edge detection algorithm [15] consists of the following steps:

1. In order to smooth the image, Gaussian filtering is applied to reduce noise effects by convolving image with Gaussian filter.
2. Image gradient magnitude and direction are computed.
3. Non-maxima suppression, according to the gradient direction, to get unilateral edge response and to preserve local maxima as these maxima correspond to the edges (The output of maxima suppression contains some local maxima which correspond to noise elements).
4. Double threshold method, in order to detect and connect edges.

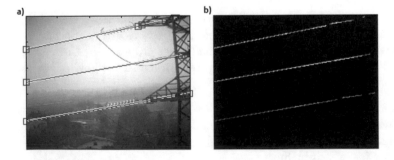

Fig. 6. (a) Detected lines after thresholding in Hough space, (b) Segmented power lines.

The results obtained by using the Canny edge detector is shown in Fig. 5a, power lines along with sharp edges of background were detected. The next step is to highlight only those edges which correspond to power lines. Hough transform is used to detect parameterized shapes through mapping each point to a new parameter space in which the location and orientation of certain shapes could be identified [16]. In this work we applied Hough transform to identify power lines, as the method identifies all straight lines in the image, maybe including roads, buildings etc. Therefore, in order to discriminate power lines from other linear object we applied clustering in the Hough space. The method usually parametrizes a line in the Cartesian coordinate to a point in the polar coordinate using the point-line duality equation:

$$x \cos \theta + y \sin \theta = \rho \qquad \rho \geq 0 \quad 0 \leq \theta \leq \pi \qquad (2)$$

Where (x, y) is the point in image in Cartesian coordinates. ρ is the perpendicular distance of the peak to the origin and θ correspond to the angle to the origin. Before detecting power lines in the Hough space, we applied the Canny edge detector to identify all edges in the images. Figure 5b highlights the detected peaks: here we filtered the three nearly perpendicular θ peaks corresponding to power lines. The detected power lines were segmented from the scene as shown in Fig. 6.

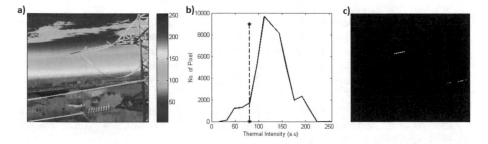

Fig. 7. (a) IR image, (b) Histogram of thermal intensity and selected threshold to highlight fault area, (c) Histogram based thresholding of IR image to identify hot spots in power lines.

4.1 Histogram Base Thresholding

At the final stage, we applied histogram based thresholding on the segmented image in Fig. 6. As both visible and infrared images are registered, therefore by using the segmentation of power lines in visible image, we mapped power lines from infrared images too. The histogram of segmented infrared image is shown in Fig. 7b. We applied thresholding on infrared thermal intensity and it results shown in Fig. 7c in highlighting hot spots present in the image. Although at this stage, we used intensity value of infrared images but in future we intend to use actual temperature value from the infrared image to classify the severity of damages in the wire.

5   Conclusions

In this work, we have identified hot spots in power transmission lines. Thermal and images in the visible spectrum have been acquired by a drone flying at a distance of approximately 10 mt from the power lines, with different cameras, near Parma, Italy in December, 2017. The acquired images from multiple sensors have different field of View, therefore, we registered both images. We used manual control point selection and applied affine transformation for image fusion. Power lines from visible images were identified using Canny edge detection Hough transform and finally, we used histogram based thresholding to identify hot spots in thermal images. The method has significantly identified hot spot from the images and in future we intend to improve this work and to use more data set to test propose method.

Acknowledgement. This work is being carried out in the framework of the Tuscany Regional Project "SCIADRO" (FAR-FAS 2014).

References

1. Alberto, G., et al.: UAVs and UAV swarms for civilian applications: communications and image. In: International Conference on Wireless and Satellite Systems, pp. 115–124. Springer (2017)
2. Li, H., et al.: Research on the infrared and visible power-equipment image fusion for inspection robots. In: 2010 1st International Conference on Applied Robotics for the Power Industry, Delta Centre-Ville Montréal, Canada, 5–7 October 2010
3. Vega, L., et al.: Power line inspection via an unmanned aerial system based on the quadrotor helicopter. In: IEEE Mediterranean Electro Technical Conference, Beirut, Lebanon, 13–16 April 2014
4. Legas, W.F., Scheeren, V.: An embedded module for robotized inspection of power lines by using thermographic and visual images. In: CARPI, ETH Zurich, Switzerland, 11–13 September 2012
5. Oliveira, J.H., et al.: Robotized inspection of power lines with infrared vision. In: 2010 1st International Conference on Applied Robotics for the Power Industry, Montréal, Canada, 5–7 October 2010
6. Wronkowicz, A.: Approach to automated hot spot detection using image processing for thermographic inspections of power transmission lines. Diagnostyka, Vol. 17, pp. 81–86 (2016)
7. Jalil, et al.: Power lines inspection via thermal and infrared imaging. In: Advanced Infrared Technology and Applications, Quebec, Canada (2017)
8. Katrasnik, J., et al.: A survey of mobile robots for distribution power line inspection. IEEE Trans. Power Delivery **25**, 485–493 (2010)
9. Rui, et al.: Registration of infrared and visible images based on improved SIFT. In: ICIMCS 2012, Wuhan, China, 09–11 September 2012
10. Zeng, L.X., et al.: SAR image registration based on affine invariant SIFT features. Opto Electr. Eng. **37**(10), 121–124 (2010)
11. Kim, K.: Survey on registration techniques of visible and infrared images. IT Convergence PRActice (INPRA) **3**(2), 25–35 (2015)

12. Hines, G., et al.: Multi-image registration for an enhanced vision system, NASA Langley Research Center, Hampton
13. Candamo, J., et al.: Detection of thin lines using low-quality video from low-altitude aircraft in urban settings. IEEE Trans. Aerosp. Electr. Syst. **45**(3), 937–949 (2009)
14. Li, Z., et al.: Towards automatic power line detection for a UAV surveillance system using pulse coupled neural filter and an improved Hough transform. Mach. Vis. Appl. **21**, 677–686 (2010)
15. Canny, J.: A computational approach to edge detection. IEEE Trans. Pattern Anal. Mach. Intell. **8**, 679–698 (1986)
16. P. V. C. Hough, Method and Means for Recognizing Complex Patterns US Patent 3,069,654, 1962, Ser. No. 17,7156 Claims

Sparse Solution to Inverse Problem of Nonlinear Dimensionality Reduction

Honggui Li[1(✉)] and Maria Trocan[2]

[1] Yangzhou University, Yangzhou 225002, Jiangsu, China
hgli@yzu.edu.cn
[2] Institut Supérieur d'Électronique de Paris, 75006 Paris, France
maria.trocan@isep.fr

Abstract. In this paper we propose a sparse solution to the inverse problem of nonlinear dimensionality reduction (NLDR), which holds potential high-performance applications in data representation, compression, generation, and visualization. Firstly, the sparse solution model of the inverse problem of NLDR is established, which consists of four components: classical NLDR, sparse dictionary learning, NLDR embedding, and sparse NLDR reconstruction. Secondly, the special sparse solution to the inverse problem of isometric feature mapping (ISOMAP), a classical NDLR algorithm, is presented. ISOMAP embedding and sparse ISOMAP reconstruction algorithms are raised, and the alternating directions method of multipliers (ADMM) is adopted to resolve the minimization problem of the special sparse solution. Finally, it is revealed by the experimental results that, in the situation of very low dimensional representation, the proposed method is superior to the state of the art methods, such as discrete cosine transformation (DCT) and sparse representation (SR), in the reconstruction performance of image and video data.

Keywords: Sparse representation · Inverse problem
Nonlinear Dimensionality Reduction · Isometric feature mapping
Data reconstruction

1 Introduction

In the current world of big data and cloud computing, efficient data representation and compression methods are pursued by researchers and engineers. Many powerful image and video coding methods have been developed, such as JPEG, AVC/H.264, HEVC/H.265, JEM/H.266, and so forth [1].

More efficient data representation and compression methods are still expected. Because nonlinear dimensionality reduction (NLDR) is capable of providing very low dimensional representation for high dimensional data, it naturally becomes a potential candidate of high efficient data representation and compression [2–7]. The classical NLDR algorithms include locally linear embedding (LLE) [2], isometric feature mapping (ISOAMP) [3], Laplacian Eigenmaps (LEM) [4], neural network dimensionality reduction (NNDR) [5], and etc. However, their inverse problems are ill-posed, undetermined, and nonlinear. The inverse problem of NLDR has almost never been

© Springer Nature Switzerland AG 2019
K. Choroś et al. (Eds.): MISSI 2018, AISC 833, pp. 322–331, 2019.
https://doi.org/10.1007/978-3-319-98678-4_33

investigated. Although support vector regression and radial basis neural network have been explored, both of them cannot offer perfect reconstruction in the situation of very low dimensional representation [6, 7].

Much prior knowledge, such as sparse representation (SR), low-rank representation (LRR), deep learning (DL), and their combinations, can be utilized to resolve the inverse problems, such as imaging, super-resolution, compressed sensing, denoising, enhancement, recovery, restoration, reconstruction, rebuilt, error concealment, and so on [8–22].

Sparse prior knowledge has been adopted in many inversion problem of imaging, such as computed tomography (CT), magnetic resonance imaging (MRI), holography, ultrasound beamforming, multimodal imaging, etc. [8–13]. Jin Liu proposes a 3D feature sparse dictionary-based reconstruction algorithm for low dose computed tomography (LDCT) imaging [8]. Ti Bai raises a 3D sparse dictionary based iterative reconstruction framework for x-ray low dose cone-beam CT (CBCT) [9]. Fei Wen presents a robust sparse reconstruction method of MRI, which exploits a generalized nonconvex penalty for sparsity inducing [10]. Xuelin Cui employs the compressed sensing sparse priors for joint reconstruction framework of CT and MRI [11]. Stijn Bettens utilizes wavelet sparsifying bases for compressive digital holography [12]. Ece Ozkan solves the inverse problem of ultrasound beamforming with sparsity constraints and regularization [13].

Low-rank prior knowledge and its combination with sparse representation have already been exploited for imaging inverse problem [14–19]. Kyungsang Kim utilizes spectral patch-based low-rank penalty for sparse-view spectral CT reconstruction [14]. Jian-Feng Cai uses low-rank matrix factorization for CBCT reconstruction [15]. Kyong Hwan Jin utilizes annihilating filter based low-rank Hankel matrix for compressed sensing and parallel MRI [16]. Lin Xu uses low-rank matrix decomposition for denoising multi-channel images in parallel MRI [17]. Sajan Goud Lingala takes advantage of sparsity and low-rank structure for accelerated dynamic MRI [18]. Anthony G. Christodoulou makes use of parallel imaging with low-rank and sparse modeling for high-resolution cardiovascular MRI [19].

Deep learning based prior knowledge and its combination with sparse representation have also been employed for imaging inverse problems [20–22]. Hu Chen brings forward a residual encoder-decoder convolutional neural network (RED-CNN) for LDCT [20]. Jo Schlemper comes up with a deep cascade of convolutional neural networks (CNN) for dynamic MRI [21]. Dufan Wu puts forward an iterative LDCT reconstruction with priors trained by K-sparse autoencoder (KSAE) [22].

This paper concentrates on sparse prior knowledge-based solution to the inverse problem of NLDR, especially ISOMAP.

The rest of this paper is organized as follows. Some theoretical fundamentals of inverse problems are recalled in Sect. 2; the proposed method is introduced in Sect. 3 and the experimental validation is conducted in Sect. 4. Finally, conclusions are drawn in Sect. 5.

2 Theoretical Background

2.1 Inverse Problem of NLDR

NLDR and its inverse problem are shown in Fig. 1, and they can also be described by the following formulas:

$$y = N(x), x = N^{-1}(y), x \in R^{H \times 1}, y \subset R^{L \times 1}, H \gg L, \tag{1}$$

where: N is the forward transformation of NLDR, such as the aforementioned classical NLDR algorithms; N^{-1} is the inverse transformation of NLDR; x is the sample in high dimensional space; y is the related sample in low dimensional space; H is the high dimension of x; L is the low dimension of y.

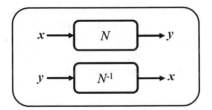

Fig. 1. NLDR and its inverse problem.

2.2 Sparse Solution to Inverse Problems

Because L is far less than H, the inversion problem of NLDR is an ill-posed and undetermined nonlinear problem. In order to resolve the inverse problem, some prior knowledge should be utilized. The prior knowledge-based solution to the inverse problem of NLDR can be depicted by the following expressions:

$$x = \arg\min_{x} \Omega(x), \; s.t. \, y = N(x), \tag{2}$$

where $\Omega(x)$ is the cost function of prior knowledge for the high dimensional sample, such as the object functions of sparse representation, low-lank representation, deep learning, and etc. If the prior knowledge is sparse representation based, it is the sparse solution to the inverse problem of NLDR, and it can be expressed by the subsequent equations:

$$x = \arg\min_{x} \|s\|_0, \; s.t. \, y = N(x), \|x - Ds\|_2^2 \leq \varepsilon, \tag{3}$$

where: D is the sparse dictionary of the high dimensional sample; s is the sparse coefficients of the high dimensional sample; ε is a given small positive constant. The cost function of sparse representation is the l_0 norm of sparse coefficients on extra condition that x can be approximately represented by the product of D and s.

3   Proposed Theoretical Framework

3.1   Theoretical Architecture

The theoretical framework of the proposed method is illustrated in Fig. 2, which comprises four components: classical NLDR (N), sparse dictionary learning (D), NLDR embedding (N_e), and sparse NLDR reconstruction (N_s).

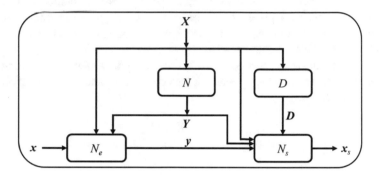

Fig. 2.   Theoretical framework of the proposed method.

The classical NLDR module, which is based on the aforesaid conventional NLDR methods, achieves the low dimensional representation of the training high dimensional data, and it can be described by the following formulas:

$$
\begin{aligned}
& Y = N(X), y_i = N(x_i) \\
& X = [x_1 \quad x_2 \quad \cdots \quad x_T], X = \{x_1, x_2, \cdots, x_T\} \\
& Y = [y_1 \quad y_2 \quad \cdots \quad y_T], Y = \{y_1, y_2, \cdots, y_T\} \\
& X \in R^{H \times T}, Y \in R^{L \times T}, x_i \in R^{H \times 1}, y_i \in R^{L \times 1}, i = 1, 2, \cdots, T
\end{aligned}
\tag{4}
$$

where: X is the matrix of the high dimensional training samples; Y is the matrix of the related low dimensional training samples; X is the set version of X; Y is the set version of Y; x_i is the i-th high dimensional training sample of X; y_i is the i-th related low dimensional training sample of Y; T is the total numbers of training samples. It is feasible to utilize the same training pair for NLDR embedding and reconstruction modules in the proposed architecture, because the same input data and classical NLDR algorithm can be adopted for each module.

The sparse dictionary learning module obtains the sparse dictionary of the training data X, and it can be depicted by the following expressions:

$$
D = D(X) = \arg\min_{D} \sum_{i=1}^{T} \|s_i\|_0, \; s.t. \sum_{i=1}^{T} \|x_i - Ds_i\|_2^2 \leq \varepsilon,
\tag{5}
$$

where: D is the learning algorithm of sparse dictionary, and D is the learned sparse dictionary.

The NLDR embedding module finds the low dimensional representation of a new testing sample, depending on the known training pair, X and Y, which are obtained by the classical NLDR algorithms. The NLDR embedding module can be replaced by any incremental NLDR algorithm. The NLDR embedding module can be described by the following formulas:

$$y = N_e(x; X, Y),$$
(6)

where: N_e is the general NLDR embedding algorithm, which obtains the low dimensional representation of a given testing sample depending on the traditional NDLR algorithms and the training pair; x is the high dimensional testing sample; y is the related low dimensional testing sample.

The sparse reconstruction module recovers the high dimensional testing sample relying on low dimensional sample and sparse dictionary, and it can be expressed by the following equations:

$$x_s = N_s(y; X, Y, D)$$
$$x_s = \arg\min_x \|s\|_0, \ s.t. \ y = N_e(x; X, Y), \|x - Ds\|_2^2 \leq \varepsilon,$$
(7)

where: N_s is the sparse NLDR reconstruction algorithm, which gains the high dimensional sample of a given low dimensional representation relying on the sparse prior knowledge and the training pair, and x_s is the high dimensional reconstruction sample.

3.2   Special Solution to ISOMAP

This paper takes ISOMAP as an example and focus on the sparse solution to its inverse problem. ISOMAP attempts to maintain the same geometric distance between neighbors in both high dimensional space and low dimensional space [3].

The embedding module for ISOMAP can be depicted by the subsequent expressions:

$$y = N_e(x; X, Y) = \arg\min_y \sum_{k=1}^{K} \left(\|y - y_{nk}\|_2 - \|x - x_{nk}\|_2 \right)^2$$
$$y_{nk} = N(x_{nk}), k = 1, 2, \cdots, K$$
$$x_{nk} \in K_n(x) \subset X \Rightarrow y_{nk} \in K_n(y) \subset Y$$
$$K_n(x) = \{x_{n1}, x_{n2}, \cdots, x_{nK}\} \Rightarrow K_n(y) = \{y_{n1}, y_{n2}, \cdots, y_{nK}\}$$
(8)

where: $K_n(x)$ is the set of K-nearest neighbors of x in set X, and $K_n(y)$ is the related set of K-nearest neighbors of y in set Y.

The sparse reconstruction module for ISOMAP can be expressed by the following equations:

$$x_s = N_s(y; X, Y, D)$$
$$= \arg\min_x \left(\|s\|_0 + \lambda \sum_{k=1}^{K} \left(\|y - y_{nk}\|_2 - \|x - x_{nk}\|_2 \right)^2 \right), s.t. \|x - Ds\|_2^2 \leq \varepsilon, \quad (9)$$

where: λ is a known positive constant.

The aforementioned minimization problem can be easily resolved by alternating directions method of multipliers (ADMM) [23]. The pseudocode of the special solution to ISOMAP is exhibited in Fig. 3.

Input: X, x
    Initialize: $Y = N(X)$, $D = D(X)$, $y = N_e(x; X, Y)$
    Resolve $x_s = N_s(y; X, Y, D)$ by ADMM
Output: x_s

Fig. 3. The pseudocode of the special solution to ISOMAP.

4 Experiments

4.1 Experiment Preparation

The experimental hardware platform is a laptop computer with 2.6 GHz CPU and 8 GB memory, and the experimental software platform is Matlab 2016a on Windows 10. The image dataset is a traditional image set, and the video dataset is the class D of the HEVC testing sequences [1]. Two conventional core methods of image and video coding are involved: sparse representation and discrete cosine transformation (DCT). The sparse representation algorithm is K-SVD, the size of sparse dictionary is 256, the sparse level is arbitrary (certainly, small fixed sparse level deserves further investigation), the number of iteration is 20, the stop criterion is a given representation error, and the algorithm of sparse coefficients estimation is orthogonal matching pursuit (OMP) [24]. The DCT algorithm is direct discrete cosine transform and dictionary learning of DCT is not adopted [25]. Two sub experiments are designed. One is the experiment of image reconstruction, and the other is the experiment of video reconstruction.

4.2 Experiment of Image Reconstruction

Self embedding and reconstruction of image are considered. Image size is 256 × 256, and each image is divided into 8 × 8 patches. The patches of an image are used as training samples and then utilized as testing samples. For the proposed method, one, two or three low dimensional representations of image patches are employed for reconstruction. For sparse representation, the most important one, two or three sparse coefficients of image patches are preserved for reconstruction. For DCT, the most important one, two or three DCT coefficients of image patches are retained for reconstruction. The experimental results of image reconstruction are shown in Table 1, which contains the reconstruction PSNRs of the proposed method, sparse representation, and DCT, in the case of low 1D, 2D, and 3D representations. The results indicate that the proposed method holds better performance than sparse representation and DCT. It is also shown by the experimental results that, if the training samples and testing samples are very similar, even the same, the proposed method can attain excellent reconstruction results. The experimental results of lenna image are also illustrated in Fig. 4.

Table 1. Experimental results of image reconstruction.

Images	Algorithms	PSNR (dB)		
		1D	2D	3D
Barbara	Proposed method	**21.5640**	**30.6138**	**43.2675**
	Sparse representation	20.1540	22.4038	24.0904
	DCT	20.6046	22.4466	24.1142
Goldhill	Proposed method	**22.7552**	**29.2620**	**41.7261**
	Sparse representation	22.1553	23.9379	25.5075
	DCT	22.6268	24.0904	25.5769
Lenna	Proposed method	**22.2545**	**32.1323**	**43.0150**
	Sparse representation	20.6448	23.4789	24.0952
	DCT	21.3308	23.6007	24.6133
Mandrill	Proposed method	**21.2782**	**25.5391**	**38.3846**
	Sparse representation	20.6219	21.2063	21.7474
	DCT	21.1779	21.8862	22.5872
Peppers	Proposed method	**20.4682**	**30.3403**	**42.4093**
	Sparse representation	19.7893	21.5121	24.2494
	DCT	20.3820	22.2148	24.3285

Fig. 4. The experimental results of lenna image, with three low dimensional representations, from left to right: original image, reconstruction image of the proposed method (PSNR = 43.0150 dB), reconstruction image of sparse representation (PSNR = 24.0952 dB), and reconstruction image of DCT (PSNR = 24.6133 dB).

It can further be discovered from Table 1 and Fig. 4 that, DCT surpasses sparse representation in reconstruction performance. This may be owing to the fact that, the implementation of DCT is not based on a trained DCT dictionary but on direct discrete cosine transform. Sparse representation is image patch set oriented and is the optimal solution to a set of image patches, so more sparse coefficients may be needed to recover a given image patch. On the contrary, DCT is image patch oriented and is the best solution to a given image patch, thus fewer DCT coefficients may be required to rebuild the image patch.

The main drawback of the image reconstruction experiment is that, the training and testing sample sets are same. On the one hand, the experimental results of image reconstruction reflect the recovery capability of the proposed method to some extent. On the other hand, the following experiment of video reconstruction employs different training and testing sample sets and can further prove the performance of the proposed method. Another disadvantage of the proposed method is that, the performance promotion is at the expense of huge computation load. However, the proposed method can be exploited by a non-real-time system, which needs high reconstruction performance.

4.3   Experiment of Video Reconstruction

Inter frame prediction is taken into account. Frame size is of 240×416 pixels and each frame is partitioned into 8×8 patches. The patches of the first frame of a video sequence are utilized as training samples and the patches of the second frame are used as testing samples.

Table 2. Experimental results of video reconstruction.

Videos	Algorithms	PSNR (dB)		
		1D	2D	3D
BasketballPass	Proposed method	**24.9477**	**26.0316**	**27.7756**
	Sparse representation	23.6037	25.2488	25.9993
	DCT	24.1969	25.3294	26.5562
BlowingBubbles	Proposed method	**23.3791**	**24.8232**	**26.4915**
	Sparse representation	21.9618	24.0896	24.6886
	DCT	22.9074	24.1142	25.0259
BQSquare	Proposed method	**18.0233**	**19.1523**	**20.6833**
	Sparse representation	17.2703	18.0921	18.7532
	DCT	17.5446	18.4825	19.6127
Flowervase	Proposed method	**29.8904**	**30.3957**	**32.2940**
	Sparse representation	28.5861	28.7340	31.0757
	DCT	28.9026	29.5711	31.1962
RaceHorses	Proposed method	**21.6374**	**23.2006**	**24.0418**
	Sparse representation	20.5837	22.0351	23.3902
	DCT	20.7119	22.0495	23.4721

The experimental results of video reconstruction are shown in Table 2, which exhibits the reconstruction PSNRs of the proposed method, sparse representation, and DCT, in the situation of 1D, 2D, and 3D representations. The experimental results manifests that the propose method possesses better performance than sparse representation and DCT. Therefore, the proposed method can be employed for inter frame prediction with high competitiveness.

5 Conclusions

This paper addresses the inverse problem of NLDR, especially the ISOMAP algorithm. In the first place, the theoretical architecture of the sparse solution to the inverse problem is built. In the second place, the special sparse solution to the inverse problem of ISOMAP is deduced. Finally, simulation experiments are devised to evaluate the performance of the proposed method, which outperforms the state of the art methods in the reconstruction of image and video data, in the case of very low dimensional representation.

In our future work, the proposed method will be testified on more image and video datasets, and more sophisticated prior knowledge will be studied for the inverse problem of NLDR.

References

1. Sullivan, G.J., Ohm, J.-R., Han, W.-J., Wiegand, T.: Overview of the high efficiency video coding (HEVC) standard. IEEE Trans. Circ. Syst. Video Technol. **22**(12), 1649–1668 (2012)
2. Sam, T.: Roweis: nonlinear dimensionality reduction by locally linear embedding. Science **290**(5500), 2323–2326 (2000)
3. Tenenbaum, J.B., de Silva, V., Langford, J.C.: A global geometric framework for nonlinear dimensionality reduction. Science **290**(5500), 2319–2323 (2000)
4. Belkin, M., Niyogi, P.: Laplacian eigenmaps and spectral techniques for embedding and clustering. Adv. Neural. Inf. Process. Syst. **14**(6), 585–591 (2001)
5. Hinton, G.E., Salakhutdinov, R.R.: Reducing the dimensionality of data with neural networks. Science **313**(5786), 504–507 (2006)
6. Zhang, C.: Reconstruction and analysis of multi-pose face images based on nonlinear dimensionality reduction. Pattern Recognit. **37**(2), 325–336 (2004)
7. Monniga, N.D., Fornberga, B., Meyerb, F.G.: Inverting nonlinear dimensionality reduction with scale-free radial basis function interpolation. Appl. Comput. Harmonic Anal. **37**(1), 162–170 (2014)
8. Liu, J., Hu, Y., Yang, J., Chen, Y., Shu, H., Luo, L., Feng, Q., Gui, Z., Coatrieux, G.: 3D feature constrained reconstruction for low dose CT imaging. IEEE Trans. Circ. Syst. Video Technol. **28**(5), 1232–1247 (2018)
9. Bai, T., Yan, H., Jia, X., Jiang, S., Wang, G., Mou, X.: Z-index parameterization (ZIP) for volumetric CT image reconstruction via 3D dictionary learning. IEEE Trans. Med. Imaging **36**(12), 2466–2478 (2017)
10. Wen, F., Pei, L., Yang, Y., Wenxian, Yu., Liu, P.: Efficient and robust recovery of sparse signal and image using generalized nonconvex regularization. IEEE Trans. Comput. Imaging **3**(4), 566–579 (2017)

11. Cui, X., Mili, L., Hengyong, Yu.: Sparse-prior-based projection distance optimization method for joint CT-MRI reconstruction. IEEE Access **5**, 20099–20110 (2017)
12. Bettens, S., Schretter, C., Deligiannis, N., Schelkens, P.: Bounds and conditions for compressive digital holography using wavelet sparsifying bases. IEEE Trans. Comput. Imaging **3**(4), 592–604 (2017)
13. Ozkan, E., Vishnevsky, V., Goksel, O.: Inverse problem of ultrasound beamforming with sparsity constraints and regularization. IEEE Trans. Ultrason. Ferroelectr. Freq. Control **65** (3), 356–365 (2018)
14. Kim, K., Ye, J.C., Worstell, W., Ouyang, J., Rakvongthai, Y., Fakhri, G.E., Li, Q.: Sparse-view spectral CT reconstruction using spectral patch-based low-rank penalty. IEEE Trans. Med. Imaging **34**(3), 748–760 (2015)
15. Cai, J.-F., Jia, X., Gao, H., Jiang, S.B., Shen, Z., Zhao, H.: Cine cone beam CT reconstruction using low-rank matrix factorization: algorithm and a proof-of-principle study. IEEE Trans. Med. Imaging **33**(8), 1581–1591 (2014)
16. Jin, K.H., Lee, D., Ye, J.C.: A general framework for compressed sensing and parallel MRI using annihilating filter based low-rank Hankel matrix. IEEE Trans. Comput. Imaging **2**(4), 480–495 (2016)
17. Lin, X., Wang, C., Chen, W., Liu, X.: Denoising multi-channel images in parallel MRI by low rank matrix decomposition. IEEE Trans. Appl. Supercond. **24**(5), 480–494 (2014)
18. Lingala, S.G., Hu, Y., DiBella, E., Jacob, M.: Accelerated dynamic MRI exploiting sparsity and low-rank structure: k-t SLR. IEEE Trans. Med. Imaging **30**(5), 1042–1054 (2011)
19. Christodoulou, A.G., Zhang, H., Zhao, B., Hitchens, T.K., Ho, C., Liang, Z.-P.: High-resolution cardiovascular MRI by integrating parallel imaging with low-rank and sparse modeling. IEEE Trans. Biomed. Eng. **60**(11), 3083–3092 (2013)
20. Chen, H., Zhang, Y., Kalra, M.K., Lin, F., Chen, Y., Liao, P., Zhou, J., Wang, G.: Low-dose CT with a residual Encoder-Decoder Convolutional Neural Network (RED-CNN). IEEE Trans. Med. Imaging **36**(12), 2524–2535 (2017)
21. Schlemper, J., Caballero, J., Hajnal, J.V., Price, A., Rueckert, D.: A deep cascade of convolutional neural networks for dynamic MR image reconstruction. IEEE Trans. Med. Imaging **37**(2), 491–503 (2017)
22. Dufan, W., Kim, K., El Fakhri, G., Li, Q.: Iterative low-dose CT reconstruction with priors trained by artificial neural network. IEEE Trans. Med. Imaging **36**(12), 2479–2486 (2017)
23. Afonso, M.V., Bioucas-Dias, J.M., Figueiredo, M.A.T.: Fast image recovery using variable splitting and constrained optimization. IEEE Trans. Image Process. **19**(9), 2345–2356 (2010)
24. Aharon, M., Elad, M., Bruckstein, A.: K-SVD: an algorithm for designing overcomplete dictionaries for sparse representation. IEEE Trans. Signal Process. **54**(11), 4311–4322 (2006)
25. Jain, A.K.: Fundamentals of Digital Image Processing, 1st edn. Prentice Hall, Englewood Cliffs (1989)

Towards Structural Monitoring and 3D Documentation of Architectural Heritage Using UAV

Danila Germanese, Giuseppe Riccardo Leone, Davide Moroni,
Maria Antonietta Pascali[✉], and Marco Tampucci

Institute of Information Science and Technologies,
National Research Council of Italy, Via Moruzzi, 1, 56124 Pisa, Italy
{danila.germanese,giuseppe.leone,davide.moroni,
maria.antonietta.pascali,marco.tampucci}@isti.cnr.it

Abstract. This paper describes how Unmanned Aerial Vehicles (UAVs) may support the architectural heritage preservation and dissemination. In detail, this work deals with the long-term monitoring of the crack pattern of historic structures, and with the reconstruction of interactive 3D scene in order to provide both the scholar and the general public with a simple and engaging tool to analyze or visit the historic structure.

Keywords: Crack quantification methodology · Crack monitoring
Photogrammetry · UAV · 3D rendering

1 Introduction

Today the cultural heritage is considered a very efficient lever to create and enhance social capital, as it has proven to have a valuable impact on economy and society. In this perspective local and national policies more and more frequently allocate resources to the cultural heritage preservation, restoration, and dissemination to the general public.

The constant growth of digital technologies, also boosted by the advent of low-cost hardware and applications, plays a role in improving the surveying, modeling, and visualization of architectural heritage. Today, the digital representation is not only used to visualize data of an architectural heritage: often complemented with simulation, the digital representation is considered by professionals and academics a fundamental aspect. Indeed, the accurate representation of an ancient structure is a reliable and accessible documentation, also used to assess the mechanical stability or to monitor specific regions, preventing critical events.

The structural deterioration of cultural heritage structures is assessed by monitoring and measuring missing or deformed structural elements, cracks and fissures. Visual inspection remains the most commonly used technique to detect damage and evaluate their progress and severity. Nonetheless, such technique may be time consuming and expensive. Moreover, in some cases, access to critical locations may be difficult.

© Springer Nature Switzerland AG 2019
K. Choroś et al. (Eds.): MISSI 2018, AISC 833, pp. 332–342, 2019.
https://doi.org/10.1007/978-3-319-98678-4_34

Apart from criticalities, the accurate monitoring of architectural structures provides, as a by-product, a set of data useful for the creation of informative models of the inspected structure, which are complex models made by geo-referenced architectural models and correlated databases of available information. Also, these models may be used to create an interactive scene of virtual reality for dissemination purpose.

The specific focus of the present work is devoted to: (i) monitor the structural health of architectural heritage via UAV; (ii) provide a clear and expressive documentation to both scholars and general public. The challenge is to acquire data using drones, hence increasing the safety of the monitoring and reducing time and cost, without loosing accuracy in measurements. Even if there are a number of promising methods in literature, the peculiarity of ancient masonry (generally showing very irregular patterns of shape and colour, or legal restrictions to the installation of sensors) poses an obstacle to the applicability of most of them. In the following, Sect. 2 provides a review of the literature about the methods used to measure and monitor crack pattern from visible images. Then we devote the remaining sections to the description of the ongoing work carried out in the framework of the MOSCARDO project, financed by the Tuscany region in the framework of the local actions devoted to the preservation and enhancement of the architectural heritage, including efficient dissemination to general public.

2   Related Works

The rich literature devoted to the structural monitoring of buildings splits into two different groups: invasive and non-invasive methods. Here we focus on those methods which are more suitable to be applied in cultural heritage: (i) Close-range photogrammetry; (ii) Marker-based structural monitoring.

Close-range digital photogrammetry is a large family of methods (see [16] for a survey). At each acquisition, the acquired images are used to produce a 3D point cloud. The assessment of the crack opening is made by comparing the point clouds generated at different dates. Such comparison may be performed in many ways, e.g.: (i) by *conventional* analysis, i.e. comparing the estimated 3D coordinates of the same points by using statistical tests, [20]; (ii) by using shape analysis techniques (matching surfaces [7] or comparing their shape signatures [3]); (iii) by comparing a specific shape parameter (the surface area associated to each crack) complemented with a bootstrap testing to detect only statistical meaningful variations in crack opening [1].

Other techniques belonging to this family aims at automatically identifying and measuring structure damages and cracks by using image-based algorithms which allow for specifically filtering out the cracking patterns. In [5], for instance, the authors refer to two image processing methods to automatically and specifically filter out the cracking patterns: the first one evaluates the color level for each pixel, in order to add more "white" or more "black" and thus making the patterns related to structural discontinuities even darker (this method may fail when the walls of the structure are not clear). The second one is based on the detection of the edges by applying a Gaussian Blur to the original image, and subsequently subtracting the filtered image from the original one again. In this work, cracks were detected and inspected by using a rotary wing octocopter micro air vehicle (MAV) and a high resolution digital camera; nonetheless, no quantitative analysis of cracks was performed.

Jahanshahi et al. [9,10] used a small cross-sectional cracks (0.4–1.4 mm) detection method based on 3D reconstruction of the scene, image segmentation and binarization to isolate the pattern related to the structural defect, and finally two classifiers (SVM and NN) trained to distinguish crack from non-crack patterns. The used approach can be applied to images captured from any distance (20 m in their experimental tests) and acquired using any resolution and focal length (600 mm in this case). Nonetheless, it is suitable for detection of anomalies over homogeneous background (for instance, over concrete). In the work presented by Niemer and colleagues [13], a system based on a commercial camera and a dedicated software (Digital Rissmess-System, DRS) was purposely developed to monitor cracks and fissures in civil structures. A cylindrical tube is fixed to the chamber which allows for a constant multi-spectral illumination. Three approaches were developed to extract crack parameters: (i) Fly-Fisher algorithm, which enables to monitor the crack over time and measure its dimensions automatically; (ii) manual measurement of the crack size at a pre-selected point and evaluation of crack profile; (iii) correlative approach, which infers crack parameters by the translation and rotation movements necessary to line up and join the two sides of the fissure.

The works of Jahanshahi [8] and Ellenberg [4] report and highlight the main challenges of cracks automatic detection in civil infrastructures performed with image-based methods. Several image processing techniques, including enhancement, noise removal, registration, edge detection, line detection, morphological functions, colour analysis, texture detection, wavelet transform, segmentation, clustering, and pattern recognition, are described and evaluated in [8]. Among the major challenges, the noise due to the edges of doors, windows, and buildings, that are sharpened when edge detection algorithms are performed.

In addition, in [4] the main problems related to the use of UAVs are reported: the environmental conditions, for instance, and the wind in particular. Moreover, the field of view of the camera, the angle of orientation of the UAV, and the GPS position must be well defined in order to perform reliable acquisitions.

It can be very useful to "mark" the most critical point of a discontinuity [19]. In the work of Nishiyama and colleagues [14], for instance, the so-called "reflective targets" were exploited. Such targets are made by glass droplets, so as to reflect the light as much as possible; they are usually positioned over the crack at points of interest. A number of images are acquired; by means of photogrammetric techniques, the coordinates of the targets can be calculated and the displacements (due to tensile and shear forces) of the two surface portions of the crack can be assessed. The main source of error is due to the calculation of the centroid coordinates. The greater the distance of the camera, the lower the accuracy with which the crack width will be calculated.

In [19], target detection is performed by Hough transform, in order to identify the geometric centers of the targets. Homography techniques are used to correct the perspective error and to identify the planar coordinates of the targets. Any displacement identified by the coordinates of the targets is used to calculate the force field along the discontinuities of interest.

Benning et al. [2] tested different structural elements of pre-stressed, reinforced and textile concrete. For the photogrammetric measurements, the surfaces were prepared by a grid of circular targets. Up to three digital cameras (Kodak DCS Pro 14n) captured images of the surface simultaneously; repeating the measurement in time intervals and calculating the relative distances between adjacent targets made it possible to monitor the cracks and discontinuities evolution. In addition, a Finite-Element-Module was developed, which simulated the test: thus, the results of photogrammetric measurements could be compared with the numeric tension calculation and iteratively improved. The markers can also be home-made; useful suggestions on their dimensions, materials, etc. can be taken from the study of Shortis et al. [17].

3   The Solution

Main goal of our work is to design and develop a robotic system able to make the inspection of ancient buildings and structures as automatic as possible, not losing in measurement accuracy. To this aim, unmanned aerial vehicles (UAV) are endowed with cameras and acquire aerial images to allow (by means of advanced image processing techniques) for a contact-less and accurate inspection of cracks and fissures, hence in a cheaper, faster, and safer way.

All the non invasive methods found in literature require high resolution cameras, good environmental and lighting conditions. The most used approaches based on image processing may be divided into two groups: marker-less and marker-based methods. Generally, methods in the first group are based on close-range photogrammetry and require the availability of feature points in the images (e.g. too homogeneous textures generally do not have enough feature points), and high quality images. Even if in literature there are some promising results, e.g. [1], we preferred to implement a marker-based approach, as the peculiarity of the considered case study of the project makes it difficult and complex to evaluate the crack opening at the required level of accuracy. In more detail, our solution was inspired by the ArUco framework, described by Salinas et al. in [12]: the authors define a dictionary of coded planar square markers and tackle the mapping problem as a variant of the sparse bundle adjustment problem, by solving the corresponding graph-pose problem; the optimization is done thanks to the sub-pixel detection of the corners of the markers placed in the scene, by minimizing their re-projection errors in all the observed frames. Such method showed to outperform the well-known Structure from Motion and visual SLAM techniques in indoor experiments.

Our preliminary experiments have been performed in a controlled setting, in order to verify the accuracy of the solution, and to define the acquisition procedure to be tested outdoor. The experiments and the results are reported in the next section. In the management of the cultural heritage, the 3D recording and documentation is a fundamental task [6] and an accurate 3D rendering is the first step towards the enhancement of cultural heritage. On the one hand the scholar may access a rich set of functionalities to explore all the digital information extracted about the building, on the other hand, the general user will enjoy an interactive survey of the building, possibly including not accessible areas, and possibly enriched with complementary information.

Various technologies can be used to build a 3D digital model [15]. Many ground system are based on LiDAR (Light Detection And Ranging) which is a remote sensing method that uses light in the form of a pulsed laser to measure ranges. Despite the LiDAR result quality, we preferred to explore less expensive solutions. A simple and lightweight monocular camera can be used for Structure from motion (SfM), the photogrammetric technique which estimate the 3D model of the structure from a sequence of different views of the object. The images could be also the still frames of a video. The number, the quality, and the resolution of the images can affect very much the time needed to obtain a full 3D reconstruction. In general, these algorithms are very time consuming, taking hours or days using a normal pc. The using of multiple parallel GPUs or of cloud computing are strongly recommended to speed up the whole process. SfM algorithm, in most of its implementations, consists of three main steps: detection of the control points, building of a dense point-cloud, and finally the surface reconstruction as a polygonal mesh. The very starting point is the feature point selection in all the images, generally obtained by applying SIFT (scale-invariant feature transform) or SURF (speeded-up robust features), very popular and performing feature detectors. Then the corresponding features are matched, hence a registration between images is provided; incorrect matches are usually filtered out with specific algorithms, e.g. RANSAC (random sample consensus). All the matched points are called control points: this set is a sparse point cloud. In the second step, using wide baseline stereo correspondence [18], a dense point cloud is built. The final step is devoted to the definition of a polygonal mesh, the final 3D model of the object. The most popular software of 3D reconstruction and some preliminary results are described in the next section.

4   Experimental Setup and Preliminary Results

The data acquisition system in made up by several optical and electronic devices mounted on the custom ISTI-CNR MAV. This drone is a Micro Air Vehicle designed and assembled at the Institute of Information Science and Technologies of the National Research Council of Italy. This drone has two flight modes: the usual free flight mode, controlled by the pilot, and the programmed flight mode, when the drone flight is based on a predefined set of GPS coordinates of waypoints. The latter modality is quite interesting, for our purpose, as it allows

Fig. 1. The ISTI-CNR drone used for this research (left) and the Ximea XiQ high speed camera used to run SLAM algorithms on the NVIDIA TX1 processor (right)

to repeat the same flight over time; hence it may support the creation of a large dataset of the site of interest over time.

In the original setting on the bottom of the drone there is a stabilized component, named gimbal, hosting a digital camera for video recording, as shown in Fig. 1. The main optical camera used for 3D reconstruction and photogrammetry is the Canon EOS M, a 18 Mega-pixel mirror-less with a sensor APS-C of 22.3 × 15 mm (aspect ratio 3 : 2). The maximum video resolution is of 1920 × 1080 pixel at 30 fps. It weighs 298 g and has dimensions 108 × 66 × 32 mm. The focal length vary in the range 18–55 mm. E.g., setting the focal length at 24 mm, and the target at 1.5 mt, the field of view will be of 1.39 m (width) and 0.93 m (height), and the pixel resolution (computed from the camera fact-sheet) will be of 0.27 mm. Beside the main device we added a lightweight camera connected to an embedded video processing unit. The aim of this extra hardware is mainly to experiment some on-line SLAM algorithm (Simultaneous Localization And Mapping) which is useful to display some real-time information on the augmented reality console of the operator. The camera is 4 MegaPixel HighSpeed USB3 Ximea XiQ[1] (Fig. 1). Its weight is only 32 g and it can acquire at 90 fps in gray scale. The embedded processor is the NVIDIA Jetson TX1[2]. It is an ARM processor couples with parallel GPU and it has an outstanding processing power for such size and consumption. It is already the de-facto standard among similar custom drone projects. It is mounted on a credit card sized carrier board from Auvidea, because the original development kit is too large and too heavy for this purpose. The connection with the ground station uses standard Wi-fi channels.

In order to assess the accuracy of the ArUco marker detection and the repeatability of such measurements, we performed some tests in our laboratory. The camera Canon Eos M has been calibrated using a ChArUco board, with a focal length of 24 mm and image dimension of 5184 × 3456 pixel. The re-projection error estimated is of 2.8 pixel. Six markers, with side length of

---

[1] www.ximea.com/en/products/usb3-vision-cameras-xiq-line/mq042cg-cm.

[2] https://developer.nvidia.com/embedded/buy/jetson-tx1.

338   D. Germanese et al.

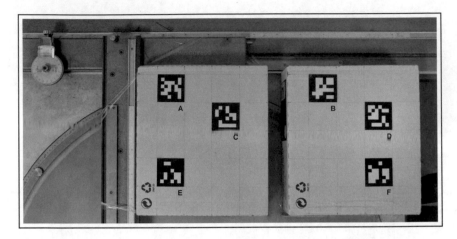

Fig. 2. Simulation of the crack opening:the coordinatographer and the six ArUco markers on it

5.5 cm, were fixed on two identical boxes (three markers on left, three on right). The left box was fixed to a mobile axis of a coordinatographer, while the right one to the table (Fig. 2). The accuracy of the coordinatographer is of 0.1 mm.

The camera acquires images at about 150 cm from the coordinatographer table. We simulated a crack opening by moving the left box along one direction far from the fixed box. The opening of the crack is performed in 10 steps: 5 by 5 mm, and 5 by 1 mm. The distance between the markers is computed following the same procedure described in [12]: a set of six frames of the same scene is acquired and at each frame the graph-pose is estimated minimizing the reprojection error in the detection of the corners of the planar markers visible. The output of the algorithm are the 3D coordinates of the markers' corners, with ids. The resulting distances between the markers' barycenters are then computed and showed in Table 1. This preliminary test pointed out that there is a need for significant improvement: the error, within 1 mm in most cases, has to be reduced. In order to reduce the error, we plan to both improve the camera calibration to subpixel accuracy and the marker mapping, by using the

Table 1. Simulation of the crack opening: three pairs of markers (A and B, C and D, E and F) moving away from each other, in five steps by 5 mm $(T_1,\ldots,T_5)$ and 5 steps by 1 mm $(T_6,\ldots,T_{10})$. All the distance values are expressed in mm.

	T_1	T_2	T_3	T_4	T_5	T_6	T_7	T_8	T_9	T_{10}
$d(A,B)$	4.76	5.56	6.19	4.35	4.01	1.05	1.32	0.41	0.86	1.87
$d(C,D)$	4.57	5.38	5.66	4.63	4.33	0.87	0.77	1.02	0.78	1.64
$d(E,F)$	4.45	5.96	5.36	4.59	4.27	1.07	1.03	0.70	0.92	1.46
Actual Δ	5	5	5	5	5	1	1	1	1	1

ChArUco diamond markers instead of the single markers. A diamond marker is a chessboard composed by 3×3 squares and 4 ArUco markers inside the white squares. The detection of a diamond marker take advantage of the known relative position of the markers in it, and it will improve the robustness and accuracy of the pose estimation in the marker mapping algorithm.

A first acquisition has been carried out flying close to an old tower in Ghezzano, near Pisa. The drone was equipped with the Canon EOS M camera. This flight was useful to verify that the ArUco markers of different dimensions (side length of 5.5 cm, 8 cm, 12 cm, and 20 cm) are detected and correctly recognized. Also, the frames extracted from the acquired video were used to compare the result of the most popular software for the 3D reconstruction. We chose to test the following software:

- **Agisoft Photoscan**[3]: it is maybe the first photogrammetric software, it is proprietary software, exploiting the CUDA parallel GPU technologies. It allows the user tuning the reconstruction parameters during the procedure to increase the quality, depending on the input data, on the user preferences, and on the computing resources.
- **COLMAP**[4]: it is an open-source software, exploiting CUDA technologies. The user can configure the reconstruction settings, but cannot interact with the middle result of the reconstruction phases.
- **Autodesk Recap Photo**[5]: it is proprietary software which exploits cloud technologies in order to perform the reconstruction remotely without burden user calculators. It processes up to 100 photos at once, and it is not possible to configure any setting relative to the reconstruction phases.

Beyond computing dense detailed models [11], we got the best result from Agisoft Photoscan because the optional interaction during the selection of the key points is very useful to process only the interesting part of the image and delete the rest. A brief result of these three stages is depicted in Fig. 3. Then, a virtual scene containing the reconstructed object has been created. The virtual scene has been realized exploiting the Unity[6] engine. The exploitation of such type of engine guarantees the easiness of navigation and, at the same time, the overall representation quality. Inside the scene, users can easily navigate around the reconstructed object and have a quick-look of all the regions of interest of the structure. In the virtual environment, structure cracks are highlighted and labelled with latest measurements. It is also possible to interact with the cracks in order to retrieve past calculated values or visualize charts representing the crack opening evolution over time.

---

[3] www.agisoft.com.

[4] colmap.github.io.

[5] www.autodesk.com/products/recap.

[6] www.unity3d.com.

Fig. 3. (a) the original aerial view from the drone; (b) the first step of the algorithm selects reliable key-points; (c) the dense point cloud is computed; (d) the final 3D model is a polygonal mesh with texture from the original frame

5   Conclusions

We believe that the accuracy of the algorithm mapping pairs of planar markers, installed along fissures and cracks of ancient buildings and structures, may be improved to enable the monitoring of the crack opening, using drones. Also, the usage of planar markers provides useful 3D information about how the two sides of a crack are moving. The testing phase showed that even if the survey carried out by the drone is detailed enough to have a 3D model usable by the expert, and enjoyable for people, the acquisition procedure, the hardware setting, and the algorithms used to assess the crack opening need to be further optimized to achieve at least the accuracy of a few tenths of a millimeter.

Acknowledgement. This work is being carried out in the framework of the Tuscany Regional Project MOSCARDO (FAR-FAS 2014). The Jetson TX1 embedded processor used for this research was donated by the NVIDIA Corporation.

References

1. Armesto, J., Arias, P., Roca, J., Lorenzo, H.: Monitoring and assessing structural damage in historical buildings. Photogram. Rec. **21**, 269–291 (2006)
2. Benning, W., Görtz, S., Lange, J., Schwermann, R., Chudoba, R.: Development of an algorithm for automatic analysis of deformation of reinforced concrete structures using photogrammetry. VDI Ber. **1757**, 411–418 (2003)
3. Cardone, A., Gupta, S., Karnik, M.: A survey of shape similarity assessment algorithms for product desing and manifacturing applications. J. Comput. Inf. Sci. Eng. **3**, 109–118 (2003)
4. Ellenberg, A., Kontsos, A., Bartoli, I., Pradhan, A.: Masonry crack detection application of an unmanned aerial vehicle. In: Proceedings of Computing in Civil and Building Engineering, ASCE 2014 (2014)
5. Eschmann, C., Kuo, C., Kuo, C., Boller, C.: Unmanned aircraft systems for remote building inspection and monitoring. In: 6th European Workshop on Structural Health Monitoring (2012)
6. Georgopoulos, A., Stathopoulou, E.K.: Data acquisition for 3D geometric recording: state of the art and recent innovations, pp. 1–26. Springer (2017)
7. Gruen, A., Akca, D.: Least square 3d surface and curve matching. ISPRS J. Photogram. Remote Sens. **59**, 151–174 (2005)
8. Jahanshahi, M.R., Kelly, J.S., Masri, S.F., Sukhatme, G.S.: A survey and evaluation of promising approaches for automatic image-based defect detection of bridge structures. Struct. Infrastruct. Eng. **5**(6), 455–486 (2009)
9. Jahanshahi, M.R., Masri, S.F.: A new methodology for non-contact accurate crack width measurement through photogrammetry for automated structural safety evaluation. Smart materials and structures **22**(3), 035019 (2013)
10. Jahanshahi, M.R., Masri, S.F., Padgett, C.W., Sukhatme, G.S.: An innovative methodology for detection and quantification of cracks through incorporation of depth perception. Machine Vis. Appl. **24**(2), 227–241 (2011)
11. Majdik, A.L., Tizedes, L., Bartus, M., Sziranyi, T.: Photogrammetric 3d reconstruction of the old slaughterhouse in budapest. In: 2016 International Workshop on Computational Intelligence for Multimedia Understanding (IWCIM) (2016)
12. Munoz-Salinas, R., Marin-Jimenez, M.J., Yeguas-Bolivar, E., Medina-Carnicer, R.: Mapping and localization from planar markers. P. Recog. **73**, 158–171 (2018)
13. Niemeier, W., Riedel, B., Fraser, C., Neuss, H., Stratmann, R., Ziem, E.: New digital crack monitoring system or measuring ad documentation of width of cracks in concrete structures. In: Measuring the Changes: 13th FIG Symposium on Deformation Measurements and Analysis (2008)
14. Nishiyama, S., Minakata, N., Kikuchi, T., Yano, T.: Improved digital photogrammetry technique for crack monitoring. Adv. Eng. Inform. **29**(4), 851–858 (2015)
15. Remondino, F., Campana, S.: 3D Recording and Modelling in Archaeology and Cultural Heritage - Theory and Best Practices. Archaeopress BAR, Oxford (2014)
16. Remondino, F., El-Hakim, S.: Image-based 3d modelling: a review. Photogram. Rec. **21**, 269–291 (2006)
17. Shortis, M.R., Seager, J.W.: A practical target recognition system for close range photogrammetry. Photogram. Rec. **29**(147), 337–355 (2014)

18. Strecha, C., Fransens, R., Gool, L.V.: Wide-baseline stereo from multiple views: a probabilistic account. In: CVPR, pp. 552–559 (2004)
19. Valença, J., Dias-da Costa, D., Júlio, E., Araújo, H., Costa, H.: Automatic crack monitoring using photogrammetry and image processing. Measurement **46**(1), 433–441 (2013)
20. Welsch, W., Heunecke, O.: Models and terminology for the analysis of geodetic monitoring observations. In: FIG 10th International Symposium on Deformation Measurements. International Federation of Surveyors, vol. 25, p. 22 (2001)

Remote Sensing for Maritime Monitoring and Vessel Prompt Identification

Marco Reggiannini[1(✉)], Marco Righi[1], Marco Tampucci[1], Luigi Bedini[1],
Claudio Di Paola[2], Massimo Martinelli[1], Costanzo Mercurio[2],
and Emanuele Salerno[1]

[1] Institute of Information Science and Technologies,
National Research Council of Italy, 56124 Pisa, Italy
marco.reggiannini@isti.cnr.it
[2] Mapsat S.R.L., Contrada Piano Cappelle 1, 82100 Benevento, BN, Italy
http://www.isti.cnr.it/

Abstract. The main purpose of the work described in this paper concerns the development of a platform dedicated to sea surveillance, capable of detecting and identifying illegal maritime traffic. This platform results from the cascade implementation of several image processing algorithms that take as input Radar or Optical maps captured by satellite-borne sensors. More in detail, the processing chain is dedicated to (i) the detection of vessel targets in the input map, (ii) the refined estimation of the vessel most descriptive geometrical features and, finally, (iii) the estimation of the kinematic status of the vessel. This platform will represent a new tool for combating unauthorized fishing, irregular migration and related smuggling activities.

Keywords: Maritime traffic monitoring · SAR sensing
Optical sensing · Ship detection · Image segmentation
Image classification · Wake detection and analysis

1 Introduction

Maritime traffic consists of more than 600.000 vessels navigating daily all over the world seas. This huge network of vehicles poses hard issues for what concerns monitoring purposes and timely countermeasures implementation. Satellite missions rotating along earth-centered orbits provide heterogeneous remote sensing data about the sea surface status on a daily basis and at different resolutions. Remote sensing data play a crucial role for vessel monitoring purposes, providing information about the vessel main features, its peculiar geometry and its kinematics.

The main goal addressed in this paper concerns the development of a software platform dedicated to sea surveillance, capable of detecting and identifying seagoing vessels. This platform will be in charge of collecting and integrating data

© Springer Nature Switzerland AG 2019
K. Choroś et al. (Eds.): MISSI 2018, AISC 833, pp. 343–352, 2019.
https://doi.org/10.1007/978-3-319-98678-4_35

made available by multi-source, multi-sensor satellite missions for specific maritime areas. High-resolution data will be provided by currently orbiting satellites such as European Space Agency (ESA) Copernicus Sentinels, ISI-IMAGESAT EROS and Italian Space Agency (ASI) COSMO-SkyMed. The data will then be processed in order to detect and identify the ships located in the area of interest and to provide estimates of meaningful ship features, such as shape and kinematics parameters.

The proposed maritime monitoring platform has been conceived as a sequential chain of procedures (Fig. 1), each addressing a specific task oriented to the extraction of meaningful information.

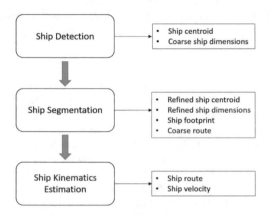

Fig. 1. Block diagram for the proposed maritime surveillance platform.

First, radar and optical images, represented in the Universal Transverse Mercator (UTM) coordinate system, are processed by a dedicated detection algorithm, inspired to state-of-the-art literature [1], that identifies potential vessels by discerning between sea background and backscattering anomalies. The output of this procedure is a set of submaps cropped from the input image so that a single target per crop is observed. These crops are then fed to a ship segmentation module [2] in charge of refining the identification of those map pixels that belong to the target. Accordingly, this operation provides additional information on the target, such as the length overall, the beam overall and the heading orientation (estimated with a 180° ambiguity, unless the velocity vector is estimated or the bow is distinguishable in the image). The morphological information provided by the latter module can be further enriched by inspecting the water surface surrounding the detected ship. Indeed, it is known that the ship's passage through the water generates a specific wake pattern whose features depend directly on the ship's kinematics [3,4]. In particular, by detecting the linear envelopes of the main wake components and performing a frequency analysis of the observable wake oscillations, it is possible to estimate, respectively, the ship's heading (univocally) and the ship's velocity.

The rest of the paper is arranged as follows: Sect. 2 concerns a detailed discussion of the adopted processing techniques. Section 3 presents an actual implementation of the discussed platform within the framework of OSIRIS, a project funded by ESA. Section 4 concludes the paper by presenting a summary of the main results and discussing future developments.

2   Processing Techniques

A main goal within the mentioned platform is to develop computational imaging procedures to process Synthetic Aperture Radar and Optical data returned by satellite sensors. We propose a system for the automatic detection and recognition of all the vessels included in a given area of interest. The maritime satellite imagery will be processed to extract visual informative features of candidate vessels and to assign an identification label to each vessel. The remaining parts of this section concern detailed descriptions of each processing stage that contribute to the fulfillment of the aforementioned tasks.

2.1   Ship Detection

The detection algorithm aims at identifying those regions in the input image data that most likely include vessels. The output of a positive detection is a subset (crop) of the input image, restricted to an Area of Interest (AoI) where only the candidate vessel is visible.

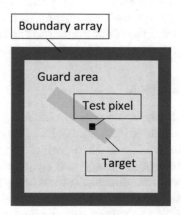

Fig. 2. CFAR detector moving window.

In case of SAR imagery, the choice of the most suitable detection method strongly depends on image features such as spatial and radiometric resolution, number of looks, transmitted and received polarization, and the like. As speed is a major requirement for prescreening, fine statistical considerations are normally overlooked to obtain fully automatic and fast detectors based on simplified statistical assumptions and a fixed maximum false-alarm rate. This is the principle of the different types of Constant False-Alarm Rate (CFAR) detectors. They normally work on single-channel intensity images, and consist in statistical tests over each pixel intensity against the intensities of a specified neighborhood. In practice, [1,5] a moving window centered on the test pixel scans the entire image. This window (Fig. 2) is formed by a test pixel neighborhood (*guard area*), which is supposed to contain any possible target that includes the test pixel, and a second neighborhood (*boundary array*), only supposed to contain background pixels, which is used to estimate the background statistics. CFAR is a type of adaptive threshold method: once the required false-alarm rate is fixed, the background probability density function $f(x)$ (usually a Gaussian), where x is the pixel value, is assumed to be parametrically known or estimated from the values in the boundary array. The probability of false alarm, PFA, is given by

$$\text{PFA} = \int_T^\infty f(x)dx \tag{1}$$

According to Eq. 1, if the test pixel value is not smaller than T, then the probability of false alarm is not larger than its pre-defined value. Thus, for each position of the moving window, the value of T obtained by solving Eq. 1 can be taken as an adaptive threshold to decide whether the test pixel belongs to a target.

By exploiting ancillary information (sensing time, ground sample distance, radiometric resolution, spacecraft position and kinematics) provided by the sensors installed aboard the host spacecrafts, the algorithm produces full radiometric and geometric resolution crop files, in geotiff format. In order to provide the subsequent processing steps with an output result with proper coordinate system, a transformation from the UTM system to the Latitude/Longitude coordinate system is finally performed.

2.2   Ship Fine Segmentation

The recognition of visual attributes within the AoI allows the user to highlight relevant informative content in the data. Imaging algorithms are applied to extract geometric and radiometric features, which provide meaningful insights about the vessel's morphology, geometry and dynamics. SAR images acquired by a satellite platform are often affected by distortions due to marine clutter, spectral leakage, or antenna sidelobes. These can mask the target image, thus hampering the possibility of evaluating the size and the behavior of the ship. An example of this type of distortions is represented in the Sentinel-I intensity image in Fig. 3, where the presence of a strong, cross-shaped artifact is apparent.

To the purpose of optimally identify the set of pixels belonging to the potential ship target, a dedicated segmentation module (SISS-Ship Image Segmentation from Space) has been developed. The main SISS features are here discussed, underlining the main differences w.r.t. previous approaches [5,6]. SISS accepts multi-sensor input data, such as SAR and optical images. In the case of SAR

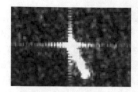

Fig. 3. Crop example of a SAR signal containing an individual ship candidate: a clearly visible cross-shaped artifact strongly affects the data quality.

images, the algorithm processes the signal looking for strong reflectors, usually appearing as bright, w.r.t. the background, connected areas. In the case of optical images, the algorithm still looks for connected regions of outliers with respect to the background statistics, with the drawback that meaningful data can only be provided during daylight and under clear sky condition.

The first step of SISS consists in building a binary shape where the target pixels are set to 1 while the remaining pixels are set to 0. This is obtained through an adaptive thresholding approach. The employed threshold is a multiple integer of the standard deviation σ, computed from the whole input image I. Hence the threshold value has the following form:

$$\sigma_{th} = k\sigma. \tag{2}$$

k is set according to the following schema:

1. $i = 1, 2, \ldots$
2. define matrix I_t as

$$I_t = \begin{cases} 1 & \text{if } I \geq i\sigma \\ 0 & \text{elsewhere} \end{cases}$$

3. consider $\bar{i}$ as the smallest i such that I_t vanishes everywhere.
4. $k = \bar{i} - 1$.

For each iteration, SISS estimates the ship footprint by simultaneously considering the signal amplitude and the interconnection between adjacent pixels. The estimated footprint also undergoes a fit process exploiting a parametric Ship Shape Model (SSM) that is defined *a-priori* by the user (e.g. an ellipse whose parameters are estimated on the image). At every iteration, the biggest connected area is computed and compared to the surface of the smallest SSM including such connected area. The algorithm finally returns the largest mask that better approximates the SSM.

After the thresholding stage, the target is isolated by clipping the image outside a mask including the ship and the possible artifacts. The subsequent step is to find the principal inertia axes by performing an eigenvalue analysis on the 2D inertia tensor, computed with respect to the barycenter of the thresholded image. Then a rectangular mask is applied around the barycenter, so as to include the whole target while cutting out part of the artifacts. By iterating this procedure, it is possible to perform a progressively refined estimation of the ship length overall (LOA), beam overall (BOA) and heading (see Fig. 4). Indeed, when no artifact is cut out anymore, we take the unit vector of the minimum inertia axis as the estimated heading, and the maximum distances between target boundary points along the principal axes as the estimated LOA and BOA. In the absence of further information, such as a detected wake or possible morphological ship features, the heading is estimated up to a 180° ambiguity.

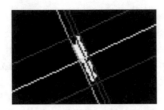

Fig. 4. Result of the ship segmentation algorithm applied to the SAR image crop in Fig. 3. The principal inertia axes and enclosing rectangle are highlighted in color.

2.3   Ship Classification and Identification

The information collected from the previous processing stages is exploited to perform a vessel classification, implemented by feeding a tailored classifier with the estimated features.

Classification and identification are not well defined in the literature. We choose to distinguish between them admitting that classification includes methods able to distinguish between a few classes of vessels, for example, fishing boats, tankers and container ships, while identification is just a finer classification, where the targets are assigned to a richer set of classes, in the limit, each including a single, well-identified vessel. Of course, more and more sophisticated features are needed to refine the classification, possibly including features that do not pertain strictly to the targets, such as wakes, and their influence in the overall performance is not independent of the chosen classifiers. For the purpose of the work described here, classification has been based on visual features that have relevant discriminative power, extracted from the captured SAR/optical images. For single-polarization, high-resolution SAR data, these can be based on simple geometric measurements. For example, the estimation of the length-to-width ratio, a standard indicator for ship structure [7], can be exploited to discriminate among target types, provided that the achieved accuracy is sufficient to get a reliable estimate.

Preliminary classification tests have been performed exploiting Automatic Identification System (AIS) data as a ground truth reference. These tests have been based on a decision-tree classifier. The availability of the ground truth allowed us to prove that the method is capable of identifying several ships categories such as fishing boats, tankers, and containers. Under these assumptions the algorithm classifies correctly in the 75% of the cases.

Aiming at increasing the performance of classification and identification methods, we have designed and developed a database system able to manage and correlate data from other sources (e.g. vessels databases) with the features extracted from the input images. By exploiting the database system, the classification and identification methods will be able to take into account any kind of additional feature extracted from both SAR and optical data, such as radiometric, polarimetric and color based features. The features extracted from new images will then be employed to perform similarity search operations to identify the possible class of the target based on the data stored into database.

2.4   Ship Kinematics Estimation

Previous research [8] provided a robust description of the physics underlying the wake pattern generation due to the ship passage through the water. Starting from this background, a main goal in the proposed maritime monitoring platform concerns the development of computational imaging procedures to provide insights about the ship kinematics through SAR imagery processing [3,4,9].

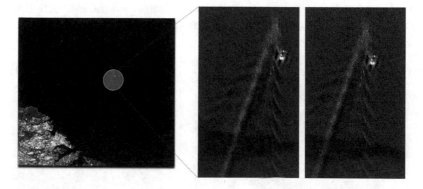

Fig. 5. Wake pattern detection. (COSMO-SkyMed Product – © ASI 2016 processed under license from ASI – Agenzia Spaziale Italiana. All rights reserved. Distributed by e-GEOS)

The wake results from multiple oscillatory components whose summation exhibits a V-shaped pattern centered on the ship route axis. The V angular aperture approximates a constant 39° value. Exploiting these observable phenomena, the route direction can be estimated by first detecting the V pattern (Fig. 5) through a Radon-transform linear detector [10], and later by identifying the wake center axis.

The wake pattern carries information about the vessel speed. Indeed, the oscillatory components observed in the external boundaries of the wake, feature wavelength values that relate to the velocity of the ship. Hence, provided the image resolution is large enough to observe these details, the Fourier analysis of a linear sample extracted from one Kelvin wake component (blue line in Fig. 6, upper left side) is performed. This allows to estimate the dominant wavelength λ (see lower left side in Fig. 6, with the spatial sampling on the right and the estimated power spectrum on the left) which is in turn employed to compute the ship's velocity v according to [9]:

$$v = \sqrt{\frac{\sqrt{3} g \lambda}{4\pi}} \tag{?}$$

where $g = 9.81 \, \text{m/s}^2$.

A second method for vessel speed estimation exploits the azimuth shift effect, a distortion which affects SAR images, causing an artificial separation between the moving ship and its wake [11]. The separation length Δ_{as} is measured directly on the SAR map (red line in Fig. 6, right side) and employed to estimate the vessel speed according to:

$$v = \frac{V_{sat} \cdot \Delta_{as}}{R_{sr} \cos \beta} \tag{4}$$

where V_{sat} is the satellite speed, R_{sr} is the slant range from satellite to the target and β is the angle between the vessel's velocity vector and the azimuth direction. Figure 6 illustrates examples of the mentioned image processing methods.

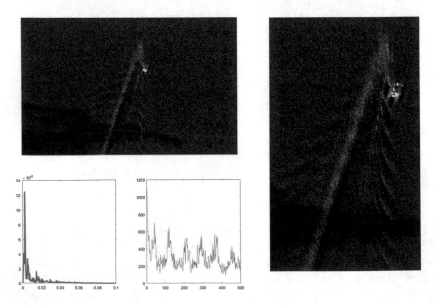

Fig. 6. External wake Fourier analysis (left) and azimuth shift (right). (COSMO-SkyMed Product – © ASI 2016 processed under license from ASI – Agenzia Spaziale Italiana. All rights reserved. Distributed by e-GEOS)

3   Implementation

OSIRIS[1] (Optical/SAR data and system Integration for Rush Identification of Ship models), an ESA project launched in February 2016, represents an effective implementation of the ideas and concepts described so far. Its main goal is the development of a platform dedicated to sea surveillance, capable of detecting and identifying illegal maritime traffic. At full operating conditions, the OSIRIS technological platform will be in charge of collecting and integrating data made available by multi-source, multi-sensor satellite missions for specific maritime areas of interest and processing the acquired data to detect and classify seagoing vessels. Crucial information for ship identification will be collected by means of popular ship monitoring systems (e.g. AIS) and by integrating additional information gathered from Open-Source Intelligence systems. The data will be finally analyzed to provide predictive models for the routes of these ships. AIS data will be provided by two different companies Astra Paging[2] (from ground based receivers) and Exact Earth[3] (from satellite constellations). A web GIS platform is currently in charge of collecting and fusing the AIS data coming from the two mentioned providers and, after a filtering stage to suppress fake or repeated records, making it available to the project operators.

4   Conclusions

A software platform dedicated to maritime traffic monitoring has been presented. The main goal of this platform is to detect and identify target vessels within a given sea surface area, which is remotely supervised by orbiting satellites such as Sentinel 1/2, CosmoSKy-Med and EROS missions. Radar and optical images represent the main input data for the described platform. These are processed by a suite of algorithms which are sequentially applied to the data returning information about (i) the ship positioning within the inspected area, (ii) the main ship geometrical attributes (length overall, beam overall, heading), and (iii) the ship kinematics.

Future developments will be devoted to improve the overall performance of the platform by enhancing the accuracy of each individual stage in the processing pipeline and devising further classification methods. We observed that with particularly noisy or distorted images, the estimates returned by the implemented segmentation module are not sufficiently robust. We are studying to establish resolution and signal-to-distortion ratio requirements that are sufficient to guarantee a specified accuracy. We noted that the result often depends significantly on the initial shape assumed, that is, on the thresholding strategy. Without relying on supplemental data or preprocessing, we are now trying to find a method to suitably tune the thresholds. Our next attempts will be to integrate the results from the same target imaged through different transmit-receive

---

[1] https://wiki.services.eoportal.org/tikiindex.php?page=OSIRIS.

[2] http://www.astrapaging.com/.

[3] https://www.exactearth.com/.

polarizations and integrate the extracted features in the classification process. Furthermore, this work is intended as a step towards a more complete ship classification method, also based on radiometric and/or polarimetric features, and a more reliable course and speed estimation, based on wake feature extraction and azimuth shift evaluation. Since wake patterns are hardly detectable in SAR maps, future developments will also be devoted to the refinement of the wake recognition process, based on the exploitation of additional information, such as the fine estimate of the vessel position as well as the theoretical constraints of this peculiar hydrodynamics problem (e.g. wake fixed angular aperture). The presented platform is being currently tested within the framework of the OSIRIS project, providing preliminary promising results from a real world testing workbench.

References

1. Crisp, D.J.: The state-of-the-art in ship detection in synthetic aperture radar imagery. Australian Government, Department of Defence, Report DSTO-RR-0272 (2004)
2. Bedini, L., Righi, M., Salerno, E.: Size and heading of SAR-detected ships through the inertia tensor. In: Proceedings of MDPI, vol. 2, no. 2. https://doi.org/10.3390/proceedings2020097. http://www.mdpi.com/2504-3900/2/2/97
3. Reggiannini, M., Righi, M.: Processing satellite imagery to detect and identify non-collaborative vessels. In: ERCIM News, vol. 108, pp. 25–26 (2017). https://ercim-news.ercim.eu/en108/special/processing-satellite-imagery-to-detect-and-identify-non-collaborative-vessels
4. Reggiannini, M., Bedini, L.: Synthetic aperture radar processing for vessel kinematics estimation. In: Proceedings of MDPI, vol. 2, no. 2. https://doi.org/10.3390/proceedings2020091. http://www.mdpi.com/2504-3900/2/2/91
5. Allard, Y., Germain, M., Bonneau, O.: Harbour Protection Through Data Fusion Technologies, pp. 243–250. Springer, Heidelberg (2009)
6. Ishak, A.B.: A two-dimensional multilevel thresholding method for image segmentation. Appl. Soft Comput. **52**, 306–322 (2017)
7. Askari, F., Zerr, B.: Automatic approach to ship detection in spaceborne synthetic aperture radar imagery: an assessment of ship detection capability using RADARSAT. Technical report SACLANTCEN-SR-338, SACLANT Undersea Research Centre, La Spezia (Italy), December 2000
8. Thomson, W.: On the waves produced by a single impulse in water of any depth, or in a dispersive medium. Proc. R. Soc. Lond. **42**, 80–83 (1887)
9. Zilman, G., Zapolski, A., Marom, M.: The speed and beam of a ship from its wake's SAR images. IEEE Trans. Geosci. Remote Sens. **42**(10), 2335–2343 (2004)
10. Gonzalez, R.C., Woods, R.E.: Digital Image Processing. Pearson Prentice Hall, Upper Saddle River (2008)
11. Eldhuset, K.: An automatic ship and ship wake detection system for spaceborne SAR images in coastal regions. IEEE Trans. Geosci. Remote Sens. **34**(4), 1010–1019 (1996)

Deep Learning Approach to Human Osteosarcoma Cell Detection and Classification

Mario D'Acunto[1], Massimo Martinelli[2(✉)], and Davide Moroni[2(✉)]

[1] Institute of Biophysics, National Research Council of Italy,
Via Moruzzi, 1, 56124 Pisa, Italy
mario.dacunto@pi.ibf.cnr.it
[2] Institute of Information Science and Technologies,
National Research Council of Italy, Via Moruzzi, 1, 56124 Pisa, Italy
{massimo.martinelli,davide.moroni}@isti.cnr.it

Abstract. The early diagnosis of a cancer type is a fundamental goal in cancer treatment, as it can facilitate the subsequent clinical management of patients. The leading importance of classifying cancer patients into high or low risk groups has led many research teams, both from biomedical and bioinformatics field, to study the application of Deep Learning (DL) methods. The ability of DL tools to detect key features from complex datasets is a fundamental achievement in early diagnosis and cell cancer progression. In this paper, we apply DL approach to classification of osteosarcoma cells. Osteosarcoma is the most common bone cancer occurring prevalently in children or young adults. Glass slides of different cell populations were cultured from Mesenchimal Stromal Cells (MSCs) and differentiated in healthy bone cells (osteoblasts) or osteosarcoma cells. Images of such samples are recorded with an optical microscope. DL is then applied to identify and classify single cells. The results show a classification accuracy of 0.97. The next step is the application of our DL approach to tissue in order to improve digital histopathology.

Keywords: Osteosarcoma cells · Deep Learning
Convolutional neural networks
Convolutional object detection systems · Cell classification

1 Introduction

Over the last decades, scientists applied different methods to detect cancer tissues at the early stage. This is because early diagnosis can facilitate the clinical managements of patients. Early diagnosis requires the ability to identify cancer tissue as small as a single cell. Classification of cancer cells is hence key research for early diagnosis and for identification of differentiation and progression of cancer in a single cell [7,11,14]. With advent of new digital technologies in the field of medicine, Artificial Intelligence (AI) methods have been applied in cancer research to complex datasets in order to discover and identify patterns and

© Springer Nature Switzerland AG 2019
K. Choroś et al. (Eds.): MISSI 2018, AISC 833, pp. 353–361, 2019.
https://doi.org/10.1007/978-3-319-98678-4_36

relationships between them. Machine Learning (ML) is a branch of AI related
to the problem of learning from data samples to the general concept of inference. The main objective of ML techniques is to produce a model, which can be
used to perform classification, prediction, estimation or any other similar task.
When a classification model is developed, by means of ML techniques, training
and generalization errors can be dealt out. The former refers to misclassification
errors on the training data while the latter on the expected errors on testing
data. A good classification model should fit the training set well and accurately
classify all the instances. Once a classification model is obtained using one or
more ML techniques, it is necessary to estimate the classifier's performance. The
performance analysis of each proposed model is measured in terms of sensitivity,
specificity, accuracy, and so-called area under the curve (AUC). Sensitivity is
usually defined as the proportion of true positives that are correctly observed by
the classifier, whereas specificity is defined as the proportion of true negatives
that are correctly identified. In turn, accuracy, that is a measure related to the
number of correct predictions, and AUC, that is a measure of model's performance, are used for assessing the overall performance of a classifier. DL is a part
of ML methods based on learning data representation. Inside DL, Convolutional
Object Detection (COD) is a recent approach to cancer analysis. In this paper,
we have applied a COD-based DL method to several differentiated samples of
cells cultured on glass slide, with the purpose to discriminate osteosarcoma cells
from MSCs (osteoblasts). The results show excellent performance with an accuracy of nearly 1. The next step will be to extend the algorithm to large populations of cells and tissues with the purpose to improve digital histopathology. In
Sect. 2 a related works are described. Section 3 describes cell culture, how was
built e handled the dataset and the network used for detection and classification. Section 4 shows the results of training and accuracy of the method applied.
Section 5 summarizes our conclusions.

2 Related Works

Image-based ML and, in particular, DL have recently shown expert-level accuracy in medical image classification, ranging from ophtalmology to diagnostic
pathology [2]. Within digital pathology, quantification and classification of digitized tissue samples by supervised deep learning has shown good results even for
tasks previously considered too challenging to be accomplished with conventional
image analysis methods [3,5,8–10,18].

Many tasks in digital pathology, such as counting mitoses, quantifying tumor
infiltrating immune cells, or tumor cell differentiation require the classification
of small clusters of cells up to single cell, if possible. For this purpose, we have
addressed our efforts to classify cultured cells with known grade of differentiation
with a supervised DL.

COD is a very recent technique of machine learning for the analysis of cancer.
A number of methods have been proposed to address the object recognition task
and many software frameworks have been implemented to develop and work with
deep learning networks (such as Caffe, Apache MXNet and many others).

Among all such methods, Google TensorFlow is currently on e of the most used framework and its Object Detection API emerged as a very powerful tool for image recognition.

In [6] a guide for selecting the right architecture depending on speed, memory and accuracy is provided.

Since our case requires the highest accuracy architecture allowable, we selected the Faster Region Convolutional Neural Network (Faster R-CNN) [12,13] that is a recent region proposal network sharing features with the detection network that improves both region proposal quality and object detection accuracy.

Faster R-CNN uses two networks: a Region Proposal Network (RPN) to generate region proposals and a detector network to discover object. The RPN generates region proposals more quickly than the Selective Search [17] algorithm used in previous solutions. By sharing information between the two networks, the accuracy is also improved and this solution is currently the one with the best results in the latest object detection competitions.

3   Materials and Methods

3.1   Cells Culture

Normal, cancerous and mixed cells were cultured on glass slides, as follows:

 (i) Undifferentiated MSCs were isolated from human bone marrow according to a previously reported method [15] and used to perform three culture strategies. MSCs were plated on glass slides inside Petri dishes at a density of 20,000 cells with 10% fetal bovine serum (FBS). The samples were cultured for 72 h, then fixed in 1% neutral buffered formalin for 10 min at 4°C.
(ii) Osteosarcoma cells. MG-63 (human osteosarcoma cell line ATCC CRL-1427) cells were seeded on 6 glass slides at 10,000 cells in Eagle's Minimum Essential Medium (EMEM) supplemented with 10% FBS and cultured for 72h; thereafter the samples were formalin fixed, as in (i).
(iii) Mixed cancer and normal cells. MSCs were plated on 6 glass slides inside Petri dishes at 10,000 cells with 10% FBS. In parallel, MG-63 cells were plated at passage 2 on 6 well tissue culture plate (50,000 cells/2 ml) in EMEM supplemented with 10% FBS. After 24 h, the MG-63 cells were harvested and seeded at 10,000 cells onto glass slides previously seeded with MSCs. After 72 h the glass slides were formalin fixed, as in (i).

At each endpoint, all the samples were fixed in 1% (w/v) neutral buffered formalin for 10 min at 4 °C.

Morphologies are visible in Fig. 1, as imaged by an inverted microscope (Nikon Eclipse Ti-E).

3.2 Data Set Collection and Annotation

Collected images have been divided into two groups, a test set and a validation set. The test set images have been labeled by the domain expert using the LabelImg Software [16].

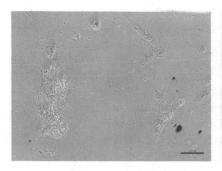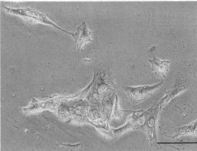

Fig. 1. Morphology of osteoblast cells, (left, 10× objective, scale bar 100 μm), and osteosarcoma cells (right, 20× objective, scale bar 50 μm).

Overall 229 objects where labeled in 48 images used for training, and 12 were used for validation, that is the size ratio of the classification network in comparison to the size of the training was 80/20.

Categories used to label were five:

- single cancer cell
- cancer_cluster
- single MSC cell
- MSC_cluster
- artifact

Image and labels have then been converted into the relative TensorFlow formats: images into the TensorFlow record, and labels into Comma Separated Values (CSV) listing: each row of the CSV contains the filename, the image dimensions, the label and the top-left and bottom-right corner of the object determined by the domain expert.

An excerpt from the above CSV file is shown in Fig. 2.

3.3 CNN for Cell Detection and Classification

We have then trained a Faster R-CNN that uses two modules, a deep fully convolutional network that proposes regions, and another module is the Fast Region Convolutional Neural Network detector using the proposed regions [12]. Specifically we selected the Inception Resnet v2 model, using TensorFlow GPU [1]. The inference graph produced has been exported and tested using new sample, that is on the validation set.

4   Results

4.1   CNN Training

A double test has been performed, due to the fact that images were obtained using two different inverted optical microscopes with grayscale and RGB, with green background density, respectively.

```
train_labels.csv ✖
    filename,width,height,class,xmin,ymin,xmax,ymax
    MG63_10x_08.jpg,1280,960,cancer_cluster,124,785,279,904
    MG63_10x_08.jpg,1280,960,cancer_cluster,7,80,600,540
    MG63_10x_08.jpg,1280,960,cancer_cluster,459,511,782,842
    MG63_10x_08.jpg,1280,960,cancer_cluster,780,150,911,557
    MG63_10x_08.jpg,1280,960,cancer,124,447,286,575
    MG63_10x_08.jpg,1280,960,cancer,563,198,653,353
    MG63_10x_08.jpg,1280,960,cancer,999,103,1091,279
    MG63_10x_08.jpg,1280,960,cancer_cluster,929,264,1211,465
    MG63_10x_08.jpg,1280,960,artifact,408,775,446,823
    MG63_10x_08.jpg,1280,960,artifact,1250,847,1277,906
    MG63_10x_08.jpg,1280,960,artifact,1111,61,1155,97
    MG63_10x_06.jpg,1280,960,cancer_cluster,115,158,570,444
    MG63_10x_06.jpg,1280,960,artifact,852,531,936,615
    MG63_10x_06.jpg,1280,960,cancer_cluster,253,493,563,940
    MG63_10x_06.jpg,1280,960,cancer_cluster,1003,387,1241,676
    MG63_10x_06.jpg,1280,960,cancer_cluster,122,547,242,690
    MG63_10x_10.jpg,1280,960,cancer_cluster,14,390,171,577
    MG63_10x_10.jpg,1280,960,cancer_cluster,122,145,513,394
    MG63_10x_10.jpg,1280,960,cancer_cluster,511,36,788,310
    MG63_10x_10.jpg,1280,960,cancer_cluster,765,508,1148,866
    MG63_10x_10.jpg,1280,960,cancer_cluster,258,604,690,954
    MG63_10x_10.jpg,1280,960,cancer_cluster,363,398,682,644
    MG63_10x_10.jpg,1280,960,artifact,194,64,282,139
    MG63_10x_10.jpg,1280,960,cancer_cluster,745,150,1279,509
    MG63_10x_10.jpg,1280,960,cancer_cluster,796,1,997,342
    MG63_10x_10.jpg,1280,960,cancer_cluster,580,417,958,509
    MG63_10x_10.jpg,1280,960,artifact,733,776,860,856
    MSC_20x_01.jpg,1280,960,MSC,283,112,664,282
    MSC_20x_01.jpg,1280,960,MSC,428,93,834,595
    MSC_20x_01.jpg,1280,960,MSC,417,598,802,863
    MSC_20x_01.jpg,1280,960,MSC_cluster,882,453,1271,960
    MSC_20x_01.jpg,1280,960,artifact,538,178,588,238
    MSC_20x_01.jpg,1280,960,artifact,234,297,411,435
    MSC_20x_01.jpg,1280,960,artifact,226,716,383,850
    MSC_20x_01.jpg,1280,960,artifact,34,178,220,286
    MSC_20x_01.jpg,1280,960,artifact,809,23,1029,72
    MSC_10x_06.jpg,1280,960,MSC,514,315,738,461
    MSC_10x_06.jpg,1280,960,MSC,129,753,272,956
    MSC_10x_06.jpg,1280,960,MSC,1096,679,1224,806
```

Fig. 2. Excerpt of CSV file

Then images produced with both microscopes have been converted into grayscale using [4] and a new training has been performed.

The training phase lasted five days for both the training sets using 300 regions proposals and a random horizontal flip has been applied to increase the number of samples and accuracy.

Both the inference graphs produced have been exported and tested for inference on new samples.

The following is an example of localization and recognition using the first graph on a RGB image.

The following is an example of the second graph localization and recognition on another gray-scale image.

The training has been performed on a personal computer equipped with a 4 cores 8 threads Intel(R) Core(TM) i7-4770 CPU @ 3.40 with 16 GB DDR3 of RAM, an Nvidia Titan X powered with Pascal, and Ubuntu 16.04 as operative system.

4.2   CNN Accuracy

Accuracy of the first trained graph on RGB images tested only on the RGB images is 0.97.

Accuracy have been also checked on all the validation set and the results are:

– 1 error
– 1 not determined

With the second graph, obtained as the first one after 106000 cycles, the accuracy is again 0.97. By checking the inference on all the validation set we obtained the following results:

– 1 not determined
– 1 false positive

Figures 3 and 4 show localized and recognized objects of the two training.

In Fig. 5 the trained graphs of the second test are shown: not only the Total Loss but also all the other region and box classification and localization graphs after 20,000 cycles tend to stabilize and then go close to zero.

On the basis of these results, the additional color layers seems not significant.

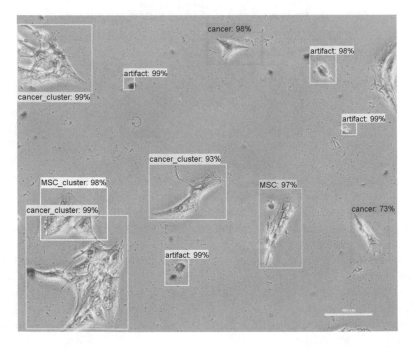

Fig. 3. An example of RGB image with localized and recognized objects under investigation. The ability of our method to discriminate between single or cell clusters is remarkable.

Localization and recognition of new images require less than one second on a personal computer with a modern Intel I7 CPU.

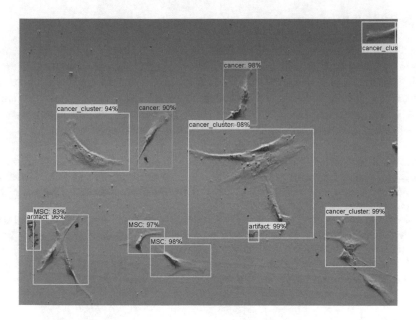

Fig. 4. An example of gray-scale image with localized and recognized objects under investigation. As in Fig. 3, the ability to discriminate between single and cell clusters is excellent.

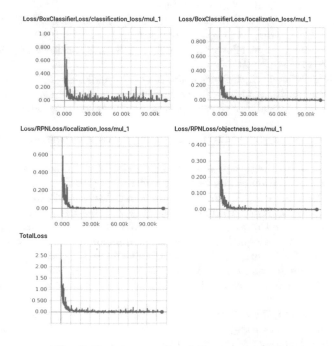

Fig. 5. Training graphs of the gray-scale test

5   Conclusions

Classification of single or small clusters of cancer cells is a crucial question for early diagnosis. In this paper, a Deep Learning approach to recognize single or small clusters of cancer cells have been presented. The Deep Learning method adopted was based on Faster-RCNN technique. The ability of such algorithm to identify and classify approximately the 100% of the investigated cells potentially will allow us to extend the method to large population cells or tissues.

Acknowledgements. This work is being carried out partially in the framework of the BIO-ICT joint laboratory between the Institute of Biophysics and the Institute of Information Science and Technologies, both of the National Research Council of Italy, in Pisa.

We would like to thank Nvidia Corporation: this work would have required an invaluable time without a Titan X board powered by Pascal won by Signals & Images Laboratory of CNR-ISTI at the 2017 Nvidia GPU Grant.

We also wish to thank Luisa Trombi, Serena Danti, Delfo D'Alessandro, from University of Pisa, for useful support with biological samples.

References

1. Abadi, M., Barham, P., Chen, J., Chen, Z., Davis, A., Dean, J., Devin, M., Ghemawat, S., Irving, G., Isard, M., et al.: Tensorflow: a system for large-scale machine learning. OSDI **16**, 265–283 (2016)
2. Bychkov, D., Linder, N., Turkki, R., Nordling, S., Kovanen, P.E., Verrill, C., Walliander, M., Lundin, M., Caj, H., Lundin, J.: Deep learning based tissue analysis predicts outcome in cllorectal cancer. Sci. Rep. **8**, 3395 (2018). https://doi.org/10.1038/s41598-018-21758-3
3. Cireşan, D.C., Giusti, A., Gambardella, L.M., Schmidhuber, J.: Mitosis detection in breast cancer histology images with deep neural networks. In: International Conference on Medical Image Computing and Computer-Assisted Intervention, pp. 411–418. Springer (2013)
4. Cristy, J.: Imagemagick website (2013). http://www.imagemagick.org/. Accessed 08 June 2018
5. Dürr, O., Sick, B.: Single-cell phenotype classification using deep convolutional neural networks. J. Biomol. Screen. **21**(9), 998–1003 (2016)
6. Huang, J., Rathod, V., Sun, C., Zhu, M., Korattikara, A., Fathi, A., Fischer, I., Wojna, Z., Song, Y., Guadarrama, S., Murphy, K.: Speed/accuracy trade-offs for modern convolutional object detectors. In: 2017 IEEE Conference on Computer Vision and Pattern Recognition, CVPR 2017, Honolulu, HI, USA, 21–26 July 2017, pp. 3296–3297 (2017). https://doi.org/10.1109/CVPR.2017.351
7. Idikio, H.A.: Human cancer classification: a systems biology-based model integrating morphology, cancer stem cells, proteomics, and genomics. J. Cancer **2**, 107 (2011)
8. Li, Z., Soroushmehr, S.M.R., Hua, Y., Mao, M., Qiu, Y., Najarian, K.: Classifying osteosarcoma patients using machine learning approaches. In: 2017 39th Annual International Conference of the IEEE Engineering in Medicine and Biology Society (EMBC), pp. 82–85 (2017). https://doi.org/10.1109/EMBC.2017.8036768

9. Mishra, R., Daescu, O., Leavey, P., Rakheja, D., Sengupta, A.: Convolutional neural network for histopathological analysis of osteosarcoma. J. Comput. Biol. **25**, 313–325 (2017)
10. Mishra, R., Daescu, O., Leavey, P., Rakheja, D., Sengupta, A.: Histopathological diagnosis for viable and non-viable tumor prediction for osteosarcoma using convolutional neural network. In: Cai, Z., Daescu, O., Li, M. (eds.) Bioinformatics Research and Applications, pp. 12–23. Springer International Publishing, Cham (2017)
11. Nahid, A.A., Mehrabi, M.A., Kong, Y.: Histopathological breast cancer image classification by deep neural network techniques guided by local clustering. BioMEd Res. Int. **2018**, 20 (2018)
12. Ren, S., He, K., Girshick, R., Sun, J.: Faster r-cnn: towards real-time object detection with region proposal networks. In: Cortes, C., Lawrence, N.D., Lee, D.D., Sugiyama, M., Garnett, R., (eds.) Advances in Neural Information Processing Systems, vol. 28, pp. 91–99. Curran Associates, Inc. (2015). http://papers.nips.cc/paper/5638-faster-r-cnn-towards-real-time-object-detection-with-region-proposal-networks.pdf
13. Ren, S., He, K., Girshick, R., Sun, J.: Faster r-cnn: towards real-time object detection with region proposal networks. IEEE Trans. Pattern Anal. Mach. Intell. **39**(6), 1137–1149 (2017)
14. Song, Q., Merajver, S.D., Li, J.Z.: Cancer classification in the genomic era: five contemporary problems. Hum. Genomics **9**, 27 (2015)
15. Trombi, L., Mattii, L., Pacini, S., D'alessandro, D., Battolla, B., Orciuolo, E., Buda, G., Fazzi, R., Galimberti, S., Petrini, M.: Human autologous plasma-derived clot as a biological scaffold for mesenchymal stem cells in treatment of orthopedic healing. J. Orthop. Res. **26**(2), 176–183 (2008)
16. Tzutalin: Labelimg. git code (2015). https://github.com/tzutalin/labelImg. Accessed 11 May 2018
17. Uijlings, J., van de Sande, K., Gevers, T., Smeulders, A.: Selective search for object recognition. Int. J. Comput. Vis. **104**, 154–171 (2013). https://doi.org/10.1007/s11263-013-0620-5. http://www.huppelen.nl/publications/selectiveSearch Draft.pdf
18. Xie, Y., Xing, F., Kong, X., Su, H., Yang, L.: Beyond classification: structured regression for robust cell detection using convolutional neural network. In: International Conference on Medical Image Computing and Computer-Assisted Intervention, pp. 358–365. Springer (2015)

Detection of Suspicious Accounts on Twitter Using Word2Vec and Sentiment Analysis

Patricia Conde-Cespedes[1]([⊠]), Julie Chavando[2], and Eliza Deberry[2]

[1] Institut Superieur d'Electronique de Paris,
28 rue Notre Dame des Champs, Paris, France
`patricia.conde-cespedes@isep.fr`
[2] Stanford University, San Francisco, CA, USA
`{chavand,eklyce}@stanford.edu`

Abstract. Twitter constantly attempts to suspend dangerous or suspicious accounts. This strategy has been questionably ineffective as users continuously recreate their accounts. As a result, there are many accounts that are left unchecked and potentially dangerous to the promotion of ideals and attacks. In this paper, we present a classification method based on sentiment analysis and word2vec to detect suspicious accounts. We evaluate our approach in real use case of main concern using data crawled directly from Twitter. Our classifier returns high accuracy in detecting suspicious accounts.

Keywords: Prediction · Classification · Twitter · Word2Vec
Sentiment analysis · SVM · Transfer learning

1 Introduction

Twitter can be seen an online news service which allows the users to interact or react (comment, like, share, etc.) some information posted by other users. Unfortunately, many accounts are created to disseminate propaganda. This propaganda on social media is being used to manipulate public opinion among the users. For instance, one can cite the works performed by the Computational Propaganda Lab at the Oxford Internet Institute, University of Oxford for the The Computational Propaganda Research Project (COMPROP) [1]. Therefore, it is important to automatically detect such accounts from the content of the tweets in order to suspend them.

Tweets therefore become extremely relevant as many voice their opinions and sometimes threats through this accessible platform. Various governmental and research organizations are on the hunt for identification of harmful accounts in order to eliminate them.

An application of big concern is the efficiently detection of suspicious accounts supporting the Islamic State of Syria and the Levant (ISIS) in Twitter. Indeed, according to research done by Twitter expert Zaman [2], the Islamic State has four goals in its terror through tweets: to spread a global threat, possess a constant stream of propaganda, have a source of recruitment, and expand the largest foreign fighter army in the world. 90% of ISIS social media is conducted through Twitter, they control about 70,000 accounts and tweet about 200,000 times per day (see also [3–5]). Twitter's ability to provide a division of labor additionally stimulates the news reporters, fanboys, recruiters, and intellectuals.

Our analysis aims to find an effective method to encounter suspicious accounts in Twitter. We apply our analysis to real data set consisting in Tweets collected during and before the terrorist attacks that took place in November 2015 in Paris. We start by identifying common words and phrases used by ISIS fanboys. Next, we perform word2vec in order to detect the words and their contexts. Finally, we apply sentiment analysis and the words studied in the previous steps to define the features to build our classifier to detect the target variable, that is, the existence or not of a suspicious account. Previously works aimed to define a classifier to automatically detect Pro-ISIS accounts, one can see for instance [6–9]. However, none of those previous studies were based on combining recently promising machine learning techniques as Word2Vec and sentiment analysis.

Section 2 presents the data used to validate our approach. Section 3 presents our approach, detailing all the aspects from the treatment of the row data to the prediction. Finally, in Sect. 4 we present some conclusions and perspectives.

2 Data Pre-processing

2.1 Data Set Description

The data set consists of tweets crawled before and after the 2015 terrorist attacks on the city of Paris, France. The tweets were crawled from November 13th, 2015 to December 2nd, 2015. We remind that The November 2015 attacks left 130 people dead and hundreds more wounded. They started in Paris on November 13 at the Bataclan Theater and continued throughout the end of the year.

The crawling step was done by the Thales Communications & Security department for the REQUEST project (http://projet-request.org). The crawling criteria were some keywords like: *#toParisWithRevenge*, *Khilafa*, *#jesuiskouachi*. Approximately 200000 tweets were crawled.

Originally, the tweets were in the form of json files. Each individual tweet had information concerning the unique identification (*"id"*), the person behind the tweet (*"screen_name"*), the location that tweet was posted from (*"location"*), the language (*"lang"*), the text of the tweet, whether it was retweeted or not (*"retweeted"*), and other information that we did not take into account in our analysis.

2.2   Language Translation

The text of the tweets were mainly in three languages English (5%), French (25%), and mostly in Arabic (70%), they were translated into English using Google Translate. Previously, all of the tweet text were encoded in utf-8 to standardize the alphabet and to ensure that all characters represented in the data set could be easily processed and later translated. In order to diminish the room for error due to translation, we used the Arabic lexicon presented in [10–12] which actually used sentimental analysis using Google Translate to process some Arabic to English. In order to further clarify some of the tweets, we also imported the Google Translate API package from GitHub [13].

2.3   Tokenization

We performed tokenization by eliminating words and phrases that would not have any impact on the anti/pro nature of the tweets such as stop words, unimportant characters, etc. Next, we performed stemming on the tokenized words. This allowed to count all instances of a given word even if it is in a plural or conjugated form. Stemmers find the stem of common words - namely it removes the morphological affixes of words. For example, the words *games, gamed, gaming* and *game* is *game*. We used the NLTK stemmer which is the word stemmer available in the Natural Language Toolkit for Python [14].

3   Data Analysis

3.1   Word Frequencies

In order to get familiar with the data set, we calculated the frequency of words and n–grams, more precisely bigrams and trigrams. An n–gram is a sequence of n adjacent elements from a string of tokens.

For instance, if the text of a tweet is the following:

"Allah is the greatest RT: #Urgent Thirty martyrs were killed and one hundred wounded in a martyrdom operation at a rally in Kano, #Cheaper".

Then, the n-grams will be the following:

n = 1: ["Allah", "is", "the", "greatest", "#Urgent", ...]
n = 2: ["Allah is", "is the", "the greatest", "greatest #Urgent",...]
n = 3: ["Allah is the","is the greatest", "the greatest #Urgent",...]

Then, for $n = 1$ we obtain single tokenized words.

For each date we considered the top 10 of n–grams that were mentioned the highest amount of times in the set of tweets. We can see these results in Fig. 1.

Notice from Fig. 1 that just after the attacks (November 14th and 15th) the there is a certain dominance of solely news sources, for instance, consider the account of *@sosparis311*, used only to raise awareness for Parisian support. We can deduce that the first days after the attacks the tweets were attempting to

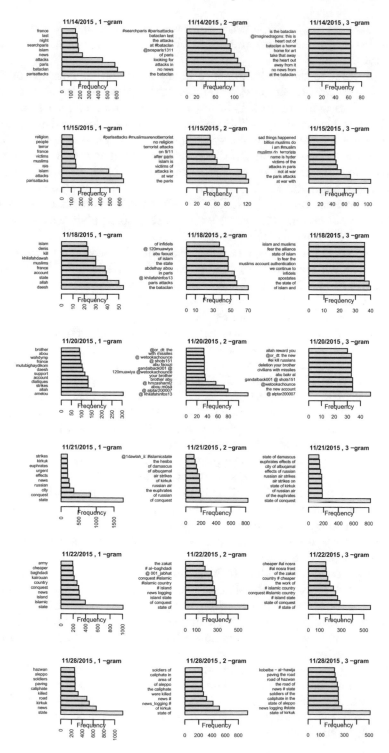

Fig. 1. *n*–grams frequencies per date.

spread awareness about the attacks. However, as time progresses onto the next day, the tweets become more indicative of opinions, both hateful toward and against ISIS.

However, notice the stark difference in the tone and connotation of the following n-grams corresponding to November 18th. #Khilafahdawah is a common hashtag used to attribute to ISIS killing of 200 children. This hashtag is tricky to attribute to suspicious pro-ISIS accounts because many used it as a spread of condemnation on the group and these specific acts. The account *@khilafahin-fos13* has been suspended on Twitter. *Abdelhay Abou* is a very active account that tweets in support of islam with phrases such as *"Lord, make the unjust ones kill each other and make the unbelievers leave them unharmed"* and *"Justice for jihadists"*. An issue presented here is, though, to make sure that the account is pro-ISIS, not pro-Islam. *@120muawiya* is another suspended account that gave rise to a pro-ISIS account *@hbebalrhman*.

As we move on the language becomes even more intense and more suspended accounts and their associates began appearing. For instance, on November 20th, *@Mutubighaydikom*, *@wahdymp* and amelou are suspended 's accounts.

Another important remark is that the familial language seems prominent as well, with *brother* and *abou* meaning *father*.

We also remarked some curiosities concerning suspicious accounts. For instance, *@alptar200007* was a belligerent account that received much attention as it is repeatedly mentioned in other tweets. For instance *@kgocGXoeSXsVpzM* tweeted *God saved you, @alptar200007*. One account, still present on Twitter and that seem suspicious is *@d175FCNkI32qHWt* who tweeted *God cursed them hypocrites*. The most glaring realization is that many of these harmful accounts have random usernames with meaningless letters. *@hmzashalm2*, *@wetookachounce*, *@shots151*, and *@or_dt* are all suspended accounts as well. In their tweets, they highlighted *@yemen66rothuan* who posted suspicious content.

Notice many of the suspended accounts appear with frequency, like *@alptar200007* and *@gandalback001*. It is important to note that although many of the frequent accounts found have been suspended by Twitter, the largest source of information resided in those who have tweeted replies or retweet the sentiments expressed by these accounts. They proved essential to our further identification of hundreds of pro-ISIS users.

The n–grams corresponding to November 21st treat mostly locations were the attacks took place, like Russian attacks. Sites such as *Albuqamal* and *Damascus* were sites of ISIS attacks while *the hesba* describes random arrests by isis. An interesting information from this set relies on the suspended account *@1dawlah_ii*. This account was eliminated because many Twitter users reported it using the tag *@gov*.

Concerning November 22nd, we found two more suspended accounts, such as *@001_jabhat* and *@smol2001*. The latter used intense language noting the same vigor that many pro-ISIS accounts convey. More Muslim ideals begin to appear as well such as *zakar*, the pillar of Islam focused on aiding the poor and *Al-baghdadi* a self-proclaimed ISIS leader.

Finally, on November 28, 2015, *Al-hawija* is a city in Iraq controlled by ISIS. This word was used several times as a hashtag to refer to the burning of civilians in this city. This highlighted belligerence in the accounts.

3.2   Classification

In this section we present our classification method used to detect suspicious accounts in our data set. The first step to perform was to label our data set. We defined the target variable as follows:

- 1: Pro-ISIS tweet.
- 0: neutral tweet.

The Tweets in our data set were interpreted and manually labeled by native speakers of Arabic and French. Out of approximately 600 labeled tweets about 60% were labeled 1 and 40% labeled 0. Besides, we observed that most likely pro-ISIS accounts were suspended.

3.3   Transfer Learning and Outside Data Sources

To perform classification, we used two outside labeled data sets from Kaggle.com:

- *Isis-Fanboys* data set [15]: a collection of tweets from accounts suspended by twitter after the 2015 Paris terrorist attacks.
- *General_Tweets* data set [16]: a combination of tweets against ISIS or of neutral sentiment collected after the Paris attacks.

We performed transfer learning from the outside sources to our data set since the outside data provided a larger labeled set to train on as shown in Fig. 2.

Therefore, all Isis_fanboys tweets were labeled 1 and all general_tweets were labeled 0.

The top word frequencies of the outside data's pro-ISIS tweets are as well as their frequencies are: *('twitter', 4705), ('island', 2800), ('account', 2484), ('suspended', 2417), ('status', 2100), ('ramiallolah', 2052), ('syria', 1791), ('breakingnews', 1447), ('killed', 1325), ('reporter', 1296), ('warrnews', 1229), ('conflict', 1107), ('salahuddin', 1067), ('ayubi', 1057), ('nidalgazaui', 1043), ('photo', 1012), ('army', 961), ('iraq', 873), ('assad', 837), ('state', 793).*

Fig. 2. Transfer learning from outside sources

Some words are often the names of leaders in the Middle East. For instance, *Salahuddin* was the name first sultan of Egypt who led the Muslim military campaign against the crusader states in the Levant, *Ayubi* was a muslim leader, and *Assad* is the name of the current Syrian president. *Nidalgazaui* is a German researcher who spreads news about terrorism & Jihadi Groups. The word *Island* refers to a famous news channel Qatari, which people on Twitter often use to relay information.

3.4   Sentiment Analysis

We used the VADER sentiment intensity analyzer from the Natural Language Processing Toolkit [17]. VADER stands for Valence Aware Dictionary and Sentiment Reasoner. The analyzer works by keeping a lexicon of words and a corresponding sentiment rating. The analyzer takes as an argument a string of text and returns four scores. The first three are the percent of positive, negative, and neutral words in the text. These three fields will thus always sum to one. The final score is a compound score. It is based on the sum of all of the lexicon ratings - having been standardized to be in the range −1 to 1. For example the word *tragedy* has a score of −3.4 as it is highly negative.

3.5   Word2Vec Approach

Word2Vec [18] is a deep learning model for computing continuous distributed representations of words. This vectorization is made in such a way that for any pair of words are mapped close to each other if they frequently appear in similar contexts, then, the calculation of similarities straightforward. Using Gensim [19], an open source Python library for natural language processing, this vectorization of words was possible.

   We identified the top 10 words in our data set that are as *similar* as possible to those the top 10 most frequent Pro-ISIS words from the *ISIS-Fanboys* data set (see Section for 3.3 this list of words). For instance these are the list of most similar words in our corpus to two keywords *Killed* and *Account* for November 18th and 20th. The similarity is also indicated.

November 18th, 2015:
Killed: [(*people*, 0.99867), (*#Bataclan*, 0.99862), (*Islam*, 0.99831), (*France*, 0.99828), (*#Ei*, 0.99818), (*Muslims*, 0.99805), (*attacks*, 0.99802), (*120Muawiya*, 0.99801)]
November 20th, 2015:
Account: [(*Abu*, 0.99897), (*120Muawiya*, 0.99872), (*brother*, 0.99861), (*Al*, 0.99848), (*new*, 0.99841), (*Support*, 0.99821), (*@KSAssa:*, 0.99796), (*brothers*, 0.99780), (*@MutuBiGhaydikom*, 0.99776), (*alptar200007*, 0.99731)]

3.6   Support Vector Machine (SVM) Classifier

In order to perform classification, we built a Support Vector Classifier (SVM). We considered 148 features chosen as follows:

- The positive, negative, neutral, and compound sentiment scores explained in Sect. 3.4.
- The other 144 features are based on a list of keywords which are identified in Sect. 3.5. Each stemmed tweet was transformed into a 144 character long array of 1's and 0's. If the nth number in the array is a 1, this indicates that the nth keyword is found in the tweet. If the nth number in the array is a 0, this indicates that the nth keyword is not found in the tweet. This array was then normalized- each number in the array of 1's and 0's was divided by the sum of all keywords found (or remained an array of 0's if there were no keywords in the tweet). This means each of the 144 features reflect a single keyword. For instance: for the tweet text:

praise be to allah peace be upon him dear brother.

The 4 words *praise, allah, peace* and *brother* being part of the keywords, the tweet will be represented by a vector of length 144, where the entries corresponding to those keywords equal 0.25 and the others are 0. The choice of these keywords was crucial for our training model.

We used a Support Vector Machine (SVM) classifier. We used a k-fold cross-validation, with p = 100, that means that for each test 100 tweets were reserved for testing and trained the algorithm on the other 47850 tweets. The resulting accuracy was 92.0%. Then we transferred. Then, we performed transfer learning to our data set. By keeping the same type of sentiment scoring and keywords, the resulting average accuracy was 87.9%.

4   Conclusion and Perspectives

In this paper we built a classifier to identify suspicious accounts in Twitter. We applied our model to a real data set consisting in Pro-ISIS and neutral tweets. We used a Support Vector Machine classifier. The model was trained on keywords and sentiment analysis scores on outside data sources. Then, we performed transfer learning to our data. The resulting accuracy was satisfactory.

We also noticed that the presence of suspended accounts led to the discovery of more pro-ISIS accounts that have not yet been eliminated by calculating word frequencies of n–grams. Additionally, the date of the Tweets was an important variable to distinguish between the spread of news and the bias that indicated the presence of dangerous users.

Some perspectives of these works are to expand on Word2Vec and improve the accuracy of the training by potentially using the words from the outside sources to build relationships. Another aspect to explore to improve our model is to perform feature selection. This leads to understand which words and sentiment analysis tools most correlated with each class.

Acknowledgment. This work was supported by REQUEST project. The authors would like to thanks Asis Zidi and Alexis Ung for labeling the Tweet texts.

References

1. Woolley, S.C., Howard, P.N.: Computational propaganda worldwide: executive summary. Project on Computational Propaganda (COMPROP), Oxford, UK, 14pp (2017). http://www.comprop.oii.ox.ac.uk
2. Zaman, K.: Isis has a twitter strategy and it is terrifying [infographic], medium, fifth tribe stories (2015)
3. Berger, J.M., Morgan, J.: The isis twitter census: defining and describing the population of isis supporters on twitter. The Brookings Project on US Relations with the Islamic World, vol. 3, no. 20, 4–1 (2015)
4. Klausen, J.: Tweeting the jihad: social media networks of western foreign fighters in Syria and Iraq. Stud. Conflict Terrorism **38**(1), 1–22 (2015)
5. Carter, J.A., Maher, S., Neumann, P.R.: #greenbirds: measuring importance and influence in Syrian foreign fighter networks (2014)
6. Saif, H., Dickinson, T., Kastler, L., Fernández, M., Alani, H.: A semantic graph-based approach for radicalisation detection on social media. In: ESWC (1). LNCS, vol. 10249, pp. 571–587 (2017)
7. Agarwal, S., Sureka, A.: Using KNN and SVM based one-class classifier for detecting online radicalization on twitter. In: Natarajan, R., Barua, G., Patra, M.R. (eds.) Distributed Computing and Internet Technology, pp. 431–442. Springer, Cham (2015)
8. Ashcroft, M., Fisher, A., Kaati, L., Omer, E., Prucha, N.: Detecting jihadist messages on twitter. In: 2015 European Intelligence and Security Informatics Conference, pp. 161–164, September 2015
9. Ferrara, E., Wang, W.Q., Varol, O., Flammini, A., Galstyan, A.: Predicting online extremism, content adopters, and interaction reciprocity. In: International Conference on Social Informatics, pp. 22–39. Springer (2016)
10. Mohammad, S., Salameh, M., Kiritchenko, S.: Sentiment lexicons for Arabic social media. In: Proceedings of 10th Edition of the the Language Resources and Evaluation Conference (LREC), Portorož, Slovenia (2016)
11. Mohammad, S.M., Salameh, M., Kiritchenko, S.: How translation alters sentiment. J. Artif. Intell. Res. **55**, 95–130 (2016)
12. Salameh, M., Mohammad, S., Kiritchenko, S.: Sentiment after translation: a case-study on Arabic social media posts. In: Proceedings of the 2015 Conference of the North American Chapter of the Association for Computational Linguistics: Human Language Technologies, Denver, Colorado, pp. 767–777. Association for Computational Linguistics (2015)
13. Aliès, A.: Google Translate API for Python. GitHub repository (2017). https://pypi.python.org/pypi?:action=display&name=mtranslate&version=1.3
14. Loper, E., Bird, S.: NLTK: the natural language toolkit. In: Proceedings of the ACL Workshop on Effective Tools and Methodologies for Teaching Natural Language Processing and Computational Linguistics. Association for Computational Linguistics, Philadelphia (2002)
15. FifthTribe: How isis uses twitter: Analyze how isis fanboys have been using twitter since 2015 Paris attacks, kaggle (2015)
16. FifthTribe: Tweets targeting isis: General tweets about isis & related words, kaggle (2015)

17. Hutto, C., Gilbert, E.: Vader: a parsimonious rule-based model for sentiment analysis of social media text. In: International AAAI Conference on Web and Social Media (2014)
18. Mikolov, T., Sutskever, I., Chen, K., Corrado, G., Dean, J.: Distributed representations of words and phrases and their compositionality. In: Proceedings of the 26th International Conference on Neural Information Processing Systems, NIPS 2013, USA, vol. 2. Curran Associates Inc. (2013)
19. Řehůřek, R., Sojka, P.: Software framework for topic modelling with large corpora. In: Proceedings of the LREC 2010 Workshop on New Challenges for NLP Frameworks, Valletta, Malta, ELRA, pp. 45–50, May 2010. http://is.muni.cz/publication/884893/en

A Hybrid CNN Approach for Single Image Depth Estimation: A Case Study

Károly Harsányi[1], Attila Kiss[1(✉)], András Majdik[1], and Tamas Sziranyi[1,2]

[1] Machine Perception Research Laboratory, MTA SZTAKI, Budapest, Hungary
{harsanyika,attila.kiss,majdik,sziranyi}@sztaki.hu
[2] Faculty of Transportation Engineering and Vehicle Engineering,
BME, Budapest, Hungary

Abstract. Three-dimensional scene understanding is an emerging field in many real-world applications. Autonomous driving, robotics, and continuous real-time tracking are hot topics within the engineering society. One essential component of this is to develop faster and more reliable algorithms being capable of predicting depths from RGB images. Generally, it is easier to install a system with fewer cameras because it requires less calibration. Thus, our aim is to develop a strategy for predicting the depth on a single image as precisely as possible from one point of view. There are existing methods for this problem with promising results. The goal of this paper is to advance the state-of-the-art in the field of single-image depth prediction using convolutional neural networks. In order to do so, we modified an existing deep neural network to get improved results. The proposed architecture contains additional side-to-side connections between the encoding and decoding branches.

Keywords: Depth estimation · Deep learning · CNN

1 Introduction

Predicting the depth of the elements of a scene has been an interesting challenge since the foundation of computer vision. It is a well known and well studied hot topic in the field of computer science. Among several techniques, some of the most known ones are using some special information from the images, for example, variations in illumination [25,29], or focus [1,4]. Another popular method in this category is the so-called Structure-from-Motion technique [20,27]. However, it is extremely difficult to gather any special information from a single image without additional knowledge about the environment. Nevertheless, finding better solutions to this problem is vital, considering that accurate depth information improves results on human pose estimation [16,26], recognition [5,17], reconstruction [13,21] or semantic segmentation [2,11].

The research was supported by the Hungarian Scientific Research Fund (No. NKFIH OTKA K-120499 and KH-126513) and the BME-Artificial Intelligence FIKP grant of EMMI (BME FIKP-MI/FM).

K. Choroś et al. (Eds.): MISSI 2018, AISC 833, pp. 372–381, 2019.
https://doi.org/10.1007/978-3-319-98678-4_38

Before neural network approaches became popular in this field, other approaches appeared based on Conditional Random Fields [8,23] (taking superpixels into consideration) and Markov Random Fields (MRF) [10,24]. Later, a very effective tool came into focus. With the help of Convolutional Neural Networks (CNNs) we became able to learn an implicit relation between depth and color pixels. Combining CNNs with CRF-based regularization via structured deep learning is a promising direction in the research [9,12].

In this paper, we introduce a fully convolutional neural network which combines the advantages of deep residual nets [6] and U-net architectures [18]. We examine the performance of this hybrid network on the NYU depth v2 dataset [21], and demonstrate that it is possible to achieve state-of-the-art accuracy without significantly increasing the depth and the number of variables. This is achieved by augmenting the CNN from the work of Laina et al. [9] with U-net-like connections between the encoding and decoding parts. This kind of augmentation is getting more and more common in the engineering society. We applied this successfully to a new domain.

2 Related Work

In the introduction, we referred some classical methods on depth estimation so, in this section, we try to focus on the latest results related to our work. Outstanding results and the possibility of new methodologies drove research towards the application of CNNs for the aforementioned problem. Classical networks, originally developed for a different task, were studied in this topic. These networks like AlexNet [7] and VGG [22] earned fame by their high level of success at the ImageNet Large Scale Visual Recognition Challenge [19]. Later, a two-scale architecture approach has appeared by Eigen et al. [3]. One year later, their work was extended to a three-scale architecture for further refinement in [2].

We have to mention two special network architectures in order to understand the motivation behind our approach. The next breakthrough in the field of neural networks was the appearance of the deep Residual Network [6] (ResNet). Training up to hundreds or even thousands of layers while achieving compelling performance became possible with it. The key idea behind ResNet is the introduction of the so-called "shortcut connection" that skips one or more layers. Thus, instead of directly fitting a desired underlying mapping, these layers can fit a residual mapping. This leads to better performance and decreases the computational complexity and the process time.

Ronneberger et al. presented a U-shaped network and a training strategy [18] that relies on the strong use of data augmentation for biomedical image segmentation. Their idea was to supplement a usual contracting network by successive layers, where pooling operators are replaced by upsampling operators. Additionally, they concatenated the corresponding feature maps between the expansive and contracting paths to enhance the information flow and consequently increase the resolution of the output.

The backbone of our network is the ResNet-based fully convolutional neural net introduced by Laina et al. [9]. We enhance this network with U-net-like lateral connections to boost its accuracy. The exact modifications are specified in Sect. 3.

3   Network Architecture

This section describes the structure of our model. For better understanding, Fig. 1 shows the building blocks we used to assemble the network.

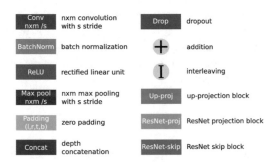

Fig. 1. The representations of the different layers used in our architecture. The Up-proj, ResNet-proj and ResNet-skip notations refer to more complex building blocks. See Figs. 2 and 3 for a detailed description. The definition for the interleaving operation can be found in [9].

The model outlined in [9] can be divided into two parts: an encoding part based on the ResNet50 architecture, and a decoding part consisting of repeated up-projection blocks (see Fig. 3). Our aim was to boost the reliability of this model without significantly increasing its depth and the number of its variables.

Inspired by U-net-like architectures, instead of deepening the model by increasing the number of layers, we enhance its interconnectedness by introducing side-to-side connections into the network flow. These lateral connections are realized by concatenating the corresponding feature maps between the encoding and decoding parts of the model. After each depth concatenation, we insert an additional 1×1 convolution in order to keep the number of input channels of the up-projection blocks intact. Thanks to these connections the network is able to translate additional information from its previous feature maps into its expansive path. The complete architecture of the proposed network is depicted in Fig. 4.

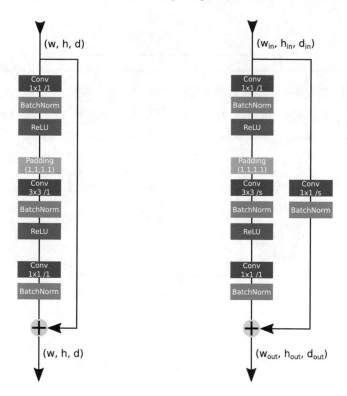

Fig. 2. (a) Residual-skip building block, (b) Residual-projection building block

4   Training

4.1   Dataset and Augmentation

We train and evaluate our model on the NYU Depth v2 dataset. This dataset captures 464 indoor scenes with an RGB camera and a Microsoft Kinect. The dataset is split into 215 testing and 249 training scenes. We extracted equally-spaced RGB-D image pairs from the training scenes of the raw dataset and ended up with roughly 48000 pairs as our training set.

Before training, we resize every RGB-D pair, to 352×264 pixels. This way the image sizes are closer to the input shape of our network and the aspect ratio stays roughly the same. Additionally, we use random online augmentation during training. Every RGB-D image pair is:

- scaled by $s \in_R [1, 1.5]$, and the depths are divided by s
- rotated by $r \in_R [-5, 5]$ degree
- randomly cropped down to 320×256 pixels
- the RGB color values are multiplied by a random $c \in_R [0.8, 1.2]^3$
- horizontally flipped with 0.5 probability.

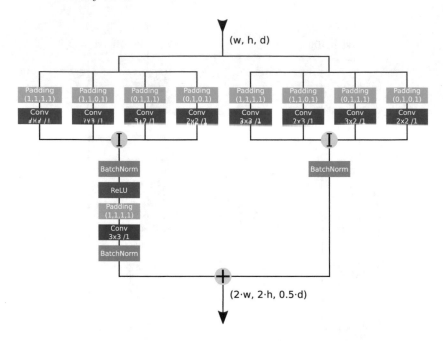

Fig. 3. An up-projection block [9]. This block is equivalent to a residual up-projection, but this version is more efficient and leads to reduced training time.

The division by s in the scaling step is necessary to preserve the world-space geometry of the scene. In order to conserve the boundaries between valid and invalid pixels on the depth images, we use Nearest-neighbor interpolation for the scaling and the rotation. After the augmentation, the size of the RGB-D pairs is 320×256 pixels. We further downscale the depth image to 160×128 pixels to match the output size of our network.

It is important to note, that the obtained training dataset contains missing and invalid depth values. We can tackle this problem by masking these invalid depth values during the loss calculation.

4.2   Loss Function

We use reverse Huber (BerHu) loss [14] as our loss function during training instead of the regularly used $\mathcal{L}_2$ loss. This concept was first introduced by Laina et al. [9]. The BerHu loss puts a higher weight on the pixels with higher residuals by applying $\mathcal{L}_2$ on them. Simultaneously, it allows smaller residuals to have a larger effect on the gradients during training due to the use of the $\mathcal{L}_1$ loss.

In order to calculate the BerHu loss $\mathcal{B}(y, \tilde{y})$ for a batch of predictions y and the ground truths $\tilde{y}$, we have to compute $c = \frac{1}{5} \cdot \max_i |y_i - \tilde{y}_i|$, where i indexes

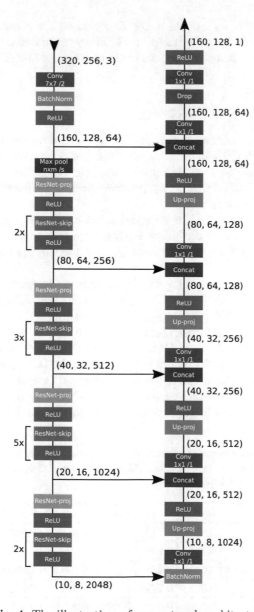

Fig. 4. The illustration of our network architecture.

over every pixel of every image in the batch y. Once c is calculated $\mathcal{B}(y, \tilde{y})$ is defined as:

$$\mathcal{B}(y, \tilde{y}) = \begin{cases} |y - \tilde{y}|, & \text{where } |y - \tilde{y}| \leq c. \\ \frac{(y-\tilde{y})^2 + c^2}{2c}, & \text{otherwise.} \end{cases} \quad (1)$$

Therefore, the BerHu loss is equal to the $\mathcal{L}_1$ norm on the pixel i where $|y_i - \tilde{y}_i| \in [-c, c]$ and equal to $\mathcal{L}_2$ outside this interval. The version defined in Eq. (1) comes from [9]. This form is favorable because it is continuous and differentiable in the switch point c.

4.3 Hyper-parameter Selection

The first half of the network is responsible for encoding the images. This part is identical to the ResNet-50 network architecture. Thus, we can initialize these layers with the ResNet-50 weights pre-trained on ImageNet [19]. The variables in the second half are initialized by random normal distribution with 0 mean and 0.001 standard deviation.

We train our network for 25–30 epochs, with a batch size of 16. We use a stochastic gradient descent optimizer with 0.9 momentum. The initial learning rate is 10^{-2}. After 10 epochs it is halved, and for the final 5–10 epochs, it is reduced to 10^{-3}.

To prevent overfitting, a dropout layer with a dropout rate set to 0.5 is inserted into the network before the final convolution. Additionally, we set a weight decay of 0.00025 for every layer in our network.

We implemented the network architecture in PyTorch [15], and trained on an NVIDIA GeForce GTX 1080 Ti Graphics Card. An entire training session took approximately 15 h. Once the model is loaded into the GPU memory, forwarding an arbitrary image through the network takes roughly 0.01 s (after resizing the picture to match the input size of the network). This means that our model is fast enough to be utilized in applications that require real-time performance.

Table 1. Quantitative comparison with state-of-the-art CNN based methods on the NYU Depth v2 dataset. In the case of RMSE, REL, and Log_{10}, lower is better. For the δ_i accuracies, higher is better.

	RMSE	REL	Log_{10}	$\delta < 1.25$	$\delta < 1.25^2$	$\delta < 1.25^3$
Eigen et al. [3]	0.907	0.215	–	0.611	0.887	0.971
Wang et al. [28]	0.745	0.220	–	0.605	0.890	0.970
Eigen and Fergus [2]	0.641	0.158	–	0.769	0.950	0.988
Laina et al. [9]	0.597	0.137	0.059	0.818	0.955	0.988
Proposed	**0.593**	**0.130**	**0.057**	**0.833**	**0.960**	**0.989**

5   Evaluation

For evaluation, we use the 654 RGB-D image pairs from the labeled test subsets of the NYU Depth v2. We resize the RGB images to 352×264 pixels, and center-crop them, to match the input size of the model. The output size of the model is 160×128 pixels, while the size of the ground truth depth maps of the test set is

Fig. 5. (a) RGB input, (b) Laina et al. [9], (c) Proposed model, (d) Ground truth

640×480. To compare the results without distortions, we resize the predictions to 582×466 pixels, and center crop the ground truth depth maps to the same size. For a fair quantitative comparison, we apply the same conversions to the output of Laina et al.'s model [9]. The rest of the data in Table 1 is based on the values reported by the authors. We computed the following error metrics on the test set, for quantitative evaluation:

- Root Mean Squared Error: $\sqrt{\frac{1}{n}\sum_{i=0}^{n}(y_i - \tilde{y}_i)^2}$
- Relative Absolute Error: $\frac{1}{n}\sum_{i=0}^{n}\frac{|y_i - \tilde{y}_i|}{\tilde{y}_i}$
- Mean Log_{10} Error: $\frac{1}{n}\sum_{i=0}^{n}|\log_{10}(y_i) - \log_{10}(\tilde{y}_i)|$
- % of pixels where $max(\frac{y_i}{\tilde{y}_i}, \frac{\tilde{y}_i}{y_i}) = \delta <$ threshold

where i iterates over every pixel of every image of the test set, and n is the number of these pixels. The comparison between our network and other models can be seen in Table 1. For qualitative results, see Fig. 5. Both the qualitative and the quantitative evaluation shows a considerable improvement compared the original architecture as well as the other state-of-the-art approaches. The trained model is publicly available at https://github.com/karoly-hars/DE_resnet_unet_hyb, along with our evaluation code.

6   Conclusion

In this paper, we studied the possibility of boosting the precision of CNNs for depth estimation by inserting lateral connections between the contracting and expansive parts of the networks. We introduced a fully convolutional network which combines the benefits of the residual shortcuts of ResNets and the side-to side feature map concatenations of U-nets. Our evaluation showed that the proposed network is able to exceed other popular CNN based depth estimation models. In the future we try to design new architectures for the problem addressed at the beginning of the paper. Our aim is to develop an architecture with less storage space requirements and smaller computational complexity in order to use the new network on mobile devices.

References

1. Alexander, E., Guo, Q., Koppal, S., Gortler, S.J., Zickler, T.: Focal flow: velocity and depth from differential defocus through motion. Int. J. Comput. Vis. 1–22 (2017)
2. Eigen, D., Fergus, R.: Predicting depth, surface normals and semantic labels with a common multi-scale convolutional architecture. In: Proceedings of the IEEE International Conference on Computer Vision, pp. 2650–2658 (2015)
3. Eigen, D., Puhrsch, C., Fergus, R.: Depth map prediction from a single image using a multi-scale deep network. In: Advances in Neural Information Processing Systems, pp. 2366–2374 (2014)
4. Grossmann, P.: Depth from focus. Pattern Recognit. Lett. **5**(1), 63–69 (1987)
5. Gupta, S., Girshick, R., Arbeláez, P., Malik, J.: Learning rich features from RGB-D images for object detection and segmentation. In: European Conference on Computer Vision, pp. 345–360. Springer, Heidelberg (2014)
6. He, K., Zhang, X., Shaoqing, R., Jian, S.: Deep residual learning for image recognition. In: Proceedings of the IEEE Conference on Computer Vision and Pattern Recognition, pp. 770–778 (2016)
7. Krizhevsky, A., Sutskever, I., Hinton, G.E.: Imagenet classification with deep convolutional neural networks. In: Advances in Neural Information Processing Systems, pp. 1097–1105 (2012)
8. Kundu, A., Li, Y., Dellaert, F., Li, F., Rehg, J.M.: Joint semantic segmentation and 3D reconstruction from monocular video. In: European Conference on Computer Vision, pp. 703–718. Springer, Heidelberg (2014)
9. Laina, I., Rupprecht, C., Belagiannis, V., Tombari, F., Navab, N.: Deeper depth prediction with fully convolutional residual networks. In: 2016 Fourth International Conference on 3D Vision (3DV), pp. 239–248 (2016)
10. Li, S.Z.: Markov random field models in computer vision. In: European Conference on Computer Vision, pp. 361–370. Springer, Heidelberg (1994)
11. Lin, G., Shen, C., Van Den Hengel, A., Reid, I.: Efficient piecewise training of deep structured models for semantic segmentation. In: Proceedings of the IEEE Conference on Computer Vision and Pattern Recognition, pp. 3194–3203 (2016)
12. Liu, F., Shen, C., Lin, G.: Deep convolutional neural fields for depth estimation from a single image. In: Proceedings of the IEEE Conference on Computer Vision and Pattern Recognition, pp. 5162–5170 (2015)

13. Long, J., Shelhamer, E., Darrell, T.: Fully convolutional networks for semantic segmentation. In: Proceedings of the IEEE Conference on Computer Vision and Pattern Recognition, pp. 3431–3440 (2015)
14. Owen, A.B.: A robust hybrid of lasso and ridge regression. Contemp. Math. **443**(7), 59–72 (2007)
15. Paszke, A., Gross, S., Chintala, S., Chanan, G., Yang, E., DeVito, Z., Lin, Z., Desmaison, A., Antiga, L., Lerer, A.: Automatic differentiation in PyTorch. Mach. Learn. Mol. Mater. (2017)
16. Pfister, T., Charles, J., Zissermann, A.: Flowing convnets for human pose estimation in videos. In: Proceedings of the IEEE International Conference on Computer Vision, pp. 1913–1921 (2015)
17. Ren, X., Bo, L., Fox, D.: RGB-(D) scene labeling: features and algorithms. In: IEEE Conference on Computer Vision and Pattern Recognition (CVPR), pp. 2759–2766 (2012)
18. Ronneberger, O., Fischer, P., Brox, T.: U-net: convolutional networks for biomedical image segmentation. In: International Conference on Medical Image Computing and Computer-Assisted Intervention, pp. 234–241. Springer, Heidelberg (2015)
19. Russakovsky, O., Deng, J., Su, H., Krause, J., Satheesh, S., Ma, S., Huang, Z., Karpathy, A., Khosla, A., Bernstein, M., Berg, A.C., Fei-Fei, L.: Imagenet large scale visual recognition challenge. Int. J. Comput. Vis. **115**(3), 211–252 (2015)
20. Schonberger, J.L., Frahm, J.M.: Structure-from-motion revisited. In: Proceedings of the IEEE Conference on Computer Vision and Pattern Recognition, pp. 4104–4113 (2016)
21. Silberman, N., Hoiem, D., Kohli, P., Fergus, R.: Indoor segmentation and support inference from RGBD images. In: European Conference on Computer Vision, pp. 746–760. Springer, Heidelberg (2012)
22. Simonyan, K., Zissermann, A.: Very deep convolutional networks for large-scale image recognition. arXiv preprint (2014)
23. Sutton, C., McCallum, A.: An introduction to conditional random fields. Found. Trends Mach. Learn. **4**(4), 267–373 (2012)
24. Szirányi, T., Zerubia, J., Czúni, L., Kato, Z.: Image segmentation using Markov random field model in fully parallel cellular network architectures. Real-Time Imaging **6**(3), 195–211 (2000)
25. Tao, M.W., Srinivasan, P.P., Malik, J., Rusinkiewicz, S., Ramamoorthi, R.: Depth from shading, defocus, and correspondence using light-field angular coherence. In: Proceedings of the IEEE Conference on Computer Vision and Pattern Recognition, pp. 1940–1948 (2015)
26. Taylor, J., Shotton, J., Sharp, T., Fitzgibbon, A.: The Vitruvian manifold: inferring dense correspondences for one-shot human pose estimation. In: IEEE Conference on Computer Vision and Pattern Recognition (CVPR), pp. 103–110 (2012)
27. Ullman, S.: The interpretation of structure from motion. Proc. R. Soc. Lond. B **203**(1153), 405–426 (1979)
28. Wang, P., Shen, X., Lin, Z., Cohen, S., Price, B., Yuille, A.: Towards unified depth and semantic prediction from a single image. In: Proceedings of the IEEE Conference on Computer Vision and Pattern Recognition, pp. 2800–2809 (2015)
29. Zhang, R., Tsai, P.S., Cryer, J.E., Shah, M.: Shape-from-shading: a survey. IEEE Trans. Pattern Anal. Mach. Intell. **21**(8), 690–706 (1999)

Accessing Multilingual Information and Opinions

Cross-Lingual Speech-to-Text Summarization

Elvys Linhares Pontes[1,2]([✉]), Carlos-Emiliano González-Gallardo[1,2],
Juan-Manuel Torres-Moreno[1,2], and Stéphane Huet[1]

[1] LIA, Université d'Avignon et des Pays de Vaucluse, Avignon 84000, France
`elvys.linhares-pontes@alumni.univ-avignon.fr`
[2] École Polytechnique de Montréal, Montréal, Canada

Abstract. Cross-Lingual Text Summarization generates a summary in
a language different from the language of the source documents. We propose a French-to-English cross-lingual transcript summarization framework that automatically segments a French transcript and analyzes the
information in the source and the target languages to estimate the
saliency of sentences. Additionally, we use a multi-sentence compression
method to simultaneously compress and improve the informativeness of
sentences. Experimental results show that our framework outperformed
extractive methods using automatic sentence segmentation, even with
transcription errors.

Keywords: Cross-Lingual Text Summarization
Multi-sentence compression · Automatic speech recognition

1 Introduction

Nowadays, audio data are part of daily life in the form of news, interviews and
conversations, whether it is on the radio or on the Internet. Manual analysis
of these data is very difficult because it requires a huge number of persons to
analyze this information in the time available. One way to analyze and acceler-
ate the data processing is Automatic Speech Summarization, which differs from
the traditional Automatic Text Summarization task [24] because there are other
problems to take into account like speech recognition errors, the lack of sentence boundaries, the wide range of sentence sizes, colloquialisms, and uneven
information distributions [3,5,21,23].

Cross-Lingual Text Summarization (CLTS) consists in summarizing a text
where the summary language differs from the original document language. This
application can be split in two sub-applications: Text Summarization (TS) and
Machine Translation (MT). Each sub-application generates outcomes with errors
and putting them one after the other may reduce the quality of cross-lingual
summaries because of the accumulation of errors.

© Springer Nature Switzerland AG 2019
K. Choroś et al. (Eds.): MISSI 2018, AISC 833, pp. 385–395, 2019.
https://doi.org/10.1007/978-3-319-98678-4_39

Recent works [25,27,30] analyzed the information of a document in both languages (source and target languages) to extract more details and identify the most relevant sentences. Following this idea, we propose a framework to realize French-to-English CLTS of transcript documents. In a nutshell, our approach first automatically segments a French transcript document and translates it into English using Google Translate. Then, we estimate the sentence relevance based on the information they contain in French and English. Similar English sentences are compressed to generate a unique, short, and informative compression. Finally, the cross-lingual summary is composed of the compression of the most relevant sentences without redundancy.

The rest of this paper is organized as follows: we make an overview of relevant work for CLTS methods and Automatic Speech Summarization in Sect. 2. Next, we detail our approach in Sect. 3. The experimental setup and results are discussed in Sects. 4 and 5, respectively. Finally, we provide our conclusion and some final comments in Sect. 6.

2   Related Work

If TS has reached a stage of maturity with well-established methods, Speech-to-Text Summarization and Cross-Lingual Summarization have their own challenges.

Speech-to-text summarization has to face three main problems: documents are not segmented into sentences, they may contain disfluencies, specific to the oral language, or they are subject to misrecognized words when using Automatic Speech Recognition (ASR). Nevertheless, it can benefit from acoustic and prosodic cues, or information about the role of speakers to determine the importance or the structure of an utterance. McKeown et al. [13] showed how the summarization approaches used in TS can be adapted to this speech-to-text task. They focused on two types of spoken sources, broadcasts news and meetings, taking advantage of acoustic, prosodic, lexical, and structural features to detect speakers' turns and overcome the difficulties that are present in spoken language.

Mrozinski et al. applied an extractive summarization approach over broadcast news stories and conference lectures [14]. In a first step, they performed sentence segmentation of the transcripts using word-based and class-based statistical language models; then during the summarization phase they selected the highest scoring sentences based on a combination of word significance score, confidence score, and linguistic likelihood.

Rott and Cerva divided their summarization system in three steps: automatic speech recognizer, syntactic analyzer, and text summarizer [22]. Sentence Boundary Detection was performed during the syntactic analysis, where they identified phrases in the recognized text using the Syntactic Engineering Tool (SET) [8]. Text summarization was performed using a TF-IDF method which selects the most informative phrases.

With regard to Cross-Lingual Summarization (CLTS), it has to deal with errors introduced by MT. CLTS was originally addressed as two separate tasks, making the information analysis in only one of the two languages [10, 17], which produces an early or a late translation scheme. In the early translation approach, the first step is to translate the source documents into the target language, the second step is to summarize the translated documents using only information of the translated sentences. The late translation approach does the reverse; first it aims to summarize the documents in its source language and then it translates them to the target language.

Further studies have considered translation quality and the information in both languages in order to generate correct and informative cross-lingual summaries. A Support Vector Machine (SVM) regression method was developed by Wan et al. in order to predict the translation quality using parse features to produce English-to-Chinese CLTS [26]. These translation quality scores were used in addition to relevance scores to select the sentences for the summaries. In order to take into account both language sides for establishing the similarity between sentences, Wan introduced two graph-based summarization methods, SimFusion and CoRank [25]. The SimFusion method was inspired by the PageRank algorithm [18] in order to calculate the relevance of sentences, where the weight arcs are defined by the linear combination of the cosine similarity of sentences in English and Chinese. The CoRank method simultaneously ranks English and Chinese sentences by incorporating mutual influences between them. The relevance of a sentence is defined by its similarity with other sentences in each language separately and between languages.

If extractive approaches are mainstream for CLTS, a few studies have been done to propose abstractive summarization. First, Yao et al. took advantage of statistical MT systems that are usually phrase-based to define relevance scores at the phrase level [28]. These scores were used to select and compress sentences simultaneously. Zhang et al. used Predicate-Argument Structures (PAS) to identify a set of concepts and facts in the source side, and their counterparts in the target side with the help of an alignment method [30]. The relevance of concepts and facts are estimated using the CoRank algorithm [25], while summaries were produced by fusing the most relevant source-side PAS elements considering their translation quality. Finally, a French-to-English cross-lingual abstractive summarization approach was proposed in [19]. This CLTS system combined multi-sentence and sentence compression methods in order to produce informative cross-lingual summaries.

3   Our Proposition

French-to-English CLTS aims to generate an English short summary that describes the main information from a French transcript document. Following the CLTS approach proposed by Linhares Pontes et al. [19], we analyze documents in the source and the target languages to select the most relevant sentences. Then, we create clusters of similar sentences to independently analyze the subjects of a document. In order to compress and to improve the informativeness of

E. Linhares Pontes et al.

the summarization, we compress the clusters composed of two or more similar sentences. Finally, the summary is composed of the compression of the most relevant sentences without redundancy.

The following subsections present the architecture of our system.

3.1 Ranking Sentences

The CoRank method jointly ranks sentences in both languages by assimilating mutual influences between them. We first translate the French sentences into English using the Google Translate system, then we use the CoRank method to estimate the informativeness of sentences (more details in [25]).

3.2 Multi-Sentence Compression

Following the idea proposed in [19], we consider the similarity in both languages to create clusters of similar sentences. Then, we use the Stanford CoreNLP tool [12] with jMWE [9] to detect Multi-Word Expressions in the English side, while the corresponding expressions were deduced on the French side with the help of the Giza++ alignment tool [16]. Among several state-of-the-art Multi-Sentence Compression (MSC) methods [1,4,15,20], we use Linhares Pontes *et al.*'s approach [19] to generate a compressed sentence from each cluster of similar sentences. This system builds compressions controlled by the presence of keywords, to increase informativeness, and 3-grams, to ensure grammaticality. Finally, the sentences of each cluster are replaced by their compression in the document.

4 Experimental Setup

We use the early translation, the late translation, and the CoRank methods [25] to evaluate the performance of our system. We adapted the SimFusion method to create the early and late translation methods. The early translation method only considers the similarity in the target language and the late translation method only considers it in the source language.

In order to avoid the generation of short compressions, we only compress sentences with at least 10 words. All systems produce summaries containing a maximum of 250 words and without redundant sentences. We consider two sentences to be similar/redundant if they have a cosine similarity value bigger than 0.5.

4.1 Dataset

The MultiLing Pilot 2011 dataset [6] is a collection of WikiNews English texts that were translated into Arabic, Czech, English, French, Greek, Hebrew and Hindi languages by native speakers. Each language version of this dataset has

10 topics where each topic is composed of 10 source texts and 3 reference summaries. Each summary has a maximum of 250 words. In this work, we use the French version of the MultiLing Pilot 2011 dataset as source language and the corresponding English version as the target language.

To our knowledge, no work has been done regarding cross-lingual summarization of transcripts generated by an ASR system. We believe this to be a good challenge given the difficulties brought by ASR transcripts. For this reason we wanted to explore this less controlled scenario and analyze the repercussions over the cross-lingual text summarization of two main problems of ASR transcripts: transcription errors and the lack of sentences.

4.2   Transcription Error Simulation

Automatic transcription performance is normally compared against one or more references using Word Error Rate (WER). This measure considers three different errors and calculates a general value indicating the quality of the transcript; the lower the value (closer to zero), the higher its quality. The three errors considered by WER (Eq. 3) are deletions, insertions, and substitutions:

$$\text{WER} = \frac{D + I + S}{N} \tag{1}$$

where D corresponds to the number of deletions, I to the number of insertions, S to the number of substitutions and N to the number of words in the reference. An ASR transcript carries all three errors at different ratios; for this controlled scenario we simulated in an isolated way each error to observe how each of them affects the performance of cross-lingual speech-to-text summarization.

We approximated WER by simulating the errors produced by ASR systems in a straightforward approach. The deletion error dataset (ASR_D) was created by choosing m words of each document randomly and by deleting them. Concerning the substitution error dataset (ASR_S), for each document we first selected a set $Y = \{y_1, ..., y_m\}$ of words randomly, then for each word w_i of the document a randomly generated decision value $v_i \in [0, 1]$ was calculated; if v_i happened to be greater than a given threshold $t = 0.5$, then w_i was replaced by y_j, this cycle was repeated until all words y_j in Y where picked. The insertion error dataset (ASR_I) followed the same procedure as ASR_D but instead of replacing w_i by y_j, y_j was placed after w_i. For all three error datasets m was calculated as:

$$m = \text{WER} \times N \tag{2}$$

where N corresponds to the length (number of words) in each original document and WER was fixed to 0.15.

4.3   Automatic Segmentation

Common ASR transcripts have no punctuation, which further complicates NLP tasks like automatic summarization. We simulated the lack of punctuation by deleting all punctuation signs inside the MultiLing Pilot 2011 French dataset (ASR_NO) and the datasets with induced transcription errors (ASR_D, ASR_S, ASR_I); then we automatically restored them. This task is known as Sentence Boundary Detection (SBD).

To restore the punctuation within the corpus we followed the best model reported by González-Gallardo and Torres-Moreno in [7]. This approach targets the segmentation problem as a classification one. It uses a Convolutional Neural Network (CNN) with Subword-level Information Vectors [2] to predict if the centered word (w_i) within a window $W = \{w_{i-(m-1)/2}, ..., w_{i-1}, w_i, w_{i+1}, ..., w_{i+(m-1)/2}\}$ corresponds to a sentence border or not.

The hidden architecture of the CNN consists of two convolutional layers with a valid padding and a stride value of one, followed by a max pooling layer and three fully connected layers with a dropout layer attached at the end. The outputs of all convolutional, max pooling and fully connected layers have a RELU activation function. The CNN was trained with a 380M words of the French Wikipedia.

Table 1 presents the automatic evaluation performed over the unpunctuated datasets. As seen from the "no boundary" class (NO_BOUND), the method has a really good performance (over 0.95 for all metrics), no matter of the type of transcription errors. Given the unbalanced nature of the data this is an expected behavior. Nevertheless for the "boundary" class (BOUND) the performance drops when trying to segment the noisy transcripts. The worst scenario corresponds to the dataset with substitution errors (ASR_S), where precision and recall present relative drops of 34% and 17% against ASR_NO.

Table 1. Results of Sentence Boundary Detection over the ASR datasets.

Dataset	Class	Precision	Recall	F1
ASR_NO	NO_BOUND	0.971	0.986	**0.978**
	BOUND	0.840	0.721	**0.776**
ASR_D	NO_BOUND	0.966	0.963	0.965
	BOUND	0.654	0.673	0.663
ASR_I	NO_BOUND	0.960	0.956	0.958
	BOUND	0.592	0.616	0.604
ASR_S	NO_BOUND	0.958	0.950	0.954
	BOUND	0.554	0.600	0.576

4.4 Automatic Evaluation

Automatic evaluation relies on comparing the information contained in the candidate summary against one or more reference summaries or the source document. The ROUGE [11] measure developed by Lin *et al.* compares the differences between the distribution of words of the candidate summary and a set of reference summaries. The comparison is made splitting into n-grams both the candidate and the reference to calculate their intersection. Standard n-grams values for ROUGE are unigrams and bigrams, both expressed as:

$$ROUGE - n = \frac{\sum_{n-grams} \in \{Sum_{can} \cap Sum_{ref}\}}{\sum_{n-grams} \in Sum_{ref}}, \tag{3}$$

where n is the n-gram order, Sum_{can} the candidate summary and Sum_{ref} the reference summary.

A third common ROUGE-n variation is ROUGE-SUγ. This ROUGE-2 variation takes into account skip units (SU) $\leq \gamma$. We considered the ROUGE-1, -2 and -SU4 measures in order to evaluate and compare our system.

5 Experimental Evaluation

Table 2 shows the ROUGE scores for each version of the MultiLing Pilot dataset. Our method outperformed the other methods for the original, ASR_NO and ASR_S dataset versions, while the CoRank method obtained the best results for other versions. As we expected, the ASR errors, introduced at the word or segmentation levels, reduced the performance of systems.

We analyzed the original dataset results as a reference to compare the performance of the systems with other dataset versions. The joint analysis of both languages generated better results. The analysis of the similarity in both languages and cross-language increased the results considerably. Finally, the addition of the compression of similar sentences to these multiple analysis of similarities achieved the best results.

The automatic segmentation process may split long sentences in two or more short sentences that can be more or less relevant to the document. In addition, these sentences are more likely to contain grammatical errors. However, the segmentation errors had little impact on the performance of systems (ASR_NO in Tables 1 and 2).

The low performance of automatic segmentation process to identify sentence bound combined with automatic speech recognition errors reduced the performance of all systems (ASR_D, ASR_I and ASR_S in Tables 1 and 2). These errors modified the structure of sentences causing large translation errors and changing the meaning of some sentences. The CoRank method achieved the best results for the deletion and insertion dataset versions; however, poor results were obtained for the substitution errors. Our approach was more stable for all kinds of ASR errors by generating cross-lingual summaries with similar ROUGE scores.

Table 2. ROUGE f-measure results for French-to-English MultiPilot 2011 dataset.

Dataset	Algorithms	ROUGE-1	ROUGE-2	ROUGE-SU4
Original	Early translation	0.4165	0.1021	0.1607
	Late translation	0.4142	0.1023	0.1589
	SimFusion	0.4173	0.1035	0.1606
	CoRank	0.4628	0.1324	0.1932
	Our proposition	**0.4724**	**0.1369**	**0.1962**
ASR_NO	Early translation	0.4115	0.0967	0.1567
	Late translation	0.4115	0.0992	0.1568
	SimFusion	0.4140	0.0981	0.1589
	CoRank	0.4608	0.1267	0.1891
	Our proposition	**0.4705**	**0.1336**	**0.1922**
ASR_D	Early translation	0.4160	0.0950	0.1566
	Late translation	0.4076	0.0896	0.1504
	SimFusion	0.4142	0.0914	0.1547
	CoRank	**0.4666**	**0.1192**	**0.1860**
	Our proposition	0.4474	0.1053	0.1711
ASR_I	Early translation	0.4027	0.0827	0.1481
	Late translation	0.3933	0.0828	0.1420
	SimFusion	0.3987	0.0814	0.1452
	CoRank	**0.4504**	**0.1089**	**0.1770**
	Our proposition	0.4481	0.1067	0.1744
ASR_S	Early translation	0.4080	0.0847	0.1495
	Late translation	0.4038	0.0834	0.1463
	SimFusion	0.4077	0.0848	0.1505
	CoRank	0.4206	0.0921	0.1584
	Our proposition	**0.4445**	**0.1072**	**0.1718**

All in all, the joint analysis of information in both languages and MSC generate more informative cross-lingual summaries. Our segmentation process kept a good quality of all summaries, i.e. all systems generated summaries with similar ROUGE scores to the original dataset. The addition of ASR errors reduced the quality of summaries of all systems because of translation and meaning errors. Our approach generated cross-lingual summaries with similar ROUGE scores for the dataset with ASR errors while the CoRank method achieved unstable results depending on the kind of errors.

6 Conclusion

We have proposed a compressive method to generate cross-lingual transcript summaries. Our framework analyzes a transcript document in French and English languages to identify the relevant information and compress similar sentences to increase the informativeness of summaries. The simulated ASR errors showed to have an impact on the performance of all systems; nevertheless, our approach achieved the best results for the original, ASR_NO and ASR_S dataset versions. Contrary to the CoRank method, our approach attained stable results for all kinds of ASR errors.

In future work, we plan to realize a manual evaluation to measure the grammaticality and the informativeness of the cross-lingual summaries. We will also use a language model or neural networks to correct grammatical errors [29] generated by ASR in order to improve the quality of transcripts and, consequently, the quality of summaries.

Acknowledgement. This work was granted by the European Project CHISTERA-AMIS ANR-15-CHR2-0001.

References

1. Banerjee, S., Mitra, P., Sugiyama, K.: Multi-document abstractive summarization using ILP based multi-sentence compression. In: 24th International Conference on Artificial Intelligence (IJCAI), IJCAI 2015, pp. 1208–1214 (2015)
2. Bojanowski, P., Grave, E., Joulin, A., Mikolov, T.: Enriching word vectors with subword information. arXiv preprint arXiv:1607.04606 (2016)
3. Christensen, H., Gotoh, Y., Kolluru, B., Renals, S.: Are extractive text summarisation techniques portable to broadcast news? In: IEEE Workshop on Automatic Speech Recognition and Understanding (ASRU), pp. 489–494 (2003)
4. Filippova, K.: Multi-sentence compression: finding shortest paths in word graphs. In: COLING, pp. 322–330 (2010)
5. Furui, S., Kikuchi, T., Shinnaka, Y., Hori, C.: Speech-to-text and speech-to-speech summarization of spontaneous speech. IEEE Trans. Speech Audio Process. **12**(4), 401–408 (2004)
6. Giannakopoulos, G., El-Haj, M., Favre, B., Litvak, M., Steinberger, J., Varma, V.: TAC2011 multiling pilot overview. In: 4th Text Analysis Conference TAC (2011)
7. González-Gallardo, C.E., Torres-Moreno, J.M.: Sentence boundary detection for French with subword-level information vectors and convolutional neural networks. ArXiv, February 2018
8. Kovář, V., Horák, A., Jakubíček, M.: Syntactic analysis using finite patterns: a new parsing system for Czech. In: Language and Technology Conference, pp. 161–171. Springer (2009)
9. Kulkarni, N., Finlayson, M.A.: jMWE: a Java toolkit for detecting multi-word expressions. In: Workshop on Multiword Expressions: From Parsing and Generation to the Real World (MWE), pp. 122–124 (2011)
10. Leuski, A., Lin, C.Y., Zhou, L., Germann, U., Och, F.J., Hovy, E.: Cross-lingual C*ST*RD: English access to Hindi information, vol. 2, no. 3, pp. 245–269, September 2003

11. Lin, C.Y.: ROUGE: a package for automatic evaluation of summaries. In: Workshop Text Summarization Branches Out (ACL 2004), pp. 74–81 (2004)
12. Manning, C., Surdeanu, M., Bauer, J., Finkel, J., Bethard, S., McClosky, D.: The Stanford CoreNLP natural language processing toolkit. In: 52nd Annual Meeting of the Association for Computational Linguistics (ACL): System Demonstrations, pp. 55–60 (2014)
13 McKeown, K., Hirschberg, J., Galley, M., Maskey, S.: From text to speech summarization. In: IEEE International Conference on Acoustics, Speech, and Signal Processing (ICASSP), vol. 5, p. v/997 (2005)
14. Mrozinski, J., Whittaker, E.W., Chatain, P., Furui, S.: Automatic sentence segmentation of speech for automatic summarization. In: IEEE International Conference on Acoustics Speech and Signal Processing Proceedings (ICASSP) (2006)
15. Niu, J., Chen, H., Zhao, Q., Su, L., Atiquzzaman, M.: Multi-document abstractive summarization using chunk-graph and recurrent neural network. In: IEEE International Conference on Communications, ICC, pp. 1–6 (2017)
16. Och, F.J., Ney, H.: A systematic comparison of various statistical alignment models. Comput. Linguist. **29**(1), 19–51 (2003)
17. Orasan, C., Chiorean, O.A.: Evaluation of a cross-lingual Romanian-English multi-document summariser. In: 6th International Conference on Language Resources and Evaluation (LREC) (2008)
18. Page, L., Brin, S., Motwani, R., Winograd, T.: The pagerank citation ranking: Bringing order to the web. In: 7th International World Wide Web Conference, Brisbane, Australia, pp. 161–172 (1998)
19. Pontes, E.L., Huet, S., Torres-Moreno, J.M., Linhares, A.C.: Cross-language text summarization using sentence and multi-sentence compression. In: Natural Language Processing and Information Systems. pp. 467–479. Springer International Publishing, Cham (2018)
20. Pontes, E.L., Huet, S., Gouveia da Silva, T., Linhares, A.C., Torres-Moreno, J.M.: Multi-sentence compression with word vertex-labeled graphs and integer linear programming. In: TextGraphs-12: The Workshop on Graph-Based Methods for Natural Language Processing. ACL (2018)
21. Pontes, E.L., Torres-Moreno, J.M., Linhares, A.C.: LIA-RAG: a system based on graphs and divergence of probabilities applied to speech-to-text summarization. In: Addendum, M.P. (ed.) Multiling CCCS (2015)
22. Rott, M., Červa, P.: Speech-to-text summarization using automatic phrase extraction from recognized text. In: International Conference on Text, Speech, and Dialogue (TSD), pp. 101–108. Springer (2016)
23. Taskiran, C.M., Pizlo, Z., Amir, A., Ponceleon, D., Delp, E.J.: Automated video program summarization using speech transcripts. IEEE Trans. Multimedia **8**(4), 775–791 (2006)
24. Torres-Moreno, J.M.: Automatic Text Summarization. Wiley, London (2014)
25. Wan, X.: Using bilingual information for cross-language document summarization. In: ACL, pp. 1546–1555 (2011)
26. Wan, X., Li, H., Xiao, J.: Cross-language document summarization based on machine translation quality prediction. In: ACL, pp. 917–926 (2010)
27. Wan, X., Luo, F., Sun, X., Huang, S., Yao, J.: Cross-language document summarization via extraction and ranking of multiple summaries. In: Knowledge and Information Systems (2018)
28. Yao, J., Wan, X., Xiao, J.: Phrase-based compressive cross-language summarization. In: EMNLP, pp. 118–127 (2015)

29. Yuan, Z., Briscoe, T.: Grammatical error correction using neural machine translation. In: Proceedings of the 2016 Conference of the North American Chapter of the Association for Computational Linguistics: Human Language Technologies, pp. 380–386. Association for Computational Linguistics (2016)
30. Zhang, J., Zhou, Y., Zong, C.: Abstractive cross-language summarization via translation model enhanced predicate argument structure fusing. IEEE/ACM Trans. Audio Speech Lang. Process. **24**(10), 1842–1853 (2016)

A Proposed Methodology for Subjective Evaluation of Video and Text Summarization

Begona Garcia-Zapirain[1], Cristian Castillo[1], Aritz Badiola[1],
Sofia Zahia[1], Amaia Mendez[1(✉)], David Langlois[2], Denis Jouvet[2],
Juan-Manuel Torres[3], Mikołaj Leszczuk[4], and Kamel Smaili[2]

[1] eVida Research Group, University of Deusto, Bilbao, Spain
`amala.mendez@deusto.es`
[2] Loria, University of Lorraine, Lorraine, France
[3] LIA, Universite d'Avignon et des Pays de Vaucluse, Avignon, France
[4] AGH University of Science and Technology Kraków, Kraków, Poland

Abstract. To evaluate a system that automatically summarizes video files (image and audio), it should be taken into account how the system works and which are the part of the process that should be evaluated, as two main topics to be evaluated can be differentiated: the video summary and the text summary. So, in the present article it is presented a complete way in order to evaluate this type of systems efficiently. With this objective, the authors have performed two types of evaluation: objective and subjective (the main focus of this paper). The objective evaluation is mainly done automatically, using established and proven metrics or frameworks, but it may need in some way the participation of humans, while the subjective evaluation is based directly on the opinion of people, who evaluate the system by answering a set of questions, which are then processed in order to obtain the targeted conclusions. The obtained general results from both evaluation systems will provide valuable information about the completeness and coherence, as well as the correctness of the generated summarizations from different points of view, as the lexical, semantical, etc. perspective. Apart from providing information about the state of the art, it will be presented an experimental proposal too, including the parameters of the experiment and the evaluation methods to be applied.

Keywords: Video summarization · Objective and subjective evaluation
Text summary

1 Introduction and Literature Review

1.1 Introduction

AMIS is an original project concerning the second call: Human Language Understanding; Grounding Language Learning. This project acts on different data: video, audio and text. We consider the understanding process, to be the aptitude to capture the most important ideas contained in a media expressed in a foreign language, which would be compared to an equivalent document in the mother tongue of a user. In other words, the understanding will be approached by the global meaning of the content of a

© Springer Nature Switzerland AG 2019
K. Choroś et al. (Eds.): MISSI 2018, AISC 833, pp. 396–404, 2019.
https://doi.org/10.1007/978-3-319-98678-4_40

support and not by the meaning of each fragment of a video, audio or text. The idea of AMIS is to facilitate the comprehension of the huge amount of information available in TV shows, internet etc. One of the possibilities to reach this objective is to summarize the amount of information and then to translate it into the end-user language. Another objective of this project is to access to the underlying emotion or opinion contained in two medias. To do this, we propose to compare the opinion of two media supports, concerning the same topic, expressed in two different languages. The idea is to study the divergence and the convergence of opinions of two documents whatever their supports. Several skills are necessary to achieve this objective: video summarization, automatic speech recognition, machine translation, language modelling, sentiment-analysis, etc. Each of them, in our consortium, is treated by machine learning techniques; nevertheless human language processing is necessary for identifying the relevant opinions and for evaluating the quality of video, audio and text summarization by the end-user.

After analysing the existing different ways of evaluating an automatic summarizer system, and taking into account the objective of the actual evaluation, it is considered the best evaluation system a combination between subjective and objective evaluation methods. It is true that there are some automatic evaluation metrics/frameworks that have demonstrated good results, like ROUGE, QARLA… But these metrics/frameworks need human-generated summaries in order to compare, so, looking for the completeness of the evaluation, we consider necessary the inclusion of both perspectives of the analysis.

Besides, it should be considered that it is complicated to analyse automatic summaries with automatic systems using the lexical and phrase-based comparison with human generated summaries, obtaining really good proved results. Indeed, the humans, obviously, are not like machines; they have the ability to really understand, take the essence of something, and express it in a different way, in different words. So, it is possible to obtain a summary generated by a machine, and a summary generated by a human, and being in essence the same, but which can be expressed in a different way. In this case an automatic system probably will not be able to see the similarity in the words, the real meaning.

So, taking into account all the expressed ideas, the subjective evaluation is considered the best way to evaluate an automatic summarizer, and, to complete the evaluation from an objective perspective, it is interesting to apply some methods/metrics to analyze the obtained summaries. Anyway, both perspectives and their methods will be presented in the present document.

1.2 Literature Review

The state of the art of the evaluation of video summaries will be presented in the following tables from 2 perspectives: image and text. So, the most relevant and interesting papers about these topics are presented, including a resume about the most important part in relation with what we are analyzing, and some information about the parameters that are proposed (Table 1).

398   B. Garcia-Zapirain et al.

Table 1. Video summary evaluation.

Paper	Abstract	Experiment
Video Abstraction: A Systematic Review and Classification [1]	**Subjective:** user studies most useful and realistic	None
*VSUMM: A mechanism designed to produce static video summaries and a novel evaluation method [2]	**Objective with user participation:** In this evaluation method, called Comparison of User Summaries (*CUS*), the video summary is built manually by a number of users from the sampled frames. The user summaries are taken as a reference to be compared with the summaries obtained by different methods. In this way, the user summaries are the reference summaries, i.e., the ground-truth. Such comparisons are based on specific metrics, which are introduced in the following paragraphs	None
*A New Method for Static Video Summarization Using Local Descriptors and Video Temporal Segmentation [3]	**Objective:** *CUS* makes a comparison between the user summary and the automatic summary. The idea is to take a keyframe from the user summary and a keyframe from the automatic video summary	None
Automatic Evaluation Method For Rushes Summary Content [4]	**Subjective:** Each submitted summary was judged by three different human judges (assessors). An assessor was given the summary and a corresponding list of up 12 topics from the ground truth	- Users: 3 - 12 topics selected from the full video by the specialist

(*continued*)

Table 1. (*continued*)

Paper	Abstract	Experiment
Video Summarisation: A Conceptual Framework and Survey of the state of the art [5]	**Objective** methods do not incorporate user judgment into the evaluation criteria but evaluate the performance of a given technique based on, for example, the extent to which specific objects and events are accurately identified in the video stream and included in the video summary	- Users: 17 The users give feedback about the content, in terms of enjoyability and informativeness by means of informal discussions.
*A Pertinent Evaluation of Automatic Video Summary [6]	**Objective** (similar to *CUS*): They propose an effective method for identifying the true matches between AT (Automatic Summary) and GT (Ground Truth User Summary) for the performance evaluation of the summarised videos. It includes the initial establishment of matched frames via two-way search followed by a consistency check where weak and false matches are eliminated	None
*Multi-video Summarization Based On Video-MMR [7]	- **Objective** (similar to *CUS*): Is meaningful to compare Video-MMR (Maximal Marginal Relevance) to human choice? In a video set, 6 videos with most obvious features were chosen. Inside 6 videos, 3 videos own the largest distances with the others in this video set, while the other 3 videos have the smallest distances	- Users: 12 - Full videos: 3 Each user selects 10 keyframes from each video

(*continued*)

Table 1. (*continued*)

Paper	Abstract	Experiment
*VERT: Automatic Evaluation of Video Summaries [8]	By borrowing ideas from ROUGE and BLEU, the authors of this paper extend these measures to the domain of video summarization. We focus our approach on the selection of relevant keyframes, as a video skim can be easily constructed by concatenating video clips extracted around the selected keyframes... The authors talk about VERT-Precision and VERT-Recall, and how they are carried out	None
VSCAN: An Enhanced Video Summarization using Density-based Spatial Clustering [9]	In this paper, a **modified version** of an evaluation method Comparison of User Summaries (*CUS*) is used to evaluate the quality of video summaries. The modifications proposed to CUS method aims at providing a more perceptual assessment of the quality of the automatic video summaries	CUS (but more complete i our opinion)

2 Experiment Design - Participants and Protocol for the Whole Integrated System

Our proposed experiment includes two different lines: subjective evaluation (questionnaires) and objective evaluation.

In order to do the evaluation as complete as possible, and obtain the best results, the problem should be approached from different perspectives, so, it has been combined information from different sources [13, 14] and developed a new way of evaluation in order to obtain better results. On each perspective there are some questions with a specific format for the answer, which can be multiple choice or ranking from 0 to 4 formats. In the multiple choice format, some specific answers will be provided in order to be selected one or some of them by the user. In the case of the scoring from 0 to 4,

Table 2. Questionnaire for the video and text summary evaluation

Criteria	Excellent	Very good	Good	Fair	Not done	Comments/Suggestions
Summary video						
Is the summary under-standable?	4	3	2	1	0	
The video doesn't con-tain any part out of context, or it does not affect to the main expressed ideas	4	3	2	1	0	
Different questions about the original video, in order to ensure that the summary contains the key ideas and the user is able to get these ideas from the summary	See below one example.					
Summary text						
Is the summary under-standable?	4	3	2	1	0	
Is it lexi-cally/grammatically cor-rect?	4	3	2	1	0	
Is it semantically cor-rect?	4	3	2	1	0	
Does it contain redun-dant information?	4	3	2	1	0	
Are the references (it, she, he...) clear? (*looking for lack of information*)	4	3	2	1	0	
Different questions about the original video, in order to ensure that the summary contains the key ideas and the user is able to get these ideas from the summary						
Summary video and summary text						
Do you think that both, the summary video and the summary text, express the same idea? (*cohesion be-tween the both formats is measured*)	4	3	2	1	0	

each number has a meaning: 0 = not done, 1 = fair, 2 = good, 3 = very good and 4 = excellent. The subjective evaluation can be done in 2 ways:

(a) Assessment of Summarized Video (image sequence) and Text

Participants: 25 per language (Arabic and French) (balanced number of men and women)
Inclusion criteria: men and women over 18 years old, with at least high school level.
Exclusion criteria: reading and writing impairment. Understanding problems.
Number of summarized video and text per user: Every user will review **a** set of 3 videos (out of the 25 prepared) with mixed topics. The test will be made of the summarized video and text version (in English) (Table 2).

(b) Assessment of the Coherence between Original and Summarized Video and Text

Participants: Four in Arabic and four in French (two men and two women for each language)
Inclusion criteria: men and women over 18 years old, with at least high school level. The participants have to be fluent in both languages.
Exclusion criteria: reading and writing impairment. Understanding problems.
Number of summarized video and text per user: one with mixed topics. One video will be selected per language for the test with original (Arabic or French) and summarized video and text version (English) (Table 3).

Table 3. Questionnaire for original and summarized video and text evaluation

Criteria	Excellent	Very good	Good	Fair	Not done	Comments/Suggestions
Original and summarized video and text						
Do you think that the provided summarizations, in video and in text, really express the main ideas of the original video?	4	3	2	1	0	
Can they be considered good summaries? - **Video** - **Text**	4	3	2	1	0	
Are they long enough to contain the main ideas? - **Video** - **Text**	4	3	2	1	0	
Do you consider it too long?	4	3	2	1	0	

Below, we propose some examples of evaluation with original videos. Specific questions are asked for each video:

The-rise-of-chemsex-on-Londons-gay-scene---**BBC-News**	**Criteria**
Which one is the main topic of the summary?	1) Homosexual parties 2) Homosexuality and STDs 3) Homosexuality, drugs and STDs
The summary says that there is not any relation between drugs and STD:	1) True 2) False
The diagnosis of HIV with drugs is related?	1) True 2) False

The assessment data analysis will consist on statistical analysis of questionnaires and the application of some machine learning techniques if possible for clusterization and comparison purposes between genders, language.

3 Results

The results regarding the questionnaires would be based in a point system, and the criteria of quality, or the different point ranges of quality level will be established depending on the total number of questions.

During 2018, all the proposed test will be carried out.

Acknowledgements. Research work funded by the Spanish Ministry of Economy, Competitiveness and Industry (Spain) conferred under the Chist-Era AMIS project.

References

1. Truong, B.T., Venkatesh, S.: Video Abstraction: A Systematic Review and Classification (2007)
2. Fontes de Avila, S.E., Brandão Lopes, A., da Luz Jr, A., de Alburquerque Araújo, A.: VSUMM: a mechanism designed to produce static video summaries and a novel evaluation method (2010)
3. Cayllahua Cahuina, E.J., Camara Chavez, G.: A New Method for Static Video Summarization Using Local Descriptors and Video Temporal Segmentation (2013)
4. Dumont, E., Bernard, M.: Automatic Evaluation Method for Rushes Summary Content (2009)
5. Money, A.G., Agius, H.: Video Summarisation: A Conceptual Framework and Survey of the State of the Art (2008)

6. Kannappan, S., Liu, Y., Tiddeman, B.: A Pertinent Evaluation of Automatic Video Summary (2016)
7. Li, Y., Merialdo, B.: Multi-video Summarization Based on Video-MMR (2010)
8. Li, Y., Merialdo, B.: VERT: Automatic Evaluation of Video Summaries (2010)
9. Mohamed, K.M., Ismail, M.A., Ghanem, N.M.: VSCAN: An Enhanced Video Summarization using Density-based Spatial Clustering (2014)
10. Molina, A., Torres-Moreno, J.-M.: The Turing Test for Automatic Text Summarization Evaluation (2016)
11. Molina Villegas, A., Torres-Moreno, J.-M., Sanjuan, E.: A Turing Test to Evaluate a Complex Summarization Task (2013)
12. Lin, C.-Y., Hovy, E.: Manual and Automatic Evaluation of Summaries (2002)
13. Hassel, M.: Evaluation of Automatic Text Summarization: a practical implementation (2004)
14. Saziyabegum, S., Sajja, P.S.: Review on Text Summarization Evaluation Method (2017)

Collection, Analysis and Summarization of Video Content

Arian Koźbiał[✉] and Mikołaj Leszczuk

AGH University of Science and Technology, 30059 Kraków, Poland
ariankozbial@gmail.com, leszczuk@agh.edu.pl
http://www.kt.agh.edu.pl/en

Abstract. Information overload is a term used to describe the difficulty of understanding when one has too much information. Information overload is one of the most common barriers in the access to e.g. video newscasts and reports. So, how a user can access and understand the overloaded information? We define the process of understanding as the assimilation of the main ideas carried by information. The best way to help and speed up understanding is summarizing the information. In this paper, we present the full scope of the summarization process, leading to a new approach for summarizing video sequences, with the special emphasis put on those with short original duration.

Keywords: Video summarization · Speech recognition
Text boundary segmentation

1 Introduction

Information overload (also known as infobesity or infoxication) is a term used to describe the difficulty of understanding when one has too much information. Information overload is one of the most common barriers in the access to newscasts and reports. Therefore, the following challenge appears: how a user can access and understand the overloaded information?

We define the process of understanding as the assimilation of the main ideas carried by information, being, e.g., recorded video. The best way to help and speed up understanding is summarizing the information. In our research we focus on summarization of newscasts and reports. Therefore, our research focuses on the most relevant information, summarizing it, and presenting to the user. This goal can be achieved by embedding its visual summarization technology into an audio-visual newscasts and reports processing system.

In this paper, we present the full scope of the summarization process. Starting from collecting videos from the public availability methods and followed by analyzing collected videos in terms of duration. Finally, we present a new approach for summarizing video sequences, with the special emphasis put on those with short original duration.

© Springer Nature Switzerland AG 2019
K. Choroś et al. (Eds.): MISSI 2018, AISC 833, pp. 405–414, 2019.
https://doi.org/10.1007/978-3-319-98678-4_41

In the related research most of video summarization processes commonly starts with Shot Boundary Detection (SBD) [9]. The video summarization process covers a number of aspects related to the analysis of the content of the video sequence, allowing for identification of meaningful characters/actors [2], shots and scenes [12] and other parts of a coherent theme [11], for presentation in a hierarchical form [1,4,16,17]. There may be hierarchical relationships between parts of one sequence of the video, as well as between different video sequences [13]. As far as generic video modality is concerned, the first notable milestone work in this area was presented by Zhang [18] (and its follow-up [19]). Authors propose an integrated and content-based solution for video parsing, retrieval and browsing. Nevertheless, specific video modalities require tailored solutions. An example of a video summarization algorithm, designed for specific video modality – video-bronchoscopy, has been provided in [10]. In this paper, we are, however, more interested in another video modality – newscasts and reports. Here, some important works include those by Maybury et al. [14], and a more recent publication by Gao et. al. [5]. Finally, a video summarization framework for newscasts and reports has been proposed by the authors of this paper as well [8].

The remainder of this paper is structured as follows. Section 2 describes the entire process of collection of video sequences. Section 3 analyzes duration of the collected videos. In Sect. 4 we present proposed approach of summarizing to image with. Sections 5 and 6 describe results and conclusions, respectively.

2 Collection of Video Sequences

One of the first challenges within the reported research was to collect various video sequences. YouTube was selected as a video provider because there are many official television channels with videos in decent quality what was one of the assumptions.

After that, appropriate televisions channels were selected. There was one simple requirements – a channel should provide video sequences being newscasts (news bulletins) or reports (documentaries) because it was one of the assumptions.

Selected YouTube channels were: Al-Arabiya, Alquds, British Broadcasting Corporation (BBC) News, Echorouk, Ennahar, Euronews, France 24, i24, Nessma, Russia Today (RT), Saudi, and SkyNews.

YouTube offers public Application Programming Interface (API) which was used to download video sequences. For each video sequence there were 5 files available: MPEG-4 Part 14 format (audio + video tracks multiplexed in 720p resolution, audio track only and video track only in 1080p resolution) as well as WebM format (audio track only and video track only in 1080p resolution).

The reason for downloading files with only audio or only video is that YouTube for video sequences with resolution higher than 720p uses DASH streams [3]. Simply it means that video and audio are not kept together.

Along with audiovisual files, for each video sequence, meta data was fetched directly to the database. Meta data contained many useful information such as name, language and duration.

3   Analysis of Video Sequence Duration

One of the most important indicator in terms of summarizing is the duration of the source material. It can affect the summarization in many ways. For example, duration of the source video sequence with constant summarization time can increase amount of shots.

Figure 1 presents the histogram which describes duration of the entire collected video sequences.

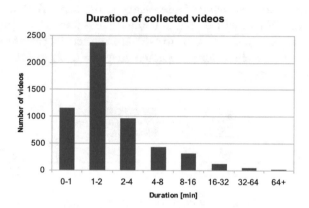

Fig. 1. A histogram of the length of the collected video sequences

The histogram shows that more than half of the collected videos sequences has duration shorter than 8 min. Almost 2500 video sequences from that has durations between 1 and 2 min. Such short duration can be source of problems which are described in the next section.

4   Summarization to Image

Short source duration is a serious challenge in terms of summarization, especially in the case described in the previous section. These short video sequences can be considered as some kind of summarization itself.

Consequently, an alternative way to summarize these short video sequences has been proposed – summarization to a static image.

The idea came from the provider service – YouTube and their video thumbnails. In these days video makers and marketers try many ways to attract viewers attention. In some cases, especially in YouTube, video makers create special thumbnails which contain the most important/interesting elements from the entire video sequence to attract attention so it can be considered as a form of summarization and it perfectly suites our case.

Besides that, currently images are one of the most important communication media. Young people create and exchange memes. Mainly for fun, but there are also research results showing that professionally generated memes were used during election campaigns and some candidates or parties earned some extra votes thanks to memes [7].

In our case, we assumed that image summarization should focus on the most important elements in the entire video sequence – **human faces** but these faces should be presented over some **background images**. It was also assumed that it would be desirable to have present the most important quotes in the form balloons. The last requirement makes it necessary to use some **Automatic Speech Recognition (ASR) module**. These three highlighted terms (**human faces, background images** and **ASR module**) are related to the three main modules of the proposed architecture (depicted in the Fig. 2) and are described in the subsequent subsections.

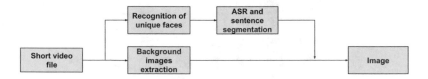

Fig. 2. Schema of the proposed solution

4.1   Recognition of Unique Human Faces

According to the previous section this module is the most important in terms of forwarding the final message to the end user.

Processing is divided into two parts – face detection and face recognition. Both modules use OpenCV Python library, which significantly speed up implementation (Fig. 3).

Face Detection. The face detection module uses Haar feature-based cascade classifiers. It is a solution based on a machine learning function. An algorithm is trained from positive and negative image examples, in this case images with faces and without faces.

Training depends of extracting features from images. These features are shown in the Fig. 4.

The features are just single values received by subtracting a sum of pixels under the white rectangle from a sum of pixels under the black rectangle. The very important term in case of the optimize recognition process is Cascade of Classifiers. Instead of testing few thousands of features they are grouped and tested one-by-one. Then if the first set of features fails the next are not tested.

OpenCV provides trained classifiers which can be used for detection of particular types of objects e.g. faces, smiles, eyes. The presented Face Detection module uses classifiers for frontal face detection.

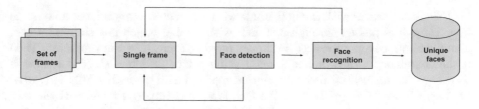

Fig. 3. Processing schema

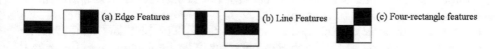

Fig. 4. Haar features

Face Recognition. The face recognition module uses Local Binary Patterns Histograms (LBPH) which is based on labeling the pixels of an image be thresholding the neighborhood of each pixel and considering the result as a binary number combined with histograms of oriented gradients descriptor.

LBPH uses four parameters:

- **radius** – mediates in building circular local binary pattern; it also represents radius around central pixel,
- **neighbors** – number of sample points used to build the circular local binary pattern,
- **grid x** – number of cells in the horizontal direction,
- **grid y** – number of cells in the vertical direction.

The LBPH implementation in OpenCV offers two main functions – a trainer and a predictor. Training the algorithm depends on passing a dataset which should contain facial images of people that we would like to recognize. Images should have the proper label, different for each person. The first step of the training depends on creating an image that describes the original image through highlighting facial characteristics points. In this step algorithm uses a concept of a sliding window which is based on the first two parameters – radius and neighbors. The second step uses the image generated in the previous step for extracting a histogram. Grid X and Grid Y parameters are used to divide the image into multiple regions and to extract histograms for them. The final histogram is obtained by concatenating histograms for each region. It represents the characteristics of the original image.

The predictor takes the image (in a form of an array) as an input argument and returns a label of the image with the closest match and a confidence indicator. The lower value of this indicator means the better match. Prediction takes place through same steps as in the training – to find the matching histogram of the input image has to be compared with histograms from the training set.

Processing starts with the first frame with a detected human face. As it is the first frame it is passed through a trainer with a label which is a simple integer. The next frames are processed by firstly passing them into the predictor. Then if the confidence indicator has a small value a face is detected. If the value is too large in means the face was not recognized and probably this is a new, unknown face. After predicting, the face is passed into the trainer. If the face was recognized it is passed with a label of the recognized face. If not, it is passed with a label indicating a new face.

After processing the entire video sequence there is a bunch of sets with grouped unique faces. It is a time to select the best face for each set. To achieve that, faces are passed to a module which is analyzing a blur level. Faces with the lowest blur level are selected as the best faces. The blur level is defined by calculating variation of the Laplacian operator. The Laplacian operator is used to measure the second derivative of an image; it means that the Laplacian operator can highlight regions where the image contains abrupt intensity changes. A variance value can be used to determine the level of edges in the image. If an image is blurred then the less edges there are. It forwards to simple conclusion – if the variance falls below a pre-defined threshold, then the image is considered blurry; otherwise, the image is not blurry [15].

4.2   Background Images

A background image is a very important component in the entire solution. It won't focus an attention but it will affect the overall perception. Let's assume that background image will be black. If there will be people with darker skin tone, the final effect of an overlay will produce a blended picture and its message will be unreadable. If colors in both cases are inverted there will be same problem.

The solution is to use (as the background) a frame from the original video, but then, there are some requirements with respect to this frame. In the ideal case the frame should not contain any faces. Nevertheless, for some kinds of source videos, it is not possible to select background frames completely without faces. In such a case, at least the selected background frame should not contain those faces which have been selected by the Unique Faces component, in order to avoid user distraction.

Before applying the background frame, it is treated with little amount of blur in order to imitate a professional "bokeh" look. In case of smoothing images, OpenCV provides a few functions. In the proposed solution Gaussian Blur was used with kernel size equal to 10% of the image dimensions.

4.3   Automatic Speech Recognition and Sentence Boundary Detection

Automatic Speech Recognition is a very hot topic these days. In solution we proposed, a third party ASR system, together with a sentence segmentation module, is used for place quotes in balloons.

The ASR system is built from two components – an acoustic model and a language model and it is based on KALDI speech recognition toolkit. Acoustic model is based on Deep Neural Network (DNN), which has input layer of 440 neurons, 6 hidden layers of 2048 neurons each. Output layer has around 4000 neurons.

The ASR system returns a set of the words appearing in the video sequence with corresponding time-codes. It's unfortunately not enough to put text in the balloons. That is why the Sentence Boundary Detection (SeBD) module is used – to combine words into sentences and detect boundaries between them. The module is built from Convolutional Neural Networks (CNN) and it is based on textual features. System was trained by subsets of Arabic, English and French Gigaword corpora [6]. SeBD module produces a Comma-Separated Values (CSV) file which contains a sentence and a time-code of appearance of the sentence in the video sequence. To fully integrate with previous modules, a time-code needs to be converted to a frame number. It is done using a Frames-Per-Second (FPS) parameter which can be easily obtained from a video sequence meta-data.

4.4   Summarization Solution

The final summarization solution is built from the previously described modules. The pipeline is shown on the Fig. 2.

The root element is the background image. It sets boundaries and size of the final image. Basically, the final image has same size as the original video sequence.

Unique faces are placed on the background image. For each selected unique face two properties are calculated – the coefficient of appearance in the entire video per person and the exact time-frame at which selected face is appearing in.

The first property is used for determining how many faces should appear on the final image. Maximum number of faces is set to three. Generally, it was empirically set that a person should appear in more than 15% of the original video's time. If there will be 3 people fulfilling the condition, their faces will be appearing in the final summarization image. If there will be more than 3 people then still 3 faces at most will be selected using ascending order of the percentage time of appearance.

The second property is used for placing the right text in balloons gathered from ASR and sentence segmentation. The text is selected be choosing the exact or the closest time-frames – same as in the selected unique faces.

5   Results

The algorithm was tested in a group of 25 randomly chosen short video sequences from category newscasts/interviews. Figure 5 presents example results.

Figure 5a and b present examples of good summarization. Proper faces was selected and dialogues was pinned to the right people.

(a) Angelina Jolie on divorce, film and Cambodia – BBC News

(b) Amber Rudd: "I have a surprise for the Brexiteers" – BBC News

(c) Donald Trump and Hillary Clinton on women's vote – BBC News

(d) Israel furious at Poland's ban on ritual slaughter – Euronews

Fig. 5. Sample summarization

Figure 5c and d are an examples of bad summarization. Both cases contains wrong selected faces. The last one doesn't even has dialogues.

Figure 5c summarize video which consist Donald Trump and Hillary Clinton speech about women's vote. In the ideal case we would expect to see these two faces in the summarization. Instead we can see two faces - one was selected correctly, second - not. In the video current USA president wore cup and his face is nearly visible. Besides that, in most shots he was standing behind. That is why algorithm didn't work correctly and instead of Donald Trump face we can see one of his supporter.

Figure 5d summarize video which contain a few interviews about decision of the Polish parliament to uphold a ban on ritual slaughter. In this kind of videos even for typical watcher it's hard to determine which speaker's opinion is the most important. In this case algorithm chose three faces which has the highest coefficient of appearance for the entire video. Unfortunately one of them shows face of the current President of the European Council Donald Tusk. His face was disposed on the protesters' banner. Furthermore lack of dialogues is the result of bilingual soundtrack – at this level it was impossible that ASR module could properly work.

6   Discussion and Conclusions

All modules related to ASR has the greatest room for improvement and extension. In many cases quotes in balloons are not related because a phrase is chosen by selecting a sentence having similar temporal localization as in the selected face so if there are two faces – one from the beginning and second from the end on the video sequence then balloons will contain completely unrelated phrases. A simple solution is to select a phrase which is time compatible with one of the face and select the second phrase which time is close to the first. There is also possibility, not to improve but to extend the solution by using social media data. In many cases people share links to the video sequences in social media by typing quotes from the video in the description field. These quotes can be considered as the most interesting phrases in the video and they can be placed inside the balloons. Besides that, it would be beneficial to have a different view hierarchy because currently all images look quite alike.

Acknowledgment. Research work funded by the National Science Center, Poland, conferred on the basis of the decision number DEC-2015/16/Z/ST7/00559.

References

1. Aghbari, Z., Kaneko, K., Makinouchi, A.: Content-trajectory approach for searching video databases. IEEE Trans. Multimedia **5**(4), 516–531 (2003). https://doi.org/10.1109/TMM.2003.819092
2. Baran, R., Rudzinski, F., Zeja, A.: Face recognition for movie character and actor discrimination based on similarity scores. In: 2016 International Conference on Computational Science and Computational Intelligence (CSCI), pp. 1333–1338 (2016). https://doi.org/10.1109/CSCI.2016.0249
3. Krishnappa, D.K., Bhat, D., Zink, M.: Dashing youtube: an analysis of using dash in YouTube video service. In: Proceedings of 38th Annual IEEE Conference on Local Computer Networks, vol. 1, pp. 407–415 (2013)
4. Fan, J., Elmagarmid, A.K., Zhu, X., Aref, W.G., Wu, L.: Classview: hierarchical video shot classification, indexing, and accessing. IEEE Trans. Multimedia **6**(1), 70–86 (2004). https://doi.org/10.1109/TMM.2003.819583
5. Gao, X., Tang, X.: Unsupervised video-shot segmentation and model-free anchor-person detection for news video story parsing. IEEE Trans. Circuits Syst. Video Technol. **12**(9), 765–776 (2002). https://doi.org/10.1109/TCSVT.2002.800510
6. González-Gallardo, C.E., Torres-Moreno, J.M.: Sentence boundary detection for French with subword-level information vectors and convolutional neural networks. arXiv preprint arXiv:1802.04559 (2018)
7. Konopka, M.N.: Rapid object detection using a boosted cascade of simple features. Przegl. Politologiczny **2**, 87–100 (2015). https://doi.org/10.14746/pp.2015.20.2.7
8. Leszczuk, M., Grega, M., Koźbiał, A., Gliwski, J., Wasieczko, K., Smaïli, K.: Video summarization framework for newscasts and reports - work in progress. In: Dziech, A., Czyżewski, A. (eds.) Multimedia Communications, Services and Security, pp. 86–97. Springer, Cham (2017)

9. Leszczuk, M., Papir, Z.: Protocols and systems for interactive distributed multimedia. In: Joint International Workshops on Interactive Distributed Multimedia Systems and Protocols for Multimedia Systems, IDMS/PROMS 2002 Coimbra, Portugal, November 26–29, 2002 Proceedings, chap. Accuracy vs. Speed Trade-Off in Detecting of Shots in Video Content for Abstracting Digital Video Libraries, pp. 176–189. Springer, Heidelberg. https://doi.org/10.1007/3-540-36166-9_16

10. Leszczuk, M.I., Duplaga, M.: Algorithm for video summarization of bronchoscopy procedures. Biomed. Eng. Online 10(1), 110 (2011). https://doi.org/10.1186/1475-925X-10-110

11. Li, S., Lee, M.C.: An efficient spatiotemporal attention model and its application to shot matching. IEEE Trans. Circuits Syst. Video Technol. 17(10), 1383–1387 (2007). https://doi.org/10.1109/TCSVT.2007.903798

12. Liu, T., Kender, J.R.: A hidden markov model approach to the structure of documentaries. In: 2000 Proceedings Workshop on Content-based Access of Image and Video Libraries, pp. 111–115 (2000). https://doi.org/10.1109/IVL.2000.853850

13. Lombardo, A., Morabito, G., Schembra, G.: Modeling intramedia and intermedia relationships in multimedia network analysis through multiple timescale statistics. IEEE Trans. Multimedia 6(1), 142–157 (2004). https://doi.org/10.1109/TMM.2003.819750

14. Maybury, M.T., Merlino, A.E.: Multimedia summaries of broadcast news. In: Proceedings of Intelligent Information Systems, IIS 1997, pp. 442–449 (1997). https://doi.org/10.1109/IIS.1997.645332

15. Pech-Pacheco, J.L., Cristobal, G., Chamorro-Martinez, J., Fernandez-Valdivia, J.: Diatom autofocusing in brightfield microscopy: a comparative study. In: Proceedings of 15th International Conference on Pattern Recognition. ICPR-2000, vol. 3, pp. 314–317 (2000). https://doi.org/10.1109/ICPR.2000.903548

16. Skarbek, W., Galiński, G., Wnukowicz, K.: Tree based multimedia indexing - a survey. In: Networked Audiovisual Media Technologies, Special VISNET Session at KKRRiT 2004, pp. 77–85 (2004)

17. Taskiran, C.M., Pizlo, Z., Amir, A., Ponceleon, D., Delp, E.J.: Automated video program summarization using speech transcripts. IEEE Trans. Multimedia 8(4), 775–791 (2006). https://doi.org/10.1109/TMM.2006.876282

18. Zhang, H.J., Low, C.Y., Smoliar, S.W., Wu, J.H.: Video parsing, retrieval and browsing: an integrated and content-based solution. In: Proceedings of the Third ACM International Conference on Multimedia, MULTIMEDIA 1995, pp. 15–24. ACM, New York (1995). http://doi.acm.org/10.1145/217279.215068

19. Zhang, H.J., Wu, J., Zhong, D., Smoliar, S.W.: An integrated system for content-based video retrieval and browsing. Pattern Recogn. 30(4), 643–658 (1997). https://doi.org/10.1016/S0031-3203(96)00109-4. http://www.sciencedirect.com/science/article/pii/S0031320396001094

An Integrated AMIS Prototype for Automated Summarization and Translation of Newscasts and Reports

Michał Grega[1]([✉]), Kamel Smaïli[2], Mikołaj Leszczuk[1],
Carlos-Emiliano González-Gallardo[3], Juan-Manuel Torres-Moreno[3,4],
Elvys Linhares Pontes[3], Dominique Fohr[2], Odile Mella[2],
Mohamed Menacer[2], and Denis Jouvet[2]

[1] AGH University, al. Mickiewicza 30, 30-059 Krakow, Poland
grega@kt.agh.edu.pl
[2] Loria University of Lorraine, Nancy, France
[3] LIA Université d'Avignon et des Pays de Vaucluse, Avignon, France
[4] Ecole Polytechnique de Montréal, Montreal, Canada

Abstract. In this paper we present the results of the integration works on the system designed for automated summarization and translation of newscast and reports. We show the proposed system architectures and list the available software modules. Thanks to well defined interfaces the software modules may be used as building blocks allowing easy experimentation with different summarization scenarios.

Keywords: Integration · Video summarization · Speech recognition
Machine translation · Text boundary segmentation
Text summarization

1 Introduction

We live in a world in which information as abundant. It is extremely easy to access all kinds of data and news. The main problem is not the availability of information, but our possibility of the digestion. This calls for effective summarization techniques, which are able to extract the most vital information and present it in an effective way. An additional problem is the language barrier. We commonly speak and understand two to three languages, while there are tens of commonly used languages in the world.

In our research within the AMIS project we solve the problem of effective summarization and translation of video newscasts and reports. We aim at reducing the length of the newscasts and reports and, at the same time, provide automatically translated subtitles to fit the users requirements. At the time of writing of this paper we offer summarization and automated translation form Arabic and French to English.

In this paper we describe the software components used in the process of summarization of newscasts and reports as well as the proposed summarization

© Springer Nature Switzerland AG 2019
K. Choroś et al. (Eds.): MISSI 2018, AISC 833, pp. 415–423, 2019.
https://doi.org/10.1007/978-3-319-98678-4_42

architectures. We have created a flexible system, based on well defined interfaces between software module which allows us to experiment with different order of the building blocks of the summarization and translation system.

The problem of video summarization has been addressed before e.g by Zhang in [18,19]. Video summarization is an important research area in medicine [11] while for newscast and reports it was investigated by Gao [6]. Our initial research on the topic has been reported in [10], while the paper describing our research on the user's requirements is available at [3].

The rest of this paper is structured as follows. Section 2 describes the Content Database. Section 3 covers the different software architectures. In Sect. 4 we present software modules used in our system and the paper is concluded in Sect. 5.

2   Content Database

In order to gather the videos we have first identified a list of controversial Twitter hash tags, such as #animalrights or #syria. Than we have extracted all twits from the Twitter service that were identified with this hashtag for a given period of time. Further on we have filtered the twits – only the twits containing valid YouTube links were passed to the next stage in which the videos were downloaded and stored into the video database.

In total we have downloaded 310 h of video from 19 TV stations in 3 languages (English, French and Arabic). The total number of 5423 videos that vary from 1 to 64 min in length. We have downloaded all available image formats.

The metadata is stored in a MySQL database and the video files are stored in the file system with paths available in the database. Each video is identified by its unique ID – videoID (that is, at the same time, the video identifier used by YouTube).

3   Scenarios and Integration

The modular construction of the AMIS system allowed us to experiment with different summarization scenarios, denoted Sc1–Sc4. The integrated scenario has a form of a Python 3.5 script that accepts as an input information on the videoID of the source video and the desired summarization length (as number of seconds or percentage of the original video). The processing is fully automated.

3.1   Scenario 1 - Sc1

Figure 1 depicts the most basic approach to newscast summarization. In this approach in the first step we perform a Video Based Summary, which is based on Shot Boundary Detection. This step results in creation o a summarization recipe. In the subsequent step we employ video processing in order to generate a video summary in the source language. After this step we perform Speech Recognition (on the summarized video) and Machine Translation to the target language. Finally, we process the subtitles in order to generate a subtitle file in the target language for the summarized video.

Fig. 1. Scenario 1 - the most basic approach to newscast summarization

3.2 Scenario 2 - Sc2

Figure 2 depicts an audio based approach to video summarization. In this scenario we generate a video summary recipe based on the audio signal, rather than visual cues as in the Sc1. The subsequent processing is identical as in Sc1. Based on the recipe file we generate a summarized video in the source language, which is than processed by the Speech Recognition module. The resulting transcription is automatically translated to the target language and finally the subtitle file is generated. At the moment of writing of this paper this scenario was not fully integrated, as work on the Audio Based Summary module was still ongoing.

Fig. 2. Scenario 2 - audio based approach to newscast summarization

3.3 Scenario 3 - Sc3

Figure 3 shows a more advanced approach to video summarization. In this architecture first Speech recognition module is employed in order to obtain the transcription in the source language. Than the transcription is translated to the target language and summarized. Based on the recipe generated by the Text Summarization module a video summary is generated. Also, a subtitle file in the target language is provided.

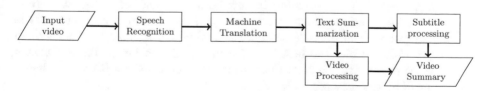

Fig. 3. Scenario 3 - text based summarization in target language

3.4 Scenario 4 - Sc4

This scenario, as depicted in Fig. 4, is a variation of Sc3. The core difference is that we perform text summarization on the source language, rather than on the target language, as in Sc3.

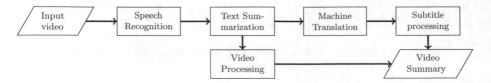

Fig. 4. Scenario 4 - text based summarization in source language

4   Available Modules

For the purpose of experimentation the whole AMIS system consists of seven separate modules. This modules, thanks to well-defined interfaces, may be connected in different configurations. This allows us to experiment with architectures in pursuit of the most effective one.

4.1   Video Based Summarization

The Video Based Summarization module role is to summarize a video based on its spatial and temporal activity. On input this module requires the video file to be summarized and the desired length of the video. The module fetches from the database a pre calculated information on the shot boundaries. It also fetches a pre calculated information on spatial and temporal activity. Based on the spatial and temporal activity the module selects most active shots, sorts them in the order of appearance in the original video and selects a number of shots to fill the required length of the summary. On the output the module provides textual description which is called 'summarization recipe' (explained later on).

4.2   Audio Based Summarization

The Audio Based Summarization module goal is to create a summary based only on the audio signal. This is a difficult task because in the audio signal there is no linguistic information (words, sentences, etc.) that would help to choose the informative parts of the video. We explore several neural architectures using deep learning in order to find the best features in the hidden layer. These features will allow us to capture the most important abstract structure (linguistic level) from a low level resource (signal).

4.3   Text Based Summarization

The Text Based Summarization module aims to select the most important information from a source transcript generated by the ASR module in order to produce an abridged and informative version of the original video. The input video transcript could be in any of the three languages involved in our research (French, English and Arabic).

We opted for an extractive summarization approach with the idea of finding the video segments that contain the most pertinent information based on the transcript. The main idea of Extractive Text Summarization (ETS) applied to video transcripts is to choose the most pertinent segments based on different criteria (information content, novelty factor and relative position) and arrange them in order of appearance to generate a shorter and informative version of original video. The Text Based Summarization module has been deployed based on ARTEX (Autre Résumer de TEXtes), an ETS system originally created by Torres-Moreno *et al.* for French, English and Spanish [17]. A Modern Standard Arabic (MSA) extension has been developed and added to ARTEX.

Sentence Boundary Detection. Before any ETS process, a Sentence Boundary Detection (SeBD) phase is needed to be performed in order to separate the unsegmented transcript produced by the ASR system.

We developed a Convolutional Neural Network (CNN) SeBD system based on textual features. In this approach, the SeBD problem is modeled as a classification task which goal is to predict if the middle word of a 5-word window is (or not) a sentence boundary [7]. During the training and testing phases we used subsets of the French, English and Arabic Gigaword corpora. The size and boundaries ratio of each corpus can bee seen in Table 1.

Table 1. Train/test Gigaword corpora

Language	Words	Boundaries	Ratio
French	587 M	56 M	9.54%
English	702 M	85 M	12.13%
Arabic	62 M	10 M	16.13%

Table 2. Performance of the AMIS SeBD

Language	Class	Precision	Recall	F1
French	NO_BOUND	0.976	0.984	0.980
	BOUND	0.838	0.768	0.801
English	NO_BOUND	0.969	0.983	0.976
	BOUND	0.856	0.762	0.806
Arabic	NO_BOUND	0.928	0.963	0.945
	BOUND	0.782	0.638	0.700

The performance in terms of Precision, Recall and F1 concerning the SeBD system over the Gigaword test datasets is shown in Table 2. For all languages, the "no boundary" (NO_BOUND) class reaches Precision and Recall scores over 0.92. Lower scores are reached for the "boundary" (BOUND) class, being the lowest the Recall for Arabic (0.638). This behavior can be explained given the sample disparity between the "boundary" and "no boundary" classes.

4.4 Automatic Speech Recognition

As the objectives of AMIS are to summarize Arabic or French videos in English and to compare the opinion of these videos with the opinion of English videos dealing with the same topic, three Automatic Speech Recognition (ASR) modules were designed. For each language, an ASR system needs at least three components: an acoustic model, a language model and a lexicon with the different pronunciations of each recognizable word. The acoustic models are based on Deep Neural Networks (DNN) - more precisely on HMM-DNN models - and their development is based on the Kaldi recipe [15]. The language models are statistical n-grams models.

For acoustic model, the French ASR uses 40 MFCC (Mel-Frequency Cepstral Coefficients) acoustic parameters calculated on 25 ms windows every 10 ms. An i-vector of size 100 is added to them. The TDNN (Time Delay Neural Network) estimates 4000 senones (contextual states of Markov models) with a network composed of 6 hidden layers. The main advantage of the TDNN is its ability to take into account a broad context to estimate the probability of senones. In our implementation, we use a context of 29 frames (290 ms). The total number of parameters is about 11 million and the 33 acoustic models were trained on a TV and radio French corpus of 250 h. The 3-gram language model contains a total of 1 million grams trained from a text corpus of 1.5 billion words and the lexicon is composed of 96000 words. The French ASR achieved a Word Error Rate (WER) of 17.2% on the Ester2 development corpus [5].

The topology of the 40 English acoustic models is the same as for French and they were trained on 212 h of TED talks (www.ted.com). The 4-gram language model contains a total of 2 million grams trained from a text corpus of 140 million words and the lexicon is composed of 98000 words. The English ASR achieved a Word Error Rate (WER) of 12.6% on the TED-LIUM development corpus [16].

For the Arabic ASR, the topology of the neural network is different: 440-dimensional input layer (40×11 fMLLR vectors), 6 hidden layers composed of 2048 nodes each and a 4264-dimensional output layer, which represents the number of HMM states. The total number of weights to estimate is about 30.6 millions. The 35 acoustic models were trained on 63 h of broadcast news after a step of parameterization with 13-dimensional Mel-Frequency Cepstral Coefficients (MFCC) features and their first and second order temporal derivatives. The 4-gram language model contains a total of 1 million grams trained from a text corpus of 1 billion words and the lexicon is composed of 95000 words. The Arabic ASR achieved a Word Error Rate (WER) of 14.4% on a test corpus [12].

4.5   Machine Translation

The Machine Translation (MT) module role is to translate an Arabic word sequence into its English corresponding one. For integration, this module needs sequence of sentences, and sentences are translated one after the other. The module outputs translated sentences.

The MT module for AMIS has been developed for the direction Arabic–English, since Arabic is considered such as the foreign language of the video to translate to a summarized video in English. The MT system has been developped using the Moses [8] and Giza++ [13] toolkits, references of the statistical phrase-based approach [2,9] in machine translation.

The statistical approach requires a parallel corpus for training the translation and language models: we chose the Arabic—English United Nation corpus [4] (UN). The training corpus is made up of 9.7 million parallel sentences extracted from UN, concerning the period from January 2000 to September 2009. This corpus has been used to train the translation model. The language model has been trained on the target language of this corpus. The vocabulary contains 224,000 words. The development and the test corpus are composed of 3,000 parallel sentences. The evaluation on the test corpus leads to a BLEU [14] of 39.

4.6   Subtitle Processing

Before presenting video sequences with translated subtitles, they have to be converted to the format acceptable commonly by video players. We have decided to choose a widely used and broadly compatible *SubRip (SubRip Text, SRT) file format* with the extension .srt. SRT files contain formatted lines of plain text in groups separated by a blank line.

A dedicated Python script converts subtitles from an internal AMIS format into the SRT format. Furthermore, the script has an ability to split long subtitles (occupying more than 2 lines of text) into a series of shorter subtitles meant to be displayed one-by-one. The maximum number of words in a single subtitle has been empirically set to 17. Any longer subtitle will be split into a required number of shorter ones and displayed for a fraction of time being relative to the subtitle length.

4.7   Video Processing

Each Summarization module produces a recipe file. This recipe file contains information on shots and their order (defined by their start and end frame number) which are supposed to be included into the video summary. Based on this information the video processing module generates the final video summary. Using the popular ffmpeg software it strips audio of the original video, splits the original video and reconstructs the summary based on the recipe.

5 Summary

In this paper we have proposed a newscast and reports summarization system architectures together with a description of the software components used in the system. We have described components that perform speech recognition, video, audio and text based summarization, video processing, machine translation and subtitle processing. The system architectures were integrated in the Python language and are currently used for experimentation on video news casts and reports summarization and translation. We foresee to work further on the system by both expanding the amount of available meta data (e.g. by incorporating face recognition [1]), modules and working further on optimizing the architecture of the system.

Acknowledgements. Research work funded by the National Science Centre, Poland, conferred on the basis of the decision number DEC-2015/16/Z/ST7/00559 under the Chist-Era AMIS project.

References

1. Baran, R., Rudzinski, F., Zeja, A.: Face recognition for movie character and actor discrimination based on similarity scores. In: 2016 International Conference on Computational Science and Computational Intelligence (CSCI), pp. 1333–1338, December 2016
2. Brown, P.F., Della Pietra, V.J., Della Pietra, S.A., Mercer, R.L.: The mathematics of statistical machine translation: parameter estimation. Comput. Linguist. **19**(2), 263–311 (1993)
3. Derkacz, J., Leszczuk, M., Grega, M., Koźbiał, A., Hernández, F.J., Zorrilla, A.M., Zapirain, B.G., Smaïli, K.: Definition of requirements for accessing multilingual information opinions. Multimedia Tools Appl. **77**(7), 8359–8374 (2018)
4. Eisele, A., Chen, Y.: Multiun: a multilingual corpus from united nation documents. In: LREC (2010)
5. Galliano, S., Geoffrois, E., Mostefa, D., Choukri, K., Bonastre, J.-F., Gravier, G.: The ester phase 2 evaluation campaign for the rich transcription of French broadcast news. In: Interspeech (2005)
6. Gao, X., Tang, X.: Unsupervised video-shot segmentation and model-free anchorperson detection for news video story parsing. IEEE Trans. Circuits Syst. Video Technol. **12**(9), 765–776 (2002)
7. González-Gallardo, C.-E., Torres-Moreno, J.-M.: Sentence boundary detection for French with subword-level information vectors and convolutional neural networks. arXiv preprint arXiv:1802.04559 (2018)
8. Koehn, P., Hoang, H., Birch, A., Callison-Burch, C., Federico, M., Bertoldi, N., Cowan, B., Shen, W., Moran, C., Zens, R., et al.: Moses: open source toolkit for statistical machine translation. In: Proceedings of the 45th Annual Meeting of the ACL on Interactive Poster and Demonstration Sessions, pp. 177–180. Association for Computational Linguistics (2007)
9. Koehn, P., Och, F.J., Marcu, D.: Statistical phrase-based translation. In: Proceedings of the 2003 Conference of the North American Chapter of the Association for Computational Linguistics on Human Language Technology-Volume 1, pp. 48–54. Association for Computational Linguistics (2003)

10. Leszczuk, M., Grega, M., Koźbiał, A., Gliwski, J., Wasieczko, K., Smaïli, K.: Video summarization framework for newscasts and reports – work inprogress. In: Dziech, A., Czyżewski, A. (eds.) Multimedia Communications, Services and Security, pp. 86–97. Springer, Cham (2017)

11. Leszczuk, M.I., Duplaga, M.: Algorithm for video summarization of bronchoscopy procedures. BioMed. Eng. OnLine **10**(1), 110 (2011)

12. Menacer, M.A., Mella, O., Fohr, D., Jouvet, D., Langlois, D., Smaïli, K.: Development of the Arabic Loria Automatic Speech Recognition system (ALASR) and its evaluation for Algerian dialect. In: ACLing 2017 - 3rd International Conference on Arabic Computational Linguistics, Dubai, UAE, pp. 1–8, November 2017

13. Och, F.J., Ney, H.: A systematic comparison of various statistical alignment models. Comput. Linguist. **29**(1), 19–51 (2003)

14. Papineni, K., Roukos, S., Ward, T., Zhu, W.-J.: Bleu: a method for automatic evaluation of machine translation. In: Proceedings of the 40th Annual Meeting on Association for Computational Linguistics, pp. 311–318. Association for Computational Linguistics (2002)

15. Povey, D., Ghoshal, A., Boulianne, G., Burget, L., Glembek, O., Goel, N., Hannemann, M., Motlicek, P., Qian, Y., Schwarz, P., Silovsky, J., Stemmer, G., Vesely, K.: The Kaldi speech recognition toolkit. In: IEEE 2011 Workshop on Automatic Speech Recognition and Understanding. IEEE Signal Processing Society, December 2011

16. Rousseau, A., Deléglise, P., Estève, Y.: Enhancing the TED-LIUM corpus with selected data for language modeling and more ted talks. In: 9th International Conference on Language Resources and Evaluation (LREC 2014), Interspeech (2014)

17. Torres-Moreno, J.-M.: Artex is another text summarizer. arXiv preprint arXiv:1210.3312 (2012)

18. Zhang, H.J., Low, C.Y., Smoliar, S.W., Wu, J.H.: Video parsing, retrieval and browsing: an integrated and content-based solution. In: Proceedings of the Third ACM International Conference on Multimedia, MULTIMEDIA 1995, pp. 15–24. ACM, New York (1995)

19. Zhang, H.J., Wu, J., Zhong, D., Smoliar, S.W.: An integrated system for content-based video retrieval and browsing. Pattern Recogn. **30**(4), 643–658 (1997)

Evaluation of Multimedia Content Summarization Algorithms

Artur Komorowski, Lucjan Janowski, and Mikołaj Leszczuk[✉]

AGH University of Science and Technology, 30059 Kraków, Poland
leszczuk@agh.edu.pl
http://www.kt.agh.edu.pl/en

Abstract. This paper provides an overview of 2 different methods for evaluation of multimedia (video) summarizations. A tag-based method focuses on tags selected for an analyzed video sequence. It checks if a summarized video sequence contains important content. An annotation method uses an annotation procedure to exact the key shots from full video sequences. Both methods used collectively allow for complex evaluation of summarizing algorithms. The paper contains descriptions and test results for both methods. The paper concludes with some suggestions for future directions.

Keywords: Evaluation · Multimedia · Summary

1 Introduction

The 21st century gives people almost unlimited access to multimedia. They can read papers, watch pictures and video materials. What is more, they can do it all without leaving their homes. They can choose what is interesting for them. Multimedia also get annotated and tagged. But a problem arises when people do not exactly know what they want to see. Life is much faster than it was before. People have to share their time with job, family, hobby. Time is important in life. Sometimes people don't want to watch or read full multimedia content because it takes too much time. Fortunately, nowadays it is possible to automatically index [1] and/or summarize [4] multimedia. Summarizing is creating a shorter version of content, still including all important data that a regular user needs. In I. Mani book from 1999 one can find: "The goal of automatic summarization is to take an information source, extract content from it, and present the most important content to the user in a condensed form and in a manner sensitive to the user's or application's needs" [6]. For example, as a result of summarization, we receive sport highlights that include only (football) goals or spectacular actions from a game. Thanks to these shortcuts, a user can see the most important elements of the full material. For example, he/she can watch highlights of ten games in time that wouldn't be even enough for watching just one full game. Sport events are only an example. Summarization of multimedia contents is used also for

K. Choroś et al. (Eds.): MISSI 2018, AISC 833, pp. 424–433, 2019.
https://doi.org/10.1007/978-3-319-98678-4_43

newscasts and reports from all over the world. In this case choosing important parts is more difficult. Several questions arise then. How to evaluate if the created summary is valuable? What does it mean that it is valuable? In a nutshell, it means that it provides all data that user may need. It has to be concise. The easiest answer for the first question is: "When a user thinks it is enough for him/her". This answer is right. But what if we don't want to make subjective tests for every multimedia summary that is made?

The main purpose of this research was to propose solutions for automatic evaluation of algorithms summarizing multilingual multimedia (video) content. Multilingualism is crucial in this case because in different languages, countries and cultures, one situation can be perceived differently. Consequently, some users may want to see different points of view.

In this paper, we show that it is possible to find methodologies that allow us to avoid subjective testing of various summarization approaches. We present 2 algorithms that can help evaluating multilingual multimedia summaries. The 1st method uses annotation technique, based on expert's selection, to create a reference summary. Afterwards, all summaries created by summarization scripts or tools, are referred to this reference. This algorithm lets checking if interesting shots from original video sequence are selected in a summary. The 2nd method is related to YouTube tags originally added to video sequence data. It allows to check another aspect – the video content. Evaluating video sequences cannot be reduced only to the visual layer. Checking the body (saved as text) of video sequence with the use of tags let us check if a summary contains all important data. Combination of these methods reduces necessity of subjective tests and lets us to evaluate summaries in a complex way.

There are some solutions for summary evaluation that we found in related research works. Obviously we are not the first group that uses annotation to create corpus for evaluating video sequences. This solution was also used with the focus on Natural Language Descriptions [3]. However, in our case, annotating was not as detailed. Instead, we complemented our method with the tag-based method. It is also necessary to mention about Recall-Oriented Understudy for Gisting Evaluation (ROUGE) [8]. We used a few of its concepts including Precision, Recall and F1 score. We used only these metrics that were needed for our method. We also refer to text summary evaluation, which was used at Document Understanding Conference (DUC) [7], and defines that the main metric is coverage. We partially use this concept while checking if all tags used in the original video sequence are "covered" in a video summary. In [6] we can also find other methods for summary evaluation that inspired us. These are intrinsic and extrinsic approaches, for which, methods for assessing informativeness are described and rated. This paper is also partially based on concepts given in [5].

The remainder of this paper is structured as follows. In Sect. 2 we describe source data used in our tests – its format, structure and preparation. Section 3 contains information about the method using an annotation procedure, based on frame selection. Section 4 introduces the 2nd algorithm that uses tags. Section 5 presents results of our tests. We also comment them discuss where, for these methodologies, there is a room for an improvement.

2   Source Data

Experiments on both evaluation methods used the same set of test of 41 video sequences. They were selected from YouTube newscast programs and reports presented in three languages: English, French and Arabic. Their durations were spread from 3 to 13 min. The idea behind selecting these video sequences was to have similar number of video sequences in three duration intervals (values in minutes): [3–6], (6–10] and (10–13]. Finally we decided to select a little bit more long video sequences (for which summarization makes more sense). For the detailed duration distribution please check Fig. 1.

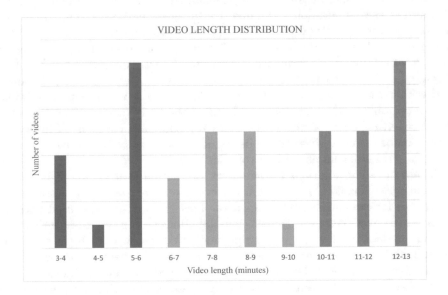

Fig. 1. Video sequence duration distribution.

All data that was processed in both methods was stored in a database. It made it easier to retrieve it for further processing.

3   Annotation Method

In this section we describe data preparation, the algorithm and results connected with the video annotation method.

3.1   Annotation Procedure

Annotation procedure allows to create reference summary based on subjective expert's frames selection. The person responsible for creating this summary had to focus on video, not audio, so annotation was done for video sequences without soundtrack. It was our intentional procedure. Selecting one frame in video were processed with our software - frame was found in video and the full shot for this frame was extracted.

We have considered different tools to be potentially used for annotating video sequences. Some of them are costly and other require very specific software configurations. Finally, we decided that we need a customizable and simple tool. Consequently, we used the VLC media player to indicate frames from shots that the expert considered interesting enough to include them in a summary.

We had to enforce a consistent procedure for annotating/extracting frames. Here the description of the extraction procedure is presented in detail:

1. Use the VLC media player to watch the video sequence. Each video sequence should be watched at least once fully before key frames are extracted.
2. Extracting frames is a simple saving selected frames to a folder by pressing a hot-key (we used S – capital s) while watching the video sequence.
3. The VLC media player has to be configured correctly so the extracted frames can be used by another tool we prepared, which extracts shots from the original video sequence, based on time of specific frame's occurrence. Here is the configuration procedure:
   (a) In the VLC media player, one needs to open the preferences window. One should go then to the video settings.
   (b) Figure 2 presents the configuration window. The settings have been set to:
      – Directory: convenient location
      – Prefix: F_$N_T_$T, where $N is the file name and $T is the frame time-code counted from the beginning of the sequence
      – Format: png

3.2   Evaluation Algorithm

We already calculated results for video sequences annotated by one person ($K = 1$). To get more specific results (to avoid subjectivity) further tests should be done by a larger pool of experts. Only one summarizing algorithm was checked also. For future tests, a more general algorithm is proposed as presented below. The annotation results in n sequences with manually chosen key shots. It means that for each sequence s_i a frame $f_{i,j}$ belongs to a chosen shot or not. More precisely since we have K people annotating videos each frame can be annotated as $f_{i,j,k}$ what means it is j-th frame of i-th sequence which was annotated by k-th person.

A summary of a sequences s_i is a set of frames $U_{i,l}$, again each frame belongs or not to summary l. To make it clearer we will denote it by u so $u_{i,j,l}$ is j-th frame of i-th sequence generated by l-th summary. Note that $u_{i,j,l}$ refers to exactly the same

Fig. 2. The VLC media player configuration window.

frame as $f_{i,j,k}$. Since both f and u are vectors of 0 and 1 values the comparison is simple detecting where both are equal to 1.

In order to describe the obtained results, we calculate Precision, Recall and F1 score for each sequence, person describing and algorithm. First for each human description and algorithm we calculate Precision (Eq. (1)), Recall (Eq. (2)) and F1 score (Eq. (3)):

$$p_{i,k,l} = \frac{\sum_j f_{i,j,k} == 1 \& u_{i,j,l} == 1}{\sum_j u_{i,j,l} == 1} \tag{1}$$

$$r_{i,k,l} = \frac{\sum_j f_{i,j,k} == 1 \& u_{i,j,l} == 1}{\sum_j f_{i,j,k} == 1} \tag{2}$$

$$F_{i,k,l} = 2\frac{p_{i,k,l} r_{i,k,l}}{p_{i,k,l} + r_{i,k,l}} \tag{3}$$

To obtain the performance metrics for all experts we calculate an average given by:

$$p_{i,l} = \frac{\sum_k p_{i,l,k}}{n_i} \tag{4}$$

where n_i is the number of experts annotating sequence i.

$$r_{i,l} = \frac{\sum_k r_{i,l,k}}{n_i} \tag{5}$$

$$F_{i,l} = 2\frac{p_{i,l}r_{i,l}}{p_{i,l} + r_{i,l}} \qquad (6)$$

A simple procedure is made to obtain the final score for an algorithm:

$$p_l = \frac{\sum_i p_{i,l}}{S_l} \qquad (7)$$

where S_l is the number of sequences annotated by algorithm l.

$$r_l = \frac{\sum_i r_{i,l}}{S_l} \qquad (8)$$

$$F_l = 2\frac{p_l r_l}{p_l + r_l} \qquad (9)$$

The best summarization algorithm will be the one with the highest F_l value but detail investigation should be made to be sure that some specific values are not influencing this result.

3.3   Experiment Results

Results that we obtained with this method are presented in Fig. 3.

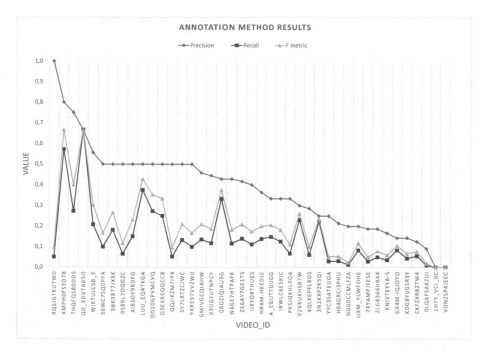

Fig. 3. Annotation method results

As we can see on Fig. 3, our Precision values are not very high. It probably is caused by 2 facts: video sequences were annotated by only one person and video sequences were annotated without an audio.

The first factor forces subjectivity of this test. One person can notice other things than group of people. He or she can choose frames that are important for him/her because of colors, connotations, perhaps even memories. An influence of this can be reduced by creating a pool of experts that independently annotate video sequences. The second factor is something that is an assumption for this method. It has its advantages like e.g. selecting frames without paying attention to spoken words, the voice's tone, etc. The disadvantage is that the expert doesn't know if any text/sound content is voiced in a given moment. To improve and rate values of metrics presented as results, it would be beneficial to test this method on other summarizing algorithms.

4   Tag-Based Method

This section presents evaluation method based on tags. It contains a data format, the scheme of the used methodology and results of our experiments.

4.1   Tag Selection

For the tag-based evaluation method, the most important data that has been used were YouTube tags. We used YouTube tags for every selected video sequence. Tags used there summarize the given newscast/report video sequence. Tags are, depending on the particular video sequences, written in different languages. Here are some examples:

```
RT, Russia Today, FSA kicks out US special forces troops,
FSA, Free Syrian Army, US special forces troops, withdraw,
kicked out, syria, war, troops, us-backed
```

4.2   Evaluation Algorithm

Every video sequence used has its own set of tags. They are in different languages. The algorithm for each video sequence is:

1. Retrieve the audio track from the original video sequence.
2. Use an Automatic Speech Recognition (ASR) system engine [2]. It is a system that is created with two parts: an acoustic model and language model. It is based on KALDI ASR toolkit. The acoustic model is based on Deep Neural Network (DNN), which has input layer of 440 neurons, 6 hidden layers of 2048 neurons each. Output layer has around 4000 neurons. ASR returns a set of words with time-codes. We use it to obtain a textual transcription of the original video sequence.
3. Retrieve tags. If a tag contains more than one word, split it. Create a set without duplicates.

4. Check which tags appear in the textual transcription of the original video sequence. Limit the set of tags to these tags that occur in the textual transcription of the original video sequence.
5. Create a summary of the original video sequence.
6. Retrieve an audio track from the recently created summary.
7. Use the ASR engine to obtain a textual transcription of the summarized video sequence.
8. Check which tags appear in the textual transcription of the summarized video sequence. Create a set of tags that occur in the textual transcription of the summarized video sequence.
9. Check the tags that occur in the summarized and the original video sequence. Calculate statistics.

4.3   Experiment Results

In this subsection results for the tag-based method are presented. In Fig. 4 one can observe the number of tags in the original video sequences and the number of tags in the summarized video sequences – for all tested video sequences.

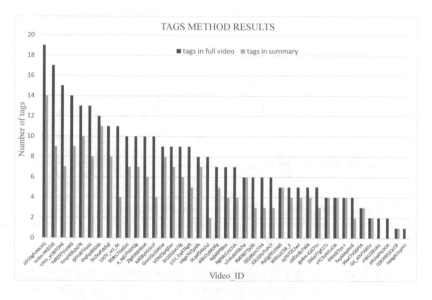

Fig. 4. Tag-based evaluation method results

As we can see, for the most of selected video sequences, the percentage of occurring tags in summaries is above 50. It means that, despite different relative lengths of the summarized video sequences for various original video sequences, they contain content that is described in tags. It is a valuable check; however, to get more reliable information, a test should be done for a larger set of video sequences.

432   A. Komorowski et al.

5   General Results and Conclusions

In Fig. 5 one can learn a comparison of a relative length of the summarized video sequences and the percent of tags contained in the summarized and in the original video sequences. The length of the summarized video sequence is based on annotation process. For the tag-based method, the length of summaries was conformed to the results from annotation method. Owing to this fact, we can evaluate if summaries created without an audio contain valuable content. In this situation summaries were created using the same algorithm.

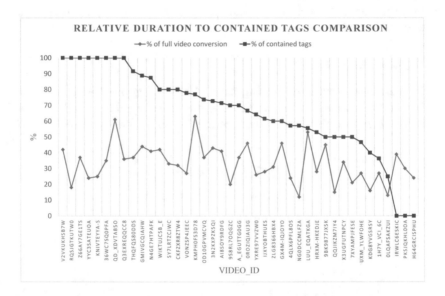

Fig. 5. Comparison of the summarization ratio and tags containing in summary

Based on data presented in Fig. 5, we can say that it is not true that if we have longer summary, it contains all tags. We see that there are video sequences that contain all of tags from the original video sequence and they are still relatively short. These results depend on many factors. According to these results, using two methods described in this paper, makes evaluation of video summaries comprehensive. It checks not only if the summary contains important shoots, but also if it contains necessary information. Our tests provide a solid base but there is some room for improvements. All tests should be performed for more video sequences. As we commented in Subsect. 3.3, annotation should be done by a larger pool of experts. It would make shots selection more objective.

Acknowledgment. I thank AMIS project and Chist-era for all support given due work with this article.

Research work funded by the National Science Center, Poland, conferred on the basis of the decision number DEC-2015/16/Z/ST7/00559.

References

1. Baran, R., Rudzinski, F., Zeja, A.: Face recognition for movie character and actor discrimination based on similarity scores. In: 2016 International Conference on Computational Science and Computational Intelligence (CSCI), pp. 1333–1338 (2016). 10.1109/CSCI.2016.0249
2. Jouvet, D., Langlois, D., Menacer, M.A., Fohr, D., Mella, O., Smaïli, K.: About vocabulary adaptation for automatic speech recognition of video data (2017)
3. Khan, M.U.G., Nawab, R.M.A., Gotoh, Y.: Natural language descriptions of visual scenes corpus generation and analysis. In: Proceedings of the Joint Workshop on Exploiting Synergies between Information Retrieval and Machine Translation (ESIRMT) and Hybrid Approaches to Machine Translation (HyTra), pp. 38–47. Association for Computational Linguistics (2012). http://www.aclweb.org/anthology/W12-0105
4. Leszczuk, M., Grega, M., Koźbiał, A., Gliwski, J., Wasieczko, K., Smaïli, K.: Video summarization framework for newscasts and reports - work in progress. In: Dziech, A., Czyżewski, A. (eds.) Multimedia Communications, Services and Security, pp. 86–97. Springer International Publishing, Cham (2017)
5. Mani, I.: Advances in Automatic Text Summarization. MIT Press, Cambridge (1999)
6. Mani, I.: Summarization evaluation: an overview (2001)
7. Nenkova, A.: Automatic text summarization of newswire: lessons learned from the document understanding conference. In: Proceedings of the 20th National Conference on Artificial Intelligence, vol. 3, AAAI 2005, pp. 1436–1441. AAAI Press (2005). http://dl.acm.org/citation.cfm?id=1619499.1619564
8. Owczarzak, K., Conroy, J.M., Dang, H.T., Nenkova, A.: An assessment of the accuracy of automatic evaluation in summarization. In: Proceedings of Workshop on Evaluation Metrics and System Comparison for Automatic Summarization, pp. 1–9. Association for Computational Linguistics, Stroudsburg (2012). http://dl.acm.org/citation.cfm?id=2391258.2391259

A Knowledge Discovery from Full-Text Document Collections Using Clustering and Interpretable Genetic-Fuzzy Systems

Filip Rudziński[✉]

Department of Electrical and Computer Engineering,
Kielce University of Technology, Al. 1000-lecia P.P. 7, 25-314 Kielce, Poland
f.rudzinski@tu.kielce.pl

Abstract. The paper presents a concept of a hybrid system consisting of two our original techniques from the computational intelligence area and its application to knowledge discovery from full-text document collection. Our first technique - self-organizing neural network with one dimensional neighborhood and dynamically evolving topological structure - aims at automatically determining the number of groups in the document collection and at grouping the documents in terms of their similarity. In turn, the main goal of our second approach - multi-objective evolutionary designing technique of fuzzy rule-based classifiers with optimized accuracy-interpretability trade-off - is to extract the most important keywords from documents and to generate classification rules which can be helpful in understanding and isolating the subjects of documents collected in the founded groups. The proposed concept may also be useful to develop systems operating in a wide area of human language understanding problems.

Keywords: Clustering · Fuzzy rule-based systems
Text document · Knowledge discovery · Information retrieval
Human language understanding

1 Introduction

Automatic clustering and classification of full-text documents without any knowledge given a priori about the meaning of their contents is still important issue in human language understanding problems. The clustering methods aim (i) to divide a collection of text documents into groups (the number of groups is unknown in advance) in such a way that all documents belonging to the same group have similar contents (topics) and (ii) to determine for each founded group the so-called prototypes of documents which represent the average knowledge about the content of documents collected in that group. In turn, the goals of classification techniques are (i) to discover the most important keywords which describe particular groups of documents obtained during the clustering and (ii) to generate the most representative class labels (unknown in advance) of those groups which help to understand the content minings (topics) of the documents. All above goals can not be

© Springer Nature Switzerland AG 2019
K. Choroś et al. (Eds.): MISSI 2018, AISC 833, pp. 434–443, 2019.
https://doi.org/10.1007/978-3-319-98678-4_44

achieve with the use of conventional clustering techniques or conventional classifiers. The major of known clustering techniques require the number of groups in advance (which is unknown in practice). The classifiers should be able to discover the knowledge from text documents in an interpretable and easy to understand form - a feature that conventional classifiers do not have.

In this paper, we propose a concept of hybrid system and its application to automatically process full-text documents, i.e., to discover groups of documents with similar content (including the number of such groups), the most important keywords describing the founded groups, as well as interpretable and easy to understand knowledge about documents collected into groups in the form of linguistic classification rules. In order to solve the above outlined tasks, the proposed system uses two novel and original techniques from the computational intelligence area, i.e., a self-organizing neural network with one dimensional neighborhood and dynamically evolving topological structure (SONN-1D) [2,3,5] as well as a multi-objective evolutionary algorithm (MOEA) (introduced in [7]) to design fuzzy rule-based classifiers (FRBCs) with optimized compromise between their accuracy and interpretability of discovered rules (see also [4,8]).

The paper is organized as follows. Section 2 outlines the proposed hybrid system and its major components (SONN-1D and FRBC). Section 3 presents the operation of our system applied to discover the knowledge from exemplary text documents. Finally, Sect. 4 concludes the results.

2   A Hybrid System for Full-Text Documents Processing

In this section we outlines three main components of the proposed system: (i) a bag of words module (BOW) which prepares the so-called bag-of-words representation of entire collection of documents in order to further processing by the next modules, (ii) SONN-1D which is able to automatically determine the number of groups and to generate the prototypes of documents, and (iii) MOEA aiming at building FRBCs with both accurate and interpretable (easy to understand) linguistic rules representing the knowledge about the content of documents.

2.1   Bag of Words Module

The BOW generates the so-called codebook (also called the vector space model [1]) from collection of L text documents:

$$D_{[L \times n]} = \{d_l\}_{l=1}^{L}, \tag{1}$$

where n is the number of terms (words) occurring in all documents. A single row $d_l = [d_{l1}, d_{l2}, \dots, d_{ln}]$ $(l = 1, 2, \dots, L)$ of the matrix D (1) is n-dimensional vector representing l-th document. We use term-frequency (TF) representation, i.e., the value of d_{li} $(i = 1, 2, \dots, n)$ indicates the occurrence count of i-th terms in l-th document (another representations are possible to use, e.g. TF-IDF [6]). Moreover, the so-called vocabulary (i.e., n-dimensional vector $v = [v_1, v_2, \dots, v_n]$)

containing the pairs v_i of i-th term and its occurrence count in all documents of the collection is generated.

Since the dimensionality of the codebook D (1) and vocabulary v is usually very high, several dimensionality-reduction techniques are applied in order to minimize the demand for computing power of the system. Very known techniques, often used at the starting point of the text document preprocessing are filtering (removing unnecessary characters from the original text like e.g., %, #, $), stemming (extracting the respective stems from words; a stem is a portion of a word left after removing its suffixes and prefixes), the stop-word removal (removing popular words which do not have identifiable meanings and therefore are of little use in various text processing tasks), and feature selection methods (removing such words that have no significant meanings from the point of view of text document processing, e.g. words that do not provide information useful for determining the similarity/dissimilarity of documents; an obvious approach is to remove words which occur in only one document or in all documents; the most advanced feature selection techniques can be also used to leave the most significant words in the codebook).

2.2 A Self-organizing Neural Network with Dynamically Evolving One-Dimensional Topological Structure

The SONN-1D with n inputs $u_1, u_2, \ldots, u_n$ and K neurons arranged in a chain (see Fig. 1 and details in [2,3,5]) with outputs $v_1, v_2, \ldots, v_K$, where $v_k = \sum_{i=1}^{n} w_{ki} u_i$, $k = 1, 2, \ldots, K$ and w_{ki} are weights connecting the output of k-th neuron with i-th input is considered. Using vector notation $v_k = \boldsymbol{w}_k \boldsymbol{u}^T$, where $\boldsymbol{w}_k = [w_{k1}, w_{k2}, \ldots, w_{kn}]$, $\boldsymbol{u} = [u_1, u_2, \ldots, u_n]$ and matrix notation $\boldsymbol{V}_{[K \times 1]} = \boldsymbol{W}_{[K \times n]} \boldsymbol{U}_{[n \times 1]}$, where $\boldsymbol{V}_{[K \times 1]} = [v_1, v_2, \ldots, v_K]^T$, $\boldsymbol{W}_{[K \times n]} = \{\boldsymbol{w}_k\}_{k=1}^{K}$ and $\boldsymbol{U}_{[n \times 1]} = [u_1, u_2, \ldots, u_n]^T$, the weights vector $\boldsymbol{w}_k$ may be treated as the prototype of l-th document represented by vector $\boldsymbol{d}_l$ from the codebook $\boldsymbol{D}_{[L \times n]}$ (1). The weights matrix $\boldsymbol{W}_{[K \times n]}$ stands for a new codebook

$$\boldsymbol{D}'_{[K \times n]} = \boldsymbol{W}_{[K \times n]} = \{\boldsymbol{w}_k\}_{k=1}^{K} \tag{2}$$

containing the prototypes of all documents ($K < L$).

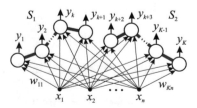

Fig. 1. The structure of self-organizing neural network with one-dimensional topological neighbourhood (the neuron chain divided into two sub-chains S_1 and S_2)

The learning process of SONN-1D lasts E epochs (E is adjusted in advance). In a single epoch, particular vectors $\boldsymbol{d}_l$ ($l = 1, 2, \ldots, L$) of the codebook $\boldsymbol{D}_{[L \times n]}$ (1) are presented exactly once to the inputs of the network using Fisher–Yates shuffle algorithm: $l = swap(T[i], T[i + rnd(L - i)])$, where $i = ((k - 1) \bmod L) + 1$, "mod" is the modulo operator, $T = [1, 2, \ldots, L]$ is a permutation of L consecutive natural numbers prepared at the beginning of a given epoch, $T[i]$ is a number placed on the position No. i in the permutation T, $rnd(n)$ returns a random integer value from $[0, n - 1]$, and $swap(a, b)$ exchanges values in a and b and then, returns the value of a.

The network aims - through the competition of neurons - to minimize the learning error $Err = \frac{1}{L} \sum_{l=1}^{L} \min_{k=1,2,\ldots,K} dist(\boldsymbol{d}_l, \boldsymbol{w}_k)$, where $dist(\boldsymbol{d}, \boldsymbol{w})$ is the Euclidean distance measure between vectors $\boldsymbol{d}$ and $\boldsymbol{w}$. The prototype $\boldsymbol{w}_k$ ($k \in \{1, 2, \ldots, K\}$) selected from the new codebook (2) for each input vector $\boldsymbol{d}_l$ represents here neuron No. k winning in the competition when $\boldsymbol{d}_l$ is presented to the inputs.

After the selection of the input vector $\boldsymbol{d}_l$, the *Winner-Takes-Most* (WTM) learning algorithm are applied to update, in j-th iteration, the weights $\boldsymbol{w}_k$ ($k = 1, 2, \ldots, K$) of both the winning neuron as well as neurons from its topological neighbourhood:

$$\boldsymbol{w}_k(j + 1) = \boldsymbol{w}_k(j) + \eta(j) N(k, v_l, j)[\boldsymbol{d}_l(j) - \boldsymbol{w}_k(j)], \tag{3}$$

where $\eta(j)$ is the learning coefficient, $N(k, v_l, j)$ is the Gaussian-type neighbourhood function:

$$N(j, v_l, k) = e^{-\frac{(j - v_l)^2}{2\lambda^2(k)}}, \tag{4}$$

where $\lambda(k)$ is the "radius" of neighbourhood (the width of the Gaussian "bell").

At the end of each learning epoch, five operations are activated (under some conditions) which dynamically modify the topological structure of the network in order to (i) adjust the number of neurons and (ii) allow the network to divide its structure (neuron chain) into parts (sub-chains) representing data clusters as best as possible (see details in [2,3,5]):

- operation 1: remove single, low-active neuron from the neuron chain,
- operation 2: disconnect the neuron chain into two sub-chains (which can be disconnected again in the course of further learning),
- operation 3: insert new neuron(s) in the neighbourhood of high-active neurons (e.g., between two directly neighbouring neurons, at the edge of (sub-)chain ended with high-active neurons, or replace of high-active neuron, accompanied by low-active neurons, by two new neurons),
- operation 4: reconnect two selected sub-chains.

It is worth to notice, that the automatically adjusting the number of neurons in the network is equivalent to automatically adjusting the number (K) of prototypes in the codebook $\boldsymbol{W}_{[K \times n]}$ (2). The prototypes belonging to the same sub-chain of the network represent a single group of the documents. The number of groups equals to the number of sub-chains.

438   F. Rudziński

2.3 Interpretability-Oriented Fuzzy Rule-Based Classifier

The last module of the system employs FRBC with n input numerical attributes $x_1, x_2, \ldots, x_n$ and an output, which is a fuzzy set over the set $Y = \{y_1, y_2, \ldots, y_c\}$ of c class labels. FRBC's knowledge base consists of R rules:

$$\textbf{IF}[x_1 \text{ is } [not]_{(sw_1^{(r)}<0)} \ A_{1,|sw_1^{(r)}|}]_{(sw_1^{(r)}\neq 0)} \textbf{AND}...\textbf{AND}$$
$$[x_n \text{ is } [not]_{(sw_n^{(r)}<0)} \ A_{n,|sw_n^{(r)}|}]_{(sw_n^{(r)}\neq 0)} \tag{5}$$
$$\textbf{THEN } y \text{ is } B_{(singl.)j^{(r)}}, \ r = 1, 2, \ldots, R,$$

where: (i) $[expression]_{(condition)}$ in (5) denotes conditional inclusion of $[expression]$ into a rule if and only if $(condition)$ is fulfilled, (ii) $|\cdot|$ returns the absolute value, and (iii) $sw_i^{(r)}$ is a switch which controls the presence/absence of the i-th input attribute in the r-th rule, $i = 1, 2, \ldots, n$ - see details in [4,8].

In general case, each input attribute x_i is represented by a_i fuzzy sets A_{ik_i}, $k_i = 1, 2, \ldots, a_i$ with trapezoidal or Gaussian-type membership functions and labelled by linguistic terms usually like, e.g., "Small" (for A_{i_1}), "Medium $1, 2, \ldots, a_i - 2$" (for $A_{i_2}, A_{i_3}, \ldots, A_{i,a_i-1}$), and "Large" (for A_{ia_i}). Class labels $y_{j^{(r)}}$ ($j^{(r)} \in \{1, 2, \ldots, c\}$) are represented by fuzzy singletons $B_{(singl.)j^{(r)}}$ characterized by membership functions defined as follows: $\mu_{B_{(singl.)j^{(r)}}}(y) = 1$ for $y = y_{j^{(r)}}$ and 0 elsewhere. In this paper, each attribute x_i is represented by one fuzzy set A_{i_1} with trapezoidal membership function and labelled "Existing" which denotes the presence of i-th term in the document. This approach allows us to discover the most important keywords representing the particular classes.

The knowledge base of FRBC is discovered from the codebook $W_{[K \times n]}$ during the learning process based on Pittsburgh-type approach (i.e., fuzzy rules and all parameters of membership functions are optimized simultaneously). Due to FRBC's accuracy and the interpretability of its knowledge base are our objectives to be optimized (complementary/contradictory objectives), we use a multi-objective evolutionary algorithm (MOEA) to design a set of FRBCs with different levels of compromise between their accuracy and interpretability.

The accuracy measure (subject to maximization) is defined as follows:

$$Q_{ACC} = 1 - Q_{RMSE}, \ Q_{RMSE} = \sqrt{\frac{1}{Kc} \sum_{k=1}^{K} \sum_{j=1}^{c} \left[\mu_{B_{(singl.)k}}(y_j) - \mu_{B'_k}(y_j) \right]^2}. \tag{6}$$

$Q_{RMSE} \in [0, 1]$, B'_k is the fuzzy-set response of system (5) for the learning data sample $x_k \leftarrow w_k$, and $B_{(singl.)k}$ is the desired fuzzy-singleton response for that sample ($\mu_{B_{(singl.)k}}(y) = 1$ for $y = y_k$ and 0 elsewhere).

The interpretability of FRBC's knowledge base is evaluated by the measure (subject to maximization):

$$Q_{INT} = 1 - Q_{CPLX}, \ Q_{CPLX} = \frac{Q_{RINP} + Q_{INP} + Q_{FS}}{3}, \tag{7}$$

where

$$Q_{RINP} = \frac{1}{R} \sum_{r=1}^{R} \frac{n_{INP}^{(r)} - 1}{n-1}, Q_{INP} = \frac{n_{INP} - 1}{n-1}, \ Q_{FS} = \frac{n_{FS} - 1}{\sum_{i=1}^{n} a_i - 1}, \ n > 1, \quad (8)$$

and $n_{INP}^{(r)}$ is the number of active input attributes in the r-th rule, n_{INP} and n_{FS} in are the numbers of active inputs and fuzzy sets in the whole system.

In [4,8] we propose original crossover and mutation operators for transformation fuzzy rules as well as we adopted some specialized crossover and mutation operators for transformation parameters of membership functions. Moreover, in [7] we propose an original generalization of very known MOEA, i.e. Strength Pareto Evolutionary Algorithm 2 (SPEA2) [9], referred to as our SPEA3, aiming at improving the spread and distribution balance of generated solutions (FRBCs). These approaches will be used in our experiments in the next section (due to limited space of the paper, they cannot be presented here in details).

3   Application to Knowledge Discovery from Text Documents

The operation of the proposed approach will be presented in this section using an exemplary collection of 476 abstracts of technical reports (published until the end of 2002) available under the name CSTR-476 in the University of Rochester repository (www.cs.rochester.edu/trs). All documents belong to one of four thematic groups: "Theory", "System", "AI", and "Robotics and Vision". It is important to notice that the knowledge about those groups in no way was used for the processing of documents by our system. It was used only to verify the obtained results.

The process will be performed in three stages aiming at (i) building a numerical model of entire document collection, that can be further processed by our system, (ii) determining the number of document groups and collects all documents for that groups in terms of their similarity (i.e., in such a way that documents with similar content are placed in the same group) and (iii) discovering the most important keywords and classification rules which can be helpful in understanding and isolating the subject of documents in particular groups.

3.1   Preparing the Numerical Model of Document Collection

Table 1 presents the dimensionality reduction process of the considered documents into the target numerical data set. First, an initial $VSM_{[L \times n_{INI}]}$ model has been built from original CSTR-476 collection, i.e., $6,752$ different terms have been extracted from all documents (overall number of occurrences of all terms in all documents equals to $41,072$). Then, stemming and stop-words removal techniques have been applied in order to reduce its dimensionality to $n_{STEM} = 4,438$ attributes (for $VSM_{[L \times n_{STEM}]}$ - model obtained after the stemming process) and then to $n_{STOP} = 4,419$ attributes (for $VSM_{[L \times n_{STOP}]}$ - model obtained after stop-words removal), respectively. Finally, the least common words have been removed (we have assumed the deletion of terms which occur less than 20 times in all documents) in order to obtain the final model ($VSM_{[L \times n_{FIN}]}$) with $n_{FIN} = 342$ attributes. The removed terms do not provide relevant information from the point of view of knowledge discovery from documents.

Table 1. The dimensionality reduction of the initial VSM for CSTR-476 collection

VSM	Number of documents	Number of attributes (terms)	Overall numbers of occurrences of all terms in all documents
$VSM_{[L \times n_{INI}]}$	476	6 752	41,072
$VSM_{[L \times n_{STEM}]}$	476	4 438	38,307
$VSM_{[L \times n_{STOP}]}$	476	4 119	26,525
$VSM_{[L \times n_{FIN}]}$	476	342	14,805

3.2 Determining the Number of Groups and Grouping of Documents

The above outlined SONN-1D is applied to determine the number of document groups in the final model $VSM_{[L \times n_{FIN}]}$, which is treated now as the learning data set for the network. Figure 2 presents the changes of the number of neurons (a) and the number of sub-chains (b) during the learning process. We use SONN-1D with two initial neurons whose weights were randomly selected from 0 to 1 as well as the learning coefficient decreasing from 0.7 to 0.01 and the Gaussian-type neighborhood function with radius decreasing from 5 to 0. After 1,000 epochs of the learning process, the network has divided its neuron chain (containing 159 neurons) into 4 sub-chains representing 4 document groups, respectively. Figure 3 shows the envelope of nearness histogram for the final route determined in the attribute space of $VSM_{[L \times n_{FIN}]}$ data set (see, e.g., [5] for the definition of that histogram). The boundaries between particular sub-chains are represented by three easy visible low values of the histogram (the highest distance between two neighboring neurons belonging to two different sub-chains).

Each sub-chain of SONN-1D represents a single group of documents with relatively similar content. The neurons of a single sub-chain constitute a multi-point prototype of documents for a given group. In turn, a single neuron of the prototype represents the most similar documents and its weight vector contains the "averaged" numbers of occurrences of all terms in that documents. Thus, the weights for all neurons can be treated as compressed and filtered data set containing the most important knowledge about the original text document collection. This data set will be used in the next stage to discover the most important keywords for particular group of documents and classification rules for those groups.

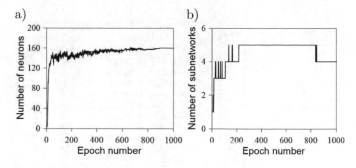

Fig. 2. The plot of the number of neurons (a) and the number of sub-chains (b) vs. learning epoch number

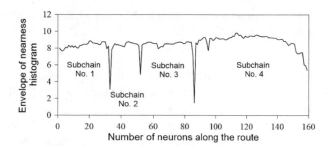

Fig. 3. The envelope of nearness histogram for the final route in the attribute space of $VSM_{[L \times n_{FIN}]}$ data set

3.3   Knowledge Discovery from Documents

At the beginning of this stage, all weight vectors of SONN-1D have been initially labelled by numbers "1", "2", "3", and "4", i.e., by number of sub-chain which contains the nearness neuron to a given input vector in the attribute space (the so-called winning neuron which wins in the competition of neurons when a given input vector is presented to the inputs of the network). After then, we applied our SPEA3 algorithm to automatically design a set of FRBCs with different levels of compromise between their accuracy and interpretability. Figure 4 shows the approximation of Pareto-optimal front of such classifiers in the objective space and their details.

Table 2 presents rules of two solutions selected from Fig. 4a and b (for simplicity, we omitted the label "Existing" of fuzzy sets). The first solution (FRBC No. 1) is characterized by the lowest complexity of rules (thus, the best interpretability of its knowledge base) and the worst accuracy (69.4% of correct decisions). This solution represents the most general knowledge about the document collection. We can distinguish 4 stems of the most important keywords (i.e., "method", "constraint", "softwar", and "design"). Keywords "method", "constraint", and "software" (in rules Nos. 1, 2, and 3) usually occur in documents mostly

a)   b)

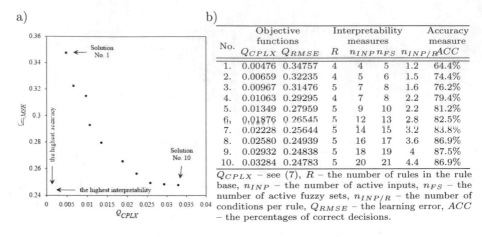

| No. | Objective functions | | Interpretability measures | | | | Accuracy measure |
	Q_{CPLX}	Q_{RMSE}	R	n_{INP}	n_{FS}	$n_{INP/R}$	ACC
1.	0.00476	0.34757	4	4	5	1.2	64.4%
2.	0.00659	0.32235	4	5	6	1.5	74.4%
3.	0.00967	0.31476	5	7	8	1.6	76.2%
4.	0.01063	0.29295	4	7	8	2.2	79.4%
5.	0.01349	0.27959	5	9	10	2.2	81.2%
6.	0.01876	0.26545	5	12	13	2.8	82.5%
7.	0.02228	0.25644	5	14	15	3.2	83.8%
8.	0.02580	0.24939	5	16	17	3.6	86.9%
9.	0.02932	0.24838	5	18	19	4	87.5%
10.	0.03284	0.24783	5	20	21	4.4	86.9%

Q_{CPLX} – see (7), R – the number of rules in the rule base, n_{INP} – the number of active inputs, n_{FS} – the number of active fuzzy sets, $n_{INP/R}$ – the number of active conditions per rule, Q_{RMSE} – the learning error, ACC – the percentages of correct decisions.

Fig. 4. The Pareto-front approximation generated by our SPEA3 (a) and interpretability and accuracy measures of solutions (b)

belonging to the original thematic groups "Theory", "Robotics and Vision", and "System", respectively. The last keyword "design" (in rules Nos. 1 and 4) usually does not occur in documents mostly belonging to groups "AI" and "Theory". The second solution (FRBC No. 9) is characterized by the best accuracy, however its rule base is much more complex and it is, to some extent, the extension of the first rule base (in fact, the rule bases of subsequent solutions from Fig. 4 are gradual extensions of the first solution's rule base; due to limited space of the paper, we cannot present all of them). This time, we can distinguish 18 stems of the most important keywords.

Table 2. Rule base of solutions (FRBCs) No. 1 and No. 9 from Fig. 4

No.	Classification rules

Solution No. 1 ($ACC = 64.4\%$):
1. **IF** method **AND** *not* design **THEN** Class "1"
2. **IF** constraint **THEN** Class "2"
3. **IF** softwar **THEN** Class "3"
4. **IF** *not* design **THEN** Class "4"

Solution No. 9 ($ACC = 87.5\%$):
1. **IF** method **AND** *not* infer **AND** *not* perform **AND** *not* design **AND** *not* goal **THEN** Class "1"
2. **IF** constraint **AND** *not* system **AND** *not* scene **THEN** Class "2"
3. **IF** softwar **AND** *not* interest **THEN** Class "3"
4. **IF** line **THEN** Class "3"
5. **IF** *not* design **AND** *not* learn **AND** *not* manipul **AND** *not* abil **AND** *not* pattern **AND** *not* softwar **AND** *not* object **AND** *not* suggest **AND** *not* large-scal **THEN** Class "4"

4 Conclusions

The concept of hybrid system which may be apply in a wide area of human language understanding problems and its exemplary application to knowledge discovery from full-text document collection have been presented. The system employs the self-organizing neural network with one dimensional neighborhood and dynamically evolving topological structure as well as multi-objective evolutionary designing technique of fuzzy rule-based classifiers with optimized accuracy-interpretability trade-off. The system is able - without any knowledge about the documents given a priori - (i) to automatically determine the number of groups in the document collection, (ii) to collect the documents in terms of their similarity, (iii) to extract the most important keywords from documents, and (iv) to generate classification rules which can be helpful in understanding and isolating the subjects of documents collected in the founded groups.

References

1. Franke, J., Nakhaeizadeh, G., Renz, I.: Text Mining: Theoretical Aspects and Applications. Physica/Springer, Heidelberg (2003)
2. Gorzałczany, M.B., Rudziński, F.: Cluster analysis via dynamic self-organizing neural networks. In: Rutkowski, L., Tadeusiewicz, R., Zadeh, L.A., Żurada, J.M. (eds.) Artificial Intelligence and Soft Computing - ICAISC 2006. Lecture Notes in Computer Science, vol. 4029, pp. 593–602. Springer, Heidelberg (2006)
3. Gorzałczany, M.B., Rudziński, F.: WWW-newsgroup-document clustering by means of dynamic self-organizing neural networks. In: Rutkowski, L., Tadeusiewicz, R., Zadeh, L.A., Żurada, J.M. (eds.) Artificial Intelligence and Soft Computing - ICAISC 2008. Lecture Notes in Computer Science, vol. 5097, pp. 40–51. Springer, Heidelberg (2008)
4. Gorzałczany, M.B., Rudziński, F.: Handling fuzzy systems' accuracy-interpretability trade-off by means of multi-objective evolutionary optimization methods - selected problems. Bull. Pol. Acad. Sci. Tech. Sci. **63**(3), 791–798 (2015)
5. Gorzałczany, M.B., Rudziński, F.: Generalized self-organizing maps for automatic determination of the number of clusters and their multiprototypes in cluster analysis. IEEE Trans. Neural Netw. Learn. Syst. **PP**(99), 1–13 (2017)
6. Leskovec, J., Rajaraman, A., Ullman, J.: Mining of Massive Datasets. Cambridge University Press, New York (2011)
7. Rudziński, F.: Finding sets of non-dominated solutions with high spread and well-balanced distribution using generalized strength Pareto evolutionary algorithm. In: Alonso, J.M., et al. (eds.) 2015 Conference on International Fuzzy Systems Association and European Society for Fuzzy Logic and Technology (IFSA-EUSFLAT-15), vol. 89, pp. 178–185. Atlantis Press, Gijón (2015)
8. Rudziński, F.: A multi-objective genetic optimization of interpretability-oriented fuzzy rule-based classifiers. Appl. Soft Comput. **38**, 118–133 (2016)
9. Zitzler, E., Laumanns, M., Thiele, L.: SPEA2: improving the strength Pareto evolutionary algorithm for multiobjective optimization. In: Proceeding of the Evolutionary Methods for Design, Optimisation, and Control, pp. 95–100. CIMNE, Barcelona (2002)

Microservices Architecture for Content-Based Indexing of Video Shots

Remigiusz Baran[1]([⊠]) ⓘ, Pavol Partila[2] ⓘ, and Rafał Wilk[3]

[1] Department of Computer Science, Electronic and Electrical Engineering
Kielce University of Technology, Kielce, Poland
r.baran@tu.kielce.pl
[2] Department of Telecommunications, VSB-Technical University of Ostrava,
Ostrava, Czech Republic
pavol.partila@vsb.cz
[3] Department of Teleinformatics, University of Computer Engineering and Telecommunications,
Kielce, Poland
r.wilk@wstkt.pl

Abstract. Three different content-based video indexing microservices dedicated to index video shots for the needs of the IMCOP Content Discovery Platform are presented in the paper. These three services as well as numerous others cooperate with each other within the IMCOP platform to describe, enrich and relate the multimedia data regarding their audio, textual and visual content. Owing to the analysis they perform, the IMCOP platform can discover, recommend and deliver the personalized multimedia content to various IMCOP's prospective recipients.

As these recipients may also require the personalized video content, services, as e.g. the presented ones, designed respectively to discriminate between characters in videos as well as text- and speech-based indexing of video shots, are absolutely essential. Goals of these services, their approaches and how they comply with objectives of the IMCOP's microservices architecture are carefully presented in the paper. Research procedures and the results of examinations that have been carried out to verify their pretty high accuracies are also reported and discussed.

Keywords: Video indexing · Text detection and recognition · Speech recognition
Face recognition · IMCOP platform

1 Introduction

Video indexing services presented in this paper have been developed as components of the IMCOP Content Discovery Platform [1]. The IMCOP system can serve, however, not only as a content discovery platform but also as a label and landmark detection tool, a reverse image search engine, an automatic generator of digital multimedia magazines, etc. To provide such a versatility and flexibility as well as a log-term agility for all its prospective applications, the IMCOP platform has been designed and implemented as microservices-based system [2]. Regarding the above, all its components, their properties, capabilities, etc. must fit the microservices architecture specifications.

K. Choroś et al. (Eds.): MISSI 2018, AISC 833, pp. 444–456, 2019.
https://doi.org/10.1007/978-3-319-98678-4_45

To achieve this goal IMCOP's services have been developed as wrapper functions that call other specialized applications or processes, as illustrated in Fig. 1.

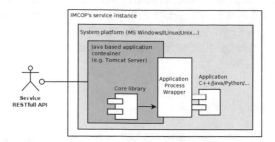

Fig. 1. Activity diagram of IMCOP's services.

As depicted in Fig. 1, IMCOP's microservices are RESTful web services with their own REST-based interfaces. Based on its RESTful API[1], services are capable to exchange messages with the IMCOP repository as well to get access to multimedia (alias media) objects – MOs (images, video clips, audio records, etc.) as return and store the results of performed analyses – the metadata (alias tags). In fact, only the descriptive metadata (including different feature descriptors) related to MOs are stored in the repository. MO objects, as such, in their original digital file formats (JPG, AVI, MP3, and so on) are stored only in special cases, namely when a long time preservation of multimedia data is required. Otherwise, only MOs' locations, given by their URLs, are preserved. A special metadata storage format, known as the Complex Multimedia Object (CMO) [3], has been invented and implemented to preserve the extracted metadata. CMO objects exist as XML documents, that are created in accordance to dedicated XSD Schema. Each MO object processed in the IMCOP system has its own CMO representation where the metadata related to it are preserved. However, the CMO objects are very capacious and they are able to store much more metadata then only the descriptive one. Relations, indicating the content-related connections between different MOs, are examples of such an extra metadata. Searching for connections between MOs is a goal of another specialized microservices.

The purpose of video indexing in the IMCOP system is to get various video materials ready for fast and efficient retrieval as well as for selecting video sequences adequately to different content-related requirements of IMCOP's prospective recipients. Microservices dedicated for that purpose are examples of IMCOP's data enrichment services, that are known also as Metadata Enhancement Services (the MES ones). Other numerous types of IMCOP's MES services are dedicated to analyze the content of still images, text objects, footage and other different forms of multimedia data. Metadata, being the results of this analysis, are related as well to various feature descriptors, including e.g. the Maximally Stable External Regions (MSER) ones [4] as to standard, e.g. MPEG-7's, descriptive schemes – labels, keywords, etc. The latter metadata types are usually the

---

[1] https://searchmicroservices.techtarget.com/definition/RESTful-API.

results of different object detection and classification schemes where the descriptors mentioned above are, in turn, the basis.

The remainder of this paper is structured as follows. Description of IMCOP's video indexing services and how they comply with the IMCOP's objectives is given in next Sect. 2. Report on their performance parameters including their accuracy with regard to tests carried out on selected video shots is presented, in turn, in Sect. 3. Short summary of the paper is given in Sect. 4.

2   Video Indexing Services

The problem of automatic content-based video indexing refers mainly to low-level audio-visual features (metadata) that can be extracted automatically to extend the video data description with respect to its audio, motion, color, shape and texture characteristics, etc. In turn, high-level features, represented for example by objects that can be identified and named within the visual content, allow to annotate video clips semantically [5]. This causes, that the so-called semantic-based approach to multimedia (video) indexing is a very "hot" topic of research, especially recently [6]. Of course, high-level metadata are usually the result of classification schemes that make their predictions on the base of low-level features [7]. Actually, extraction of high- and low-level metadata is essential not only for content-based video indexing and retrieval but also for quite other applications, including e.g. fighting against child pornography [8].

A vast variety of application dependent and thematically differentiated video contents that may be processed and indexed in the IMCOP platform lead to the problem of their proper annotation, which should be made with regard to application/user semantic requirements. According to the concept of microservices architecture, each IMCOP service should be internally coherent, independent from other services, as small as possible and should run in its own process within a specific end-to-end domain while maintaining the capability to communicate with other processes[2]. Regarding the above, the metadata enhancement tasks in the IMCOP platform have been designed as the scenarios where different and precisely specified MES services are involved in a scheduled manner. For example, with reference e.g. to [9], video indexing tasks can be divided between several MES microservices consecutively responsible for: Shot Boundary Detection (SBD), abstraction of video shots [10] and feature extraction. Of course, the IMCOP platform is capable to carry out many altered scenarios where different MES services (based on various approaches) may be involved. Numerous distinct and specialized MES microservices, managed in that way, create the Data Enrichment Engine (DEE) of the IMCOP platform. In general, DEE's microservices are provided in the cloud, regarding the overall computing model depicted in Fig. 2.

---

[2] https://docs.microsoft.com/en-us/dotnet/standard/microservices-architecture/architect-micro-service-container-applications/service-oriented-architecture.

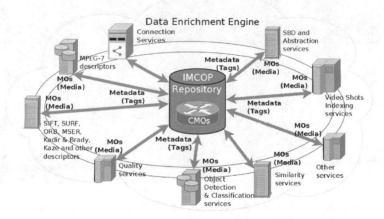

Fig. 2. Data enrichment engine (DEE) of the IMCOP platform.

Three different video indexing microservices have been developed, so far, to analyze the content of video shots and index them in the IMCOP platform. They can be applied to video shots and other basic as well as more complex video sequences where the SBD step can be omitted or (because of application requirements) is not needed. In particular, they have been dedicated to:

- identify faces and assign them to different classes (distinct characters/persons) regarding the similarity scores between them (namely, *facerv* service),
- detect and recognize text objects (written words or sentences) within the natural content of video frames (namely, *textdrv* service),
- recognize selected spoken words or phrases (sentences) on the base of applied speech recognition toolkit (namely, *speechrv* service).

Specific characteristic of each of the IMCOP's video indexing services will be given in the following sub-sections of Sect. 2. However, there can be drawn a generalized scheme with regard to which each of the above services is carried out - see Fig. 3[3].

---

[3] The flowchart presented in Figs. 3, 4, 6 and 8 have been drawn under inspiration of the Fuji Xerox Video Indexing Technology website: https://www.fujixerox.com/eng/company/technology/production/multimedia/talkminer.html.

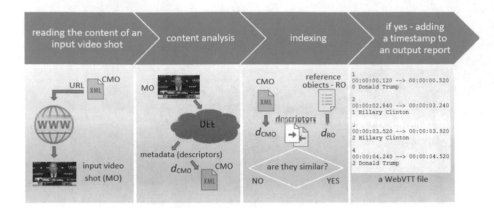

Fig. 3. A generalized flowchart of IMCOP's video indexing services.

2.1   The *facerv* Service

The *facerv* service has been designed to detect, extract and finally group the human faces with respect to their mutual similarities. In general, the *facerv* service extracts and performs every n-th frame of an input video shot ($1 \leq n$). Extracted frames F_i, where i = 1, 2, 3 ..., are searched for human faces, wherein recognized Viola-Jones algorithm [11] for face detection and localization is utilized within this context. Sub-images with detected faces are extracted as distinct pictures and converted to their grayscale variants - I_{ij}, where j = 1, 2, 3 ... (more than one face image can be extracted from performed frame). Next, each I_{ij} picture is normalized (scaled and rotated) to get finally the single face image of fixed size - N_{ij}, with face localized in its center and with the eye line horizontally aligned. Normalization procedure applied in that stage has been described in details in [12]. After that, Oriented FAST and Rotated BRIEF (ORB) [13] features - interest points (**IP**) are extracted and ORB descriptors (**D**) corresponding to these interest points are calculated for each N_{ij} image, wherein:

$$IP = \left\{ \left(x_{IP_k}, y_{IP_k}, s \right) \right\}_{k=1}^{k_{max}} \tag{1}$$

where x_{IP_k} and y_{IP_k} are the vertical and horizontal coordinates of k-th interest point (k = 1, 2, ..., k_{max}), at a given scale s in multiscale space of ORB transformation, and

$$D = \left\{ d_k \right\}_{k=1}^{k_{max}} \tag{2}$$

where $d_k = \left\{ d_{kq} \right\}_{q=1}^{q_{max}}$ is the q-dimensional descriptor of the k-th interest point.

Regarding the above, the face descriptor FD corresponding to a given N image can be defined as follows:

$$FD = (KP, DM) = \left\{ \left(\left(x_{IP_k}, y_{IP_k}, s \right), d_k \right) \right\}_{k=1}^{k_{max}} \tag{3}$$

Similarity measure of two distinct face images N_{ij} and N_{mn}, represented by descriptors FD_{ij} and FD_{mn}, respectively, is calculated as follows:

$$d\left(FD_{ij}, FD_{mn}\right) = \frac{m_{max}}{k_{max_{ij}}}$$

(4)

where m_{max} is the size of matrix known as the matcher object[4], where distances between matched descriptors [12], in sense of the HAMMING norm, are stored.

Finally, all faces that have been detected, extracted and carried out, are clustered to distinct clusters, where similar faces are grouped. An overall flowchart of the *facerv* inner process is depicted in Fig. 4. As *facerv* is the process that goes on in time, consecutive clusters, referred to distinct persons (dissimilar anchormen, actors, characters, etc.) appearing in successive moments of time in the video shot, are also determined/updated in a successive manner. Consecutively created clusters are enumerated in ascending order, wherein the cluster with #0 is the cluster that contains the first extracted face as well as all other faces that have been marked, in the next moments of time, as similar to the first one. However, the first face dissimilar to these from the cluster #0, creates, in turn, cluster #1. Remaining faces, extracted successively while the indexing process proceeds, can be included either into the cluster #0 or cluster #1. Nevertheless, they can be also parts of other clusters with higher numbers.

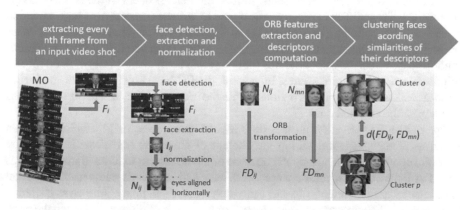

Fig. 4. An overall flowchart of the *facerv* service.

An example of the *facerv* report (selected rows of an output WebVTT file), obtained as a result of indexing of a sample video shot – "VSs01" [14], is depicted in Fig. 5a. Presented report shows that clusters returned by the *facerv* service are not labeled automatically. Labels, e.g. names of the characters represented by particular clusters (as given for instance in Fig. 5b), must be assigned manually, by prospective IMCOP's end users. Actions like that, taken in the post-processing steps, are parts of the social recommendation layer of the IMCOP platform.

---

[4] https://docs.opencv.org/3.3.0/dc/dc3/tutorial_py_matcher.html.

a)
```
1
00:00:00.120 --> 00:00:00.520
0

2
00:00:02.840 --> 00:00:03.240
1

3
00:00:03.520 --> 00:00:03.920
2

4
00:00:04.120 --> 00:00:04.520
3
```

b)
```
1
00:00:00.120 --> 00:00:00.520
0 Donald Trump

2
00:00:02.840 --> 00:00:03.240
1 Hillary Clinton

3
00:00:03.520 --> 00:00:03.920
2 Hillary Clinton

4
00:00:04.240 --> 00:00:04.520
3 Donald Trump
```

Fig. 5. An exemplary *facerv* report in its original form (a) and after labels assignment (b).

Figure 5 shows that one and the same character, appearing in the video shot, may refer to several clusters. It may happen because of many factors, including e.g. facial expression, emotions, etc., that vary while the indexing process proceeds.

2.2 The *textdrv* Service

The *textdrv* service is dedicated to detect, extract and finally recognize text present in natural content of an input video shots. In fact, the inner process of the *textdrv* service is composed of two different processes designed to text regions detection and extraction that are called in parallel. One of them is an OpenCV-based application while the other one is our own implementation that has been written from scratch. Both applications mentioned above can be categorized as the connected component (CCs) method for text regions detection in natural scenes. In general, the CC-based methods focus on localization and extraction of uniform regions within the image content, where the text regions are amongst the selected ones.

The our own application utilizes the MSER feature detector [4], which has been proven as pretty successful within the context of the scene text detection by many scientific reports [15]. This is mainly because of its robustness to affine transformations as well as light and viewpoint changes. It detects and extracts a number of co-variant stable connected components of some (gray-level) sets of an image, known as Extremal Regions (ERs). ERs regions include as well text as non-text objects. To reach the goal, regions containing non-text objects must be rejected. Remaining regions (probably the text ones) are the MSER regions of interest – Q. To solve this problem, a set of especially designed filters is applied. Actually, our application (known as the MSER-based one) is an extension of an approach presented in details in [16]. Above report specifies all the applied filters, including two the novel ones referred to the height and the relative positions between regions, respectively. However, to improve the accuracy of the *textdrv* service, an additional filter has been added. The goal of this filter is to reject these regions that are not in line regarding thickness of their edges, as proposed in [15].

The OpenCV-based implementation[5], in turn, refers to the scene text detection approach, known as the Class Specific Extremal Regions (CSER) [17], where CSER

---

[5] https://docs.opencv.org/3.0-beta/modules/text/doc/erfilter.html.

regions (*R*) are generalization of the MSER ones. In general, the main difference between the CSER approach and the MSER one is a learned sequential classifier that is used (in the CSER case) to select suitable, not necessarily maximally stable (but class specific) ERs, instead of a set of specialized filters (as in the MSER-based approach). The CSER classifier is a two-stage-classifier that uses different sets of descriptors, including area, perimeter, bounding box, Euler number, etc. (in the first stage) and hole area ratio, convex hull ratio, the number of outer boundary inflexion points (in the second stage) to finally classify the ERs into character and non-character categories. The CSER detector is robust to blur, illumination, color and to partial occlusions. It is also affine invariant.

As in the case of *facerv* service, the *textdrv* one extracts and performs every n-th frame of an input video shot ($1 \leq n$). Text objects (written words, sentences or strings) returned by MSER- and CSER-based processes (T_{MS} and T_{CS}, respectively) are then recognized using the Tesseract OCR engine[6] and verified by checking them up with the dictionary. Finally, verified objects are compared against the reference text objects (alias dictionary words). Reference text objects (given by supervisory/management services or by the user) are the objects that are searched for when indexing a video shot.

An overall flowchart of the *textdrv* inner process is depicted in Fig. 6.

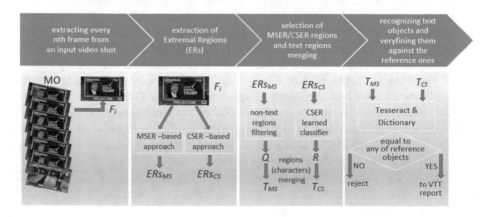

Fig. 6. An overall flowchart of the *textdrv* service.

An exemplary output report of the *textdrv* service, obtained as results of indexing sample video shot – "VSs02" [14], is depicted in Fig. 7a. View of the *textdrv* service user interface with reference text objects ("Dictionary words") set just before the test has started is depicted, in turn, in Fig. 7b.

---

[6] https://opensource.google.com/projects/tesseract.

a)   b)

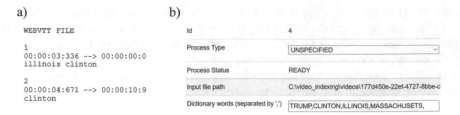

Fig. 7. Exemplary *textdrv* results of indexing video shot "VSs02" [14].

As depicted in Fig. 7a, the *textdrv* service is capable to report presence of two (or even more) text objects in the same period of time.

2.3   The *speechrv* Service

The *speechrv* MES service wraps a speech recognition application that has been based on Kaldi [18] toolkit, which is an open-source toolkit for automatic speech recognition (ASR). Deep neural networks used for training the acoustic models (AMs) make Kaldi pretty advanced and state-of-the-art toolkit for speech recognition. Kaldi also benefits greatly from the Weighted Finite-State Transducers (WFST) model that is used to decode the speech. Core Kaldi tool for creating, optimizing and searching the WFST model is the OpenFST library. Both the above mentioned merits significantly reduce the complexity of ASR systems based on Kaldi toolkit[7]. Systems like that usually apply a mixed Gaussian Mixture - Hidden Markov model (GMM-HMM) to recognize the speech. They also use Mel-Frequency Cesptral Coefficients (MFCC) transformation (MFCC descriptors – d_{MFCC}) [19], which is one of the most successful methods for speech parametrization. The goal of speech parametrization is to reduce the negative aspects of speech variability, including e.g. phonetic or pronunciation differences among various speakers that depends on their age, gender, voice, and so on. Because of this, using the MFCC transformation improves significantly robustness of ASR systems, especially in the case of newscasts, television debates, etc. Thus, the MFCC features were also our choice.

The *speechrv* service has been designed to recognize and mark in time, in analyzed audio-video sequence, selected spoken words and short sentences, set as the reference speech objects. The *speechrv* process starts from extracting an audio (speech) signal – A, from an input video shot (MO). Signal A is then carried out by the Kaldi ASR system. To train the AM model as well as to prepare the language model (LM) for *speechrv* service, an open English language speech corpus, taken from Voxforge[8], has been used. An amount of data represented by speech recordings downloaded within this corpus was about 1.8 GB. Following the Kaldi implementation, to train the data the MFCC-GMM-HMM model has been applied. Speech objects recognized and returned by the Kaldi ASR system – ω, are finally compared to reference speech objects. The overall flowchart of the *speechrv* ASR system is depicted in Fig. 8.

---

[7] http://kaldi-asr.org/.

[8] http://www.voxforge.org/.

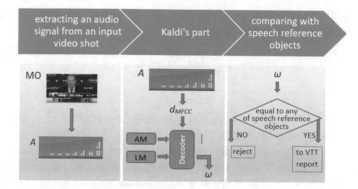

Fig. 8. An overall flowchart of the *speechrv* service.

The inner speech recognition process of the *speechrv* service has been implemented with the use of the OpenFST ver. 1.3.4.

An exemplary output report of the *speechrv* service, obtained as results of indexing sample video shot – "VSs03" [14], is depicted in Fig. 9a. View of the *speechrv* service user interface just before the test has started is depicted, in turn, in Fig. 9b.

a)

```
WEBVTT FILE

1
00:00:00.270 --> 00:00:05.160
JERSEY

2
00:00:05.260 --> 00:00:09.899
JERSEY
```

b)

Id	2
Process Type	SPEECH_RECOGNITION
Process Status	PROCESSED
Input file path	/home/imcop/Dokumenty/speech/upload/57e765d3-7fc0-4b5d-ae9
Dictionary words (separated by ',')	VIRGINIA,OREGON,JERSEY

Fig. 9. Exemplary *speechrv* results of indexing video shots "VSs03" [14].

As depicted in Fig. 9b, the reference speech objects that have been searched for when sample video shot "VSs03" was indexed, contain three selected words. However, a WebVTT file presented in Fig. 9a contains only one from these words. In fact, it is accurate, because words Virginia and Oregon have not been spoken in "VSs03" shot.

3 Experiments and Accuracy Report

To verify the accuracy of video indexing services presented in this paper, a set of test video shots have been prepared. As the basis for building the test set, a number of selected Youtube video sequences, related to the United States presidential election of 2016, has been downloaded and segmented into short video shots. Test video shots should be short (no longer than 30 s) because of limited computational resources that can be currently allocated for presented services in the DEE cloud. Examples of test video shots as well as links to the their source video sequences can be found in [14].

To verify the efficiency of the proposed approaches the True Positive Rate (TPR) measure, defined as follows, has been applied:

$$TPR = \frac{TP}{TP + FN} 100\% \tag{5}$$

where:

TP – Is the number of correctly tagged moments (true positive alias hit) while

FN – is the number of missed moments in test shots (false negative alias miss).

Experiments we have carried out consisted, in general, in fixing the reference objects (text or speech objects to be searched for) in the cases of *textdrv* and *speechrv* services, as depicted in Figs. 7b and 9b, respectively (reference objects are not required in the case of *facerv* service) and careful analysis of timestamps in output reports. Purpose of this analysis was to classify the timestamps as the correctly tagged moments (*TP*) or the missed ones (*FN*) wherein all properly detected and recognized text and speech objects (where properly means also accurately with respect to periods of time given in time-stamps) have been classified as the correctly tagged moments. In turn, all these fragments of tested video shots where a given reference object was present but it has not been recorded in a WebVTT file have been classified as the missed moments.

In the case of *facerv* service as the correctly tagged moments have been classified only these records of VTT files that were related to the same single human face (of the same character/person/newscaster/etc.) present on video shot throughout the entire period of time given in a timestamp. Moments in time where not detected and recorded faces were present, have been categorized as the missed ones.

Regarding the above, the True Positive Rates we obtained during the experiments carried out with presented video indexing services, with respect to the number of 200 extracted and selected test video shots, were as follows:

- for *facerv* service – **98.1%**,
- for *textdrv* service – **71.3%** (This is the cumulative value being the result of combining two scene text detection approaches. When applied apart, the TPR is: 70% - for the MSER-based approach and 65% - for the CSER-based one),
- for *spechrv* service – **90.5%**.

There can be indicated some reasons why the TPR measure, obtained in the case of *texdrv* service, was pretty low. The main reason is that scene text detection is still a challenge. Variety of text objects distortions that are usually observed in natural scene images lead to many wrong detections of ERs regions in CC-based approaches. Very destructive in this context are perspective distortions, including e.g. 3-D rotations, that are very often used as special effects in modern TV newscasts, especially during the election nights. They were probably the reasons of many missed moments during our examinations. Other *FN* cases could have been caused by multicolored graphics that are also pretty often used in such TV shows to emphasize the text.

In turn, the TPR rate of *speechrv* service strongly depends on speech corpus that have been used to train its Kaldi ASR component. As given in Sect. 2.3, the Voxforge English language speech corpus have been used during presented research. The main reason why we have chosen the Voxforge corpus, instead e.g. Kaldi's English Broadcast News recipes[9], is that it is open and can be accessed without any fees.

4 Summary

Three different content-based video indexing services dedicated to index video shots for the needs of the IMCOP Content Discovery Platform [1] have been presented in the paper. They have been designated, respectively, to trace and classify the faces detected within the video content as well as to search this content for given reference text and speech objects. Two of the presented services have been based on novel approaches, described in details in [12] (*facerv*) and [16] (*textdrv*). In the case of *textdrv* service, the novel approach, known as the MSER-based one, has been combined with an OpenCV CSER-based implementation. However, as shown is Sect. 3, the novel MSER-based approach was more accurate than the CSER-based one – with respect to performed tests. The main advantages of presented services and the IMCOP's microservices architecture are their flexibility and versatility as well as pretty high accuracy.

References

1. Baran, R., Dziech, A., Zeja, A.: A capable multimedia content discovery platform based on visual content analysis and intelligent data enrichment. Multimed. Tools Appl., 1–15 (2017). https://doi.org/10.1007/s11042-017-5014-1
2. Wolff, E.: Microservices: Flexible Software Architectures. Addison-Wesley, Boston (2016)
3. Baran, R., Zeja, A.: The IMCOP system for data enrichment and content discovery and delivery. In: Proceedings of the 2015 International Conference on Computational Science and Computational Intelligence, Las Vegas, USA, pp. 143–146 (2015)
4. Matas, J., Chum, O., Urban, M., Pajdla, T.: Robust wide-baseline stereo from maximally stable extremal regions. Image Vis. Comput. **22**(10), 761–767 (2004)
5. Bloehdorn, S., et al.: Semantic annotation of images and videos for multimedia analysis. In: Gómez-Pérez, A., Euzenat J. (eds.) The Semantic Web: Research and Applications. ESWC 2005. LNCS, vol. 3532, pp. 592–607. Springer, Heidelberg (2005)
6. Budnik, M., et al.: Learned features versus engineered features for semantic video indexing. In: 13th International Workshop on Content-Based Multimedia Indexing, Prague, pp. 1–6 (2015)
7. Leszczuk, M., Grega, M.: Prototype software for video summary of bronchoscopy procedures with the use of mechanisms designed to identify, index and search. In: Piętka, E., Kawa, J. (eds.) Information Technologies in Biomedicine. Advances in Intelligent and Soft Computing, vol. 69, pp. 587–598. Springer, Heidelberg (2010)
8. Grega, M., et al.: Multimed. Tools Appl. **68**(1), 95–110 (2014)
9. Zhang, H.J., Wu, J., Zhong, D., Smoliar, S.W.: An integrated system for content-based video retrieval and browsing. Pattern Recognit. **30**(4), 643–658 (1997)
10. Leszczuk, M., et al.: Video summarization framework for newscasts and reports – work in progress. In: Dziech, A., Czyżewski, A. (eds.) MCSS 2017, CCIS, vol. 785, pp. 86–97. Springer, Cham (2017)

---

[9] https://github.com/kaldi-asr/kaldi/tree/master/egs/hub4_english/s5.

11. Viola, P., Jones, M.: Rapid object detection using a boosted cascade of simple features. In: Proceedings of the 2001 International Conference on Computer Vision and Pattern Recognition (CVPR), Kauai, HI, USA, vol. 1, pp. 511–518. IEEE (2001)
12. Baran, R., et al.: Face recognition for movie character and actor discrimination based on similarity scores. In: Proceedings of the 2016 International Conference on Computational Science and Computational Intelligence (CSCI), pp. 1333–1338. IEEE, Las Vegas (2016)
13. Rublee, E., et al.: ORB: an efficient alternative to SIFT or SURF. In: 13th International Conference on Computer Vision (ICCV), pp. 2564 2571. IEEE, Barcelona (2011)
14. http://research.wstkt.pl/?page_id=88
15. Chen, S.S., et al.: Robust text detection in natural images with edge-enhanced maximally stable extremal regions. In: Proceedings of the 18th International Conference on Image Processing, Brussels, pp. 2609–2612. IEEE (2011)
16. Baran, R., Partila, P., Wilk, R.: Automated text detection and character recognition in natural scenes based on local image features and contour processing techniques. In: Karwowski, W., Ahram, T. (eds.) IHSI 2018, AISC, vol. 722, pp. 42–48. Springer, Cham (2018)
17. Neumann, L., Matas, J.: Real-time scene text localization and recognition. In: Proceedings of the 2012 International Conference on Computer Vision and Pattern Recognition (CVPR), Providence, RI, USA, pp. 3538–3545. IEEE (2012)
18. Povey, D., Ghoshal, A., Boulianne, G., et al.: The Kaldi speech recognition toolkit. In: Proceedings of the Workshop on Automatic Speech Recognition and Understanding. IEEE, Big Island (2011)
19. O'Shaughnesssy, D.: Invited paper: automatic speech recognition: history, methods and challenges. Pattern Recognit. **41**(10), 2965–2979 (2008)

Intelligent Audio Processing

Spatial Audio Scene Characterization (SASC)

Automatic Classification of Five-Channel Surround Sound Recordings According to the Foreground and Background Content

Sławomir K. Zieliński$^{(\boxtimes)}$

Faculty of Computer Science, Białystok University of Technology,
Białystok, Poland
s.zielinski@pb.edu.pl

Abstract. Spatial audio becomes increasingly popular in domestic and mobile multimedia applications. Evaluating quality of experience (QoE) of such applications requires the development of algorithms capable of identification and quantification of perceptual characteristics of spatial audio scenes. This paper introduces a method for the automatic categorization of surround sound recordings using a criterion based on the distribution of foreground and background audio content around a listener. The principles of the method were demonstrated using a study in which a corpus of 110 five-channel surround sound recordings was computationally classified according to the two basic spatial audio scene categories. In order to develop the proposed method a novel metric, representing spatial audio characteristics, was identified. Moreover, five machine learning algorithms, including neural networks, random forests and support vector machines, were employed and their performance compared. According to the obtained results, the proposed method was capable of categorization of surround sound recordings reaching accuracy of 99%.

Keywords: Spatial audio · Multimedia systems · Machine learning
Quality of experience

1 Introduction

The modern multimedia systems provide the means of surrounding listeners with sound exhibiting three-dimensional characteristics. Spatial sound is nowadays commonly used in home cinema systems, virtual reality applications, computer games, TV broadcasts, and it has recently been introduced to one of the most popular video-sharing Internet services. There are many popular technologies used for transmission and rendering of spatial audio, including 5.1-channel standard, 22.2-channel format, Ambisonics, binaural audio techniques, and spatial audio object-based coding [1–3].

Auditory images reproduced by spatial audio systems can either be assessed using high-level attributes, such as *basic audio quality* [4], *spatial audio quality* [5], *overall listening experience* [6], or by means of low-level attributes, including *envelopment*, *scene width* and *localizability* [7]. In the latter case, more detailed information

© Springer Nature Switzerland AG 2019
K. Choroś et al. (Eds.): MISSI 2018, AISC 833, pp. 459–469, 2019.
https://doi.org/10.1007/978-3-319-98678-4_46

regarding the performance of spatial audio systems could be acquired, which is essential in terms of quality of experience (QoE) assessment.

The psychoacoustical studies aiming to define a taxonomy of low-level spatial audio attributes were pioneered by such researchers as Berg and Rumsey [8] and extended, among others, by Le Bagousse et al. [9]. Over the past fifteen years, the scholars reached a considerable consensus regarding a common lexicon of low-level spatial audio attributes [7, 9, 10]. Rumsey introduced a scene-based paradigm, whereby the auditory images could be characterized using geometrical properties of sound scenes, e.g. *source distance, ensemble depth* and *scene depth* [11]. This approach was simplified by Zieliński et al. [12], who proposed to describe spatial audio scenes using two basic categories, depending on the spatial distribution of the foreground and background audio content. Such simplified categorization was employed by Conetta et al. [5] to characterize audio material selected for the listening tests. This simplified categorization was also adopted in this study (see Table 1).

Table 1. Taxonomy of basic spatial audio scenes (inspired by Rumsey [11]).

Acoustic scene	Distribution of audio content across channels
Foreground-background (FB)	Front loudspeakers reproduce predominantly foreground audio content (identifiable, important and clearly perceived audio sources), whereas the rear loudspeakers reproduce only background audio content (room response, reverberant, unimportant, unclear, ambient, and "foggy" sounds)
Foreground-foreground (FF)	Both front and rear loudspeakers reproduce foreground content. This scene may refer to the audio impression where a listener is surrounded by an ensemble of musicians or simultaneously talking speakers

The aim of this study is to introduce a machine learning-based method for characterization of basic spatial audio scenes, according to the distribution of foreground and background audio content. The developed method is currently limited to the classification of scenes of five-channel surround sound recordings. The ITU-standardized [1] five-channel surround sound loudspeaker layout is illustrated in Fig. 1.

The proposed method could be used as a "building block" of a quality of experience (QoE) model for multimedia applications incorporating spatial audio. It could also be exploited by researchers as a tool for automatic selection of audio material for subjective audio quality tests. Moreover, the method could be applied for the purposes of semantic search and retrieval of information from spatial audio recordings.

The paper offers the following contributions: (1) it introduces a method for the automatic categorization of basic spatial audio scenes according to a foreground and background audio content distribution; (2) it introduces a new metric conducive for differentiation between the basic audio scenes, (3) it compares the performance of the five machine learning-based algorithms applied to classify surround sound recordings using the aforementioned approach.

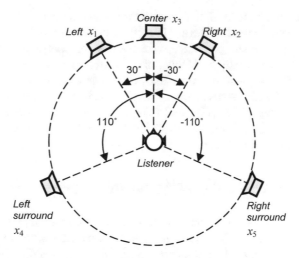

Fig. 1. Loudspeaker layout of the five-channel surround sound system conformant to the ITU-R BS.775 Recommendation [1].

This study extends and builds on our earlier work [13], which was predominantly concerned with the feature extraction and identification of the best topology for the integration of multichannel audio information (*early fusion* versus *late fusion* schemes). In comparison with our previous contribution, this paper introduces a new metric designed to differentiate between the basic spatial audio scenes. While in our previous study only a single machine learning algorithm was employed (random forest), in this work five machine learning algorithms were employed and compared, yielding much improved results. The classification accuracy between the previous and the present study was enhanced from 84% up to 99%, respectively.

2   Related Work

While several objective methods aiming to assess the quality of spatial audio using high-level attributes have already been developed [12, 14, 15], only a few studies have been devoted to the objective characterization of spatial audio in terms of the low-level attributes. George et al. [16] developed a so-called "Envelometer", that is the algorithm capable of quantification of envelopment sensation evoked by spatial audio scenes. Their method was intended for the five-channel surround sound systems. More recently, Francombe et al. [17] proposed another method for the quantification of envelopment, compatible with the object-based spatial audio format. However, to the best of the author's knowledge, no work has been done towards the objective characterization of spatial audio using other low-level attributes than envelopment, which constituted the motivation underlying this work.

Despite the nomenclature similarity, the purpose of the proposed method is distinctively different compared to the other state-of-the-art "machine-listening" approaches. Three growing fields can be distinguished in the literature in this respect, namely, acoustic event recognition (AER), acoustic scene classification (ASC), and computational auditory scene analysis (CASA) [18, 19]. The aim of the AER is to identify the source of an auditory event (e.g. voice, vacuum-cleaner, footsteps), whereas the purpose of the ASC is typically to identify the environment in which the sound was recorded (e.g. office, kitchen, metro station). CASA is predominantly concerned with isolating individual audio sources from complex auditory scenes. In contrast to the above three approaches (AES, ASC, and CASA), the purpose of the proposed method of spatial audio scene characterization (SASC) is to identify and/or quantify basic properties of spatial audio scenes.

3 Corpus and Annotation of Audio Recordings

A corpus of 110 five-channel surround sound excerpts (5.0 format [1]) was gathered for the purpose of this work. The recordings were extracted from the commercially available DVD recordings. The mean duration of the acquired audio samples was equal to 11.6 s with a standard deviation of 7.2 s. The recordings represented a broad range of genres, including classical music, pop music, jazz, and movies. The recordings were sampled at a 48 kHz rate with a 16-bit resolution and stored as uncompressed multi-channel audio files. During the selection of the recordings, attention was paid that each excerpt represented a single spatial scene (either *FB* or *FF*).

Each recording in the corpus was manually annotated by this author using either an *FB* or *FF* label. The corpus of audio recordings was slightly imbalanced, since 57 recordings represented the *FB* scene (52%), whereas the remaining 53 recordings were annotated as exhibiting the *FF* scene (48%). There were also some differences between the recording genres in terms of the spatial scenes. Most of the *FB* excerpts were represented by such genres as pop music, classical music, film music, movies, and jazz. In the case of the *FF* recordings, the most frequent categories were as follows: pop music, rock, sound effects, electronic music, and blues.

As mentioned above, in the present study the audio recordings were manually annotated by a single researcher (the author). One cannot exclude a possibility that other persons (e.g. experienced sound engineers) would create a slightly different division into *FB* and *FF* content. While the present approach was deemed to be acceptable at this stage of the work, in the future studies it is planned to involve more researchers in the annotation procedure. Moreover, in order to increase the validity of the annotation task, formal subjective tests are also envisioned.

4 Feature Extraction

In this Section, a brief overview of the feature extraction procedure is provided. Due to a space limitation, only a new metric (*theta*) is explained in more detail. For further information regarding the parametrization of spatial audio, an interested reader is referred to our previous paper [13], which was solely devoted to the topic of the feature extraction of the five-channel surround sound recordings.

The input audio signals were divided into time-frames of 42.6 ms in duration (2048 samples) with a 50% overlap factor. In addition, the signals were split in frequency using a gammatone filter-bank with the 24 ERB-spaced filters. The center frequency of the first and the last filter was equal to 100 Hz and 16 000 Hz, respectively.

Prior to extracting the metrics, three types of down-mix signals were derived. The first one was calculated as an unweighted sum of all signals: $x_{all} = x_1 + x_2 + ... + x_5$, where x_n represented the n–th channel signal (see Fig. 1). The second one was obtained by summing the three front channel signals: $x_{front} = x_1 + x_2 + x_3$, whereas the last one was estimated as a sum of the rear channel signals: $x_{back} = x_4 + x_5$. Then, for each time-frame and every frequency band, two types of metrics were normally computed: *front-to-back* and *all-to-front difference metrics*. Front-to-back difference metrics were calculated as $m(x_{front}) - m(x_{back})$, whereas all-to-front difference metrics were derived as $m(x_{all}) - m(x_{front})$. The symbol m represents a given signal descriptor, such as, for instance, loudness or zero-crossing. The difference metrics calculated across time-frames and frequency bands were summarized using the mean value, standard deviation and the range.

4.1 Theta Metric

A directional metric *theta* was designed for the purpose of scene classification. It could be interpreted as an angle of incidence (azimuth) of a dominant sound source for a given time-frame and frequency band. It was defined using the following equation: $\theta = \arg(v)$, where

$$v = \sum_{n=1}^{5} v_{rms,n} \exp(j\theta_n). \tag{1}$$

The term $v_{rms,\,n}$ in the above equation denotes the rms value of the n-th channel signal x_n, while $j^2 = -1$. The angle θ_n is the azimuth of the n-th loudspeaker with respect to the line crossing the listener and the center loudspeaker (Fig. 1). An example of the distribution of the directional vector v, for the two selected audio recordings exhibiting the *FB* and *FF* scenes respectively, is shown in Fig. 2. As can be seen, the distribution of the angle of incidence (*theta*) for the *FF* scene is broader, compared to the *FB* one.

4.2 Remaining Metrics

All the metrics extracted for the purpose of this study could be divided into six categories: *spatial*, *spectral*, *cepstral*, *binaural*, *temporal* and *loudness-based*. They were outlined below.

The introduced above *theta* metric, *percentage variance* of the PCA components of the surround sound signals [16], *spatial cepstrum* [19], and *PCA centroid* [13] could be classified as spatial features. In addition, the following spectral descriptors, commonly used for the music information retrieval applications [20], were also included in the study: *centroid, spread, brightness, high-frequency content, crest, decrease, entropy, flatness, irregularity, kurtosis, skewness, roll-off, flux* and *spectral variation*.

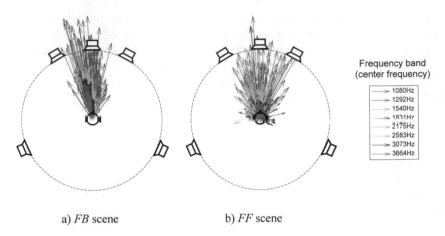

a) *FB* scene   b) *FF* scene

Fig. 2. Examples of the distribution of the directional vector for the two selected audio recordings representing *FB* and *FF* scenes, respectively. For clarity, the results are limited to eight frequency bands (out of 24).

The standard mel-frequency cepstral coefficients (MFCCs) [18] were calculated for the down-mixed signals x_{all}, x_{front} and x_{back}, respectively. Then, the Euclidean distances between these coefficients were estimated. Moreover, the popular binaural audio descriptors were also incorporated in the study, including interaural level difference (*ILD*), interaural time difference (*ITD*) and interaural coherence (*IC*) [2]. Furthermore, the following temporal metrics were included in the study: *zero crossing*, *crest factor* and *cross-correlation coefficient* between the rate maps. Finally, the loudness-based features were also incorporated in the study. The loudness of audio signals was estimated using the method developed by Glasberg and Moore [21].

5   Machine Learning-Based Classification

The following five algorithms were compared in terms of their ability to classify the spatial scenes: *k*-nearest neighbors algorithm (*k*-NN), logistic regression with a least absolute shrinkage and selection operator (lasso [22]), random forest, neural network, and support vector machine (svm). They were selected since they represent typical methods currently applied to the task of audio classification [18].

The parameters of the classification algorithms used for the final model were as follows. For the *k*-NN method, the value of *k* was set to 7. As far as the lasso regression is concerned, the parameters alpha and lambda were set to 0.1 and 624.97×10^{-6}, respectively. The random forest consisted of 500 trees and a number of signal metrics randomly sampled as candidates at each split was set to 2. The single-hidden-layer neural network was used with five hidden units and a weight decay equal to 0.1. For the support vector machine, the radial basis function kernel was used, with the values of the parameters sigma and *C* (cost) set to 0.1023013 and 0.25, respectively.

In typical machine learning experiments, the database is normally split into two subsets used for training and validation, respectively. However, when the data set is small, resampling techniques are applied instead [22]. Since the database used in this study contained only 110 items, the latter approach was adopted in this study in order to evaluate the performance of the machine learning-based algorithms. To this end, a standard method of a 10-fold cross-validation repeated 10 times was applied.

6 Results

A dataset containing the 166 metrics, overviewed earlier in Sect. 4, was employed to undertake the classification task. The classification results of the initial model are presented in the middle row in Table 2. The results obtained for the initial models could be regarded as satisfactory since an accuracy achieved for the lasso regression and the random forest methods exceeded 90%. However, considering that the all 166 metrics were used in the initial models, there was a strong chance of model overfitting. Therefore, it was decided to reduce the set of metrics.

Table 2. Classification Results (Accuracy). Results in brackets represent standard deviation.

Model	k-NN	Lasso	Random forest	Neural network	svm
Initial	0.785 (0.10)	0.9048 (0.09)	0.9045 (0.08)	0.890 (0.09)	0.885 (0.10)
Final	0.906 (0.08)	0.999 (0.01)	0.923 (0.08)	0.983 (0.04)	0.9128 (0.08)

The procedure employed to reduce the number of metrics was undertaken iteratively, by means of a backward stepwise selection technique [22]. Initially, all 166 metrics were included in the model, and then in each iteration the least useful metric was discarded. The procedure was performed separately for each classification algorithm. The best results across the methods were obtained for approximately 20 metrics, as indicated in Fig. 3. Further reduction of a number of metrics caused a slight deterioration of a classification accuracy. Another interesting observation that could be made by visual inspection of the curves plotted in Fig. 3 is that while random forests and svm algorithms were relatively robust to the variation in the number of retained metrics, the remaining three algorithms (k-NN, lasso and neural network) exhibited an effect of an improved performance for the reduced subset of metrics.

In summary, for the database at hand, the lasso regression and neural network outperformed the remaining three algorithms. Their accuracy averaged across the cross-validation folds reached a level of 99% and 98%, respectively (see bottom row in Table 2).

Figure 4 shows the ten the most important metrics selected in the final model for the lasso regression. They were ranked in the figure according to their normalized importance (the most important metric as 100%). Front-to-back *loudness difference* proved to be the most important descriptor (loudness_fb). The range of the *theta* metric was identified as the second most important metric (theta_r), while the range of the *all-to-front difference* of the *theta* metric was ranked as the third most important metric

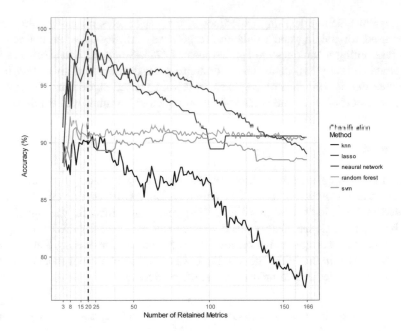

Fig. 3. Accuracy of the models as a function of a number of metrics retained.

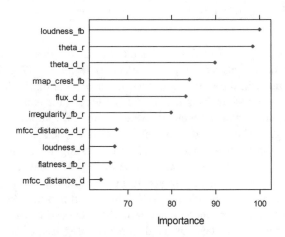

Fig. 4. Top ten most important metrics selected in the final model for the lasso regression.

(theta_d_r). The remaining metrics constituted a combination of the temporal, spectral, cepstral and loudness-based features. These results indicate that the proposed *theta* metric played an important role in the classification of the basic spatial audio scenes.

7 Conclusions and Future Work

This paper introduces a method for the automatic categorization of surround sound recordings according to a distribution of a foreground and background audio content around a listener. Moreover, it introduces a new feature, namely the *theta* metric, which proved to be instrumental in differentiation between the two basic spatial audio scenes.

The following machine learning-based algorithms were employed in the developed audio scene classifier and their performance compared: *k*-nearest neighbors, logistic regression with a least absolute shrinkage and selection operator (lasso regression), random forest, neural network, and support vector machine. According to the obtained results, the algorithms based on the lasso logistic regression and the neural network outperformed the remaining classification algorithms. The classification accuracy reached with the two above algorithms was equal to 99% and 98%, respectively.

It is difficult to compare the proposed method against the existing state-of-the-art machine-listening algorithms since the work presented in this paper can be considered as an attempt to embrace a new area of research, rather than to improve the existing techniques. Moreover, the currently presented study was concerned with the automatic classification of the basic spatial audio scenes (*FB* and *FF*); a scenario which has not been incorporated in the traditional machine-listening algorithms yet.

In contrast to the modern multimedia content classification algorithms [18], which are typically trained and validated using two separate datasets, respectively, the developed method was cross-validated jointly exploiting a relatively small corpus of audio recordings. This approach, in the opinion of the author, was deemed acceptable to prove the concept, however, it might not be sufficient to generalize the empirical observations regarding the most prominent features and the achieved accuracy of the method. Training and validating the method on two separate (and extended in size) audio corpora were left to future work.

While the scope of this study was limited to a simple dichotomous categorization of spatial audio scenes, more categories will be considered in future methods, including those accounting for elevated audio images. Moreover, considering already high popularity of binaural audio recordings in the publically available Internet repositories, it is intended that the next version of method would be tailored to the binaural audio format. Such method could be applied to automatic audio tagging of binaural audio recordings. It could also be applied for semantic search and retrieval of thereof. For example, one may consider searching the Internet for soundtracks exhibiting a prominent foreground audio content 'behind-the-head' or 'above-the-head' of a listener.

Finally, it is planned to incorporate more audio attributes in future algorithms, allowing for a more comprehensive characterization of spatial audio scenes. The recent research reports on a consensus vocabulary regarding spatial audio assessment could serve as examples of such attributes [7, 9, 10]. For instance, Francombe et al. [23, 24] recently discovered that the listeners' preference of the modern spatial audio systems is governed by such attributes, among others, as *envelopment* (*immersion*), *horizontal width*, *spatial naturalness*, *surrounding*, *spatial clarity*, *sense of space*, and *spatial movement*. Therefore, in line with the above body of research, such attributes should be considered for incorporation in future classification algorithms. It is hoped that this paper will prompt other researchers to undertake further studies in the area of a machine learning-based spatial audio scene characterization (SASC).

Acknowledgments. This work was supported by a grant S/WI/3/2013 from Bialystok University of Technology and funded from the resources for research by Ministry of Science and Higher Education.

References

1. ITU-R Rec. BS.775: multichannel stereophonic sound system with and without accompanying picture. International Telecommunication Union, Geneva, Switzerland (2012)
2. Blauert, J.: The Technology of Binaural Listening. Springer, New York (2013). https://doi.org/10.1007/978-3-642-37762-4
3. Sugimoto, T., Nakayama, Y., Komori, T.: 22.2 ch audio encoding/decoding hardware system based on MPEG-4 AAC. IEEE Trans. Broadcast. **63**(2), 426–432 (2017). https://doi.org/10.1109/tbc.2017.2687699
4. ITU-R Rec. BS.1116: methods for the subjective assessment of small impairments in audio systems. International Telecommunication Union, Geneva, Switzerland (2015)
5. Conetta, R., Brookes, T., Rumsey, F., Zieliński, S., Dewhirst, M., Jackson, P., Jackson, P., Bech, S., Meares, D., George, S.: Spatial audio quality perception (part 1): impact of commonly encountered processes. J. Audio Eng. Soc. **62**(12), 831–846 (2014). https://doi.org/10.17743/jaes.2014.0048
6. Walton, T.: The overall listening experience of binaural audio. In: Proceedings of the 4th International Conference on Spatial Audio (ICSA), Graz, Austria (2017)
7. Zacharov, N., Pedersen, T., Pike, C.: A common lexicon for spatial sound quality assessment–latest developments. In: Proceedings of the 8th International Conference on Quality of Multimedia Experience (QoMEX), Lisbon, Portugal (2016). https://doi.org/10.1109/qomex.2016.7498967
8. Berg, J., Rumsey, F.: Spatial attribute identification and scaling by repertory grid technique and other methods. In: Proceedings of the 16th International AES Conference, On Spatial Sound Reproduction, Rovaniemi (1999)
9. Le Bagousse, S., Paquier, M., Colomes, C.: Categorization of sound attributes for audio quality assessment–a lexical study. J. Audio Eng. Soc. **62**(11), 736–747 (2014). https://doi.org/10.17743/jaes.2014.0043
10. Lindau, A., Erbes, V., Lepa, S., Maempel, H.-J., Brinkman, F., Weinzierl, S.: A spatial audio quality inventory (SAQI). Acta Acust. U. Acust. **100**, 984–994 (2014). https://doi.org/10.3813/aaa.918778
11. Rumsey, F.: Spatial quality evaluation for reproduced sound: Terminology, meaning, and a scene-based paradigm. J. Audio Eng. Soc. **50**(9), 651–666 (2002)
12. Zieliński, S., Rumsey, F., Kassier, R., Bech, S.: Development and initial validation of a multichannel audio quality expert system. J. Audio Eng. Soc. **53**(1/2), 4–21 (2005)
13. Zieliński, S.K.: Feature extraction of surround sound recordings for acoustic scene classification. In: Rutkowski, L., Scherer, R., Korytkowski, M., Pedrycz, W., Tadeusiewicz, R., Zurada, J. (eds) Artificial Intelligence and Soft Computing, ICAISC 2018. LNCS, vol. 10842. Springer, Cham (2018). https://doi.org/10.1007/978-3-319-91262-2_43
14. Härmä, A., Park, M., Kohlrausch, A.: Data-driven modeling of the spatial sound experience. In: Proceedings of the 136th AES Convention, Berlin (2014)

15. Conetta, R., Brookes, T., Rumsey, F., Zieliński, S., Dewhirst, M., Jackson, P., Jackson, P., Bech, S., Meares, D., George, S.: Spatial audio quality perception (part 2): a linear regression model. J. Audio Eng. Soc. 62(12), 847–860 (2014). https://doi.org/10.17743/jaes.2014.0047

16. George, S., Zieliński, S., Rumsey, F., Jackson, P., Conetta, R., Dewhirst, M., Meares, D., Bech, S., George, S.: Development and validation of an unintrusive model for predicting the sensation of envelopment arising from surround sound recordings. J. Audio Eng. Soc. 58 (12), 1013–1031 (2010)

17. Francombe, J., Brookes, T., Mason, R.: Determination and validation of mix parameters for modifying envelopment in object-based audio. J. Audio Eng. Soc. 66(3), 127–145 (2018). https://doi.org/10.17743/jaes.2018.0011

18. Stowell, D., Giannoulis, D., Benetos, E., Lagrange, M., Plumbley, M.D.: Detection and classification of acoustic scenes and events. IEEE Trans. Multimed. 17(10), 1733–1746 (2015). https://doi.org/10.1109/tmm.2015.2428998

19. Imoto, K., Ono, N.: Spatial cepstrum as a spatial feature using a distributed microphone array for acoustic scene analysis. IEEE/ACM Trans. Audio Speech Lang. Process. 25(6), 1335–1343 (2017). https://doi.org/10.1109/taslp.2017.2690559

20. Peeters, G., Giordano, B.L., Susini, P., Misdariis, N., McAdams, S.: The timbre toolbox: extracting audio descriptors from musical signals. J. Acoust. Soc. Am. 130(5), 2902–2916 (2011). https://doi.org/10.1121/1.3642604

21. Glasberg, B.R., Moore, B.C.J.: A model of loudness applicable to time-varying sounds. J. Audio Eng. Soc. 50(5), 331–342 (2002)

22. James, G., Witten, D., Hastie, T., Tibshirani, R.: An Introduction to Statistical Learning with Applications in R. Springer, London (2017). https://doi.org/10.1007/978-1-4614-7138-7

23. Francombe, J., Brookes, T., Mason, R.: Evaluation of spatial audio reproduction methods (part 1): elicitation of perceptual differences. J. Audio Eng. Soc. 65(3), 198–211 (2017). https://doi.org/10.17743/jaes.2016.0070

24. Francombe, J., Brookes, T., Mason, R., Woodcock, J.: Evaluation of spatial audio reproduction methods (part 2): analysis of listener preference. J. Audio Eng. Soc. 65(3), 212–225 (2017). https://doi.org/10.17743/jaes.2016.0071

Automatic Clustering of EEG-Based Data Associated with Brain Activity

Adam Kurowski[1], Katarzyna Mrozik[1], Bożena Kostek[2(✉)], and Andrzej Czyżewski[1]

[1] Faculty of Electronics, Telecommunications and Informatics, Multimedia Systems Department,
Gdansk University of Technology, G. Narutowicza 11/12, 80-233 Gdansk, Poland
[2] Faculty of Electronics, Telecommunications and Informatics, Audio Acoustics Laboratory,
Gdansk University of Technology, G. Narutowicza 11/12, 80-233 Gdansk, Poland
bokostek@audioacoustics.org

Abstract. The aim of this paper is to present a system for automatic assigning electroencephalographic (EEG) signals to appropriate classes associated with brain activity. The EEG signals are acquired from a headset consisting of 14 electrodes placed on skull. Data gathered are first processed by the Independent Component Analysis algorithm to obtain estimates of signals generated by primary sources reflecting the activity of the brain. Next, the parameterization process is performed in two ways, i.e. by applying Discrete Wavelet Transform and utilizing an autoencoder network. The resulting sets of parameters are then used for the data clustering and the effectiveness of correct assignment of data into adequate clusters is checked. It occurs that the performance of wavelets- and autoencoders-based parametrization is similar, however in several cases, autoencoders allowed for obtaining a higher mean distance and lower standard deviation than distances provided by the wavelet-based method. Moreover, a supervised classification of signals is performed as a form of benchmarking.

Keywords: EEG signal · Brain activity
Independent Component Analysis (ICA) · Human-Computer Interfaces (HCI)
Data clustering · Deep learning

1 Introduction

Monitoring of EEG signals may be a valuable source of information for the purpose of biofeedback [1–3]. Either a person involved in a task performed could observe her/his state, which may be associated with a desired mental state, getting into such a state or training the ability to stay in such a state for a desired period of time despite existing distractors or a task-controller may survey an improvement in certain cognitive aspects. The mentioned objectives constitute research directions performed over several decades, but recently gained new capabilities [4, 5], however automatization of such processes is still far from being fully achieved, thus this should be further researched. Moreover, information about changes of the mental state of the subject may be utilized for investigating the mood of patients with brain injuries for whom Brain-Computer Interface (BCI) often may be the only modality allowing for communication with other people.

© Springer Nature Switzerland AG 2019
K. Choroś et al. (Eds.): MISSI 2018, AISC 833, pp. 470–479, 2019.
https://doi.org/10.1007/978-3-319-98678-4_47

The objective of this study was to acquire electroencephalographic (EEG) data from six subjects performing three tasks. To distinguish between time periods related to tasks performed by the subjects two types of parametrization were utilized, namely: wavelet- and autoencoder-based algorithms. Then, unsupervised clustering employing k means, spectral and Ward hierarchical algorithm [6–8] is executed and the results of the autoencoder-based clustering are compared with the outcomes achieved with the use of vectors of parameters calculated by Discrete Wavelet Transform (DWT). In both cases the EEG signal was pre-processed by employing Independent Component Analysis (ICA) as a blind source separation algorithm in order to extract estimates of primary sources of EEG signals. Some future directions were also outlined, e.g. taking into account influence of personal interests of subject in activities (e.g. specified kind of music) to be performed, fatigue and mood of the person participated in the test, or level of engagement in performed tasks on the assignment of data to different clusters.

2 Signal Processing

2.1 Signal Acquisition

The Emotiv EPOC headset was used for the signal acquisition. The device collects raw signals from 14 electrodes assigned according to the 10–20 placement standard (electrode placement on the scalp defines a set of standard positions that results can be related to): AF3, F7, F3, FC5, T7, P7, O1, O2, P8, T8, FC6, F4, F8 and AF4. To process these signals several scripts were written in the Python environment extended with libraries provided by the headset manufacturer. The electrodes used in this type of the headset are saline-based ones, thus the sampling rate was set to 128 S/s. Data acquired from the headset were stored in the csv data format and they were processed with the use of a separate set of scripts, also written in the Python programming language, which will be further explained in Sect. 2.2. Six subjects of age between 20 and 30 participated in tests. Each test session consisted of three tasks performed:

- relaxing with closed eyes - during this task execution the subjects were asked to relax in order to induce mental states associated with such an activity,
- watching a music video - music genre chosen for this task was "folk metal" due to its engaging and lively character of this music style which creates an opportunity to induce mental states opposite to those induced in the preceding task,
- playing a logic game, the game used at this stage of the experiment was Netwalk logic computer game, in which a player has to rotate elements of the board in such manner, that there would be a connection between a single central element called "server" and multiple peripheral elements called "computers".

Each task lasted for approximately 5 min, therefore the total session duration was equal to 15 min.

2.2   Data Processing

The starting point of data processing was splitting signals into a set of 15 seconds-long frames, with the sampling rate of 128 S/s, entailing the length of 1920 samples per single channel of a single frame. First, signals obtained from the electrodes of the headset were processed by the ICA algorithm in order to perform blind source separation, which estimates primary sources of signals located on the surface of the brain. The implementation of the algorithm requires the same number of estimated source signals as the number of the input signals. Therefore, each frame contains data associated with all 14 channels of the ICA-processed EEG data. These estimates were then used for calculating parameters for the data clustering stage.

Two types of feature extraction methods were used, i.e. Discrete Wavelet Transform and an autoencoder neural network belonging to deep learning algorithms. The feature extraction process was carried out according to the following assumptions:

- wavelet transform-based extraction of parameters, which represent mean values and variances of all possible levels of discrete wavelet transform (DWT),

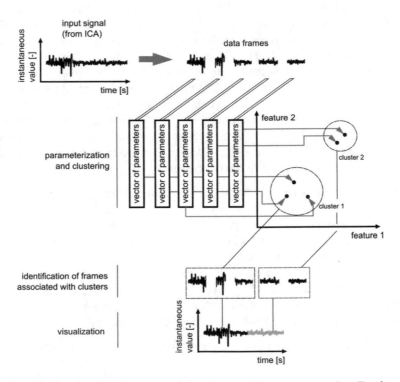

Fig. 1. Graphic visualization of the overall algorithm used for data processing. For the purpose of visualization the decision space was shown as if it had two dimensions associated with two features. In our research the number of features was 1513 in the case of the wavelet-based and 1120 in the case of the autoencoder-based feature extraction.

- obtained features calculated with the use of encoding part of the autoencoder neural network, which was trained with the use of segmented signals provided by the ICA algorithm output.

Data processing and clustering are shown in Fig. 1. All algorithms, including processing of signals gathered from the EEG acquisition headset, were implemented in the Python environment. Several libraries were employed: NumPy, SciPy and PyWavelets, Scikit-learn, Theano and Keras were used as machine and deep learning libraries [9–12]. The frame contains wavelet transform-based parameters associated with all 14 channels of the EEG headset. In the case of parameters extracted by autoencoder networks – an algorithm is capable of creating "custom" features related with each channel of the input signal. Therefore the classification algorithm takes into consideration both spectral and spatial characteristics of the EEG signal measured.

Resulted from the wavelet- and autoencoder-based feature vectors are then analyzed by three selected clustering algorithms, i.e.: k means, spectral and Ward hierarchical algorithm, which are used for the unsupervised assignment of frame-related vectors of parameters to one of k clusters. Each of the investigated algorithms performs an assignment of data into clusters in a different way:

- k means algorithm minimizes the sum of the squared distances between k centroids of clusters and feature-related points in their neighborhood [8],
- spectral clustering is based on the extraction of eigenvalues of data before dividing them into clusters with another secondary algorithm (i.e. k means) [6],
- Ward hierarchical clustering performs an assignment to classes by iterative merging of most similar clusters of data. The similarity measure is a sum of squared distances between points of clusters, however the result of calculation is used for making decision to which clusters they should be merged. The algorithm execution ends when there are only k clusters left [7, 13].

Different principles of operation of chosen algorithms makes it more likely that one of them will be suitable to the structure of the data obtained in experiment than others. The number of k clusters referring to carried out study was set to 5. Though arbitrarily chosen, it was justified by the number of the tasks performed by each subject during the experiment associated with expected three states, i.e. relaxing, listening to music and playing a game, and two additional states, which may be associated with some unexpected events like distraction of subjects or noise present in the input data.

In the case of the wavelet-based processing, coefficients of discrete wavelet transform were computed for the maximum possible level of decomposition. For each level of decomposition, mean and variance were calculated in order to decrease the number of generated features. Wavelets used were from the three families, namely: Coiflets 1 and 2, Daubechies 1, 2 and 9 and Symlet 9. The selection of both the basis mother wavelet and its order was dictated by the critical literature review results [14, 15].

When applying machine learning to feature extraction, a special type of an artificial neural network was used, i.e. an autoencoder neural network. Autoencoders, in addition to data compression, can perform pre-training of deep learning models [16, 17]. The network in autoencoder contains two parts: encoder function $h = f(x)$ and decoder function $r = g(h)$, where: x denotes an input, h refers to the hidden layer. The aim of the decoder function is to produce a reconstruction of the original function $r = g(f(x)) = x$ with the least possible amount of deformation, but at the same time with compressed representation. The network may also be utilized for the extraction of n features associated with vector of N samples. In the autoencoder-based feature extraction, signals obtained from ICA were normalized using standard score, which is represented by the following formula:

$$z = \frac{x - x'}{\delta_x} \tag{1}$$

where z is normalized vector of features, x is unnormalized vector of features, x' refers to the mean value of vector x, and δ_x is standard deviation of such vector.

Subsequently, signals were split into frames of the same length and the number of channels and then a training phase of an autoencoder started. Each single output signal from the ICA has its own autoencoder trained, therefore it was necessary to train 14 autoencoder networks in the process of feature extraction. Illustration of this process is shown in Fig. 2. Therefore, the resulting parameters provide an effect of nonlinear dimensionality reduction performed on data returned by the ICA-based processing stage. This approach has already been used in machine learning-based studies and results obtained were comparable or better than ones obtained with PCA [18].

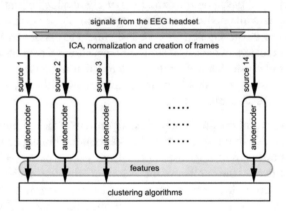

Fig. 2. The autoencoder-based feature extractor used in the experiment.

The encoder consisted of six layers of neurons. The following sizes of neuron layers were used: 640, 420, 320, 240, 160 and 80 neurons. This was the best performing set of hyperparameters over a number of tested sets of values in terms of unsupervised classification outcomes. The decoder was designed analogously, with the reversed order of neuron layers. The hyperparameters were selected on the basis of the number of trials and selection of best results of clustering performed on their basis. The outcomes allowing for a better distinction between time periods associated with performing particular exercises were considered as the better ones. Resulting from this analysis, for each frame, 80 features were calculated by a single autoencoder instance. Next, the encoder parts of the trained networks were used for the computation of a feature vector associated with each frame of the analyzed EEG signal, which in turn can be used for data clustering. The resulting vector of parameters consisted of 14 concatenated l vectors calculated for each output signal generated by ICA.

The diagram of operations performed in the case of both types of feature extraction investigated is shown in Fig. 3. Each vector is associated with one time frame of the EEG signal. Such an association make it possible to connect a certain moment of time with a given vector of parameters. Therefore, after the clustering stage, each time frame was associated with one class assigned by the clustering algorithm. As seen in Fig. 3, three types of clustering algorithms, mentioned earlier, were used in the final stage of data processing in order to compare their performance and ability to detect different types of data clusters present in the analyzed sets of feature vectors.

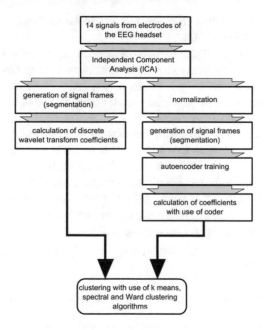

Fig. 3. Feature extraction processes with the use of ICA, two types of feature extraction: wavelet transform-based features (left side of the diagram) and encoder-based feature extractor (right side of the diagram), and three clustering methods (k means, spectral and Ward).

3   Experiment Results

A graphic representation of signals obtained by the ICA algorithm output is not always suitable because of the complexity of the acquired results. Often, the result of classification obtained for a time frame associated with a task completed consisted of a sequence of several mental states switched on and off. Though, this type of visualization makes it possible to evaluate roughly the performance of the clustering algorithm, there is a need for more objective evaluation of results. Therefore, another way of data presentation – an estimate of the probability density of states associated with each exercise was proposed. It allows for quantitative analysis of occurrence frequency of clusters as a result of the EEG data unsupervised classification. This estimate was then treated as a vector in a k-dimensional space. As the number of classes provided by clustering algorithms was set to 5, the k is also equal to 5. Therefore, each task performed by the subject was associated with one vector representing the estimate of probability of each cluster-related state. Next, distances between each pair of such vectors were calculated. The Euclidian distance was used in the process of assessing the mean spatial separation of clusters:

$$d = \sqrt[2]{\sum\nolimits_{i=1}^{k} d_i^2}, \tag{2}$$

where d denotes distance between clusters and d_i is the ith distance from k distances, which are possible to be calculated.

The mean value was calculated for each of three distances obtained for each subject (see Table 1). The Ward clustering algorithm was not capable of producing satisfactory results, therefore, results for only k means and spectral algorithms are contained in Table 1, as the Ward algorithm provided almost no separation between classes. Values were rounded up to two decimal places.

Results obtained from the unsupervised classification of the gathered EEG signals allows for drawing a conclusion that it is possible to associate each activity performed by the person involved in the experiment with a probability vector. The probability vector is a vector in the n-dimensional decision space and can be treated as a kind of *fingerprint* of a state of an EEG signal gathered from the subject. Data from Table 1 were further processed to obtain four more metrics shown in Table 2, where $\bar{d}$ denotes the mean Euclidean distance between vectors associated with each exercise, δ_d refers to the standard deviation of the mean distance, d_{max} and d_{min} denote maximal and minimal distance, consecutively.

Table 1. Mean distance between vectors associated with clusters related to tasks performed by subjects.

Subject ID	Method							
	Wavelets		Autoencoder-based (50 epochs)		Autoencoder-based (100 epochs)		Autoencoder-based (200 epochs)	
	k means	Spectral	k means	Spectral	k means	Spectral	k means	Spectral
1	0.05	0.22	0.04	0.17	0.2	0.15	0.09	0.08
2	0.4	0.53	0.89	0.2	0.41	0.47	0.53	0.38
3	1.06	0.77	0.63	0.46	0.76	0.56	0.06	0.56
4	0	0.14	0.73	0.19	0.7	0.29	0.17	0.29
5	0.02	0.12	0.1	0.14	0.18	0.28	0.11	0.08
6	0	0.3	0.66	0.39	0.12	0.16	0.13	0.18

Table 2. Statistical parameters calculated from intermediate parameters presented in Table 1.

Method								
	Wavelets		Autoencoder-based (50 epochs)		Autoencoder-based (100 epochs)		Autoencoder-based (200 epochs)	
	k means	Spectral	k means	Spectral	k means	Spectral	k means	Spectral
$\bar{d}$	0.26	0.35	0.51	0.26	0.40	0.32	0.18	0.26
δ_d	0.42	0.26	0.35	0.13	0.28	0.17	0.17	0.19
d_{max}	1.06	0.77	0.89	0.46	0.76	0.56	0.53	0.56
d_{min}	0	0.12	0.04	0.14	0.12	0.15	0.06	0.08

The mean distance and its variance between probability vectors vary depending on parameters used for the feature extraction process. In the case of the machine learning-based features extracted by an autoencoder network there is also dependence on the number of epochs of the training algorithms. The least value was associated with the largest separation distance provided by the k means algorithm. In some cases - autoencoders allowed for obtaining a greater value of the mean distance and smaller standard deviation than ones obtained for the wavelet-based algorithm. It is also worth mentioning that in some cases k means algorithm was not able to separate time periods associated with each task.

Parameters calculated for four scenarios considered, i.e. obtained with the use of wavelets and autoencoders with an increasing number of training algorithm epochs, were used in supervised classification of tasks performed by subjects. Acquired signals were split into 339 frames assigned to three classes, that were connected to each of three exercises, 226 of them were randomly assigned to the training set, 113 were used for validation stage. Data were normalized according to the formula (1). The neural network used for the purpose of classification consisted of 1120 input neurons, 100 neurons in the hidden layer, and 3 neurons in the output. Classes associated with exercises were encoded in the one-hot encoding manner. Therefore, n classes are associated with n vectors consisting of n-1 zeros and one value equals 1, which is placed in the unique position in the vector, allowing for identification of the associated class. A plain back-propagation algorithm was used for the training the neural network. Vectors of features were randomly assigned to the training and validation sets. Also, the process of training and assessing the performance of the network was repeated 100 times and then repeated in order to evaluate mean efficiency of the classifier based on each of the feature sets investigated. Results of classification are shown in Table 3 in the form of mean efficiency and standard deviation of the classifier efficiency.

Table 3. Results of classification of signals gathered in the experiment with the use of artificial neural network expressed by classification error rate.

	Wavelet-based	Autoencoder, 50 epochs	Autoencoder, 100 epochs	Autoencoder, 200 epochs
Mean value	0.495	0.536	0.526	0.573
Standard deviation	0.041	0.038	0.039	0.044

The lowest mean value of the mean efficiency was obtained for the wavelet-based classification feature set. In the case of random classification, the efficiency of classification tended to be 1/3. Therefore, in each case the classifier performed better than random classification. All algorithms based on a set of autoencoders performed better than the one based on the wavelet-derived parameters. Values of the standard deviations are similar in each case. The best performance was obtained for the system trained for 200 epochs. It should be noted that clustering algorithms performed worse for the last mentioned scenario, therefore additional experiments with the use of other clustering algorithms applied to a larger group of subjects should be done in order to further investigate the dependence between the distance of clusters obtained in the process of unsupervised clustering algorithms and the supervised classification methods with the use of artificial neural networks.

4   Conclusions

Results obtained show, that it is possible to perform the unsupervised clustering of the EEG signals based on feature vectors derived from the wavelet-based analysis and autoencoder neural networks. The proposed architecture of an autoencoder-based feature extractor allowed for obtaining similar Euclidean distances as the ones provided by the wavelet-based feature extractor, however it should be noted, that the number of epochs of the training algorithm has a significant influence on the performance of this kind of the feature extraction algorithm.

Resulting feature vectors were clustered with the use of various clustering algorithms and the outcomes of each algorithm were different. Also, the satisfactory performance was achieved for only two algorithms: k means and spectral. Achieved distances were significantly smaller for the Ward clustering algorithm and they were close to zero, providing almost no separation between classes. Moreover, in some situations, a series of consecutive signal frames were not consistently associated with one cluster. Therefore, for each set of EEG signal frames associated with various tasks performed by subjects, a vector of probabilities of classes assigned to frames by each algorithm was calculated. Such vectors, treated as vectors in the n-dimensional space can be used for further analyses and for the analysis of the state of the EEG signal gathered from the subject. More research should be performed in order to determine how an assignment to different clusters can be connected to such factors as personal interests of the subject in activities (e.g. specified kind of music) to be performed, fatigue and mood of the person participated in the test, or level of engagement in performed tasks. Such a goal may be achieved by extending the range of performed activities, through the increasing the number of participants and by the use of some additional signal analysis techniques. An example of such a technique may be the use of convolutional neural networks and data augmentation methods to obtain features better suited for the type of the processed data.

Acknowledgments. The project was funded by the National Science Centre on the basis of the decision number DEC-2014/15/B/ST7/04724.

References

1. Sun, J.C.-Y., Yeh, K.P.-C.: The effects of attention monitoring with EEG biofeedback on university students' attention and self-efficacy: the case of anti-phishing instructional materials. Comput. Educ. **106**, 73–82 (2017). https://doi.org/10.1016/j.compedu.2016.12.003
2. Vernon, D.J.: Can neurofeedback training enhance performance? An evaluation of the evidence with implications for future research. Appl. Psychophysiol. Biofeedback **30**(4), 347–364 (2005)
3. Loo, S.K., Makeig, S.: Clinical utility of EEG in attention-deficit/hyperactivity disorder: a research update. Neurotherapeutics **9**, 569–587 (2012)
4. Ilyas, M.Z., Saad, P., Ahmad, M.I., Ghani, A.R.I.: Classification of EEG signals for brain-computer interface applications: performance comparison. In: 2016 International Conference on Robotics, Automation and Sciences (ICORAS), 5–6 November (2016). https://doi.org/10.1109/icoras.2016.7872610
5. Nicolas-Alonso, L.F., Gomez-Gil, J.: Brain computer interfaces, a review. Sensors **12**(2), 1211–1279 (2012). https://doi.org/10.3390/s120201211
6. Ng, A., Jordan, M., Weiss, Y.: On spectral clustering: analysis and an algorithm. In: Advances in Neural Information Processing Systems, pp. 849–856. MIT Press (2001)
7. Murtagh, F., Legendre, P.: Ward's hierarchical agglomerative clustering method: which algorithms implement Ward's criterion? J. Classif. **31**, 274–295 (2014). https://doi.org/10.1007/s00357-014-9161-z
8. Harrington, P.: Machine Learning in Action. Manning Publications Co., Greenwich (2012)
9. Jones, E., Oliphant, T., Peterson, O. et al.: SciPy: open source scientific tools for Python, Internet website of the Python library. http://www.scipy.org/. Accessed Apr 2018
10. PyWavelets: Internet website of the Python library. http://pywavelets.readthedocs.io. Accessed Apr 2018
11. Theano: Internet website of the Theano Python machine learning library. http://deeplearning.net/software/theano. Accessed Apr 2018
12. Keras: Internet website of the Python machine learning library. https://keras.io/. Accessed Apr 2018
13. Zaki, M., Meira, W.: Data Mining and Analysis: Fundamental Concepts and Algorithms. Cambridge University Press, New York (2014)
14. Al-Qazzaz, N.K., Bin Mohd Ali, S.H., Ahmad, S.A., Islam, M.S., Escudero, J.: Selection of mother wavelet functions for multi-channel EEG signal analysis during a working memory task. Sensors **15**, 29015–29035 (2015)
15. Saeid, S., Chambers, J.A.: EEG Signal Processing. Wiley, Chicester (2007)
16. Gluge, S., Bock, R., Wendemuth, A.: Auto-encoder pre-training of segmented-memory recurrent neural networks. In: ESANN 2013, Bruges, Belgium (2013)
17. Saroff, A. Musical audio synthesis using autoencoding neural networks. In: Proceedings of the International Computer Music Conference, Athens, Greece (2014)
18. Wang, W., Huang, Y., Wang, Y., Wang, L.: Generalized autoencoder: a neural network framework for dimensionality reduction. In: IEEE Conference on Computer Vision and Pattern Recognition Workshops, Columbus, pp. 496–503 (2014). https://doi.org/10.1109/cvprw.2014.79

Comparative Analysis of Spectral and Cepstral Feature Extraction Techniques for Phoneme Modelling

Gražina Korvel[1(✉)], Olga Kurasova[1], and Bożena Kostek[2]

[1] Institute of Data Science and Digital Technologies, Vilnius University,
Akademijos str. 4, 04812 Vilnius, Lithuania
grazina.korvel@mii.vu.lt
[2] Audio Acoustics Laboratory, Faculty of Electronics,
Telecommunications and Informatics, Gdańsk University of Technology,
G. Narutowicza 11/12, 80-233 Gdansk, Poland

Abstract. Phoneme parameter extraction framework based on spectral and cepstral parameters is proposed. Using this framework, the phoneme signal is divided into frames and Hamming window is used. The performances are evaluated for recognition of Lithuanian vowel and semivowel phonemes. Different feature sets without noise as well as at different level of noise are considered. Two classical machine learning methods (Naive Bayes and Support Vector Machine) are used for classifying each problem, separately. The experiment results show that cepstral parameters give higher accuracies than spectral parameters. Moreover, cepstral parameters give better performance compared to spectral parameters in noisy conditions.

Keywords: Parameter extraction · Machine learning · Spectrum
Cepstrum

1 Introduction

One could argue that there are many aspects of speech analysis and modelling that are already well understood. However, it is also clear that there is much that we still do not know. Speech modelling is still a difficult unsolved problem. The phoneme mathematical models are tools for describing speech. In order to create a mathematical model, it is important to find an acoustic description of speech as well as speaker emotional states and sentiments. This paper is related to the speech acoustic parameters extraction for speech modelling. The machine learning methods are used for evaluation of extracted features.

In the speech technology applications, the researchers focus attention on the spectral speech parameterization techniques. The spectral features like: spectral centroid, skewness, kurtosis, spread are extracted [1]. The popular implementations of different cepstral analysis approaches are as follows: mel frequency cepstral analysis, linear prediction cepstral analysis, perceptual linear prediction cepstral analysis [2]. Similar to the representation of a signal in the form of spectrum, time domain

© Springer Nature Switzerland AG 2019
K. Choroś et al. (Eds.): MISSI 2018, AISC 833, pp. 480–489, 2019.
https://doi.org/10.1007/978-3-319-98678-4_48

(temporal) features such as: energy of the signal, maximum amplitude, zero crossing rate, minimum energy, maximum correlation coefficient are well-known [3]. The correlation between temporal and spectral characteristics are thoroughly reviewed by Fastl and Zwicker [4]. An overview of the constantly growing and defining the field, introducing typical applications, presenting exemplary resources, and sharing a unified view of the chain of processing is provided by Schuller et al. [5]. There are also some standard speech feature sets, like Geneva minimalistic acoustic parameter set (GeMAPS) [6] or Praat features [7].

In this paper, a phoneme parameter extraction framework based on spectral and cepstral parameters is proposed. Spectral parameters are obtained as frequency response of the all-pole Linear Time Invariant (LTI) Filter. In the cepstral domain, a signal is transformed to mel scale. As opposed to widely used MFCC calculation technique, the power spectrum for cepstral coefficient calculation is used in this research study. The scale of filters is liner up to 1 kHz. Above this frequency, the scale becomes logarithmic.

Speech recognition based on feature vectors is generally performed using well-known methods such as Support Vector Machines [8], Artificial Neural Networks [9], Nearest Neighbors [10], Naive Bayes [11], and Hidden Markov Model [12]. Some scientists have created hierarchical classifiers by combining existing methods [13]. In our study, we use two classical machine learning algorithms to compare classification accuracy. The first of them is the Naive Bayes classification method, and the second one - Support Vector Machine (SVM). A desirable feature of the Naive Bayes classifier is that it only requires a small amount of training data to estimate means and variances of the variables necessary for classification. The advantage of SVMs is that they are particularly suited to distinguish between objects of different class memberships.

Checking if the extracted parameters do not lose the characteristics of the phonemes in the presence of noise is another important aspect. Therefore, the recognition process is performed with clean speech (without noise) and for speech data at various signal-to-noise (SNR) ratios.

2 Parameter Extraction Techniques

The speech signal is pre-processed before features are extracted. Usually, in the speech recognition process, features based on the short-term spectrum are used [14]. In this research study, the speech signal is divided into N sample frames (N is an integer power of 2) as shown in Fig. 1.

It can be seen in Fig. 1, that an overlapping applies to frames. Then the frames are windowed with the Hamming window:

$$w(n) = 0.54 - 0.46\cos\left(\frac{2\pi n}{N-1}\right) \tag{1}$$

where $n = 0, \ldots N - 1$.

For each frame, a set of parameters is calculated. Two feature extraction techniques are described in this Section.

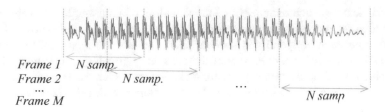

Fig. 1. Speech signal dividing into frames.

2.1   Extraction of Cepstral Features

The Mel-Frequency Cepstral Coefficients (MFCCs) convert information of the speech signal into cepstral coefficients. The difference from the real cepstral is that a nonlinear scale, based on human hearing, is used. The process of MFCC coefficients extraction constitutes a part of the algorithm prepared. The detailed description of this algorithm is explained below:

Input:
$x(n)$ - a short-time interval of the analyzed speech signal, where $n = 0, ... N - 1$.

Output:
MFCC coefficients

The steps of the algorithm:

Step 1. The short-time power spectrum is calculated.
Step 2. The filter bank is constructed.
Step 3. The logarithm is calculated.
Step 4. The Discrete Cosine Transform (DCT) is applied.

The MFCC feature extraction begins with calculating the power spectrum of the speech signal $x(n)$. The power spectrum is given by the following formula:

$$PS(n) = \left| \frac{1}{N} \sqrt{(X(k))_{re}^2 + (X(k))_{im}^2} \right|^2 \tag{2}$$

where $X(n)$ are the Fourier transform coefficients of the speech signal $x(n)$ $(n = 0, \ldots, N - 1)$, re means a real part and im – an imaginary part.

In the second step, triangle bandpass filters are constructed over the frequency range from the lower to the upper frequency. The scale of filters is linear up to 1 kHz. Above this frequency, the scale becomes logarithmic. The relation between frequency of speech and mel scale is defined by the formula:

$$Mel = 2595\log\left(1 + \frac{f}{700}\right). \tag{3}$$

The mel spectrum is obtained by multiplying the power spectrum coefficients by the triangular filter coefficients. A log magnitude of mel spectrum is calculated in order to obtain the real cepstrum. DCT applied to the transformed mel frequency coefficients produces a set of cepstral coefficients [15]. The MFCC are calculated as follows:

$$c_j = \sum_{i=0}^{M-1} m_i \cos\left(\frac{\pi j(i - 1/2)}{M}\right) \tag{4}$$

where m_i are the log filterbank amplitudes, M is the number of filers, $j = 0, \ldots, K$, K is the number of cepstral coefficients.

As a result, the K coefficients of MFCC are obtained for each frame. The mean values of frames are used.

2.2   Extraction of Spectral Features

Speech signal can be expressed as a time varying linear filer [16]. However, shorter intervals of this signal can be modeled as a time invariant system. In this paper, we use all-pole LTI Filter:

$$H(z) = \frac{1}{1 + a_1 z^{-1} + a_2 z^{-2} + \ldots + a_p z^{-p}} \tag{5}$$

A filter is constructed based on the Linear Prediction (LP) technique. According to this technique, the speech signal can be approximated as a linear combination of its p previous samples:

$$\hat{x}(n) \approx a_1 x(n - 1) - a_1 x(n - 2) - \ldots - a_p x(n - p) \tag{6}$$

where $\hat{x}(n)$ is the predicted signal, $a = [1, a_1, \ldots, a_p]$ are a set of the coefficients.

The algorithm for Linear Predictive Coding (LPC) coefficients extraction is given below:

   Input:

$x(n)$ - the short-time interval of the analyzed speech signal $(n = 0, \ldots N - 1)$

   Output:
LPC coefficients

   Steps of the algorithm:

Step 1. The autocorrelation analysis
Step 2. The LPC analysis

First of all, the autocorrelation method to find the filter coefficients is used. The autocorrelation function can be described by the following formula:

$$r(j) = \sum_{i=0}^{N-1-j} x(i)x(i+j) \qquad (7)$$

where $j = 0, \ldots p$.

As a result of Step 1, we obtain $(p+1)$ autocorrelation coefficients. In the next step, the LPC analysis is performed. In order to obtain LPC coefficients, the Yule-Walker equations are solved:

$$\begin{bmatrix} r(0) & \cdots & r(1-p) \\ \vdots & \ddots & \vdots \\ r(p-1) & \cdots & r(0) \end{bmatrix} \begin{bmatrix} a_1 \\ \vdots \\ a_p \end{bmatrix} = \begin{bmatrix} -r(1) \\ \vdots \\ -r(p) \end{bmatrix} \qquad (8)$$

The recursive Levinson Durbin algorithm is used to convert the autocorrelations into LPC parameters. According to this algorithm, equations, which are given below, are solved recursively for $i = 1, \ldots, p$ [17]:

$$E^{(0)} = r(0) \qquad (9)$$

$$k_i = \left(r(i) - \sum_{l=1}^{i-1} \alpha_l^{(i-1)} r(i-l) \right) / E^{(i-1)} \qquad (10)$$

$$\alpha_i^{(i)} = k_i \qquad (11)$$

$$if\ (i > 1)\ then\ \alpha_j^{(i)} = \alpha_j^{(i-1)} - k_i \alpha_{j-1}^{(i-1)}, j = 1, \ldots, i-1 \qquad (12)$$

$$E^{(i)} = \left(1 - k_i^2 \right) E^{(i-1)} \qquad (13)$$

When $i = p$, the set of LPC parameters is given by formula:

$$\alpha_k = \alpha_k^{(p)} \qquad (14)$$

where $k = 1, \ldots, p$.

As a result p coefficients of LPC are obtained for each frame. The obtained values are used for the filter design, described by Eq. (14) and then the frequency response of the filter is calculated. In this research study, we use logarithmic frequency response data, obtained by the formula:

$$h_j = 20 \cdot log10(abs(fr)) \qquad (15)$$

where fr is the frequency response vector, $j = 1, \ldots, N$ (N is the length of the transform). The mean values of the frames are used as spectral parameters.

3 Data Analyzed

An experiment is performed on the Lithuanian speech database, created during the LIEPA (Services controlled by the Lithuanian Speech) project [18]. The database is adapted for this specific research. Information about audio data is given in Table 1.

Table 1. Parameters of audio data

Format	wav
Sampling frequency	22 kHz
Quantification	16 bits
Number of channels	1

The recordings consist of 100 h of words, phrases and sentences, different speakers. The consonant phonemes were extracted manually, thus this tedious process limited creating a larger set for the analysis. The analyzed phonemes are given in Fig. 2.

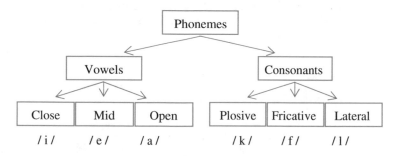

Fig. 2. Vowel phoneme grouping.

As we see from Fig. 2, the analyzed phonemes are divided into 6 classes. The goal of this study is to determine which of the parameters described above (cepstral or spectral) give higher classification accuracy.

4 Technique of Phoneme Classification

Due to the fact that the set of samples is not very large, and it is important to estimate the true error rate of a given classifier, the cross validation method is used. We use k-fold cross validation technique [19]. The data are partitioned into k equally sized segments, and k iterations are performed. Within each iteration, a different fold of the data is employed to test a model and the remaining $k - 1$ folds to train this model.

In the experiment, two classical machine learning algorithms to compare classification rates are used. First of them is the Naive Bayes classification method, based on

Bayes theory [20]. It provides the probability of each attribute set belonging to a given class. The test data are assigned to the class with the maximum class probability.

The second classification algorithm employed is Support Vector Machine [21]. SVM algorithm builds a model from training data that assigns test data to the appropriate class.

5 Experiment Results

In the experiment, 600 utterances (100 for each phoneme) were considered. The data are divided into two segments: the first one employed to train a model and the second one to test this model. As mentioned before, the 10-fold cross validation is used.

Before the extraction of features is performed, the signal pre-processing is carried out. The frame length is equal to 128 samples, an overlap constitutes 50% of the segment length.

In the experiment, 13 cepstral ($c_k, k = 1, ..13$) coefficients are used. This number of coefficients is usually used in scientific research [22]. An observation reveals that only part of spectral coefficients is useful for separation of the consonant classes. That is why only the first 12 spectral coefficients ($h_k, k = 1, ..12$) were chosen as representative features. The graphical representation of these features is given in Figs. 3 and 4, respectively.

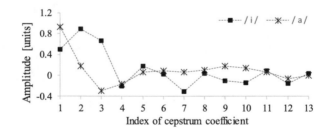

Fig. 3. Cepstrum parameters of the phoneme /a/ and phoneme /i/.

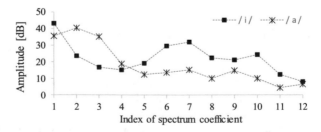

Fig. 4. Spectrum parameters of the phoneme /a/ and phoneme /i/.

Additionally, both their variances and first-order derivative (Δ) are considered. Additional tests were executed on the same feature vectors, but in the presence of noise. We add random white noise to the phoneme signals before feature extraction. The signal-to-noise ratio (SNR) of the noisy signal is equal to 20 dB, 10 dB, and 5 dB.

The average results of phoneme classification for spectral parameters are presented in Table 2. The classification accuracies, obtained based on cepstral parameters, are given in Table 3.

Table 2. The results of classification for 6 phoneme classes, based on spectral parameters

		Naive Bayes			SVM		
		h_k	$h_k + vh_k$	$h_k + \Delta h_k$	h_k	$h_k + vh_k$	$h_k + \Delta h_k$
Without noise	Mean	91.8%	90.2%	**95.0%**	82.0%	87.0%	**90.2%**
	STD	5.469	2.987	3.143	6.421	6.025	6.453
SNR = 20 dB	Mean	86.3%	85.3%	**92.3%**	80.8%	87.8%	**88.5%**
	STD	5.019	2.919	3.063	5.679	4.846	4.997
SNR = 10 dB	Mean	84.8%	85.6%	**89.6%**	82.0%	88.7%	**88.0%**
	STD	4.041	4.318	3.752	5.432	5.432	4.831
SNR = 5 dB	Mean	84.2%	85.3%	**90.2%**	83.0%	**90.8%**	87.2%
	STD	5.569	4.360	4.544	4.431	5.285	5.446

Table 3. The results of classification for 6 phoneme classes based on cepstral parameters

		Naive Bayes			SVM		
		h_k	$h_k + vh_k$	$h_k + \Delta h_k$	h_k	$h_k + vh_k$	$h_k + \Delta h_k$
Without noise	Mean	94.8%	93.3%	**95.2%**	83.5%	83.5%	**90.8%**
	STD	3.282	3.768	3.282	5.636	5.636	4.795
SNR = 20 dB	Mean	94.8%	93.5 V	**95.3%**	84.3%	84.3%	**90.3%**
	STD	3.639	3.805	3.123	4.098	4.527	5.317
SNR = 10 dB	Mean	95.0%	94.3%	**95.3%**	86.5%	86.8%	**90.8%**
	STD	4.006	3.784	3.833	4.191	4.264	4.321
SNR = 5 dB	Mean	**95.5%**	94.7%	**95.5%**	87.0%	87.0%	**92.2%**
	STD	3.338	4.289	4.585	4.216	4.831	3.338

The accuracy highlighted in bold font (see Tables 2 and 3) indicate the feature type with the highest accuracy for each classifier.

6 Conclusions

Phoneme parameter extraction framework based on spectral and cepstral parameters has been proposed in this paper. Our intention was to check which parametrization method (spectral or cepstral) works better under conditions in which noise is present. In the experiment, we have compared the performance of these feature extraction methods for vowel and consonant (/i/, /e/, /a/, /k/, /f/, /l/) classification. In order to evaluate classification accuracy, two methods, namely SVM and Naive Bayes were used.

The highest accuracy for the signal without noise added (95.2%) has been achieved for cepstral features plus their first order derivative in the case of the Naive Bayes method, while for the SVM classifier the highest accuracy was equal to 90.8% for the same parameters. Interestingly, for cepstral features both Naive Bayes and SVM classifiers returned higher accuracies in the case of SNR = 5 dB. Moreover, even in the presence of higher SNR still accuracy was higher for the Naïve Bayes method. These results are statistically significant, but still this may be an indication that cepstral parameters work better in the presence of noise.

In order to show how the method deals in the presence of noise, various noise levels were tested. It has been observed that cepstral parameters give a relatively robust performance compared to spectral parameters in noisy conditions. The cepstral parameters kept similar classification accuracy at different noise levels, while classification accuracy based on spectral parameters decreased in the presence of noise.

In the future research, we will test these features on the entire set of Lithuanian phonemes and in addition compare them to the results obtained with English and Polish phonemes [23] to create mathematical models based on these features. This way we will increase their applicability to more robust phoneme synthesis due to noise-resistant features.

Acknowledgment. This research is funded by the European Social Fund under the No 09.3.3-LMT-K-712 "Development of Competences of Scientists, other Researchers and Students through Practical Research Activities" measure.

References

1. Korvel, G., Kostek, B.: Examining feature vector for phoneme recognition. In: Proceeding of IEEE International Symposium on Signal Processing and Information Technology, ISSPIT 2017, Bilbao, Spain (2017). https://drive.google.com/open?id=1ugidXH_qNO9LRWTnkJU6Mbrk4AfTonGw
2. Eringis, D., Tamulevičius, G.: Improving speech recognition rate through analysis parameters. Electr. Control Commun. Eng. **5**(1), 61–66 (2014)
3. Kostek, B., Piotrowska, M., Ciszewski, T., Czyzewski, A.: Comparative study of self-organizing maps vs subjective evaluation of quality of allophone pronunciation for non-native english speakers. In: Audio Engineering Society Convention, p. 143 (2017)
4. Fastl, H., Zwicker, E.: Psychoacoustics: Facts and Models. Springer Series in Information Sciences, 3rd edn. Springer, Heidelberg (2007)
5. Schuller, B.R., Steidl, S., Batliner, A., Burkhardt, F., Devillers, L., Müller, C., Narayanan, S.: Paralinguistics in speech and language - state-of-the-art and the challenge. Comput. Speech Lang. **27**(1), 4–39 (2009)
6. Eyben, F., Scherer, K.R., Schuller, B.W., Sundberg, J., Andre, E., Busso, C., Devillers, L.Y., Epps, J., Laukka, P., Narayanan, S.S., Truong, K.P.: The geneva minimalistic acoustic parameter set (GeMAPS) for voice research and affective computing. IEEE Trans. Affect. Comput. **7**(2), 190–202 (2016)

7. Boersma, P.: Praat, a system for doing phonetics by computer. Glot Int. **5**(9/10), 341–347 (2002)
8. Matsumoto, M., Hori, J.: Classification of silent speech using support vector machine and relevance vector machine. Appl. Soft Comput. **20**, 95–102 (2014)
9. Bojanic, M., Crnojevic, V., Delic, V.: Application of neural networks in emotional speech recognition. In: 2012 11th Symposium on Neural Network Applications in Electrical Engineering (NEUREL), pp. 223–226. IEEE (2012)
10. Sadjadi, S.O., Pelecanos, J., Zhu, W.: Nearest neighbor based i-vector normalization for robust speaker recognition under unseen channel conditions. In: Fifteenth Annual Conference of the International Speech Communication Association, Singapore, pp. 1860–1864 (2014)
11. Kotsiantis, S.B.: Supervised machine learning: a review of classification techniques. Informatica **31**, 249–268 (2007)
12. Metallinou, A., Katsamanis, A., Narayanan, S.: A hierarchical framework for modeling multimodality and emotional evolution in affective dialogs. In: IEEE International Conference on IEEE Acoustics, Speech and Signal Processing (ICASSP), pp. 2401–2404 (2012)
13. Vasuki, P., Aravindan, C.: Improving emotion recognition from speech using sensor fusion techniques. In: IEEE Region 10 Conference, TENCON, pp. 1–6 (2012)
14. Rabiner, L., Juang, B.H.: Fundamental of Speech Recognition. Prentice Hall, Upper Saddle River (1933)
15. Rao, K.S., Vuppala, A.K.: Speech Processing in Mobile Environments, Springer Science & Business Media, Heidelberg (2014)
16. Pyž, G., Šimonytė, V., Slivinskas, V.: Developing models of lithuanian speech vowels and semivowels. Informatica **25**(1), 55–72 (2014)
17. Rabiner, L.R., Schafer, R.W.: Introduction to Digital Speech Processing. Now Publishers Inc., Breda (2007)
18. LIEPA Homepage. https://www.raštija.lt/liepa/about-project-liepa/7596
19. Refaeilzadeh, P., Tang, L., Liu, H.: Cross-validation. In: Encyclopedia of Database Systems, pp. 532–538 (2009)
20. Ghosh, J.K., Delampady, M., Samanta, T.: An Introduction to Bayesian Analysis: Theory and Methods, 1st edn. Springer Science Business Media, LLC, New York (2006)
21. Palaniappan, R., Sundaraj, K., Sundaraj, S.: A comparative study of the svm and k-nn machine learning algorithms for the diagnosis of respiratory pathologies using pulmonary acoustic signals. BMC Bioinf. **15**, 1–8 (2014)
22. Hegde, S., Achary, K.K., Shetty, S.: Feature selection using fisher's ratio technique for automatic speech recognition. Int. J. Cybern. Inf. (IJCI) **4**(2) (2015). https://doi.org/10.5121/ijci.2015.4204
23. Korvel, G., Kostek, B.: Voiceless stop consonant modelling and synthesis framework based on MISO dynamic system. Arch. Acoust. **3**(42), 375–383 (2017). https://doi.org/10.1515/aoa-2017-0039

Selection of Features for Multimodal Vocalic Segments Classification

Szymon Zaporowski[✉] and Andrzej Czyżewski

Faculty of Electronics, Telecommunication and Informatics,
Multimedia Systems Department, Gdańsk University of Technology,
Narutowicza 11/12, 80-233 Gdańsk, Poland
smck@multimed.org
http://www.multimed.org

Abstract. English speech recognition experiments are presented employing both: audio signal and Facial Motion Capture (FMC) recordings. The principal aim of the study was to evaluate the influence of feature vector dimension reduction for the accuracy of vocalic segments classification employing neural networks. Several parameter reduction strategies were adopted, namely: Extremely Randomized Trees, Principal Component Analysis and Recursive Parameter Elimination. The feature extraction process is explained, applied feature selection methods are presented and obtained results are discussed.

Keywords: Automatic speech recognition · Face motion capture
Neural networks

1 Introduction

Currently, there is a significant increase in the popularity of solutions related to automatic speech recognition (ASR). The main method of classification and detection used in this field is machine learning. The vast majority of these solutions have been known for years and thoroughly described. However, there is a lack of work on a decidedly deeper approach and attempts to classify speech elements at the allophonic level. Approaching the problem in this way is definitely more difficult, because the allophone articulation is a complex process. Especially, the dependence on the preceding sounds, context, short duration time and individual features requires a change in the approach popular in the literature. Therefore, it was decided to use Facial Motion Capture (FMC) as the main information carrier related to the articulation of allophonic sounds and to test the suitability of such an approach. Another solution that has been described in the paper is the fusion of audio parameters, namely MFCC parameters and LPC coefficients.

The use of FMC causes a number of difficulties, especially related to the placement of markers on the faces of speakers and the need for an uniform lighting of the speaker's entire face. The literature describes the detection of

© Springer Nature Switzerland AG 2019
K. Choroś et al. (Eds.): MISSI 2018, AISC 833, pp. 490–500, 2019.
https://doi.org/10.1007/978-3-319-98678-4_49

lip movement using a video image [1, 2]. Currently, there is no problem with determining the region of interest (ROI) or to make further image processing [3], but the issue is the amount of data limited by practical reasons, namely by the lack of possibility to record a large number of speakers with reflection markers mounted on their lips. Automatic recognition algorithms, such as neural networks deep learning are trained with use of data collected from hundreds or thousands of speakers. Another problem is positioning the speaker face in perpendicular to the camera. Each change in the camera's angle of view yields shifts, which in the case of a 2D image significantly affects the obtained results. An application of FMC with markers arranged in 3D space on the speaker's face, allows for obtaining coordinates in 3 dimensions. In this way the problem with the perspective of view is significantly reduced.

The use of FMC as a lip tracking system has not been described in the literature, also no descriptions for parameterizing data for such FMC applications could be found elsewhere. The authors and their co-workers (Szwoch, Cygert, Ciszewski) submitted for publishing early works pertaining this problem, however at the time this article is written, those papers were not printed, yet. It should be mentioned that the problem of redundant data was spotted. The data vector for each speaker contains a specific amount of parameters. Meanwhile, not all parameters contained in the vector serve efficiently as carriers of information significant from the classification point of view. This means that some parameters are not needed or may be treated by the classifier as noise, what in turn, may deteriorate the accuracy of the classification. Hence, it was decided to find the most optimal way to reduce feature vectors without decreasing the accuracy of classification.

2   Data

2.1   Recording Sessions

The recordings were made in an adapted room with the use of acoustic dampers to reduce unwanted sounds. Two microphones, a lavalier and a super-directional microphone with an external recorder were used to record the audio signal. The signal was recorded with sampling rate 48 kHz/s and 16 bit resolution. Six Vicon Vue cameras were used to capture the FMC data, while the video was recorded using a single Vicon reference camera (120fps) and a digital camera (50fps) plus a sports camera (240fps). Before every use the system was calibrated due to fact that it is needed to measure the positions of the markers in physical dimensions. 32 markers were used, 20 of which were placed on speaker's inner lips. 4 of them were placed on a special cap, that was point of reference in stabilizing the FMC image in post-processing. The system of coordinates was oriented in such manner, that the speaker's face was set in parallel to the XY plane, with Y axis coordinates towards the forehead and chin, and with the X axis running towards the ears. However the Z axis directed towards the cameras.

The recording sessions were divided into two days, data from 6 speakers were acquired. Each speaker pronounced 300 words or short expressions. Only selected vowels and diphthongs were used for the presented study. This is a result of the fact that they entail robust imaging in the FMC system, thus with the help of phonetic specialist it was decided to use only consonants. It is known by phonology principles that they had visible articulatory features, which are easier to notice employing FMC recordings. The list created contains 300 items that constitute a set of English pure vowels and consonants that are placed in various consonantal context. The recorded subjects were native speakers speaking with varying pronunciation (withing Standard Southern British English variety), one of them is of Polish origin, working as a professor of English phonology. The received classes are highly imbalanced in numbers as it can be seen in Table 1.

Table 1. Number of different vocalic segments for each speaker

Segment	Quantity	Segment	Quantity
ae	60	I	39
e	38	A	36
ei	29	O	26
i:	26	ai	22
eu	18	a:	17
u:	13	3:	10
o:	10	U	5
au	5	aiE	1
oi	1	ir	1
eE	1	ie	1

During the recording session few problems were encountered. The first problem was the room lighting. FMC cameras are highly sensitive to light change. The light in the visible spectrum can cause reflection artifacts creation, that affects FMC recording and may create false marker images. Therefore with use of infrared system it was not possible to record FMC data efficiently without any source of artificial light. Moreover, we will be not able to record any high quality video that was crucial as a reference for phonetic specialist to detect and to label specific vowels. So we were forced to use both celling light and spotlight directed towards a speaker's face. Another problem were different mouth shapes of each speaker. That caused a necessity to use a bit different placing of markers for every speaker. It can be seen in Fig. 1. As a result we needed to find a solution for a normalization of data for every speaker. This problem is further discussed in Sect. 4.

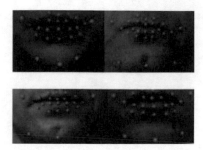

Fig. 1. Different lips shapes requiring differentiated marker placement

2.2 Data Acquisition

The prepared recordings were used by specialists in the field of phonetics for manual labeling of allophonic sounds. Due to the lack of possibility to automate of the process, the labeling took about a week of manual processing for each speaker. This is one of the biggest disadvantages of such an approach, because the acquisition of more data is a very time-consuming process and it does not allow the use of a decision algorithm based on deep learning. Moreover, duration differences of the spoken excerpts were problematic. Segments can differ substantially in their duration e.g. less than 40 ms for pure vowels to almost 400 ms for diphthongs. Since obtained classes were highly imbalanced, the experimenters were forced to reject from classification all classes with a number of elements smaller than 5.

3   Feature Extraction

3.1   Wide-Height

The first attempt to parametrize FMC signal was to create a simple, two-parameter approach. As a parameter speaker's open mouth width and height (W-H) were used. In Fig. 2 the main idea is illustrated. Further simple neural network classifier that is described in Sect. 4 was used. Purpose of this NN was to classify the extracted parameters.

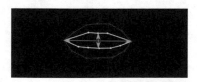

Fig. 2. Illustration of "W-H" idea of parameterization

In the pilot experiment a simple geometrical parametrization was used. Results presented in Table 2 were not satisfying, therefore another approach to feature extraction described in next section was adopted.

Table 2. Accuracy of classification with "W-H" parametrization

Speaker id	Accuracy
1	39.6
2	33.4
3	38.5
4	20.1
5	29.4
6	37.2

3.2 Grid - Shifting

A more advanced parametrization was developed employing Cartesian coordinates to shift position of markers. The first step is to determine the reference frame, the next one is to apply normalization of coordinates to the range x = [−3, 3] y = [−1, 1], according to the model, (0, 0) in the middle of the speaker's mouth. Subsequently, there are calculated frame shift factors that are applied to all frames. For each frame a feature vector is created that represents the difference between standardized coordinates of the current and of the reference frame. In the end one can get 40 parameters for one frame that means there are 20 markers with 2 parameters for each (x, y coordinates).

In Fig. 3 one can see the whole grid for the given range with markers inside the grid. This parametrization were used in examples and results shown below.

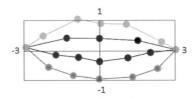

Fig. 3. Illustration of the Grid-shifting idea of feature extraction

4 Classification with Neural Network

For years, there has been a significant increase in interest in neural networks, especially in the context of image or speech recognition. Over the last decade, the popularity of the "black-box" approach with the use of deep learning has become the foundation for the development of many commercial systems for the voice and face detection. Increase of the computing power and the use of graphics cards for the acceleration of parallel computations has contributed to the creation of some more complex topologies of neural networks for the processing of huge amounts of data. Meanwhile, experiments employing FMC data provide a not particularly rich source of data. Therefore, in case of the classifier used for the presented research, a simple feed forward architecture was used. A neural network using one hidden layer consisting of 80 neurons with a Relu activation function and softmax function on the last layer, which size is determined by the size of the classes being the object of classification was used. The network was implemented with use of Keras library in the Python programming language. For the training the SGD algorithm with the learning rate 0.03 was used and also a categorical cross entropy loss function has been applied. Data were preprocessed using standard procedure of zero-centering and normalization to [−1, 1] interval. For the evaluation the 10-fold cross validation were used where for each fold the model with the best training loss has been chosen. Then this model was used for the test. The average values of all folds are presented. Results of the neural network-based classification employing grid-shifting parametrization varied between 35.1% up to 63.1% for recorded speakers.

5   Multimodal Approach

From our previous work, it is known that FMC data can enhance vowel recognition in the multi-modal approach. However, the question remains whether and how the accuracy of classification in the multimodal approach will change with significantly reduced feature vector. Standard MFCC features which are commonly used for speech representation according to [4,5] including allophone analysis [6] were used for the parametrization.

Speech was sampled at 48 kHz/s. The MFCC parameters were calculated for 512 samples of speech frames (with an overlap of 50%) and then the average and the variance values were calculated, which results in 40-dimensional vector of parameters. The decision algorithm using the audio signal only is based on the MFCC coefficients forming the input feature vector. Combined with FMC parameters vector, final dimension of data vector size is 80 for the multimodal approach. Results achieved for this approach are presented in Table 3.

Table 3. Results of classification employing grid-shifting parametrization and NN classifier

Speaker ID	FMC	Audio	Multimodal
1	63.1	76	86.3
2	47.1	80.3	86.3
3	61.4	86	92.6
4	35.1	84.9	85.4
5	52.1	75.6	78.5
6	44.6	82	85.4

6   Parameter Set Reduction

6.1   Feature Importance Assessed with ERT

The first presented algorithm for feature vector reduction is based on the
Extremely Randomized Trees (ERT) algorithm [7]. The concept derives from
the random forest providing a combination of tree predictors such that each tree
depends on the values of a random vector sampled independently and is also
characterized with the same distribution for all trees in the forest. The error
connected with generalization for forests converges to a limit as the number of
trees in the forest grows. The generalization error of a forest of tree classifiers
depends on the correlation between trees in the forest and on the strength of the
individual trees in the whole set [8,9].

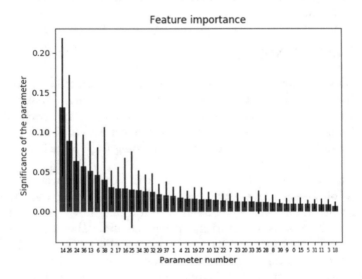

Fig. 4. Feature importance of every parameter in FMC data vector

Above algorithm was implemented with use of scikit python library [10]. ERT
was used with 10 estimators, entropy criterion, minimum sample split equal two,
minimum samples per leaf equal one, maximum number of features equal twenty
and with balanced class weight to balance an uneven number of examples for
classes of classified vocalic segments. Warm start and bootstrap were not used.

Moreover, with the use of Extremely Randomized Trees it was possible to
obtain a detailed information pertaining most important parameters, as it is
shown in Fig. 4 in terms of the classification for each speaker. In this way also
the role of markers as the key points was defined, what is seen in Fig. 5.

Table 4. Most important parameters for each speaker obtained with Extremely Randomized Trees

Speaker 1	Speaker 2	Speaker 3	Speaker 4	Speaker 5	Speaker 6
36	26	6	14	6	14
14	14	36	36	14	26
26	36	16	6	34	24
38	16	34	26	16	36
28	8	9	38	4	13
18	6	4	2	28	6
32	28	32	16	22	28
3	12	14	32	8	2
6	37	17	25	32	17
2	2	28	35	38	16

Having the knowledge about marker enumeration and remembering that every even number parameter plus zero defines the x-axis shift and every parameter with odd number represents y-axis shift, it is easy to determine which markers are important in the parameterization process. Ten most important parameters for each speaker are presented in Table 4. The most influential values are 6, 14, 24, 26, 34, 36. That corresponds x-axis shifts for markers number 4, 8, 13, 14, 18, 19 in Fig. 5. A very small number of components connected with y-axis shift in 10 most important parameters is an interesting observation.

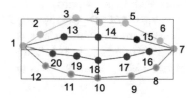

Fig. 5. Order of markers in the feature vector with the use of grid-shifting parametrization

As is seen in Table 5, the number of parameters vary depending on the speaker. The results are in most cases similar to those earlier obtained ones employing the entire (not reduced) parameter vector. Consequently, it turned out that the dimension of data was reduced more than twice with no significant impact on classification results.

Table 5. NN classification accuracy after ERT-based feature vector dimension reduction

Speaker ID	FMC	No. of param	Audio	No. of param	Multimodal	No. of param
1	57.14	13	67.43	14	81.43	27
2	48.86	18	75.71	14	85.14	24
3	62	18	82.29	13	90	24
4	34	18	78.29	15	83.14	25
5	51.47	15	79.12	16	78.53	27
6	46.57	16	80.57	14	82.57	27

6.2   PCA-Based Data Reduction

Principal Component Analysis is one of most known and used algorithm to reduce data dimension [11,12] The following goals of PCA are in-line with the tasks related to the compression of feature vectors:

- extract the most important information from the data table;
- compress the size of the data set by keeping only this important information;
- simplify the description of the data set;
- analyze the structure of the observations and the variables.

PCA was used to reduce dimension of our gathered data with minimal loss of accuracy in classification. Our approach was determined by the fact that FMC data have some redundancy and part of data has noise. It has quite big impact on classification results. It was possible to reduce dimensions of all data vectors four times, that means both FMC and audio vectors got 10 most important parameters and multimodal vector got 20 parameters. Accuracy of classification corresponding to reduced feature vector dimensions is shown in Table 6. As can be observed, in Audio and Multimodal approach accuracy of classification is slightly worse. However, in the FMC vector there is no bigger change in results.

Table 6. NN classification accuracy after PCA-based feature vector dimension reduction

Speaker ID	FMC	Audio	Multimodal
1	58	58.29	74.86
2	46	71.42	80.57
3	59.43	73.71	84.29
4	35.14	78	76
5	51.18	71.47	75.88
6	44.86	74.86	81.43

6.3   Recursive Parameter Elimination

Given an external estimator that assigns weights to features, the aim of Recursive Feature Elimination (RFE) is to select features through a recursive reduction of feature set content. First, an estimator is trained on the initial set of features and the importance of each feature is obtained for given attributes. Then, the least important features are pruned in the current set of features. That procedure is recursively repeated on the pruned set until the desired number of features is reached [13].

RFE model from scikit library was used [10], with the same dimension as in PCA (10 parameters for MFC and audio, 20 for the multimodal approach).

Results are presented in Table 7.

Table 7. NN classification accuracy after RFE-based feature vector dimension reduction

Speaker ID	FMC	Audio	Multimodal
1	55.14	65.71	82
2	48.86	76	84.86
3	54.86	81.14	88.29
4	32.57	78.29	83.71
5	49.71	73.82	80.29
6	47.14	79.14	85.14

As is seen from the table, the results are similar to the ones obtained for the full-size feature vector. The dimension of the vector was reduced 4 times, however any increase in accuracy for FMC was not achieved. It is worth noting, however, that for all speakers above 80% accuracy of the classification was obtained for the case of the multimodal approach.

7   Conclusions

The aim of the presented experiments was to find out whether the reduction of parameters in face motion capture and audio data can be used for improving vocalic segment recognition accuracy. The accuracy of vocalic segments classification with the use of 3 methods described in the previous chapter was measured for this purpose. That was done also in 3 ways: by using motion capture only, through the employment of audio features only and finally, by merging the data from both modalities. The obtained results do not allow to determine definitively, which algorithm of parameter reduction is the most effective one. Each of them works best for some data subsets, RFE being most effective for the multimodal approach, and Feature Importance ERT working best for FMC data and for audio. PCA-based feature vector dimension reduction performed worst of all 3 tested algorithms. An interesting observation is the fact that an important effect on the FMC classification is caused by shifts of FMC marker images along the X axis. This may occur due to the fact that with opening the mouth, the markers located near the right and left end of the lips cause the largest shift with respect to other markers and the center of the mouth, regardless of whether the spoken phoneme requires a wide or a narrow mouth opening. This observation may suggest designing of a new pattern of markers and testing its effectiveness upon new recording data processing.

Further plans related to the classification of vocalic segments is changing the classifier to HMM-based one. This step would require a significant rebuilding of the hitherto approach to speech representation, because a dynamic approach to the process of speech articulation is required in this case.

Acknowledgement. Research sponsored by the Polish National Science Centre, Dec. No. 2015/17/B/ST6/01874.

References

1. Cox, S., Harvey, R., Lan, Y., Newman, J., Theobald, B.: The challenge of multi-speaker lip-reading. In: International Conference on AuditoryVisual Speech Processing (2008)
2. Pei, Y., Kim, T., Zha, H.: Unsupervised random forest manifold alignment for lipreading. In: IEEE International Conference on Computer Vision, pp. 129–136 (2013)
3. Jachimski, D., Czyzewski, A., Ciszewski, T.: A comparative study of English viseme recognition methods and algorithms. Multimedia Tools Appl. (2017). https://doi.org/10.1007/s11042-017-5217-5
4. Eringis, D., Tamulevičius, G.: Modified filterbank analysis features for speech recognition. Baltic J. Modern Comput. **3**(1), 29–42 (2015)
5. Zheng, F., Zhang, G., Song, Z.: Comparison of different implementations of MFCC. J. Compu. Sci. Technol. **16**(6), 582–589 (2001)
6. Kostek, B., Piotrowska, M., Czyżewski, A.: Comparative study of self-organizing maps vs. subjective evaluation of quality of allophone pronunciation for non-native english speakers. In: Audio Engineering Society Convention 143 (2017)
7. Geurts, P., Ernst, D., Wehenkel, L.: Extremely randomized trees. Mach. Learn. **63**, 3 (2006). https://doi.org/10.1007/s10994-006-6226-1
8. Louppe, G., Wehenkel, L., Sutera, A., Geurts, P.: Understanding variable importances in forests of randomized trees. Adv. Neural Inf. Process. Syst. **26**, 431–439 (2013)
9. Svetnik, V., Liaw, A., Tong, C., Culberson, J.C., Sheridan, R.P., Feuston, B.P.: Random forest: a classification and regression tool for compound classification and QSAR modeling. J. Chem. Inf. Comput. Sci. **43**(6), 1947–1958 (2003)
10. Pedrosa, F., Varoquaux, G., Gramfort, A., Michel, V., Thirion, B., Grisel, O., Blondel, M., Prettenhofer, P., Weiss, R., Dubourg, V., Vanderplas, J., Passos, A., Cournapeau, A., Brucher, M., Perrot, M., Duchesnay, É.: Scikit-learn: machine learning in Python. J. Mach. Learn. Res. **12**, 2825–2830 (2011)
11. Bro, R., Smilde, K.: Principal component analysis. Anal. Methods, 2812–2831 (2014)
12. Abdi, H., Williams, L.J.: Principal component analysis. Wiley Interdisc. Rev. Comput. Stat. **2**(4), 433–459 (2010)
13. Mao, Y., et al.: Accelerated recursive feature elimination based on support vector machine for key variable identification. Chin. J. Chem. Eng. **1**(14), 65–72 (2006)

Innovations in Web Technologies

Weight Adaptation Stability of Linear and Higher-Order Neural Units for Prediction Applications

Ricardo Rodriguez-Jorge[1]([⊠]), Jiri Bila[2], Jolanta Mizera-Pietraszko[3], and Edgar A. Martínez-Garcia[1]

[1] Institute of Engineering and Technology, Universidad Autónoma de Ciudad Juárez, Partido Romero, 32310 Ciudad Juárez, Mexico
`ricardo.jorge@uacj.mx`
[2] Czech Technical University in Prague, Technická 4, 166 07 Praha 6, Prague, Czech Republic
[3] Institute of Mathematics and Computer Science, Opole University, Opole, Poland

Abstract. This paper is focused on weight adaptation stability analysis of static and dynamic neural units for prediction applications. The aim of this paper is to provide verifiable conditions in which the weight system is stable during sample-by-sample adaptation. The paper presents a novel approach toward stability of linear and higher-order neural units. A study of utilization of linear and higher-order neural units with the foundations on stability of the gradient descent algorithm for static and dynamic models is addressed.

Keywords: Linear neural units · Higher-order neural units · Stability analysis

1 Introduction

Artificial neural networks were developed in the 1940's. Dealing with nonlinearities and uncertainties has been of major interest to artificial neural networks (ANNs). Due to their inherent approximation capabilities, the ANN has been extensible used in modeling of high complex nonlinear dynamic systems. These architectures can approximate spatio-temporal data by providing to the input of the network past inputs and past outputs. However, ANNs have been naturally studied with some problems about local minima, overfitting issue, and generalization capability [1–4].

One of the main applications of the ANNs is in the prediction field. Prediction has been considered as great interest to know what is unknown in the future. These ANNs have been applied to predict time series (a set of values of an attribute sensed in regular periods). Some applications of prediction by neural networks are: predicting the weather in a period, behavior of the merchandise sale level or maintaining the stock, financial reasons prediction, prediction catastrophic events into electrocardiograms, prediction of breathing patterns, and so on [5, 6].

Measurements of real-world systems usually have influences of several external processes, which can generate long-range correlations and non-stationarity.

© Springer Nature Switzerland AG 2019
K. Choroś et al. (Eds.): MISSI 2018, AISC 833, pp. 503–511, 2019.
https://doi.org/10.1007/978-3-319-98678-4_50

A non-stationary time series is usually characterized by time varying variance, mean, or even both. Thus, the learning process should then capture the current data behavior (dynamics) to the predictive model for improved prediction accuracy [7, 8].

Quadratic neural unit (QNU) and cubic neural unit (CNU) are known as type of higher-order neural unit (HONU), and comparing with linear neural unit (LNU), QNU and CNU provide good quality of nonlinear approximation. Basic concepts of learning and adaptation have been reviewed in [9].

This paper presents the sample-by-sample adaptation of linear, and higher-order neural units in their static and dynamic architectures, with error back-propagation and explores the gradient descent method. The implementation of linear and higher-order neural units is performed using a novel weight adaptation stability of the predictive models.

This paper is organized as follows: in Sect. 2, the static neural models are described, as well as the gradient descent method for linear and higher-order static neural models in sample-by-sample adaptation. Section 3 discusses the dynamic neural models for prediction. Section 4 discusses the weight system stability of the linear and higher order neural models. Section 5 discusses the results obtained after applying the weight update stability of the models. Finally, conclusion remarks and future work are described in Sect. 6.

2 Static Neural Models for Prediction

2.1 Linear Neural Unit

Linear neural units are simplest and computationally efficient, especially when the input data contain relatively higher number of inputs. LNUs yields the neural output as shown in Eq. (1).

$$\tilde{y}(k+h) = \sum_{i=0}^{n} w_i \cdot x_i = \mathbf{w} \cdot \mathbf{x}(k) \tag{1}$$

where, $\mathbf{w}$ denotes the row vector of all neural weights and $\mathbf{x}(k)$ represents the vector of inputs. The weight updates are directly calculated for a linear predictive model as presented next.

$$\Delta w_i(k) = -\frac{1}{2} \cdot \mu \cdot \frac{\partial e^2(k+h)}{\partial w_i} \tag{2}$$

where μ is the learning rate, $e(k+h)$ is the prediction error, and $\Delta w_i(k)$ is an adaptive weight increment of $i - th$ weight. The updates of all weights can be in their simplest form, i.e., without momentum or regularization term, as shown in Eq. (3).

$$w_i(k+1) = w_i(k) + \Delta w_i(k) \tag{3}$$

2.2 Quadratic Neural Unit

QNU is known as a type of HONU [2], and comparing with the traditional neural networks, QNU provide good quality of nonlinear approximation [6]. In addition, the mathematical structure is relatively comprehensible due to the minimum number of neural parameters. Basic concepts of learning and adaptation has been reviewed in [8]. The static quadratic neural unit used for prediction is shown in Fig. 1.

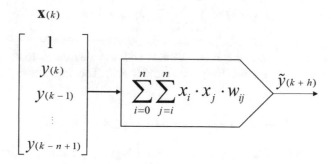

Fig. 1. Static quadratic neural unit for prediction.

In the case of the QNU static architecture, a quadratic operation using an $(1+n) \times (1+n)$ upper triangular weight matrix $\mathbf{W}$ is performed. This operation is presented in Eq. (4).

$$\tilde{y}(k+h) = \sum_{i=0}^{n} \sum_{j=i}^{n} x_i \cdot x_j \cdot w_{ij} = \mathbf{x}^{\mathrm{T}} \cdot \mathbf{W} \cdot \mathbf{x} \tag{4}$$

where the vector $\mathbf{x}$ denotes the $(1+n) \times 1$ vector with bias $x_0 = 1$, x_i denotes individual inputs (scalars), and h stands for the prediction horizon.

The static QNU is adapted every $k-th$ sample by means of sample-by-sample adaptation as presented in Eq. (5).

$$w_{ij}(k+1) = w_{ij}(k) + \Delta w_{ij}(k) \tag{5}$$

where the weight update is performed by gradient descent as shown in Eq. (6).

$$\Delta w_{ij}(k) = -\frac{1}{2} \cdot \mu \cdot \frac{\partial e^2(k+h)}{\partial w_{ij}} \tag{6}$$

where μ is the learning rate, and $e(k+h)$ stands for the error, which is defined as in Eq. (7).

$$e(k+h) = y(k+h) - \tilde{y}(k+h) \tag{7}$$

According to the sample-by-sample adaptation $\tilde{y}(k+h)$ stands for the neural output, $y(k+h)$ is the real value.

2.3 Cubic Neural Unit

A CNU is a type of HONU, the input-output relation of a static CNU is given as in Eq. (8)

$$\tilde{y}(k+h) = \sum_{i=0}^{n}\sum_{j=i}^{n}\sum_{l=j}^{n} x_i \cdot x_j \cdot x_l \cdot w_{i,j,l} \qquad (8)$$

where $\mathbf{x}$ and $\tilde{y}(k+h)$ are the input vector and the neuron output, respectively; $\mathbf{W} = \{w_{i,j,l} : i,j,l \in n\}$ are the weights of the CNU. The input $\mathbf{x}(k)$ is defined by Eq. (9).

$$\mathbf{x}(k) = [\, 1 \quad y(k) \quad y(k-1) \quad \cdots \quad y(k-n+1)\,]^{T} \qquad (9)$$

where n is the total number of samples of the real signal $y(k)$ and ^{T} stands for transposition. The CNU calculates the output neuron $\tilde{y}(k+h)$ when the input vector $\mathbf{x}$ is provided. The CNU learns by applying the back-propagation technique as a learning rule to adjust the weight matrix $\mathbf{W}$. The structure of the CNU for prediction is shown in Fig. 2.

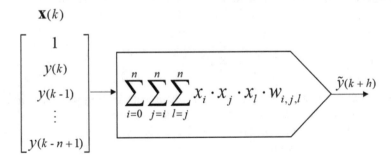

Fig. 2. Cubic neural unit architecture for prediction.

The back-propagation technique as a learning rule is applied in a sample-by-sample adaptation. It follows the convergence criterion of the training performance, which is given as the sum of squared error in each iteration time, for N training samples as shown in Eq. (10).

$$Q(epoch) = \sum_{k=1}^{N} e(k+h) \qquad (10)$$

where $e(k+h) = y(k+h) - \tilde{y}(k+h)$.

The neural weights adaptation $w_{i,j,l}(k+1)$ in the new adaptation time of the model is presented in Eq. (11).

$$w_{i,j,l}(k+1) = w_{i,j,l}(k) + \Delta w_{i,j,l}(k) \tag{11}$$

The increment of each individual weight with the gradient descent rule is described in Eq. (12).

$$\Delta w_{i,j,l}(k) = \mu \cdot e(k+h) \cdot \frac{\partial \tilde{y}(k+h)}{\partial w_{i,j,l}} \tag{12}$$

3 Dynamic Neural Models for Prediction

Towards the goal of predicting data from a system with the biologic neural characteristics, a mathematical model named dynamic linear neural unit (D-LNU), dynamic
quadratic neural unit (D-QNU), and dynamic cubic neural unit (D-CNU) have been
developed [1, 2]. The D-QNU and D-CNU are considered a type of dynamic higher-
order neural unit (D-HONU). D-HONU have been capable of processing systems that
present linearities and non-linearities on spaces of continuous or discrete time.
The QNU model is considered as a special class of polynomial neural networks. The
works of [1, 2, 6] have presented QNUs indicating the characteristics of the neural unit
of providing a faster response comparing to a controller by linear states feedbacks as
well as with controllers applied to nonlinear systems, unstable systems, and unknown
non-linear dynamic systems.

3.1 Dynamic Quadratic Neural Unit

The D-QNU use a RTRL learning method and can be implemented in discrete or
continuous real-time. In Eq. (13) the notation of D-QNU with RTRL is shown, where
$\tilde{y}(k+h)$ is the predicted neural output and ^{T} represents the transpose of the vector $\mathbf{x}$,
which contains the feedback neural output values and the signal from the system.

$$\tilde{y}(k+h) = \left[\sum_{i=0}^{nx-1} \sum_{j=0}^{nx-1} x_j \cdot w_{i,j} \cdot x_i \right] = (\mathbf{x}^T \cdot \mathbf{W} \cdot \mathbf{x}) \tag{13}$$

The upper triangular matrix of weights $\mathbf{W}$ with neural bias $w_{0,0}$ is defined as

$$\mathbf{W} = \begin{bmatrix} w_{0,0} & w_{0,1} & \cdots & w_{0,nx-1} \\ 0 & w_{1,1} & \cdots & w_{1,nx-1} \\ \vdots & \vdots & \ddots & \vdots \\ 0 & \cdots & 0 & w_{nx-1,nx-1} \end{bmatrix} \tag{14}$$

Equation (15) is the column of the augmented vector **x** of the neural input, $y(k)$ is the real value, n is the number of real values that feeds the neural input, k is the variable that describes the discrete time and h is the prediction horizon.

$$\mathbf{x} = \begin{bmatrix} 1 \\ \tilde{y}(k+h-1) \\ \tilde{y}(k+h-2) \\ \vdots \\ \tilde{y}(k+1) \\ y(k) \\ y(k-1) \\ \vdots \\ y(k-n+1) \end{bmatrix} \tag{15}$$

In the architecture of the D-QNU, $1/z$ represents the values of feedback, which are fed to the augmented vector **x**.

An error function is defined in such a way that, in each learning step indicates how close the solution is to minimize the error function (mean squared error function) of the D-QNU model; $\frac{1}{N}\sum_{k=1}^{N} e(k+h)^2$ in each prediction time $k+h$. In Eq. (16), $w_{i,j}(k+1)$ stands for neural weights adaptation in the new adaptation time of the model, which are the sum of the neural weights increment $\Delta w_{i,j}(k)$ from the adaptation of the model, with individual weights $w_{i,j}$ of each k discretized value from the system.

$$w_{i,j}(k+1) = w_{i,j}(k) + \Delta w_{i,j}(k) \tag{16}$$

The increment of the neural weights with RTRL is described in (17), where μ is the learning rate that determines in which proportion the neural weights are updated and the velocity of the learning process, $e(k+h)$ is the error in each discrete time $k+h$ that multiplies the partial derivative of the neural output with respect to the weights $w_{i,j}$ described in Eq. (18).

$$\Delta w_{i,j}(k) = \mu \cdot e(k+h) \cdot \frac{\partial \tilde{y}(k+h)}{\partial w_{i,j}} \tag{17}$$

$$\frac{\partial \tilde{y}(k+h)}{\partial w_{i,j}} =$$
$$\frac{\partial (\mathbf{x}^T \cdot \mathbf{W} \cdot \mathbf{x})}{\partial w_{i,j}} = \left(\frac{\partial \mathbf{x}^T}{\partial w_{i,j}} \cdot \mathbf{W} \cdot \mathbf{x} + \mathbf{x}^T \cdot \frac{\partial \mathbf{W}}{\partial w_{i,j}} \cdot \mathbf{x} + \mathbf{x}^T \cdot \mathbf{W} \cdot \frac{\partial \mathbf{x}}{\partial w_{i,j}} \right) \tag{18}$$

In Eq. (19), $\mathbf{jxw}_{i,j}$ is described as the partial derivative of the augmented vector $\mathbf{x}$ of the neural inputs with respect to the neural weights.

$$\frac{\partial \mathbf{x}}{\partial w_{i,j}} = \mathbf{jxw}_{i,j} =$$

$$\left[0 \quad \frac{\partial \tilde{y}(k+h-1)}{\partial w_{i,j}} \quad \frac{\partial \tilde{y}(k+h-2)}{\partial w_{i,j}} \quad \cdots \quad \frac{\partial \tilde{y}(k+1)}{\partial w_{i,j}} \quad 0 \quad \cdots \quad 0 \right]^T \tag{19}$$

Meanwhile, in Eq. (20) the representation of a full matrix is shown, which is a convenient adaptive representation of a D-QNU, where $\mathbf{r\Delta W}(k)$ is a vector with all increased weights and $\mathbf{jxw}_{i,j}$ is the Jacobian matrix evolving recurrently [9].

$$\mathbf{r\Delta W}(k) = \mu \cdot e(k+h) \cdot$$
$$\left((\mathbf{jxw}_{i,j}(k)^T \cdot \mathbf{W} \cdot \mathbf{x})^T + \mathbf{r}\{\mathbf{x}^T \cdot \mathbf{x}\} + \mathbf{x}^T \cdot \mathbf{W} \cdot \mathbf{jxw}_{i,j}(k) \right) \tag{20}$$

where $\mathbf{r}\{\mathbf{x}^T \cdot \mathbf{x}\}$ represents all the combination of the input $\mathbf{x}$. The Jacobian matrix is updated on a recurrent basis for the adaptation of the model.

4   Weight Adaptation Stability

The weight system to be influenced during the weight adaptation stability can be expressed in matrix form for the static neural model, even linear, quadratic or cubic neural unit, and the learning rates will be used individually for each weight as shown in Eq. (21).

$$\mathbf{colW}(k+1) = \mathbf{colW}(k) + \mathbf{M} \cdot e(k+h) \cdot \mathbf{colx} \tag{21}$$

where $\mathbf{M}$ is $(nw \times nw)$ main diagonal matrix of individual learning rates. Accordingly, the individual learning rates corresponds to the main diagonal of the matrix $\mathbf{M}$ while the other values remain in zero.

Let us express the weight adaptation system using the introduced notation by substituting the definition of the error, as shown in Eq. (22).

$$\mathbf{colW}(k+1) = \mathbf{colW}(k) + \mathbf{M} \cdot (y(k+h) - \tilde{y}(k+h)) \cdot \mathbf{colx} \tag{22}$$

After several hand-work step-by-step, and considering the weight state vector $\mathbf{colW}$, the representation of the weight dynamics system can be rewritten as shown in Eq. (23).

$$\mathbf{colW}(k+1) = \mathbf{colW}(k)(1 - \mathbf{M} \cdot \mathbf{colx} \cdot \mathbf{rowx}) + \mathbf{M} \cdot y(k+h) \cdot \mathbf{colx} \tag{23}$$

where **1** denotes $(nw \times nw)$ identity matrix and $\mathbf{A} = (\mathbf{1} - \mathbf{M} \cdot \mathbf{colx} \cdot \mathbf{rowx})$. Next, the stability criterion is applied using Eq. (24).

$$Max|eig(\mathbf{A})| \leq 1 \tag{24}$$

Then, the weight system is stable (contractive) at time *epoch* if Eq. (24) is true. However, when the stability condition (non-expansiveness) of weight system is not satisfied, the learning rates are recalculated by random vector $\boldsymbol{\delta}$ of $(1 \times nw)$ dimension and the vector of learning rates are slightly modified as shown in Eq. (25).

$$\boldsymbol{\mu} = \boldsymbol{\mu} \cdot \left(1 - \frac{\boldsymbol{\delta}}{5} \right) \tag{25}$$

This is performed to maintain the condition (24). The division of the random vector rate $\boldsymbol{\delta}$ by, for example: the number 5, allows maintaining the learning rate very close to the original values. Next, after recalculating the new learning rates, the maximum absolute eigenvalues are evaluated again according to the condition Eq. (24). In case that the new learning rate matrix influence the stability condition, i.e. Eq. (24) is satisfied, the neural unit is updated. Otherwise, the learning rate matrix **M** is changed again until at least non-expanding autonomous part of weight dynamic update system is obtained.

For the dynamic neural model (D-LNU, D-QNU or even D-CNU) the weight update stability can be given using the previous expressions by fundamental gradient descent in RTRL.

5 Discussion

According to the adaptation of each individual weight as shown in the linear and higher-order models, all weights might appear as linear, therefore the adaptation weights can be transformed into a state-space form. Therefore, the vector **colW** has been introduced to perform the weight adaptation stability; which corresponds to the transformation of the weight matrix **W**, for higher-order neural units, or the vector **w** for linear neural units, i.e., **colW** contains all the weights ordered from every row of the upper-triangular weight matrix, for the higher-order neural units. In the case of **colx**, stands for a column vector of input multiplications, for the case of HONUs, and **rowx** stands for $\mathbf{colx}^T$.

6 Conclusions

This paper presented the linear and higher order neural units mathematical notations and their architecture as well. In addition, their stability evaluation approach has been introduced to ensure the stability of the weight update system of static and dynamic neural models applied in prediction applications. The approach of the stability analysis presented is based on the eigenvalues of state-space representation of the weight update dynamic system for gradient descent adaptation rule.

Acknowledgements. The project is supported by a research grant No. DSA/103.5/16/10473 awarded by PRODEP and by Autonomous University of Ciudad Juarez. Title - Detection of Cardiac Arrhythmia Patterns through Adaptive Analysis.

References

1. Rodríguez Jorge, R.: Lung tumor motion prediction by neural networks. Ph.D. thesis. Czech Technical University in Prague, Czech Republic (2012)
2. Gupta, M.M., Jin, L., Homma, N.: Static and Dynamic Neural Networks. Wiley, U.S.A (2003)
3. Bach, F.: Breaking the curse of dimensionality with convex neural networks. J. Mach. Learn. Res. **18**, 1–53 (2017)
4. Rodriguez, R., Vergara Villegas, O.O., Cruz Sanchez, V.G., Bila, J., Mexicano, A.: Arrhythmia disease classification using a higher-order neural unit. In: 2015 Fourth International Conference on Future Generation Communication Technology (FGCT), pp. 1–6. IEEE, July 2015
5. Rodríguez, R., Mexicano, A., Bila, J., Ponce, R., Cervantes, S., Martinez, A.: Hilbert transform and neural networks for identification and modeling of ECG complex. In: 2013 Third International Conference on Innovative Computing Technology (INTECH), pp. 327–332. IEEE, August 2013
6. Cancino, E., Rodriguez Jorge, R., Vergara Villegas, O.O., Cruz Sánchez, V.G., Bila, J., Nandayapa, M., Israel, P., Soto, A., Abad, A.: Monitoring of cardiac arrhythmia patterns by adaptive analysis. In: The 11th International Conference on P2P, Parallel, Grid, Cloud and Internet Computing (3PGCIC-2016) (2016)
7. Mironovova, M., Bíla, J.: Fast fourier transform for feature extraction and neural network for classification of electrocardiogram signals. In: 2015 Fourth International Conference on Future Generation Communication Technology (FGCT), pp. 1–6. IEEE, July 2015
8. Rodriguez, R., Bukovsky, I., Bila, J.: Weight Adaptation Stability of Static Quadratic Neural Unit. Ústav Přístrojové a řídicí techniky, pp. 83–88. Czech Technical University in Prague (2011)
9. Rodríguez Jorge, R., Martínez García, E., Mizera-Pietraszko, J., Bila, J., Torres Córdoba, R.: Prediction of highly non-stationary time series using higher-order neural units. In: Xhafa, F., Caballé, S., Barolli, L. (eds.) Advances on P2P, Parallel, Grid, Cloud and Internet Computing, 3PGCIC 2017. Lecture Notes on Data Engineering and Communications Technologies, vol. 13. Springer, Cham (2017)

Visual Analysis of Differential Evolution Algorithms

A. Mexicano-Santoyo[1], R. Rodríguez-Jorge[2(✉)], A. Abrego[1,3], M. A. Jiménez[1],
R. Zúñiga-Treviño[1], and Edgar A. Martínez-Garcia[2]

[1] Instituto Tecnológico de Cd., Victoria, Mexico
mexicanoa@gmail.com
[2] Institute of Engineering and Technology, Universidad Autónoma de Ciudad Juárez,
Partido Romero, 32310 Ciudad Juarez, Mexico
ricardo.jorge@uacj.mx
[3] Universidad Autónoma de Tamaulipas Cd., Victoria, Mexico

Abstract. In this article a web tool which contributes to the visual analysis of the Differential Evolution (DE) algorithms is presented. The tool provides a graphic interface with 8 views that allows understanding the underlying process of the algorithm. The tool has a library which extracts data from DE algorithms and its main feature is that the functions of the library can be embedded in the code of any DE algorithm to be analyzed. To validate the tool, three DE algorithms: *DE/Rand/1/bin*, *DE/best/1/bin*, and *JADE* and three test functions: *Sphere*, *Rosenbrock*, and *Rastrigin* have been used, which produced a total of 234 different tests, all of them performed successfully. The tool can allow to experts to analyze algorithms, particularly DE algorithms, and it can contribute to improve such algorithms or in generating new strategies that can emerge from the analysis of the extracted information.

Keywords: Differential evolution algorithms · Algorithm analysis · Web tool

1 Introduction

Setting the parameters of an evolutionary algorithm to solve a combinatorial optimization problem can be a simple task. However, analyzing the results to discover when the process is efficient enough is a non-trivial task. The representation of individuals, the choice of operators, and the parameter configuration are factors that influence the computation speed and the quality of the result [1]. The development of visualization tools to understand the inherent mechanisms of evolutionary algorithms is of great importance to understand the underlying process and the impact caused by the initial configuration of the parameters [2, 3]. Some tools developed for the analysis of evolutionary algorithms can be found at [4–8].

In this work, a web tool that contributes to the visual analysis of the performance of differential evolution algorithms (DE) is presented. The purpose of the tool is to provide the researcher with a graphical interface that allows understanding, in general, the operation of the DE algorithms, which are a class of evolutionary algorithms, and provide a notion of what is happening within it. In contrast to the related work, the tool was developed to be useful in the analysis of the DE algorithm which use a coding with floating

© Springer Nature Switzerland AG 2019
K. Choroś et al. (Eds.): MISSI 2018, AISC 833, pp. 512–521, 2019.
https://doi.org/10.1007/978-3-319-98678-4_51

point values unlike evolutionary algorithms which in genetic algorithm uses a bit string for representing the solutions. Furthermore, the tool has a library which extracts data from any variant of DE algorithms by using a library whose functions can be embed into the code of any specific DE algorithm in order to be analyzed and evolutionary operators can be visualized in order to observe the variations. Another important contribution is that, the presented tool is web and can be accessed anywhere.

The rest of the article is organized as follows: Sect. 2 shows the description of the tool developed for the analysis of DE algorithms; Sect. 3 shows the conducted tests to prove the proper functioning of the tool, and Sect. 4 shows the conclusions and future work.

2   Tool for Analyzing Differential Evolution Algorithms

In this section the web tool for the visualization of the DE algorithms is described, this tool has been developed using the R language, since according to [9], R is one of the most popular environments for the analysis and visualization of data and allows the application to be hosted on a remote server to be acceded through the use of a web browser. The graphical interface of the tool provides views with 2D graphs and information in tabular form, where statistical data are presented as: average, minor value, higher value, and standard deviation of the analyzed information. Additionally, the tool has interactive 3D graphics that have been created with the WebGL technology, property of the Institut de Radioprotection et de Sûreté Nucléaire [10]. The tool uses a generic library implemented in C++ that captures the necessary data to perform the generation of the visualizations inspired by [11], even modern variants with autoadaptive parameters.

2.1   Algorithm Behavior Menu

The menu "Algorithm behavior" allows visualizing the information corresponding to each of the generations produced during the execution of the DE algorithm. The views presented by this menu are *Convergence diagram*, *Best individuals* and *Autoadaptive parameters*. As an example, only some views of the tool are shown, especially those that differ from other implementations.

Convergence Diagram. The "Convergence diagram" view presents the objective value of the best individual at each generation, this view allows observing how the algorithm approaches the result until it reaches a point where it converges.

Best Individuals. The "Best individuals" view presents the characteristics of the best individual in each generation during the execution of the algorithm. Figure 1 shows a screenshot of the tool with the view "Best individuals". Figure 1a presents the graphic representation of the best individuals where the axis "Dimensions" represents the characteristics of the individuals and the axis "Individuals" stands for the best individual of each generation. This graph allows observing the evolution of the best individual through the execution of the algorithm. In the lower left part (Fig. 1b) there is a control that allows selecting the data that is loaded in the view and a scroll bars that allow controlling the amount of data that will be displayed in the tabular form (Fig. 1c). This feature allows the user the manipulation of the range of values to analyze, according to their need.

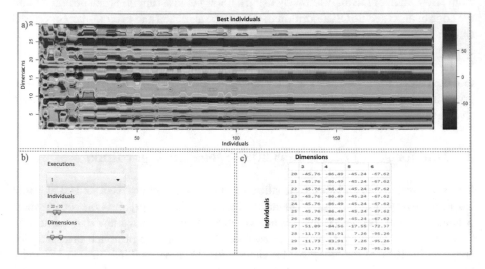

Fig. 1. Best individuals view.

Autoadaptive Parameters. Some variants of the DE algorithm contain parameters that are adjusted in the course of the algorithm in order to generate more accurately individuals, for example: JADE, jDE, SaDE, among others. In this view the behavior that

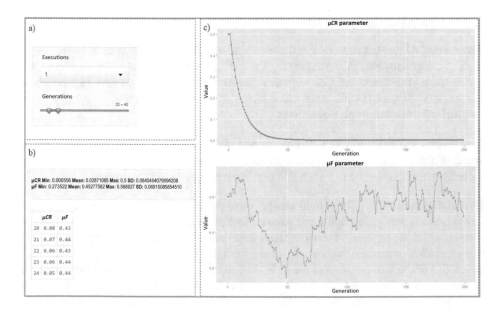

Fig. 2. Autoadaptive parameters view.

the autoadaptive parameters μF (mean of the mutation factor) and μCr (average of the crossing probability) takes during the execution of the algorithm is presented. This view allows observing the trends of the function used to adjust the auto-adaptive parameters. Figure 2 shows the "Autoadaptive Parameters" view of the tool, where the user can select the information of an execution (Fig. 2a) and the information shown in tabular form (Fig. 2b). In addition, Fig. 2c shows graphically the course of the μF and μCr parameters.

2.2 Algorithm Status Menu

The "Algorithm Status" menu contains views as: "Visualization of individuals", "Variables of individuals", "Objective values" and "Evolutionary operators", which present the information of each generation of the algorithm, allowing the visualization of every individual of any generation.

Visualization of Individuals. Through the "Visualization of individuals" view, the dimensions of all the individuals of a particular population can be visualized, it allows having an idea of the distribution of the population and its closeness to the solution. Figure 3b shows this view, where the "Individuals" axis represents the individuals that form the population in a specific generation, the "Dimensions" axis represents the characteristics of the individuals, and the "Value" axis corresponds to the value of each dimension, which, in the example, are defined in a range of $[-100, 100]$ because the characteristics of the individuals oscillate between these values. The graph can be rotated, zoomed in or zoomed out by the user. In Fig. 3a, the "Generations" scroll bar allows to see the graph of the population corresponding to a specific generation and the "Individuals" and "Dimensions" scroll bars allow controlling the number of individuals and dimensions, respectively, which are also visualized in tabular form.

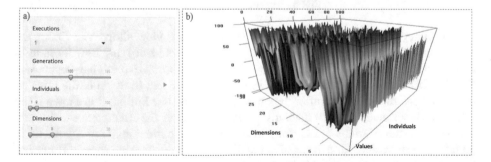

Fig. 3. Visualization of individuals view.

Variables of Individuals. This view provides an overview about the distribution of individuals, the distances between them and the concentration of characteristics in one or more regions.

Objective Values. The "Objective values" view allows observing how the fitness of a given population is distributed. In the graph "Objective values" (Fig. 4-superior) the axis "Objective value" represents the aptitude of the individual and the axis "Individuals" corresponds to each of the individuals of the population in the given generation. In the graph "Quantification of objective values" (Fig. 4-below) the distribution of the objective values is visualized and helps to determine how many individuals of the population have a good aptitude and how many do not have it

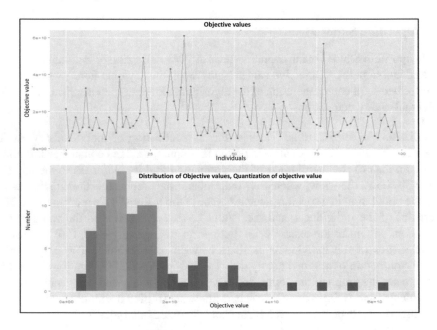

Fig. 4. Objective values view.

Evolutionary Operators. The "Evolutionary Operators" view allows visualizing the formation of individuals in a given generation and allows identifying how these individuals are affected by the Mutation, Crossing, and Selection evolutionary operators. Figure 5-below shows the tabular representation of a current individual, the mutated individual and the individual after the operation crosses. In addition, it is shown if the individual is selected, that is, if the individual resulting from the cross passes to the next generation. In the "Variables of the individuals" graph the current individual (solid line), mutated (line in dashes), and the resultant of the cross (dotted line) are shown. However, the user can select any specific individual of any generation or execution.

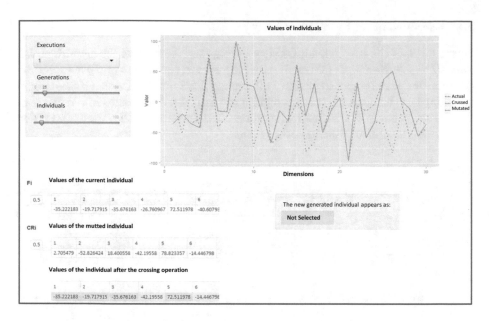

Fig. 5. Evolutionary operators view.

3   Experimental Results

The experimental environment has been established under the following conditions: computer with Intel core i7 2.50 GHz processor; 16 GB of RAM; 64-bit Windows 8.1 operating system; R version 3.1.2, Google Chrome web browser version 39.0 and the tool developed. Data extraction has been performed using three different DE algorithms: *DE/Rand/1/bin*, *DE/best/1/bin* and *JADE*, implemented in C/C++. The parameters of the three algorithms were initialized using the same values *NP* (number of the population) = 100, *D* (dimensions) = 30, *F* (mutation factor) = 0.5, *CR* (probability of crossing) = 0.5; and using a number of generations = 300. The three algorithms have been run using three test functions widely used for the evaluation of optimization algorithms, these functions are: *Sphere* a well-known simple function with a single global minimum, *Rosenbrock* or *Banana* function that has a valley that descends slowly by the origin of the function which makes it difficult to find its global minimum, and *Rastrigin* function, this function has a global minimum that is quite difficult to find, because the function contains many local minima. All the functions were obtained from the *Congress on Evolutionary Computation 2005*, published in the work [12].

As an example of the use of the tool, Sects. 3.1 and 3.2 show the views of the *Best Individuals* and the *Convergence Diagrams* when observing the behavior for three different algorithms.

3.1 Execution of Best Individual View

Figure 6 shows the best individuals view during the execution of the algorithms *DE/ Rand/1/bin*, *DE/Best/1/bin* [13], and *JADE* [14]. As can be seen in Fig. 6 each algorithm performs the search for the solution in the range determined between [100, −100]. In addition, it can be observed that the algorithms perform a different evolution that leads them to obtain three different optimal solutions. In the case of the *DE/Best/1/bin* version, the algorithm stabilizes faster than the others.

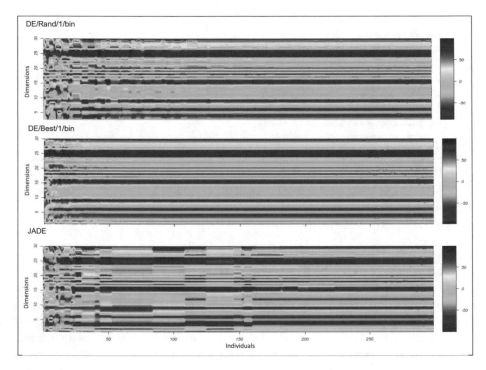

Fig. 6. Visualization of the best individuals during for algorithms *DE/Rand/1/bin*, *DE/Best/1/ bin*, and *JADE*.

3.2 Execution of Convergence Diagram View

Figure 7 shows the convergence diagram of the algorithms: *DE/Rand/bin/1*, *DE/Best/1/ bin*, and *JADE*. In the figure, the red line corresponds to the average of the objective values of the population and the blue line is the best individual. It can be observed as in the Variables of the best individual view that the *DE/Best/1/bin* algorithm converges more quickly than the *DE/Rand/1/bin* and *JADE* algorithms.

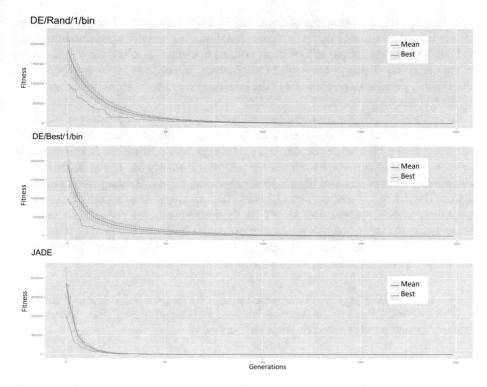

Fig. 7. Visualization of the convergence diagram for algorithms *DE/Rand/1/bin*, *DE/Best/1/ bin*, and *JADE*.

4   Conclusions

The main contribution of this work is the development of a tool to support the experimental analysis of DE algorithms. The tool was developed in R and contributes to the Visual Analysis of the behavior of the DE algorithms. The tool helps to understand the intermediate processes of the algorithm and the evolution of individuals until finding a solution. The tool is composed of three elements: (a) the library, which allows extracting the data from executions of any DE algorithm implemented in C/C++; (b) the repository of files generated by the library, which contains the data of the executions of the experiments carried out, and (c) the web graphical interface, which enables the user to interact with the graphs and information in tabular form, of the stored information previously obtained.

In order to test the correct functioning of the tool, three DE algorithms were implemented in C/C++ (*DE/Rand/1/bin*, *DE/best/1/bin* and *JADE*), each algorithm was executed with three test functions used to evaluate search and optimization algorithms (*Sphere*, *Rosenbrock* and *Rastrigin*). The three algorithms have been run twice with the three test functions using the same input configuration ($NP = 100$, $D = 30$, $F = 0.5$, $Cr = 0.5$ and $G = 300$) making a total of 18 executions, which produced a set of 36 files with the data extracted from the experimentation. The information of the executions has been extracted using the library, which uses functions that are embedded into the source code of the implementations of the DE algorithms to extract the data from the execution of the algorithm. Through the use of the graphical interface it has been possible to observe in a visual and correct manner the information extracted from the execution of the evolutionary algorithms.

The experimental tests show the generality of the tool since, regardless of the algorithmic implementation, it has been possible to extract the necessary data from each algorithm for its later visualization. Using this type of visualization, it is feasible to compare the results between iterations to decide which strategy can be the most optimal. The development of this tool has allowed supporting experts in analysis of DE algorithms. The use of this tool may contribute to the improvement of algorithms or in the generation of new strategies that arise from the analysis of extracted information. For future work, the use of techniques belonging to the field of "visual data mining" is recommended to obtain more significant graphic representations from the large amount of information obtained during the experimentation.

Acknowledgements. The project is supported by a research grant No. DSA/103.5/16/10473 awarded by PRODEP and by Autonomous University of Ciudad Juarez. Title - Detection of Cardiac Arrhythmia Patterns through Adaptive Analysis.

References

1. Hart, E., Ross, P.: GAVEL - a new tool for genetic algorithm visualization. IEEE Trans. Evol. Comput. **5**(2), 335–348 (2001)
2. Wu, H.-C., Sun, C.-T., Lee, S.-S.: Visualization of evolutionary computation processes: from the perspective of population. In: Proceedings of the Fifth World Congress on Intelligent Control and Automation, pp. 2077–2081. IEEE Xplore, Hangzhou (2004)
3. Kerren, A., Egger, T.: EAVis: a visualization tool for evolutionary algorithms. In: IEEE Symposium on Visual Languages and Human-Centric Computing, pp. 299–301. IEEE Xplore, Texas (2005)
4. Lutton, E., Fekete, J.D.: Visual analytics and experimental analysis of evolutionary algorithms. INRIA Rapport de recherche 7605 (2011)
5. Bullock, S., Bedau, M.A.: Exploring the dynamics of adaptation with evolutionary activity plots. Artif. Life **12**(2), 193–197 (2006)
6. Pohlheim, H.: Visualization of evolutionary algorithms - set of standard techniques and multidimensional visualization. In: Proceedings of the 1st Annual Conference on Genetic and Evolutionary Computation, vol. 1, pp. 533–540. Morgan Kaufmann Publishers Inc., San California (1999)
7. Lotif, M.: Visualizing the population of meta-heuristics during the optimization process using self-organizing maps. In: IEEE Congress on Evolutionary Computation, pp. 312–319. IEEE Xplore, Beijing (2014)
8. Mach, M., Zetakova, Z.: Visualising genetic algorithms: a way through the labyrinth of search space. In: Sincak, P., Vascak, J., Kvasnicak, V., Pospichal, J. (eds.) Intelligent Technologies-Theory and Applications, pp. 279–285. IOS Press, Amsterdam (2002)
9. The R Foundation: Comprehensive R Archive Network. The R Project for Statistical Computing (2015). http://www.r-project.org/index.html

10. Institut de Radioprotection et de Sûreté Nucléaire, WebGL in Shiny. http://trestletechgithub.io/shinyRGL/. Accessed 2013
11. Pérez, J., Alvarado, L., Almanza, N., Mexicano, A., Zavala, C.: A graphical visualization tool for analyzing the behavior of metaheuristic algorithms. In: Proceedings of ICITSEM, Dubai, UAE, pp. 120–124 (2014)
12. Suganthan, P.N., Hansen, N., Liang, J., Deb, K., Chen, Y.P., Auger, A., Tiwari, S.: Problem Definitions and Evaluation Criteria for the CEC 2005 Special Session on Real-Parameter Optimization. Nanyang Technological University, Singapore (2005)
13. Price, K.V., Storn, R.M., Lampinen, J.A.: Differential Evolution: A Practical Approach to Global Optimization. Springer, New York (2005)
14. Zhang, J., Sanderson, A.C.: JADE: adaptive differential evolution with optional external archive. IEEE Trans. Evol. Comput. 13(5), 945–958 (2009)

PCA Algorithms in the Visualization of Big Data from Polish Digital Libraries

Grzegorz Osinski[1]([⊠]), Veslava Osinska[2], and Piotr Malak[3]

[1] College of School and Media Culture, Torun, Poland
grzegorz.osinski@wsksim.edu.pl
[2] Nicolaus Copernicus University, Torun, Poland
[3] Wroclaw University, Wroclaw, Poland

Abstract. The visualization of large data sets from Polish digital libraries requires proper preparation of a comprehensive consolidated data set. Differences in the organizational systems of digital resources, and other factors affecting the heterogeneity of distributed data and metadata, require the use of clustering algorithms. To achieve this goal, the authors decided to use the PCA method and compare it with k-means results. PCA fulfills the condition of efficient size reduction for multidimensional data but is largely sensitive to deviations and differences in stochastic distributions. To eliminate the problem of noise in the input data, the deterministic model in the form of the Langevin function was used first. This leads to the "flattening" of the distribution of factors influencing the data structure. Due to such an approach, the most relevant categories to information systems were distinguished and Polish digital libraries were visualized.

Keywords: Digital libraries · Object type · PCA · k-means clustering

1 Introduction

The development of digital libraries is a response to the constantly growing amount of digital knowledge in particular communities. Polish digital libraries do not have a long history: the first ones were founded in 2001, while in 2008, all nineteen libraries hosted about 130,000 items in total. In 2016, Polish users could access even two million records stored in more than 120 libraries. Increase of the number of possessed resources was slow, but systematic. It is worth to note that the rapid rise of the number of both resources and digital centers emerged in 2010, when the national aggregator like Europeana – the Federation of Digital Libraries (FDL) - has been developed [1].

Digital libraries vary in terms of size and content, technical infrastructure, including user interface, as well as target audience. Such differentiation is caused by different organization forms of DLs. They may be created upon an agreement of a few institutions or by single entities. Despite the considerable interest of Polish researchers in the

This research is sponsored by National Science Center (NCN) under grant: 2013/11/B/HS2/03048/Digital knowledge structure and dynamics analysis by means of visualisation

© Springer Nature Switzerland AG 2019
K. Choroś et al. (Eds.): MISSI 2018, AISC 833, pp. 522–532, 2019.
https://doi.org/10.1007/978-3-319-98678-4_52

problematics of digital libraries, the questions referring to the high heterogeneity of the source data, as well as quality and functionality assessment, have not been sufficiently studied, and thus they are poorly described in the subject literature [1, 2].

This is not surprising, as digital libraries store various types of objects, e.g. old books, ancient chronicles, and other archival materials mixed with the newest novels, poetry or even scientific documents. The process of digitization is determined by many factors: the specificity of the collection, copyright issues, the equipment available in the digitization unit, funds provided for that purpose, the policy of managing institution, the state of the analogue resources, and, finally, scientific demand [3]. Digitization of resources is also often performed in an ad hoc manner, especially in the very beginning phase of DL formation. This all causes that data in digital libraries of millions of records have an inhomogeneous structure, grow quickly but not systematically. These features allow to consider them as big data, which is a novelty in the current research. In subject literature, DL data were applied rather for statistical description, not for structure studies [4–6]. The authors have attempted to systematize the distributed description of Polish digital libraries and their resources. Metadata, such as the type of document in the dataset containing over 1.6 million records, was used, and next, clustering algorithms, such as PCA and k-means, were applied.

2 Parallel Research Problem: Data

The Federation of Digital Libraries is a national aggregator of digital resources, gathering metadata from Polish cultural institutions, including digital libraries of various types and backgrounds. Actually, FDL provides open access to over four million objects and this collection is constantly growing. It is often noticed that digital libraries store documents of various types, not only those with academic context, e.g.: official letters, leaflets and brochures, lists of exhibitions etc. The FDL use the Dublin Core standard for object description and for searching tasks. There are fifteen main fields as advances option available in web interface. From our research perspective, the most valuable fields are: *dc:Title*, *dc:Type*, *dc:Author* and *dc:Date*.

Since FDL range is not limited to documents with scientific content, it proved difficult to deliver one of the project's goals – filtering humanity papers. Metadata, provided in the XML format, which are built with Dublin Core compatibility, however, do not provide credible basis of domains selection. Authors have assumed that *dc:type* values, describing publication type, will be useful in the filtering task, allowing to detect at least scientific documents. But, the great number of errors made by librarians during filling this field out made it more complicated. Instead of real documents types, one can also find there, for example: the title, the physical size of book, author's name or publish date. Another problem in the type description is a lack of any standardization of linguistic forms like singular or plural, different language use (Polish, German or English words interchangeably) and polysemy. Generally, the lack of controlled or prompt vocabulary used during filling metadata fields list has been qualified as a common difficulty in providing searchable descriptions of digital objects. This situation results in excessive fragmentation of resources typology on one hand and reduced precision of search on the other. Another troublesome data refer to the date field *dc:date* which reveals a lot of inaccuracies and multiple formats [7, 8].

A big diversity of types among the set of libraries has been observed. One of instance, WBC storing 222,521 records uses 1,594 unique description for types of documents. Figure 1A presents the quantity of types; one can observe the curve has a long tail, but it is not distributed according to a power law (Fig. 1B). We can talk about the type variation, which is generated rather by human (mistakes) than randomly. The majority of units do not exceed 100 types, but this is still too many for further considering and classification needs.

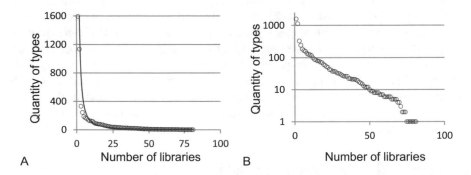

Fig. 1. Quantity of types distribution among the number of libraries (a); in logarithmic scale (b).

The most frequent type in 60%–90% of digital libraries is journals. In all datasets 70% of records refer to the journals. Next significant identified item in types set was newspaper(s) (Fig. 3). It can be explained that the digitization policy largely depends on the copyright law, which does not apply to the pre-WWII press. Such big variety of types is useless in most studies on digital libraries and needs significant reduction. Beside the problem of categorization, the question *what is the optimal number of types in current collection which is growing?* has appeared.

3   Methodology

The processing stage involved a matrix construction where columns present all documents' types and the rows are digital sources ($N = 120$). The number of columns is much more than 1594 (number of types) because the types do not coincide in different libraries. For convenience, only the types, which describe 99% of objects for each library, have been selected for analysis; the rest of data was rejected. The sorting from the largest number of types in particular library (24) to the smallest one (2) results in the matrix similar to triangular (Fig. 2). This action and also the use of both synonyms and unified lexical form with regard to the selected types allow to produce the final types-libraries matrix with dimensions of 20 × 75. This natural way we reduced the variety of types with a minimum loss of information, because no aggregation of data type was made.

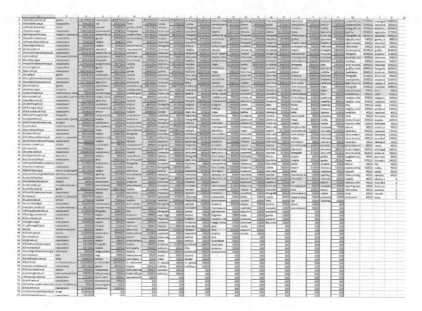

Fig. 2. Construction of triangular matrix: libraries in rows; library types in columns sorted according to frequency; selected types for analysis are highlighted.

For the types-libraries matrix, we have applied selected statistical methods such as PCA and k-means clustering. The second one was used to verify obtained key categories. The obtained results of types clustering were compared with an expert's categorization based on substantive criteria [7].

These criteria describe: (1) type of organization (for example regional or institutional) and (2) the range of leading subjects of collection (i.e. special, concerning narrow subject; church, technical, and universal with diverse collection). An assumption: institutional libraries have a wide thematic scope attends this division. This way the libraries were labeled by the first letter of this type: "r", "k" (church), "t", "u", "s". Additional information on localization was included to labels to analyze whether or not the topological separation of institutions matches the geographical one.

4   Results

The above mentioned, twenty types (columns) are: journals, newspapers, archival (archives), articles, books, thesis (typescripts), maps, graphics (images), photography, flyers (brochures), old prints, catalogues, lists, official documents, real objects, sculpture, music prints, multimedia, elaborations (statistical, didactic, professional). The quantitative distribution of objects among the most frequent types (>1%) is presented in Fig. 3. This graph relates to all data in corpora. It would seem rejection of the first item – journals, as the most common category, will produce more comprehensive background for further analysis. But factually, journals type does not occur as single category; all considered libraries have coupled more than two types and these combinations form the goal of the analysis.

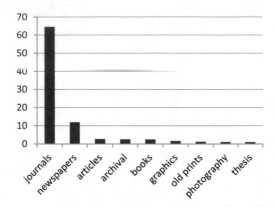

Fig. 3. Percentile distribution of types frequency in whole database.

Considering the research data in terms of a complex system, we assumed that the large variance of types was not caused by randomness, but rather diffusion processes between different categories of knowledge, and therefore we decided to apply the deterministic model [9].

We have applied particular function to flatten the very strong differences in the steep distribution of types frequency (Fig. 1a). It is difficult to find a function characterized by a relatively long segment of plateau, thus matching to existing condition has been performed heuristically. Finally we referred to phenomenological thermodynamics and for normalization Langevin function was used [10].

$$f(x) = a_0 + C\left((c_0)cothyp(x - x_c) - \frac{1}{x - x_c} \right) \qquad (1)$$

Where hyperbolic cotangent is defined in complex number space as:

$$cothyp(x) = \frac{e^x + e^{-x}}{e^x - e^{-x}} \qquad (2)$$

where is x – the number of types for particular digital library ($x \geq 2$), a_0 (offset), x_c (center) and C (amplitude) – are fittings parameters, empirically matched during numerical simulation. In current experiment the values were given: $a_0 = 20.34$, $x_c = 2.42$, $C = 1574$.

To reduce the dimension of the data vector determined by the number of DL types, the PCA was used [11]. The best PCA results of main components' variability were observed for a covariance matrix of initial data normalized by the Langevin function. The first two factors correspond with 78% of the variance. It means that representation based on these two factors is a good quality projection of initial 20 dimensional data.

The scree plot in Fig. 4 illustrates that eight factors can be considered as significant to represent dataset.

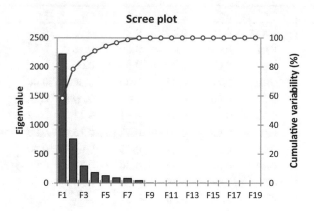

Fig. 4. Percentile distribution of PCA components as a new types frequency.

To reduce 20-dimensional structure of data we used primarily PCA for Pearson's correlation matrix with rotation of components [12]. Output first two factors contained about 67% of the whole variability ($\alpha < 5\%$). These components cover largely the first two types, which needs to introduce normalization as well as some weights towards the number of types. Table 1 shows how output PCA components are composed from initial variables.

Although journals is the largest category in whole database, it is does not represented in a new data space. Journals' combination with articles forms the first factor (F1).

Books are projected on a second factor – F2. Newspapers turn into a separate category (F6). The documents such as elaborations: statistical, educational or lists are connected within factor F3. All photography objects relate to separate component (F4), but graphics (pictures, images) together with the maps and physical objects constitute the F7. F5 factor consists mostly of old prints and audio books, which means that those both types are similarly popular in DL. There are also thesis works, which form the factor F8.

If we compare Table 1 with the chart in Fig. 3, we find the same number of mostly similar categories describing our dataset, i.e. eight. Essential differences concerned archival documents which ranked 4th in initial data distribution (Fig. 3) and were not represented in PCA. And inversely, audio prints, elaborations (lists), catalogues, maps of physical objects that influenced the output components space, did not exist among significant input data. Despite these differences, it could be noted that the largest categories played the key roles in types spectrum representation, but PCA created their combinations which were impossible to detect by human.

Observations space formed by the first two main components (Fig. 5) allows us to see hidden similarities of units (digital libraries) and verify the obtained pattern with a categorization framework performed by the expert. The data are coded by the type letter and the residence city of digital library.

Table 1. Contribution of initial variables to the first eight PCA components (fragment of greater table)

	F1	F2	F3	F4	F5	F6	F7	F8
journals	0,84	0,15	0,01	0,00	0,00	0,00	0,00	0,00
newspapers	0,01	0,00	0,03	0,00	0,26	0,67	0,02	0,00
archival	0,00	0,00	0,00	0,00	0,00	0,00	0,00	0,00
articles	0,77	0,21	0,02	0,00	0,00	0,00	0,00	0,00
maps	0,01	0,00	0,01	0,00	0,09	0,02	0,67	0,03
graphics	0,01	0,01	0,02	0,01	0,11	0,03	0,81	0,01
photography	0,02	0,00	0,10	0,71	0,15	0,00	0,00	0,00
catalogoues	0,01	0,20	0,44	0,25	0,09	0,00	0,01	0,00
objects	0,00	0,01	0,02	0,00	0,11	0,03	0,80	0,01
books	0,00	0,58	0,38	0,00	0,03	0,00	0,00	0,00
oldprints	0,00	0,16	0,00	0,03	0,31	0,25	0,18	0,02
thesis	0,05	0,00	0,00	0,00	0,05	0,01	0,00	0,89
elaborations	0,01	0,19	0,43	0,26	0,10	0,00	0,01	0,0
audio books	0,01	0,15	0,00	0,03	0,32	0,26	0,19	0,03

Data distribution resembles "Y" letter, but it is convenient for further communication to number quarters of observation space by digits I-IV. We can observe that in quarter II there is a cluster unifying the types, excluding universal one ("u" prefix). The same cluster groups among others all church digital libraries ("k"). Moreover, in this quarter there is the largest concentration of special type ("s"). Quarter I is a place of almost all universal and regional libraries. This means there are not so big differences between them like an expert has assumed. Journals dominance in collections unifies these libraries. These common features for both universal and regional digitization centers spread partly on the data located in quarters III, IV. No more practical interpretation for data distributed in III and IV quarters was possible. Some outsiders like Koszalin digital library ("t_Koszalin") provide unique, extremely profiled technical literature. Contrary to expectations, spatial mapping of data does not reflect any geographical distances or importance rankings but the results can be taken into consideration for improving professional categorization.

Another method – k-means clustering was used to verify outlined by PCA significant categories. Statistical results of the 20 variables and weighted data clustering are shown in Table 2. The algorithm was tested by the number of classes from 6 to 8, but the smallest within-class variance (14%) and the same time the largest – for between-classes variance (86%) was obtained for 8. Twenty initial categories that have grouped into eight classes largely resembled main PCA components contributions (Table 1).

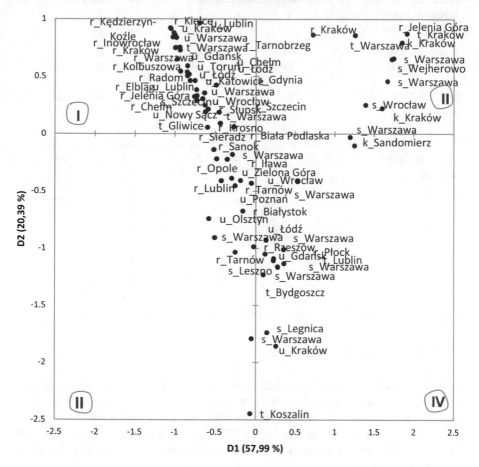

Fig. 5. Observations map formed by PCA components.

The primary difference is that articles' class is separated from others, meanwhile both journals and articles constitute one PCA component. The question: *does the dominant type – journals have to be distinct category or should it be integrated with another significant one for digital era – articles* can evoke wide discussion among professionals, who have to manage and arrange digital resources in libraries.

Table 2. Results of k-means clustering for dataset variables, i.e. types.

Generated class	1	2	3	4	5	6	7	8
Objects	1	1	9	1	2	2	1	3
Sum of weights	1	1	9	1	2	2	1	3
Within-class variance	0	0	2498	0	2274	1336	0	5666
Min. distance to centroid	0	0	9	0	7	7	0	13
Averege instance to centroid	0	0	16	0	7	7	0	24
Max. distance to centroid	0	0	68	0	7	7	0	34
	journals	news-papers	Archives, maps, photography, flyers, lists, official	articles	graphic objects	catalogues, elaborations	books	old prints, thesis, sound

5   Conclusion and Discussion

Metadata of digital libraries is a valuable source of data that can be analyzed using machine learning techniques [7]. But there is a problem of metadata quality, caused mostly by humans. However, using Dublin Core, digital libraries do not always follow uniform standards and normalization in data. There are different terms used for the same type of digital documents not only varying between different libraries, but also within meta data of one library [4, 5].

This multi variety of data types and field descriptions in digital libraries causes difficulties in the construction of a relevant typology of resources, and consequently, reduced precision in searching and exploring objects. Visual analysis of type set (Fig. 1) proved the human nature of its large diversity. Such a big variety of types is useless in the most studies on digital libraries and needs significant reduction [6]. Naturally, we simultaneously focused on seeking the optimal number of types in digital libraries, which can be helpful in their systematics. The results of statistical technique and the unsupervised learning algorithm, such as PCA and k-means clustering, show that eight categories are sufficient to describe the variety of current digital collection.

The question aroused in the current study concerns whether the used techniques are necessary for managing the digital collections in libraries. There are documented more than 1500 object types in Polish digital libraries; the number of the latter exceeds 100; there is an expert's categorization based on both organizational and specialization criteria. All these properties can provide the basic characteristics of digitized knowledge in Poland. But if we look, first of all, at the benefits to librarians, the presented approach becomes reasonable. PCA selected components can enable librarians to create a new classification as well as to manage particular types of objects.

If DL managers map the data set to a given optimal number of types, the redundancy of types and a huge amount of human mistakes while entering data can be minimized. Qualitative grouping made by experts only fractionally coincide with statistical clustering (Fig. 5). The results show the separate clustering for special types of DLs as well as universal and regional libraries feature the journals dominance in their collections.

Librarians' skepticism to quantitative results can be confronted with a new question: does the leading type – journals - have to be a distinct category or should it be integrated with another significant one for digital era – articles. This issue should evoke a wide discussion among professionals, who have to manage and arrange digital resources in libraries.

The authors still continue these studies as the combination of qualitative and quantitative approaches; they apply other valuable metadata such as a date, title and abstracts [8, 13]. The challenge to systematize digital knowledge by improving typology corresponds with semantic Web current needs. This can be a milestone in developing the ontology-based automatic searching systems in digital libraries.

References

1. Mazurek, C., Werla, M.: Federacja Bibliotek Cyfrowych – studium przypadku. In: Janiak, M., Krakowska, M., Próchnicka, M. (eds.) BIBLIOTEKI CYFROWE, pp. 225–239. SBP, Warszawa (2012). (in polish)
2. Werla, M.: Metadane dokumentów w bibliotekach cyfrowych (2010). http://lib.psnc.pl/Content/284/CPI-Werla.pdf. Accessed 18 Apr 2018. (in polish)
3. Calhoun, K.: Exploring Digital Libraries: Foundations, Practice, Prospects. Neal-Schuman, Chicago (2014)
4. Costello, L.: Title, description, and subject are the most important metadata fields for keyword discoverability. Evid. Based Libr. Inf. Pract. 11(3), 88–90 (2016)
5. Xie, I., Matusiak, K.: Discover Digital Libraries: Theory and Practice. Elsevier, Amsterdam (2016)
6. Zavalina, O.L.: Complementarity in subject metadata in large-scale digital libraries: a comparative analysis. Cataloging Classif. Q. 52(1), 77–89 (2014)
7. Osinska, V., Matusiak, K., Kowalska, M., Malak, P., Bednarek-Michalska, B.: Distribution of date elements and its relationship to the types of digital libraries. J. Librarianships Inf. Sci. 11 (2017)
8. Osinska, V., Malak, P.: Maps and mapping in scientometrics. In: Goralska, M., Wandel, A. (eds.) Tools and Methods for Analysing the Scientific Literature and Readers, pp. 59–73. WUW, Wrocław (2016)
9. Schramm, P., Oppenheim, I.: Properties of noise correlation functions of Langevin-like equations. Phys. A Stat. Mech. Appl. 137(1–2), 81–95 (1986)
10. Abrahams, E., Keffer, F.: Langevin Function. AccessScience. McGraw-Hill Education (2014)
11. Abdi, H., Williams, L.J.: Principal component analysis. Wiley Interdisc. Rev. Comput. Stat. 2(4), 433–459 (2010)

532   G. Osinski et al.

12. Egghe, L., Leydesdorff, L.: The relation between Pearson's correlation coefficient r and Salton's cosine measure. JASIST **60**(5), 1027–1036 (2009)
13. Osinska, V., Osinski, G., Komendzinski, T.: Altmetrics and visualisation – the complementary tools for analysing scientific collaboration and behaviour on researchgate. Cult. Educ. **4**(110), 105–121 (2015)

Recognition of the Pathology of the Human Movement with the Use of Mobile Technology and Machine Learning

Kazimierz Frączkowski[✉] and Sandra Łaska

Faculty of Computer Science and Management,
Wroclaw University of Science and Technology,
ul. Łukasiewicza 5, 50-370 Wrocław, Poland
Kazimierz.fraczkowski@pwr.edu.pl

Abstract. The work covers the research results related to the use of mobile devices for measuring human physical activity related to human gait whose quantitative measurement can be used to diagnose and monitor progress in therapy. Change in human gait may be a symptom of ongoing disease changes of various etiologies and occur during human life. Walking disorders can be caused by both nervous system function and bone, muscle or joint structures. Examples of diseases that can change normal gait for a pathological gait are Parkinson disease, multiple sclerosis, muscular dystrophy and other. One of the directions of studying this problem is the use of image analysis systems but continuous development of mobile technologies and machine learning techniques allows the use of human motion recognition, including human gait, using sensors in smartphones, such as accelerometer and gyroscope. The work demonstrated the possibility of using this kind of technologies to support health problems in health care.

Keywords: Human movement recognition · Gait analysis · Machine learning
Mobile technology · Pathological gait

1 Introduction

Gait disorders have a significant impact on the decrease in quality and comfort of life. Gait disorders are not only a natural consequence of the aging process of a human being but it can be a factor indicating the presence of many specific diseases. Among the most common causes of abnormal gait are neurological conditions, orthopedic problems and medical conditions [8]. Currently, a large group of patients with gait disorders are people affected by Parkinson disease (7–10 million people) and patients after a stroke (15 million people every year). In the case of these patients, continuous gait monitoring and analysis is very important for the evaluation the results of rehabilitation. Gait analysis is a systematic study of human locomotion involving the measurement, description and assessment of features characterizing the way of human movement. Development of mobile technologies has made recording and gait analysis much easier than using existing, standard methods based on camera-based motion recording. With the development of micro-electromechanical systems technology, the dimensions of these

© Springer Nature Switzerland AG 2019
K. Choroś et al. (Eds.): MISSI 2018, AISC 833, pp. 533–541, 2019.
https://doi.org/10.1007/978-3-319-98678-4_53

devices have been significantly reduced, which facilitated their application and applications. The development of advanced technologies has made the smartphone a great
tool that allows to control many vital functions. Thanks to built-in components (GPS,
magnetometer, accelerometer, gyroscope) - it became a kind of Wearable Ambulatory
Monitor. In the case of Parkinson disease thanks to motion sensors, it is possible to
observe parameters related to walking, body posture, leg movement, hand movement,
body tremor, fall, dyskinesia and bradykinesia. Gait monitoring and analysis also plays a
key role in the rehabilitation of patients after stroke, the most common consequence of
which is motor disability. An important role in the case of systems for diagnosing gait
pathology and dysfunctions are algorithms based on machine learning techniques. Data
acquisition from motion sensors and the use of machine learning techniques can recognize the first symptoms of pathological gait.

2 Related Work

Monitoring and human movement analysis is an area of interest for many researchers.
Information that can be obtained through monitoring and analyzing systems can be
used in many areas of life, such as rehabilitation, health care [3] sports or monitoring
the physical activity of older people - e.g. fall detection [2]. In most of the work related
to gait analysis, smartphone was used as a motion sensor. Kwapisz [6] used the
accelerometer built into the smartphone and an application based on the Android
system that monitors the daily physical activity of users. The main purpose of this
application was to recognize on the basis of data from the accelerometer the appropriate
types of human physical activity: walking, running, climbing stairs, going down stairs,
standing and sitting. In work Bao and Intille [1], 5 biaxial accelerometers, placed in
different parts of the body (wrist, forearm, hip, thigh, foot), have been used so as to be
able to recognize about twenty different daily human activities. The recognition and
prediction of activity was based on the Bayesian classifier. The results of this work
indicate that the human thigh is the most effective place for the sensor (phone or the
accelerometer itself) in order to distinguish different physical activities. In work Zheng
and Ordieres-Mere [11] the device registering the movement was a smartwatch placed
on the user's wrist. In this case, the human movement monitoring system is based on:
registering the movement using a smartwatch, sending the collected data to the
application on the mobile phone via Bluetooth, then sending data to the server and in
the final stage analyzing the collected data. In work [10] used machine learning
approach for Parkinson disease gait classification. As a machine classifiers used
Artificial Neural Network (ANN) and Support Vector Machine (SVM) in distinguishing gait pattern. As features in classifying Parkinson disease used spatiotemporal,
kinematic and kinetic gait parameters. In work [7] machine learning techniques was
tested on two pathological groups - Huntington's disease and post-stroke subjects. Use
of SVM classifier allowed to properly discriminate abnormal gait patterns with an
accuracy of 90,5%.

3   Characteristics of Normal and Pathological Gait

Human gait should generate the least possible energy losses and this situation is possible when the movement of the center of mass moves along a path similar to a straight line. While walking, it is possible to distinguish certain movements of body parts referred to as determinants or determinants of gait such as: pelvic rotation, tilting the pelvis, bending the knee, movement in the knee joint. The dysfunction or lack of any of the gait determinants increases the energy expenditure during movement. There are a number of methods to measure, describe and analyze gait. Research methods often include:

– measurement of spatio-temporal parameters: gait speed, step length, step frequency,
– measurement of kinematic quantities: trajectory of selected points of the moving body, measurement of angles between body segments (angles in human joints), velocity and acceleration of segments,
– measurement of dynamic quantities: forces and moments of forces acting on the joints during gait, reactions of forces arising in the joints.

The basic assumptions adopted in the analysis of human motion are based on the rules of rigid body physics. It is assumed that the segments of the body are treated as rigid elements, which means that the distance between any two points of the body is constant during movement.

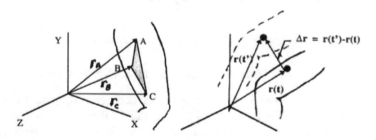

Fig. 1. Markers (A.B, C) defining the rigid body segment and displacement vectors [9]

Based on Fig. 1, the distance AB, BC, CA is constant during movement, according to the equations:

$$r_A = X_A(t)\hat{I} + Y_A(t)\hat{J} + Z_A(t)\acute{K} \tag{1}$$

$$r_B = X_B(t)\hat{I} + Y_B(t)\hat{J} + Z_B(t)\acute{K} \tag{2}$$

$$r_C = X_C(t)\hat{I} + Y_C(t)\hat{J} + Z_C(t)\acute{K} \tag{3}$$

Where:

r_A, r_B, r_C - displacement vectors

Î, Ĵ, Ḱ - unit vectors

X, Y, Z - coordinates of points (markers)

The linear velocity of the point can be described by as follows:

$$v = Lim\Delta t \rightarrow 0 \frac{\Lambda r}{\Delta t} = \frac{dr}{dt} \tag{4}$$

Relative speed of points (A and B):

$$v_{B/A} = v_B - v_A \tag{5}$$

Rotational speed of point (ω):

$$v_{B/A} = \omega \times r_{B/A} \tag{6}$$

Using the results of experimental studies, gait assessment can be performed using qualitative and quantitative analysis or numerical indicators. During gait quantitative analysis, pathological gait parameters (e.g. step length, walking speed) are compared with the standard gait parameters. Qualitative analysis is a comparison of registered values and a comparison with normal walking norms. Any deviations from the pattern allow the determination of a specific pathology of gait.

4 Technology Platform and Test Environment

4.1 Requirements for Sensors for Measuring Human Movement

Accelerometers used to monitor and evaluate human activity must meet the conditions for accurate recording of the frequency and amplitude of the acceleration of a human movement. The frequencies and amplitudes of the acceleration signal associated with typical human motion are relatively small. More dynamic human activities such as running or jumping, during which the acceleration of movement is greater, generate higher frequencies and amplitudes. The place on the human body in which the acceleration was measured also has a huge impact on the value of acceleration. The smallest value of acceleration is noted when placing the sensor on the head, the highest acceleration values concern the placement of the sensor on the lower parts of the body such as legs or feet. When moving at natural speed, frequencies generated in the upper body are 0.8–5 Hz, in the lower part (feet) are sometimes even about 60 Hz (while running). The average frequency value for daily activities is around 20 Hz, including 5 Hz for walk. The desired range of acceleration amplitude values are values from −12 g to 12 g and frequency up to 30 Hz.

4.2   Supervised Machine Learning in Human Gaits Recognition

Machine learning is an area of knowledge focused on the design, development and evaluation of systems capable of learning from data. Machine learning systems are based on predicting future activities based on past experience. In machine learning, statistical methods are used in the construction of mathematical models, because the main purpose of machine learning is inference based on a sample. In supervised machine learning there is a set of input data called a training set, usually shown as an x input vector pair x and its target y value. In gait identification, the input vector is the signal from the accelerometer, while the target value is the current label corresponding to each sample. In supervised machine learning, the input variable and the resulting output variable are known, by finding a relationship between input and output data, it is possible to predict output data for data not already belonging to the training set. In this case each example is a pair consisting of an input vector and a desired output value. The training data is analyzed by the machine learning algorithm and, based on the obtained functions, can be used to map new examples. The purpose of supervised machine learning is to build a model for evidence-based prediction in the presence of uncertainty. Patterns from data are identified by adaptive algorithms, which makes it possible to learn from observations. In order to solve a specific problem with the use of supervised machine learning, follow a specific pattern. The first step is to determine the type of training data. In the case of gait analysis, training data will be measured thanks to the sensor's acceleration and angular velocity of the moving body. The training set must be representative for the use of functions in the real world. The next stage is the representation of the input features of the learning function. The accuracy of the learning function closely depends on how the input data is represented. The input object is usually in the form of a vector, containing a series of functions that describe the object. The number of features should not be too large, but the amount of information should be sufficient enough to accurately predict the output.

4.3   Classification

Classification in machine learning enables finding a mapping of unknown data in a set of predefined classes thanks to the use of a model called a classifier. The classification process consists of three stages. The first stage is the construction of a classifier - model. The second stage is the process of testing the model. The last stage is the prediction of unknown data using the classifier. In the case of classification, building a model is the most important issue. A part of the collected data, can be used to build the model - training set. The training set is composed of a list of features and the class-decision values assigned to these features. The classifier is learned from the training set, assigning to each vector decision value a class that is the output value of the model. In the testing stage, the accuracy of the constructed model is verified. The classifier can be used to classify new data when its accuracy is acceptable - third stage. Classification in machine learning can be solved using many algorithms - techniques based on logic, neural networks and statistical methods. In work used algorithm based on logical methods: decision tree and k nearest neighbours classifier. In the decision tree, each internal node divides space into two or more subspaces according to a certain discrete function of the value of the input attribute (Splitting). The method of decision tree algorithm is based on the classification of objects by sorting them, based on the values of features. Objects are classified by going from the root to the leaves of the tree, according to test results along a specific path. Tree leaves represent all classes that can be assigned to the input observations. The tops represent the test on the attribute and the arcs the test result. The decision tree divides the training set to the moment when there are data belonging to one class in a given partition or the classifying data prevails.

Characteristics of the model (specification of the classification problem):

X - feature space

C - set of classes

$c:X \rightarrow C$ - perfect classifier for X

$D = \{(x_1, c(x_1)), \ldots, (x_n, c(x_n))\} \subseteq X \times C$ - set of examples

The main goal of the k nearest neighbours classifier is to create a multi-dimensional space of features in which each dimension corresponds to a different feature. The feature space filled in initially is the points of the training data set. An unmarked vector of x features and a set of marked classifier feature vectors seek to find k nearest neighbors of the input vector. The vector x is assigned to the most common class among the remaining neighboring classes. The performance of the classifier in the k nearest neighbours depends on: k (the number of considered neighbors) and the metric distance used to search for k nearest neighbors. The similarity between two objects can be determined using, for example, Euclidean metrics.

Specification of the classification problem:

x_i - sample from the training set

x - sample from the test set

ω - the true class for the training set

$\hat{\omega}$ - the predicted class for the test set

Ω - total number of all classes ($\omega, \hat{\omega} = 1, 2, \ldots, \Omega$)

As a rule, the 1-nearest neighbor, the predicted class of the test sample x is equal to the true class ω of its nearest neighbor, where m_i it is the nearest neighbor x, if the distance is: $d (m_i, x) = \min_j \{d (m_j, x)\}$

Characteristics of the model:

Input parameters: $P = \{<x_i, f (x_i)>\}$ where: P - set of objects, f - is the target function, x_q - object for classification

Output parameters: the value of $f (x_q)$ - the decision class.

5   Methods

5.1   Data Acquisition

Gait recording was carried out using a smartphone with built-in sensors: accelerometer type MEMS LIS302DL and gyroscope type MEMS L3G4200D. The obtained data is the acceleration measured along three axes x, y, z and angular velocity measured along the y axis. The sampling frequency of data was 60 Hz. Data acquisition was carried out for two cases of smartphone location on the thigh and ankle. The training data was collected during recording different types of gait: normal gait and four different abnormal gait.

5.2   Data Preprocessing - Filtration and Feature Extraction

For supervised machine learning and classification process applied raw data. Data filtration was performed as a preliminary process of data analysis related to spatio-temporal gait parameters. The main purpose of data filtration is to separate the component of the gravitational acceleration being a constant component of the accumulated acceleration data and to reduce the noise resulting from the dynamics of human motion. The classification process uses features based on the absolute value of the measured acceleration: mean, standard deviation, kurtosis, mean absolute difference, percentile 25, percentile 75.

5.3   Classification and Determination Spatio-Temporal Gait Parameters

In the classification process applied supervised machine learning techniques: decision tree classifier and k nearest neighbors classifier (4.3). To show accuracy of classification used confusion matrix.

Spatio-temporal parameters of human gait were determined on the basis of acceleration and angular velocity. The human gait parameters determined in the work: number of steps in 20 s time interval determined on the basis of a function locating the local maxima of the acceleration data, steps frequency, linear speed, distance, step length.

6   Results

The aim of the study was to compare the accuracy of applied machine learning algorithms in pathology gait recognition and to compare spatio-temporal gait parameters depending on the sensor location.

7   Conclusions

Using a smartphone as a mobile device that measures human movement is a good solution for experimental research. The results obtained in the work are satisfactory but it is worth noting that studies of human movement in the clinical conditions require the use of specialized mobile sensors for this purpose, as well as many other additional equipment that can be used to test many other aspects related to the human movement - not only spatio-temporal parameters such at this work. The accuracy of classification for pathological gait recognition using decision tree classifier and k nearest neighbors classifier was slightly greater for decision tree classifier (Tables 1 and 2). The determined spatio-temporal parameters of gait clearly show differences in the value between normal and pathological gait. The acceleration and angular velocity values are

Table 1. Confusion matrix for true class for decision tree and k nearest neighbours (TC- true class, PC - predicted class, TPR - true positive rate, FNR - false negative rate)

Normal TC	783 45.0%	251 14,4%	346 19,9%	197 11,3%	163 9,4%	Normal TC	399 22,9%	279 16,0%	600 34,5%	304 17,5%	158 9,1%	22,9% 77,1%
Abnor mal 1 TC	193 9,5%	1234 60,5%	404 19,8%	104 5,1%	105 5,1%	Abnor mal 1 TC	95 4,7%	1055 51,7%	658 32,3%	131 6,4%	101 5,0%	51,7% 48,3%
Abnor mal 2 TC	247 12,1%	333 16,3%	1313 64,4%	37 1,8%	110 5,4%	Abnor mal 2 TC	118 5,8%	227 11,1%	1566 76,8%	45 2,2%	84 4,1%	76,8% 23,2%
Abnor mal 3 TC	292 14,3%	187 9,2%	98 4,8%	1292 63,3%	171 8,4%	Abnor mal 3 TC	216 10,6%	208 10,2%	145 7,1%	1304 63,9%	167 8,2%	63,9% 36,1%
Abnor mal 4 TC	291 14,3%	332 16,3%	397 19,5%	213 10,4%	807 39,6%	Abnor mal 4 TC	214 10,5%	280 13,7%	587 28,8%	254 12,5%	705 34,6%	34,6% 65,4%
Per TC	Norma l PC	Abro mal 1 PC	Abnor mal 2 PC	Abnor mal 3 PC	Abnor mal 4 PC	Per TC	Norma l PC	Abnor mal 1 PC	Abnor mal 2 PC	Abnor mal 3 PC	Abnor mal 4 PC	TPR/F NR

Table 2. Confusion matrix for predicted class for decision tree and k nearest neighbours (TC-true class, PC - predicted class, PPV - positive predictive value, FDR - false discovery rate)

Normal TC	399 38,3%	279 13,6%	600 16,9%	304 14,9%	158 13,0%	Normal TC	783 43,4%	251 10,7%	346 13,5%	197 10,7%	163 12,0%
Abnor mal 1 TC	95 9,15%	1055 51,5%	658 18,5%	131 6,4%	101 8,3%	Abnor mal 1 TC	193 10,7%	1234 52,8%	404 15,8%	104 5,6%	105 7,7%
Abnor mal 2 TC	118 11,3%	227 11,1%	1566 44,0%	45 2,2%	84 6,9%	Abnor mal 2 TC	247 13,7%	333 14,2%	1313 51,3%	37 2,0%	110 8,1%
Abnor mal 3 TC	216 20,7%	208 10,2%	145 4,1%	1304 64,0%	167 13,7%	Abnor mal 3 TC	292 16,2%	187 8,0%	98 3,8%	1292 70,1%	171 12,6%
Abnor mal 4 TC	214 20,5%	280 13,7%	587 16,5%	254 12,5%	705 58,0%	Abnor mal 4 TC	291 16,1%	332 14,2%	397 15,5%	213 11,6%	807 59,5%
Per PC	Norma l TC	Abnor mal 1 TC	Abnor mal 2 TC	Abnor mal 3 TC	Abnor mal 4 TC	Per PC	Norma l TC	Abnor mal 1 TC	Abnor mal 2 TC	Abnor mal 3 TC	Abnor mal 4 TC
PPV/F DR	38,3% 61,7%	51,5% 48,5%	44,0% 56,0%	64,0% 36,0%	58,0% 42,0%	PPV/F DR	43,4% 56,6%	52,8% 47,2%	51,3% 48,7%	70,1% 29,9%	59,5% 40,5%

significantly higher when the sensor is placed on the ankle and this is consistent with the distribution of the acceleration values measured on the human body that grows from head to foot (Table 3).

Table 3. Spatio-temporal normal (N) and abnormal (A) gait parameters

Parameter	Sensor on thigh	Sensor on ankle
Average acceleration [m/s^2]	N:10,16 A:10.06	N: 11,5 A:10,43
Acceleration std. deviation [m/s^2]	N:1,36 A:1.4	N: 2,54 A:2.46
Min acceleration [m/s^2]	N:4,84 A:2,15	N:0,99 A:1,99
Max acceleration [m/s^2]	N:16,42 A:19,92	N:26,04 A:33,46
Average angular velocity [rad]	N:0,81 A:0.26	N:0,81 A:0,22
Angular velocity std. deviation [rad]	N:0,85 A:0,73	N:1,44 A:0,85
Min angular velocity [rad]	N:−1,39, A:−2.39	N:−,00 A:−2.48
Max angular velocity [rad]	N: 3,72 A:9,78	N:6,29 A:4.69

References

1. Bao, L., Intille, S.S.: Activity recognition from user-annotated acceleration data. In: International Conference on Pervasive Computing. Springer, Heidelberg (2004)
2. Bellazzi, R., et al.: Web-based telemedicine systems for home-care: technical issues and experiences. Comput. Methods Programs Biomed. **64**(3), 175–187 (2001)
3. Berndt, R.D., et al.: SaaS-platform for mobile health applications. In: 2012 9th International Multi-Conference on Systems, Signals and Devices (SSD). IEEE (2012)
4. bin Abdullah, M.F.A., et al.: Classification algorithms in human activity recognition using smartphones. Int. J. Comput. Inf. Eng. **6**, 77–84 (2012)
5. Kotsiantis, S.B., Zaharakis, I., Pintelas, P.: Supervised machine learning: a review of classification techniques. Emerg. Artif. Intell. Appl. Comput. Eng. **160**, 3–24 (2007)
6. Kwapisz, J.R., Weiss, G.M., Moore, S.A.: Activity recognition using cell phone accelerometers. ACM SIGKDD Explor. Newsl. **12**(2), 74–82 (2011)
7. Mannini, A., et al.: A machine learning framework for gait classification using inertial sensors: application to elderly, post-stroke and huntington's disease patients. Sensors **16**(1), 134 (2016)
8. Pirker, W., Katzenschlager, R.: Gait disorders in adults and the elderly. Wien. Klin. Wochenschr. **129**(3-4), 81–95 (2017)
9. Soutas-Little, R.W.: Motion analysis and biomechanics. J. Rehabil. Res. Dev. **12**(2), 49–68 (1998)
10. Tahir, N.M., Manap, H.H.: Parkinson disease gait classification based on machine learning approach. J. Appl. Sci. **12**(2), 180–185 (2012)
11. Zheng, X., Ordieres-Meré, J.: Development of a human movement monitoring system based on wearable devices. In: The International Conference on Electronics, Signal Processing and Communication Systems (ESPCO 2014) (2014)

A Multimedia Platform for Mexican Cultural Heritage Diffusion

Francisco de Asís López-Fuentes[✉] and Juan Alejandro Ibañez-Ramírez

Universidad Autónoma Metropolitana-Cuajimalpa, 01871 Mexico City, Mexico
flopez@correo.cua.uam.mx

Abstract. New Web technologies that allow greater interaction among users have emerged in the Internet. In such a way that now it is common for users to participate actively by writing, publishing, exchanging or sharing digital content over the Internet. This fact has allowed the emergence of web communities, web services, social networking services, wikis, blogs, etc. that are developed on a specific topic. This paper presents a multimedia web platform for the creation, edition and dissemination of historical cultural contents. Security strategies are considered in the design and implementation of the interactive platform. A digital platform with this purpose is very important due to the great cultural heritage of Mexico. We evaluate the operation of our platform with cultural information related with different historic stages of México such as prehispanic, colonial and independent. Our proposal seeks that the population has knowledge of cultural and historical contents in order to continuing preserve their cultural heritage.

Keywords: Distributed application · Multimedia · Internet · Web 2.0

1 Introduction

Culture brings together general knowledge such as beliefs and customs that shape the way in which an organized society interacts [1]. To preserve the knowledge that we have about a culture, it is necessary that this knowledge be transmitted so that society as a whole acquires, values, assimilates and continues to enrich them. Mexico has an extensive and rich culture, and this valuable cultural heritage needs to be preserved and disseminated among the population in order to strengthen its identity as a nation and generate awareness of the importance of continuing to preserve its cultural heritage. The Internet has grown exponentially in recent years and has allowed information to be available to almost all people, since ICTs (Information and Communication Technologies) and distributed systems [9] have allowed people to be connected and communicated at all times no matter where they are. This new scenario presents a great potential to spread the culture through new digital platforms such as social networks, wikis, blogs, etc.

Digital platforms emerge as ideal scenarios for social interaction, since these are entities that allow the transmission of knowledge, store or create new digital contents, or all these possibilities at the same time [1]. In this paper, we present a multimedia web system for the creation, edition and dissemination of historical cultural content of Mexico. This multimedia platform is displayed on a secure infrastructure authenticated

© Springer Nature Switzerland AG 2019
K. Choroś et al. (Eds.): MISSI 2018, AISC 833, pp. 542–551, 2019.
https://doi.org/10.1007/978-3-319-98678-4_54

based on the Kerberos protocol [2]. Our work mainly pursues two objectives. The first objective has been fixed to disseminate quality information about our tangible and intangible historical heritage among the population. Thus, most of society can access to these data, which are generated directly by researchers specialized in the subject. The second objective aims to create a dissemination web system that generates a direct bridge between the researchers and the population, which guarantees that the information disclosed comes from authorized researchers on the subject and promotes collaboration among researchers.

This paper presents an architecture for developing and implementing a digital platform that allows managing multimedia contents in order to disseminate the cultural heritage of Mexico. Currently, there are already different digital platforms related to culture issues such as the Portal of Culture of Latin America and the Caribbean [6], the Digital Culture Center [7] or the CIP (CDMX Heritage Information Center) [8] to name a few. However, the purpose of these portals is different from the proposal presented in this project, as well as the vision for the future, since it seeks to produce specific technological development for cultural heritage. Main benefits of this project with respect to other platforms are to encourage active collaboration between specialists from different areas, to generate a platform that securely contains content with reliable and quality information, through authentication controls of authors and mechanisms that guarantee the no redundancy of information.

The rest of this paper has the following organization. In Sect. 2, the design of the application is presented, where the parts of which the platform is composed are described. Section 3 presents the progress of the implementation of the proposed model. Our conclusions are given in Sect. 4.

2   Design

In this section, we introduces some points related to design of our interactive media platform. For the design is very important to consider the following objectives:

1. To disseminate quality information about our tangible and intangible historical heritage among the population, in a way that most of society can have access to these data, which are generated directly by researchers specialized in the subject.
2. Create a dissemination web system that generates a direct bridge between researchers and the population, which guarantees that the information disclosed comes from authorized researchers in the field and promotes collaboration among researchers.

The system must have the following basic functions:

1. Allow access to the system to those researchers who have previously registered, through a series of validations based on the kerberos security model. In such a way that researchers can:
   • Generate new content
   • Edit content already published on the platform (collaborate).
2. Allow free access to users in general to consult the topics or articles published on the platform.

A major challenge in the design of the platform is that a fine adjustment has to be made between the usability [5] part and the security, in such a way that greater security does not imply having a more rigid system from the user's perspective [3, 4]. Likewise, it should be avoided that a system that provides greater usability to the user makes the security of the system more vulnerable. In order to maintain effective access control to the platform, an authentication mechanism has been designed, which is based on the kerberos model, which consists of a general three-step validation. Taking into account that each message pass will have to be done sending the encrypted information; additionally, the validations will be done in a transparent way for the user, keeping the access and use of the system simple. The steps to perform to access the platform are:

1. Validation of the user using a password on a web server
2. Generation of an access ticket
3. Generation of a request to access the content server

For the design of the platform, it is considered to allow access to the system to those researchers who have previously registered, through a series of validations based on the Kerberos security model [10, 11]. Operation of authentication mechanism is illustrated in Fig. 1.

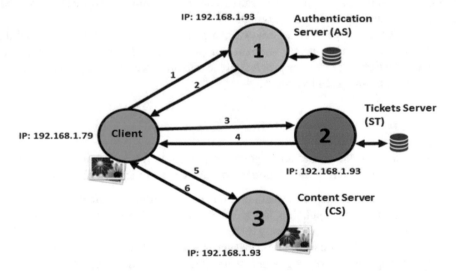

Fig. 1. Authentication scheme for access to interactive platform [12].

This authentication mechanism realizes the following steps:

- Step 1: The client application sends its user credentials to the authentication server (node 1).
- Step 2: The authentication server validates the credentials and responds with a message, which contains a ticket to access the ticket server (TGS).
- Step 3, the client application sends a message with the ticket obtained (TGS) in step 2 together with the desired content ID to the ticket server (node 2).

- Step 4: If the content is in the same domain (local content), a message with a ticket (TKC) is generated to access the content server.
- Step 5: The client application sends the obtained ticket (TKC) to the content server (SC) (node 3),
- Step 6: Content server validates the received data and proceeds to send the requested file to requested client.

Using this platform, the researchers can generate new content or edit a content already published on the platform. Free access to users in general should also be allowed to consult the topics or articles published on the platform. The content will be organized into categories according to the type of information or topic. Thus, we can have the following content organization:

- Museums
- Archeological zones
- Monuments.
- Artistic pieces
- Bibliographic collection.
- Historical dates.
- Traditions and customs

To offer a multimedia experience the description of the content will be accompanied with multimedia elements such as:

- Videos
- Images
- Audios.

In each article, researchers and users in general will have information about author of the article, category and historical stage, date of creation and date of last edition.

In order to avoid redundancy in the stored elements and guarantee non-thematic redundancies, three types of controls are planned:

1. Verification and validation of articles assisted by a group of expert administrators in specific categories of the platform.
2. Automatic semantic analysis of the textual content of each article.
3. Analysis of multimedia files using machine learning techniques.

Finally, once any incidence of repeated content is detected, the author will be offered options to modify or collaborate on an article that has already been created in order not to hinder the creation and enrichment of the content hosted on the platform. Figure 2 shows the UML diagram for the design of the digital platform.

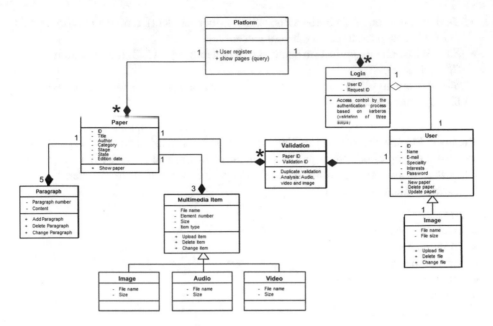

Fig. 2. UML diagram for the design of the platform.

3   Prototype

A basic implementation of the digital platform has been carried out and evaluated in our laboratory with some cultural test contents. Figure 3 presents the design of the main page of the platform. Prototype implementation considers the following steps:

1. Web support: HTTP protocol (Hypertext Transfer Protocol) is used because the application will run in a web browser, the interfaces are designed using the Axure RP V6 prototyping tool in its trial version. This tool allows integrating images, text and links to other interfaces to finally generate a website composed of HTML (Hypertext Markup Language) and CSS style sheets (Cascading Style Sheets).

2. Frontend: for the final implementation of views, we intend to use the JavaScript language to give dynamism or a certain level of interactivity and perform some functions such as the prior validation of user data (ID and password) through regular expressions as mentioned above.

3. Backend: for the dynamic creation of pages with content at the request of the client on the server side, the JSP pages (Java Service Pages) will be used. The Apache Tomcat server will respond to the user's requests. With respect to the implementation of user access control, there will be three servers that will perform each of the three validation steps described above in the access control section of the platform. Requests and sending of access control messages will be made by sending encrypted messages.

4. Data Persistence: to guarantee the persistence and validation of user data in the first step of access control within the authentication server, MySQL will be used. In the same way, to store the information of the articles generated in the platform by the users,

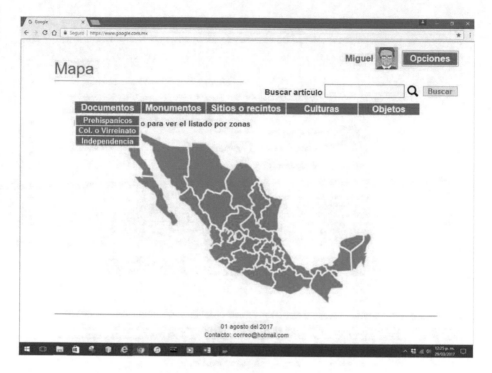

Fig. 3. Main page of the platform.

the same database administrator will be used only within the content server's execution environment.

Operation of our platform is shown in Figs. 4, 5, 6 and 7. For example, Fig. 4 shows a user profile called "Miguel", who is an archaeologist. Profile shows his interests too. Authentication mechanism allows that information upload by Miguel can be authenticated by other specialists. On the other hand, a user can investigate information about different cultures in different times and states of Mexico. Figure 5 shows a case for a monument during prehispanic period in the state of Mexico, specifically about the Teotihuacan cultural. Figures 6 shows a case for a colonial monument in the city of Guanajuato, while information about Mexican muralism is shown in Fig. 7. All these multimedia contents has been upload by the user of name Miguel. All contents are of the multimedia type because audio, images and video are combined.

Fig. 4. A user profile.

Fig. 5. A multimedia content for a prehispanic monument in the state of Mexico

Fig. 6. An example with a colonial monument in the city of Guanajuato

Fig. 7. A case about Mexican muralism during the independent stage in Mexico City

Preliminary results

The exploration and investigation of this type of systems constitutes a great development challenge; since, based on the experience acquired through the design phase and having carried out a detailed analysis considering different factors, such as usability and security, it generates as a result the certainty of knowing that the development of a secure system it does not have to mean sacrificing usability. On the other hand, we know that the handling of large volumes of information must involve robust controls against redundancies, so because of studying the needs for the operation of the platform, we consider it very important to develop automatic analysis functions for syntax and semantics, as well as the analysis of multimedia elements, doing research and developing automatic learning algorithms that are efficient in the consumption of computing resources and do not imply a large processing time.

A preliminary prototype has been shown to a group of expert who collaborate in the Thematic Network of Digital Technologies for the Dissemination of Cultural Heritage (Red Temática en Tecnologías Digitales para la Difusión del Patrimonio Cultural) [13]. The multiple comments from these experts in Mexican culture will allow us to improve the scope of the system. Finally, an important experience obtained from the social point of view is that the design and implementation of this platform can constitute an important advance when combining technological development with specifically cultural objectives with a vision of impact and social benefits in the future.

4   Conclusions

The great development of the Internet has allowed the emergence of online digital cultural platforms as an important scenario where diverse actors can interact to share information. These platforms are driven from different sectors, both public and private. Spreading the heritage and protecting the cultural wealth of society is vital to strengthen your identity. However, various regions of Mexico find it difficult to disclose their own cultural heritage due to lack of knowledge of these platforms or limited access to these technologies. This project aims to create a community digital platform to disseminate the historical cultural heritage where experts from different topics can contribute to enrich the contents and the general public can access them. Our work is in progress, and we currently working on the one hand with the authentication part to ensure the accuracy of the contents, but also on the part of usability in such a way that the user has a friendly interaction with the system. As future work, different aspects are considered from general aspects related to the structure of the platform, to individuals related to aspects of content such as their sources and usefulness. Also technical aspects related to data storage, data organization and search in the system are considered.

References

1. Túñez-López, M., Chillón-Álvarez, A.: Difusión de la cultura en Internet: mapa mundial de las plataformas online. Fonseca J. Commun., 123–149 (2010)
2. Neumann, B.C., Ts'o, T.: Kerberos: an authentication service for computer networks. IEEE Commun. Mag. **32**(9), 33–38 (1994)
3. Yee, K.-P.: Aligning security and usability. IEEE Secur. Priv. **2**(5), 48–55 (2004)
4. Garfinkel, S.: Design principles and patterns for computer systems that are simultaneously secure and usable. Ph.D. thesis, Massachusetts Institute of Technology (2005)
5. Alshamari, M.: A review of gaps between usability and security/privacy. Int. J. Commun. Netw. Syst. Sci. **9**(10), 413–429 (2016)
6. Portal de la cultura de America Latina y el Caribe. http://www.lacult.unesco.org/home/indice_new.php. Accessed 28 April 2018
7. Centro de Cultura Digital http://www.centroculturadigital.mx/. Accessed 28 April 2018
8. Centro de Información del Patrimonio CDMX (CIP). www.patrimonio.cdmx.gob.mx. Accessed 28 April 2018
9. López-Fuentes, F.A.: Sistemas Distribuidos, UAM, Unidad Cuajimalpa, pp. 1–203 (2015)
10. MIT Kerberos Consortium: The Role of Kerberos in Modern Information Systems, pp. 1–53 (2008)
11. MIT Kerberos Consortium: Why is Kerberos a credible security solution? pp. 1–13 (2008)
12. Ibañez-Ramírez, A.: Sistema de autenticación para acceso a datos distribuidos basado en Kerberos. UAM-Cuajimalpa, Technical Report (PT-III) (2018)
13. INAH-Red Temática en Tecnologías Digitales para la Difusión del Patrimonio Cultural. http://www.redtdpc.inah.gob.mx/. Accessed 4 June 2018

Authentication Mechanism to Access Multiple-Domain Multimedia Data

Juan Alejandro Ibañez-Ramírez and Francisco de Asís López-Fuentes[(✉)]

Universidad Autónoma Metropolitana-Cuajimalpa, 04871 Mexico City, Mexico
flopez@correo.cua.uam.mx

Abstract. Today most data is stored in remote sites that are usually distributed systems. The increase of virus, trojans or ramsomware has led to the implementation of security mechanisms to protect the access to data and guarantee their privacy and integrity. Important security issues to be considered in a computer system are the authentication and access control. Authentication provides proof of identity, while access control limits the actions or operations that a legitimate user of a computer system can perform. However, both authentication and access control are strongly related, because access control assumes that the authentication of the user has been previously successfully verified. Likewise, strong authentication methods requires to integrate cryptography techniques. This paper introduces and evaluates an authentication mechanism to access distributed data in a multi-domains environment. Authentication mechanism based on Kerberos uses encryption of messages and data. Interoperability between domains allows clients and servers that belong to different domains to offer data or services among them for previously authenticated users.

Keywords: Security · Authentication · Access control · Kerberos

1 Introduction

Multimedia content has had an explosive growth in recent years. Information and communication technologies (ICTs) have allowed people to be connected and communicated at all times no matter where they are. Most generated and exchanged data by the persons are multimedia data such photos, video or music. This ubiquity requires that the information systems can be distributed in different places and that the data can also be accessed from anywhere. However, distributed systems operating in an open communication environment are susceptible to different security attacks such as virus, trojans or ramsomware [6]. Security refers to logical and physical procedure measures aimed at the prevention, detection and correction of cases of misuse, as well as the characteristics that a computer system must have to resist attacks. The most common attacks on security in a computer system are attacks of interception, modification or fabrication of messages. One of the main objectives in information security is to implement access controls, which integrate policies and criteria that indicate under what circumstances access to the resources of a system should be granted. The basic access control mechanisms are [2, 3, 5]:

© Springer Nature Switzerland AG 2019
K. Choroś et al. (Eds.): MISSI 2018, AISC 833, pp. 552–561, 2019.
https://doi.org/10.1007/978-3-319-98678-4_55

- *Integrity* serves to prevent improper or unauthorized changes to the content of the information.
- *Confidentiality* is the mechanism that allows the concealment of resources to unauthorized entities. The use of cryptography helps in the task of preserving confidentiality.
- *Authentication* offers mechanisms that allow a correct identification of the origin of the message, ensuring that the entity is not false.
- *Authorization* is the mechanism that determines if an entity once authenticated is authorized to obtain access to the requested resource.

On the other hand, access control policies in a security system must be implemented to help establish who will have access to system resources, who owns the resources and what permissions users will have over them. Distributed systems operating within an open communications environment, are susceptible to interception, modification or message-making attacks. This paper aims to solve the problems related to access control to ensure the security of distributed data in different sites by integrating the Kerberos authentication control model [1] in a distributed environment. In order to reach this goal, the proposed model aims to cover the following features: operation within a network environment, resistance against password guessing, safeguarding the data against a false request, offering a unique user authentication service, and guaranteeing the encryption of messages and data.

The rest of this paper is organized as follows. A briefly description about related works are presented in Sect. 2. In Sect. 3, we introduce our architecture and present its main components to be implemented. Then, we explain the implementation of our prototype in Sect. 4. In Sect. 5, we evaluate the performance of our prototype in a local network and the results obtained from it. Our paper concludes in Sect. 6.

2   Related Work

Some of the issues to be resolved by the current distributed systems consist of the need to guarantee effective and secure access to information. For this type of scenario, the authors in [9] propose the use of the Kerberos model. After reviewing the previous proposal, it can be detected that a possible improvement to this system would be the use of a more current encryption algorithm instead of DES. On the other hand, Vandyke [13] raises the issue of user authentication for the transfer of files using the SFTP (Secure Shell File Transfer Protocol) protocol through a program developed by them. This program creates an encrypted communications tunnel using the client-server model to establish remote connections. Among the features of this program are that communication is generated through the SSH (Secure Shell) protocol, the system offers SSL (Secure Sockets Layer) communication encryption and not of files in the source and that is paid. Another relevant issue for a security system is the control of user authentication. To achieve this goal in [11] the authors propose using the Kerberos protocol to offer an authentication service that allows access to a resource hosted in the cloud. In addition, authors modify the original model by adding a Distributed Authentication protocol, which allows the system to perform a pre-authentication. In relation to the issue of file transfer using the Kerberos model for access control to a file download server, Al-Ayed and Liu [10] propose using the FTP (File Transfer Protocol) protocol and not using SSL to encrypt the connections, since they consider that the use of Kerberos is a more robust solution. Authors also suggest using a learning machine based on the Markov model to detect intrusions. In order to offer a light version of the Kerberos model to solve the problem of authentication and authorization of digital rights, Zhang et al. [12] formulate a redesign of the Kerberos model with the intention of reducing the load of each node of the system.

3 Architecture

The architecture for our proposed authentication mechanism is described in this section. A basic model is explained as well as the distributed model. The proposed authentication model takes as reference the Kerberos model, which is a credible and functional model for the following reasons [4]:

- It has been widely used, tested, studied and supported by a broad community of developers.
- Kerberos meets the requirements of modern distributed systems, since from the beginning it was designed to work within open communications environments.
- The architectural model is solid and functional, which has allowed the evolution of the model for easy integration with different systems.
- The Kerberos model is currently in operation and integrated into several systems, such as Apache Hadoop, Ad-hoc Networks, or Open Source OS and is an integral part of the current information technology infrastructure.

The authentication mechanism model consists of three nodes which are located in a single network domain or distributed in different domains. Figure 1 shows a scenario for a domain. In this case there are three servers: authentication (AS), tickets (TS) and content (CS). Each server exchanges messages with the client. The user will be able to transparently download the data through a client node which will establish a connection with the system, and the system will in turn be responsible for carrying out the validations and links to download the distributed content. All communications or pass messages between nodes will have to be processed by an encryption function, in order to avoid to send data as clear text.

Authentication mechanism operates as follows. In step 1, the client application sends its user credentials to the authentication server (node 1). Then, the authentication server validates the credentials and responds with a message (step 2), which contains a ticket to access the ticket server (TGS). In step 3, the client application sends a message with the ticket obtained (TGS) in step 2 together with the desired content ID to the ticket server (node 2). If the content is in the same domain (local content), a message with a ticket (TKC) is generated to access the content server (step 4). Then, the client application (step 5) sends the obtained ticket (TKC) to the content server (SC) (node 3), which validates the received data and proceeds to send the requested file (step 6) to requested client.

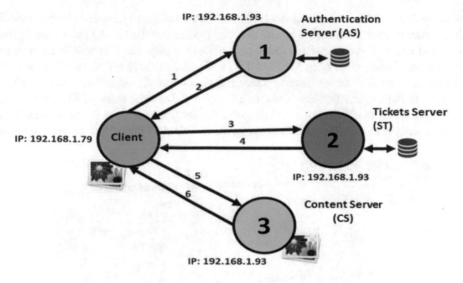

Fig. 1. Authentication scheme for a domain [15]

Figure 2 shows a scenario with multiple domains. In this case, authentication between domains is done through the authentication servers of each domain. Each domain has a single database, but with knowledge of other domains. Thus, a client can access content in other domains, but this access must do so through the authentication server of local domain. In a scenario with multiple domains, the authentication mechanism considers the following protocol. Step 1 and 2 are same as a domain. If in step 3, where the ticket obtained in the authentication server (TGS) and the content ID that you

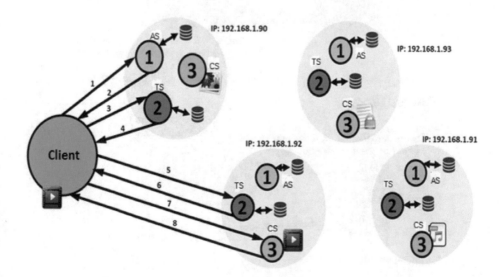

Fig. 2. Authentication scheme for a multiple domains [15]

want to download are sent, the ticket server validates that the requested content is stored
in another domain, responds with the data for redirecting the client application to the
external ticket server (step 4); in this way the client application proceeds to connect to
the external ticket server (step 5) so that this one I sent a ticket (TKC) to be able to access
the external content server (step 6). Once the client application has the ticket (TKC),
request to the external content server (step 7), it validates the ticket (TKC) and sends
the content that the client requested (step 8). All the above is done by the system in a
transparent manner, so that the client only enters their credentials the client application

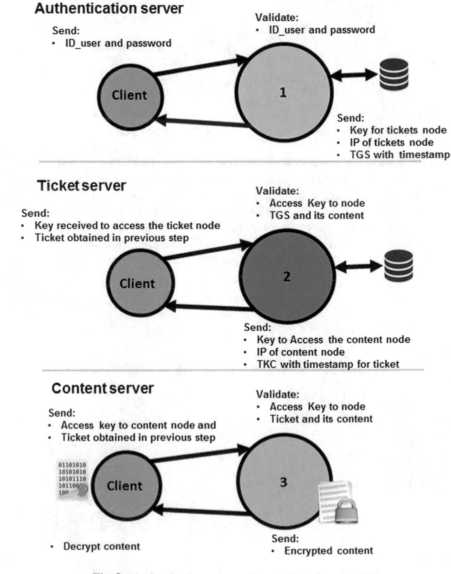

Fig. 3. Authentication scheme for a multiple domains [15]

and requests a content. Figure 3 shows communication and sending of messages proposed in each stage of communication. Likewise, the validations made by each of the nodes of the system are indicated in a general way. A problem in authentication to be considered is associated with the lifetime of a message in the grant ticket. In such a way that [7]:

- If it is too short, then it is requested repeatedly by the password.
- If it is too long, then there is a greater opportunity for a repeat attack.
- The threat is that an opponent steals the ticket and uses it before the time expires.

In the communication model of Fig. 3 this time-of-life problem is considered for each ticket that is issued. In such a way that it can be determined that whoever presents the ticket is the same customer for whom the ticket was issued. An important part of the operation of this project is the interoperability between domains. This feature allows that clients and servers located in different organizations or domains can offer services to each other for previously authenticated users.

4 Implementation

A prototype using proposed authentication scheme based on Kerberos has been implemented in our lab. The basic system has three nodes, which could be localized in a single domain or distributed in different domain. A user can download the content in a transparent way via a client node, which establishes connection with the system. Then, the authentication mechanism validates the user and allows to download the content from the content server. Each domain has a single database, but with information about other domains. The communications between nodes is implemented with sockets for Linux OS using the TCP/IP protocol. TCP/IP protocol is used in order to offer retransmission against the packet loss. All transmission of messages between nodes are encrypted. In this prototype, an encrypted function is responsible of this task. Thus, a content is encrypted by the content source and it will be decrypted in the client node by the end user.

Our prototype is implemented using threads in order to attend different user in a parallel way. Each node in the system is a separate application which can be under a single network domain (IP) or distributed under a different domain; inclusive, each of the nodes could be contained within several containers (e.g.: Docker or Linux Container). Our implementation assumes that all the clocks within each environment where each of the nodes is running are synchronized. Therefore, a clock synchronization service has not been implemented. Additionally each node was programmed to be able to serve different clients in a parallel way making use of threads that attend all the requests of each client. For the implementation a communication protocol was designed for the interaction with each of the servers, this protocol is based on the kerberos model [1]. In this protocol, each message is composed of several data to be validated by each of the servers. An important feature of the system is that messages sent between servers are generated with a length of 128 bytes, of which, depending on the case, 64 bytes correspond to the tickets generated by the authentication server (TGS) and by the server.

Ticket holder (TKC). It is important to clarify that the TGS is encrypted by the authentication server and later added to the message to compose a chain of 128 bytes, thus being protected by two rounds of encryption, the first to generate the TGS of 64 bytes and the second to encrypt the message of 128 bytes to be sent. The AES (Advanced Encryption Standard) algorithm used to encrypt the multimedia data in the system. We use the AES version for 256 bit-keys for block and key size in the CBC block encryption mode (Cipher Block Chaining Mode). This version is used instead of the AES version for 128 bit-key because a larger size gives the system greater strength against possible attacks and at the same time obtain a fast encryption and low memory consumption. The handling of keys is done in programming time in each node being defined in a static way, similarly, the initialization vector for the block encryption is established statically in programming time.

5   Evaluation and Analysis

Our authentication mechanism is evaluated against several scenarios. Password strength is evaluated against two basic types of cracking methods which are the brute force attack and the dictionary attack. The encryption key length is 256 bits, so a brute force attack is unfeasible, because to decode the password it is necessary to make 2^{256} combinations. In other words, 1.1×10^{77} combinations are required. Therefore, a supercomputer capable of performing operations at the level of 1 Peta would need [14], in the best case, billions of years to decipher any message, TGS or TKC. To avoid a brute force attack by using passwords dictionaries, the system contemplates several policies such as use of passwords and ID's of users with a length of 10 characters, mandatory use of combinations of uppercase, lowercase, symbols and numbers for the credentials of registered users in the system, or constant change of passwords.

We use Wireshark [8] to audit the sent packets and corroborate the sending of encrypted messages within the system. Wireshark is a free software used to analyze the packets sent in a network with multiple protocols. Figure 4 shows the capture of two messages sent by a previous version of the system, in which the encryption function was not integrated. This was done to highlight the importance of encrypting network traffic, since it can be easily captured.

On the contrary, Fig. 5 shows the capture of two encrypted packets, these messages are sent by the system with the integrated encryption function to verify that the content sent in the TCP packet is protected by the encryption function. In this way, the privacy and integrity of the messages and files sent is guaranteed.

An important part for the functionality and protection of the data stored in the system is to guarantee the functionality and validation of the Tickets issued by the authentication and ticket servers. In order to test the system protocol and ticket validations with respect to the data sent to each server, the following tests were generated for the following cases:

- Attempt to use a false TGS
- Cloning of the client's IP and attempt to use a TGS
- Attempt to use a TKC to obtain content

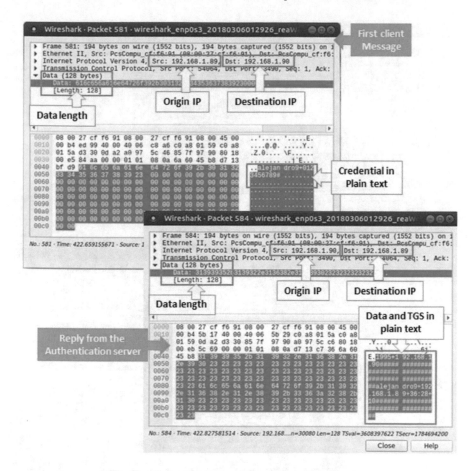

Fig. 4. Message capture without encryption function

To protect the system during its operation against interceptions or modifications of the generated tickets, a life time was defined for the use of these tickets. This time of life is embedded within each ticket and is protected by encryption.

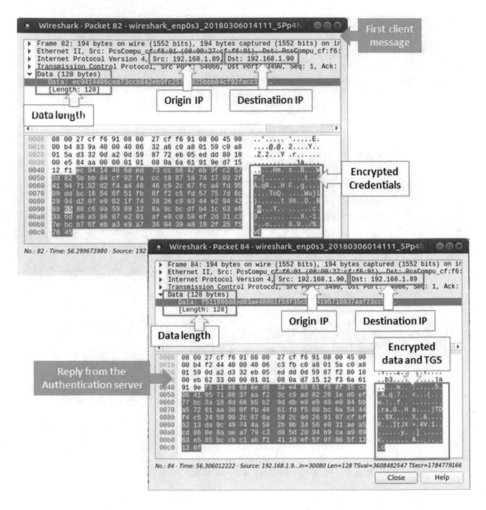

Fig. 5. A capture of encrypted messages.

6   Conclusions

The rapid development and increasing complexity of computer applications that are currently deployed over communication networks have generated an increasing demand in terms of security and privacy on the part of users. This has generated a great technological challenge and the need to build secure systems. In this paper an authentication system for distributed data based on Kerberos model is presented. A distributed authentication system based on a widely tested model is important to transparently access high-value data dispersed and stored at remote sites. This gives to user the impression of location and uniqueness of the data.

The contribution of our work lies in the design and implementation of an authentication system, based on an operating scheme based on interoperability between different domains. The mechanism evaluated offers a control of secure access to data, integrates a proven symmetric encryption algorithm for messages and files, and storage in databases for its operation. The proposed mechanism guarantees the privacy and integrity of data and messages within an open communications environment, it is undoubtedly an effort to contribute to meet the current cybersecurity demands of the information society.

Our solution proposed in this work can be addressed in different directions. As future work, we can integrate clock synchronization services and encryption key management for messages and files. Also, we can explore the impact of Kerberos in the Internet of things in order to study how security and privacy can be guaranteed in open network environments.

References

1. Neumann, B.C., Ts'o, T.: Kerberos: an authentication service for computer networks. IEEE Commun. Mag. **32**(9), 33–38 (1994)
2. López-Fuentes, F.A.: Sistemas Distribuidos, UAM, Unidad Cuajimalpa, pp. 1–203 (2015)
3. MIT Kerberos Consortium: The Role of Kerberos in Modern Information Systems, pp. 1–53 (2008)
4. MIT Kerberos Consortium, Why is Kerberos a credible security solution? pp. 1–13 (2008)
5. Bishop, M.: Introduction to Computer Security, Pearson Education, Inc. pp. 2–3, 97 (2005)
6. Symantec: Informe sobre las amenazas para la seguridad en Internet de 2017. Junio-2017 (2017). https://www.symantec.com/es/mx/security-center/threat-report
7. Stallings, W.: Fundamentos de Seguridad en Redes Aplicaciones y Estándares, Pearson Educación, pp. 28, 31–32, 35, 106, 109 (2004)
8. Wireshark. https://www.wireshark.org/download.html. Accessed 6 May 2018
9. Hojabri, M., Venkat, R.K.: Innovation in cloud computing: implementation of Kerberos version 5 in cloud computing in order to enhance the security issues. In: IEEE International Conference on Information Communication and Embedded Systems (ICICES), India (2013)
10. Al-Ayed, F., Liu, H.: Synopsis of security: using Kerberos method to secure file transfer sessions. In: IEEE International Conference on Computational Science and Computational Intelligence, USA (2016)
11. Liu, Y., Li, Z., Sun, Y.: Distributed authentication in the cloud computing environment. In: LNCS, vol. 9532. Springer International Publishing Switzerland (2015)
12. Zhang, N., Wu, X., Yang, C., Shen, Y., Cheng, Y.: A lightweight authentication and authorization solution based on Kerberos. In: IEEE Advanced Information Management, Communicates, Electronic and Automation Control Conference (2016)
13. VanDyke Software Inc. Secure File Transfer with SSH (2008)
14. Arora, M.: How secure is AES against brute force attacks. EE Times **5**(7). https://www.eetimes.com/document.asp?doc_id=1279619. Accessed 4 June 2018
15. Ibañez-Ramírez, A.: Sistema de autenticación para acceso a datos distribuidos basado en Kerberos. UAM-Cuajimalpa, Technical Report (PT-III) (2018)

Video Game Development Methods
and Technologies

Accident Prevention System During Immersion in Virtual Reality

Dawid Połap$^{(\boxtimes)}$, Karolina Kęsik, and Marcin Woźniak

Institute of Mathematics, Silesian University of Technology, Kaszubska 23,
44-100 Gliwice, Poland
{Dawid.Polap,Marcin.Wozniak}@polsl.pl, Karola.Ksk@gmail.com

Abstract. More and more development under the sign of reality modification brought glasses that do not require connection to the phone, and all calculations can be performed on a connected computer. An additional advantage is the increase in the user's perception through the introduction of motion sensors on the market. However, it is worth noting that immersion into the virtual world requires an initial selection of the area where the user can move. In fact, various obstacles can appear. To remedy this, we suggest to improve the immersion system by using camera sensors and a classifier that can determine if anything has changed in the current area.

Our proposal has been described in detail, then created, implemented and tested on several examples. The obtained results were discussed on the wider use in the increasingly frequent use of mixed reality.

Keywords: Virtual reality · Image processing · k-nearest neighbor
Warning system

1 Introduction

The rapid development of technology is mainly caused by two factors. The first one is the continuous competition of companies on the market, which can be illustrated by an example of leading graphic card producers such as Nvidia or AMD. Both companies are in a fierce battle for over a dozen years – when one of them presents its latest product to the market, the second one is forced to offer something at least slightly better in a few months. This type of struggle to stay on the graphics card market is additionally determined by demand. Demand is created by players who increasingly require not only efficient hardware but also games. It is games that are the second leading direction causing more and more technological development. The increasing gameplay, more realistic graphics and feelings are the demands that gamers make.

Nowadays, game production does not only concern the console or computer market but also the mobile. Especially the mobile application is fairly quickly being expanded due to the introduction of augmented or virtual reality technology. Virtual immersion gets rid of mobile equipment in favor of embedded

© Springer Nature Switzerland AG 2019
K. Choroś et al. (Eds.): MISSI 2018, AISC 833, pp. 565–573, 2019.
https://doi.org/10.1007/978-3-319-98678-4_56

technology in goggles. The development of specific technologies has many applications in the aforementioned games, education or even medicine. Another important brand is augmented reality, which gained its initial popularity by introducing the game *Pokemon Go*. Through the use of navigation, gyroscope and camera in smartphones, the possibility of catching classic creatures was introduced, and this caused that young people left the houses for fresh air [11]. This type of solution also has its drawbacks, because kids are focused on phones and do not always notice threats around.

More and more technological possibilities have meant that people can play games anywhere and anytime. These types of possibilities increase the ability to coordinate between sight and hand, concentration or even teach. It is not difficult to notice that such activities have, however, many disadvantages. An example is the augmented reality, where players do not notice the world around them, which is notable for numerous accidents. These types of problems are not solved only by controlling the players, but by using surveillance applications connected to various sensors.

1.1 Related Work

The game titled *Pokemon GO* has a large impact on modern technologies, which, despite the large number of advantages, also has some drawbacks, because kids are focused on phones and do not always notice threats around. In [13], the authors proposed a system of informing about the imminent threat in real time. The technology has found a lot of use in education, where it does not need complex mechanisms to show some action, but only to put on glasses and analyze the displayed effects [1]. Of course, this technology still needs to be improved because it has many flaws and imperfections. The main subject of continuous development are scientific works such as [2], where the authors presented the idea of near-eye varifocal augmented reality or [8] where the framework based on distributed visual slam was shown.

Similarly, the virtual reality has found its application in medicine for patients after the stroke [5], or as an aid in disease simulations [16]. The effects of research on the impact and burden of this technology on modern phones are presented in [10]. Again in [17], the authors shows the creation of a virtual keyboard, which can be influenced by users' fingers. The solutions not dedicated to the chosen application, which can be assimilated, also have a significant impact. An example is the automatic detection of the playing field and the extraction of dominant colors in sports games [4], or a novel approach to the group of agents [15]. All the indicated research works at their bases use various achievements that were not dedicated to the game. Particular attention should be paid to optimization algorithms [9] or fuzzy logic [12]. These works find their application in image analysis, processing and description of data from various sensors [6,14].

In this paper, we present our research on facilitating immersing in virtual reality by preventing unprecedented events such as the appearance of an obstacle during the game.

2   Real-Time Warning System

Current software for virtual reality means wearing glasses that cover the entire viewpoint of the user. Full immersion means transferring consciousness into a created world. Immersion can take place in three different ways – sitting, standing and in a certain area where the user can move. The problem occurs in the last case – when the user has a certain area on which he can move. The unconscious user can take a step forward or backwards where an obstacle may occur during the immersion and have tragic consequences.

The assumption of glasses in the case of Microsoft's software requires marking the area of moving with some restrictions. More precisely, the requirement is a minimum size of the area, which is given as a rectangle with dimensions of 1.5×2 meters. In this area, the child can throw the toy, and the unknowing user can stumble over it. To prevent this, we propose a system to control this area and inform the user about any changes.

2.1   Architecture of the System

In order to propose the architecture of such a system, it is worth paying attention to some result of too urgent alerting about the problem. A quick alarm about the obstacle can stop the immersion, and it would not have to affect the game itself. This causes that the system should analyze the user's movements and decide if it will affect on his movement. Breaking the game too quickly may result in discouraging to the technology.

The Video from Camera. The proposed solution uses a camera that is set on the ceiling of the room where the player is immersed in a virtual world. The obtained image is composed of a series of frames (average 24 frames per second), to simplify the number of calculations, we suggest taking one frame every second. To locate new objects, we need to analyze and compare whole frames. Unfortunately, a simple comparison pixel by pixel is too burdensome for the computer, so we suggest searching for key points of the image and comparing them. The advantage of this solution is the stable position of the camera, which records one area without making any movements. In this purpose, SURF (*Speeded Up Robust Features*) algorithm is used [3]. It is based on Hessian which indicates local changes on the analyzed fragment of the image. Hessian matrix is defined as

$$H(x,\omega) = \begin{bmatrix} L_{xx}(x,\omega) & L_{xy}(x,\omega) \\ L_{xy}(x,\omega) & L_{yy}(x,\omega) \end{bmatrix}, \tag{1}$$

where the individual cells of the matrix are the convolution of the image I with the second derivative of the Gaussian ($g(\omega)$ is the Gaussian kernel), which can be defined as

$$L_{xx}(x,\omega) = I(x) \frac{\partial^2}{\partial x^2} g(\omega), \tag{2}$$

$$L_{yy}(x,\omega) = I(x)\frac{\partial^2}{\partial y^2}g(\omega), \tag{3}$$

$$L_{xy}(x,\omega) = I(x)\frac{\partial^2}{\partial xy}g(\omega), \tag{4}$$

where $I(x)$ is an integral image, where x is understood as point in the center of the analyzed grid what is calculated as

$$I(x) = \sum_{i=0}^{i\leq x}\sum_{j=0}^{j\leq y}I(x,y). \tag{5}$$

The idea of these algorithm is to analyze non-maximal-suppression of determinant the matrix presented in Eq. (1). All the extremes are understood as a key–points. Mentioned matrix determinant can be depicted as

$$det(H_{approximate}) = L_{xx}(x,\omega)L_{xy}(x,\omega) - (wL_{xy}(x,\omega))^2, \tag{6}$$

where w is a certain weight.

2.2 Motion Analysis on Video

Take two adjacent video frames (according to the previous assumptions, they are separated by one second). Each picture can be identified with a set of found points. In the absence of any movement, all points will overlap (within a few pixels). Of course, it may happen that the light will fall differently, therefore if there is a point and there is no other in its vicinity with a radius of 10 pixels, it should be removed. If the points coincide, they will be removed from both sets. If there is only a player in the area, all unique points will belong to him. This situation is certain at the beginning of the action. At this point, we start detecting and classifying objects having the first two frames, all the repeating points are deleted. The rest belong to the user, so four points must be found that will be the most distant points in relation to the center, i.e. a set of four points is found in the following way

$$\{(\min(x_1),y_1),(x_2,\min(y_2)),(\max(x_3),y_3),(x_4,\max(y_4))\}. \tag{7}$$

A defined set allows for defining an object on the image that can be identified with the user. In the next frame, there is no re-search for the position of the player with the given location of these four points above.

Of course, having another frame and a list of found points, we remove those that repeat. Then it is necessary to analyze all points using the k-nearest neighbors algorithm [7] to find the player's position. Algorithm is based on the probability estimation of the x observation belonging to the k class, what is defined as

$$\hat{p}(k|x) = \frac{1}{K}\sum_{i=1}^{n}I(\rho(x,x_i)\leq\rho(x,x^{(k)}))I(y_i = k), \quad k = 1,\ldots,L, \tag{8}$$

where $\rho(\cdot)$ is a given metric (in our case Euclidean), and $x^{(k)}$ is K-th as to the distance to the point from the learning sample x. These classifier could be defined as

$$\hat{d}_{KNN}(x) = \arg\max_k \hat{p}(k|x). \tag{9}$$

If the point is assigned by the algorithm to the object (here, we have only one object – player defined by the Eq. (7) and consider its movement by a maximum of 10% in each direction within one second), we remove it from the list. The remaining points will be considered as an emerging obstacle. In the case, when the object is close to player (as closeness we mean 20% of the image length), the player was informed about the problem.

The proposed solution is presented in Fig. 1 and it can be combined with goggle equipment and simply interrupt their operation or, in case of an obstacle, alert the user with a sound signal – this proposal is simpler to create because there is no need to influence the operation of virtual reality app.

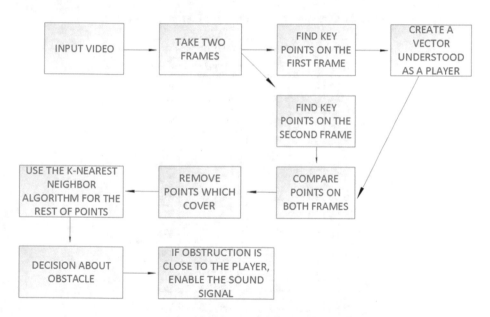

Fig. 1. Visualization of the system's operation.

3   Experiments

We tested our solution on the basis of adjusting neighborhood coefficients and achieved accuracy for different cases. We considered two – when the game does not require a lot of moves and a game that requires movement in a certain area. In our research Dell Visor Virtual Reality Headset with controllers were used. Additionally, the smartphone was suspended on the ceiling of the room. The phone recorded the video and sent selected frames to the computer. The computer recalculated and in the event of a problem, a loud signal was turned on.

Fig. 2. An example of two frames away by one second with a gently moved camera stand by a child.

Table 1. Obtained results due to different parameters.

Without obstacle				With obstacle			
k	r	d [%]	Alarm	k	r	d [%]	Alarm
2	2	5	Yes	2	2	5	No
4	2	5	Yes	4	2	5	No
6	2	5	Yes	6	2	5	Yes
8	2	5	No	8	2	5	Yes
2	4	10	Yes	2	4	10	Yes
4	4	10	Yes	4	4	10	No
6	4	10	No	6	4	10	Yes
8	4	10	No	8	4	10	Yes
2	6	15	No	2	6	15	Yes
4	6	15	No	4	6	15	Yes
6	6	15	No	6	6	15	Yes
8	6	15	No	8	6	15	Yes
2	8	20	No	2	8	20	Yes
4	8	20	No	4	8	20	Yes
6	8	20	No	6	8	20	Yes
8	8	20	No	8	8	20	Yes

The solution was tested depending on three parameters – the size of neighborhood k, the number of radius r from key-point that were deleted if they do not have any neighbors and the distance d between the obstacle and the player (in previously section, we mentioned at most 20% of the length of the frame which is calculated as $\sqrt{wh}$ where w is the width and h the height) (Table 1).

We tested our proposition due to different parameters like $k \in \{2, 4, 6, 8\}$, $r \in \{2, 4, 6, 8\}$, and $d \in \{5\%w, 10\%w, 15\%w, 20\%w\}$ where w is the width of the frame. In case of alarming when the obstacle is near the user – between $15 - 20\%$ of the length of the frame – the obtained results are correct for more than half of the tests. For each results (for given parameters), 5 experiments were made and depending on where the majority went, this is the answer in the column *Alarm*. Based on the obtained results, it can not be unambiguously stated that this is the ideal solution. The problem may be the classifier that makes decisions based on the Euclidean metric. Better results could be obtained using a more complex design. However, it is worth noting that even with this selection, good efficacy and correctness for such a solution can be obtained what can be seen using confusion matrix presented in Fig. 3.

In Fig. 2, two frames with found key-points are presented. The video frames are made by the camera placed on the stand quite low compared to the user's height. It is worth noting the arrangement of key points – if they are removed under the assumptions and method given in the previous section, they result in their absence, and thus the absence of any possible obstacle.

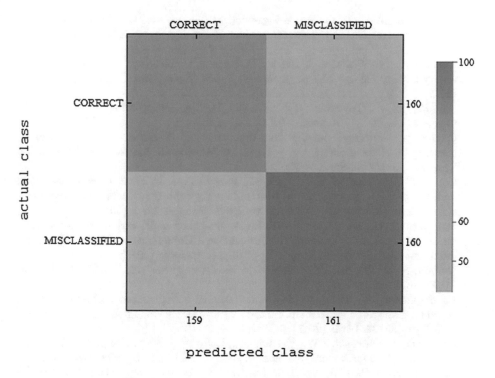

Fig. 3. Confusion matrix for obtained results.

4   Conclusions

In this paper, we proposed a system informing the user immersed in the virtual reality about the emerging obstacle. We have tested this type of solution on several examples and the analysis of the results indicates that the expectations of the work are fulfilled. This type of solution is needed because different situations can take place. Especially if children play around. However, it is worth noting that the system is not perfect. The especially chosen classifier is quite naive in this case. In the future, we will test a similar action by using other classifiers and try to increase the effectiveness of the proposed system.

Acknowledgement. Authors acknowledge contribution to this project of the "Diamond Grant 2016" No. 0080/DIA/2016/45 from the Polish Ministry of Science and Higher Education and the Rector pro-quality grant No. 09/010/RGJ18/0033 at the Silesian University of Technology, Poland.

References

1. Akçayır, M., Akçayır, G.: Advantages and challenges associated with augmented reality for education: a systematic review of the literature. Educ. Res. Rev. **20**, 1–11 (2017)
2. Akşit, K., Lopes, W., Kim, J., Shirley, P., Luebke, D.: Near-eye varifocal augmented reality display using see-through screens. ACM Trans. Graph. (TOG) **36**(6), 189 (2017)
3. Bay, H., Ess, A., Tuytelaars, T., Van Gool, L.: Speeded-up robust features (surf). Comput. Vis. Image Underst. **110**(3), 346–359 (2008)
4. Choroś, K.: Automatic playing field detection and dominant color extraction in sports video shots of different view types. In: Multimedia and Network Information Systems, pp. 39–48. Springer (2017)
5. Cogné, M., Violleau, M.H., Klinger, E., Joseph, P.A.: Influence of non-contextual auditory stimuli on navigation in a virtual reality context involving executive functions among patients after stroke. Ann. Phys. Rehabil. Med. (2018)
6. Dobrowolski, G., Byrski, A., Siwik, L.: Agent-based multi-variant crisis handling strategies for SCADA systems. In: International Conference on Multimedia Communications, Services and Security, pp. 61–71. Springer (2015)
7. Dudani, S.A.: The distance-weighted k-nearest-neighbor rule. IEEE Trans. Syst. Man Cybern. **4**, 325–327 (1976)
8. Egodagamage, R., Tuceryan, M.: A collaborative augmented reality framework based on distributed visual slam. In: 2017 International Conference on Cyberworlds (CW), pp. 25–32. IEEE (2017)
9. Kacprzyk, J., Owsinski, J.W., Viattchenin, D.A.: A new heuristic possibilistic clustering algorithm for feature selection. J. Autom. Mobile Rob. Intell. Syst. **8** (2014)
10. Lai, Z., Hu, Y.C., Cui, Y., Sun, L., Dai, N.: Furion: engineering high-quality immersive virtual reality on today's mobile devices. In: Proceedings of the 23rd Annual International Conference on Mobile Computing and Networking, pp. 409–421. ACM (2017)
11. LeBlanc, A.G., Chaput, J.P.: Pokémon go: a game changer for the physical inactivity crisis? Prev. Med. **101**, 235–237 (2017)

12. Nowicki, R.K., Starczewski, J.T.: A new method for classification of imprecise data using fuzzy rough fuzzification. Inf. Sci. **414**, 33–52 (2017)
13. Połap, D., Kęsik, K., Książek, K., Woźniak, M.: Obstacle detection as a safety alert in augmented reality models by the use of deep learning techniques. Sensors **17**(12), 2803 (2017)
14. Sieminski, A., Kozierkiewicz, A., Nunez, M., Ha, Q.T.: Modern approaches for intelligent information and database systems (2018)
15. Świechowski, M., Kacprzyk, J., Zadrozny, S.: A novel game playing based approach to the modeling and support of consensus reaching in a group of agents. In: 2016 IEEE Symposium Series on Computational Intelligence (SSCI), pp. 1–8. IEEE (2016)
16. Tyrrell, R., Sarig-Bahat, H., Williams, K., Williams, G., Treleaven, J.: Simulator sickness in patients with neck pain and vestibular pathology during virtual reality tasks. Virtual Reality, 1–9 (2017)
17. Wu, C.M., Hsu, C.W., Lee, T.K., Smith, S.: A virtual reality keyboard with realistic haptic feedback in a fully immersive virtual environment. Virtual Reality **21**(1), 19–29 (2017)

Design and Development of "Battle Drone" Computer-Based Trading Card Game (CTCG)

Reza Andrea[1]([✉]) [iD], Nursobah[1] [iD], and Marek Kopel[2] [iD]

[1] STMIK Widya Cipta Dharma, Samarinda 75123, Indonesia
{reza,nursobah}@wicida.ac.id
[2] Faculty of Computer Science and Management,
Wroclaw University of Science and Technology, Wybrzeze Wyspiaskiego 27,
50-370 Wroclaw, Poland
marek.kopel@pwr.edu.pl

Abstract. One of the most popular game types is card strategy game (collectible card game), which is a game of collecting and trading cards games. The card pictures vary from cartoon, fantasy to magical hero characters. This game can also be played by custom game rules and individual strategies. This means it is not only limited to collecting the cards. Collectible card games are usually played manually without a computer system. That is why "Battle Drone" was created. This is a card strategy game that has been computer-based. Battle Drone is a type of game focused on card selection strategies and chance of victory. Development methods are based on game development life cycle consisting of analysis, game design, implementation, and testing. A computer-based card strategy game that uses a barcode system to summon monsters and use ability from game card into console game.

Keywords: Battle Drone · Trading Card Game · Computer-based

1 Introduction

Trading card games are usually only played manually without a computer system. Players only use picture cards for which they create rules themselves. Players only use primitive devices such as calculators, coins, dice to perform calculations (e.g. to count Live point, Damage, etc.) in the game. The game card (trading card) is connected to the console game so it can be played in console or arcade machine (computer-based) which requires an intermediary system that can be played practically, fast and automatically. The technique of barcode system can be used here.

The creation of Battle Drone is inspired by an *"Animal Kaiser"* game released by Namco Bandai Games Inc. at the end of 2008 which also uses a barcode system. The name "Battle Drone" which means the fighting monster robot from another dimention.

© Springer Nature Switzerland AG 2019
K. Choroś et al. (Eds.): MISSI 2018, AISC 833, pp. 574–585, 2019.
https://doi.org/10.1007/978-3-319-98678-4_57

2   Research Procedures

Trading Card Game (TCG) is a game based on both luck as well as skill. It shows off the skills in organizing strategy in choosing the type and ability of the cards used in the right conditions by relying on luck to come out as the winner.

The Battle Drone Computer-based Trading Card Game (CTCG) is developed using the basic game development life cycle [7]. The steps are as follows:

1. Analysis.

At this stage the game concepts, such as the goal or rules of this game, are further developed

2. Game Design

Graphic design process in multimedia include text, sound, animation and image. The selection of the right elements in graphic design can optimize the process and make results fit in the game application. In accordance with the design, the text will use attractive colors, sound in WAV format, images in JPEG format, and animation in SWF format. Game rules will also greatly help users understand how to play.

3. Implementation

This stage serves to implement the design into an application using a programming language.

4. Testing

This stage will test and examine the system to see if the application is running as expected.

3   Game Concept and Design

Battle Drone Computer-based Trading Card Game (CTCG) is a computerized card game included in the CCG (Collectible Card Game) category. This card game uses special rules and the cards are not only playable but can also be collected.

The game trains the competitiveness and luck of the players in strategizing the right card to build a deck, so they can make the right combination to come out as the winner.

3.1   Game Rules

In *Battle Drone CTCG*, the rules are as follows:

1. Duel game must be played by two players
2. In single duel, each player must use a deck consisting of 1 Drone card, 1 Attribute card, and 3 different Gear cards
3. In match duel, each player must use a deck consisting of 1 Drone card, 1 Attribute card, 3 different Gear cards, and a backup deck containing 1 Attribute card, and 3 Gear cards

4. A player cannot use the same Gear card in one game
5. Gear cards can only be used or activated in every even round in a match (Example: in round 2, round 4, round 6, and so on)
6. In match duel, each player can replace their Attribute and Gear card at the beginning of the game, but cannot replace the Drone card
7. The game will end if one player can make the Drone Health point of their opponent became 0 (zero), or both of their Drone Health point became 0 (zero) at the same time.

3.2   Gameplay Concept

When the game is launched for the first time, the game will ask the player to choose the match type (single duel or match duel), then the first player (P1) input Drone card to summon it into the game followed by the second player (P2). They use barcode scanner on arcade machine to summon it (see Fig. 1).

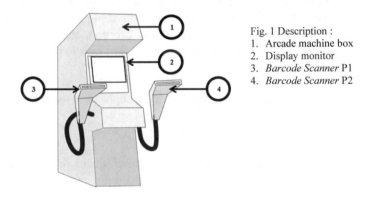

Fig. 1 Description :
1. Arcade machine box
2. Display monitor
3. *Barcode Scanner* P1
4. *Barcode Scanner* P2

Fig. 1. Arcade Machine Design

Then the first player (P1) input Attribute card that will change the Drone element in the game followed by the second player (P2), if one player wrongly inputs an attribute that is not in accordance with the Drone, the game will end and the opponent wins. Once both Drone players have an element input the Battle Stage will begin. In Stage 1 neither player is allowed to enter Gear card (Gear card unavailable), P1 and P2 input attack power bonus with randomized numbers through keyboard input then Battle Stage Calculation can be done by program. If there is no Drone whose player Health point (HP) become 0, then the program flow will return to Battle Stage. In the next Battle Stage, if the player is allowed to input Gear card then the program will go into Menu Gear input card. Once one or both players have input Gear card then the program will activate the player's Gear card effect and the Battle stage will start again.

If there has been one Drone whose player Health point dropped down to 0 (defeat) then the winning player will get a score of 1 (WIN = 1), If the game type is single duel then the player who get the score WIN = 1 will come out as the winner and the program will displays Battle Result. But if the game type is match duel then the program will return to the beginning, into the next round of matches. In the next round, the player cannot change their Drone card. Players only can replace the Attribute and Gear cards. In match duel the player must get the score WIN = 2 to become a winner.

3.3 Character and Game Environment Design

Designing the characteristics of game art is very important, because the style of characters, figures, visualization of the game effects supports the story of the game [5]. This research covers the design of the character of drone monster and game environment, including battle arena, rider character (drone card user), and etc. Drone is a monster in Battle Drone game. Drones are monster summoned by Drone cards in the battle arena (blue print of Drones can be seen in Fig. 2 on left image), and "Radenz" (see Fig. 2 in middle) is a human character from another dimensional world of Battle Drone. He is one of the Drone riders who will teach players how to play Battle Drone CTCG. Battle arena design that became the arena where fight Drone, it designed with the theme of the dimensions of Battle Drone World. Design of the Battle Drone's battle arena shown in Fig. 2 (right).

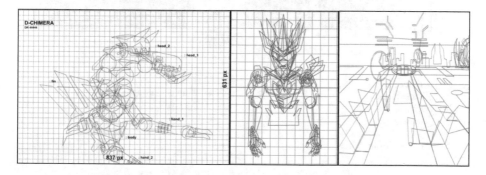

Fig. 2. Blue print of character and game environment

4   Results and Discussion

The results of the implementation are the real form of collection cards and the Battle Drone Computer-based Trading Card Game (CTCG) application, which can be played with arcade machine and barcode system.

4.1   Game Cards

There are three types of cards used in the *Battle Drone Computerized Trading Card Game* (CTCG). They are:

1. *Drone Card* – Fighting Monster Card

    Drone Monster card is the main card in this game that can be connected to a computer/ arcade machine using the printed barcode on the card. The card can be scanned by a barcode scanner to bring it up automatically in the battle arena. Each Drone card has a HLT (life points), ATK (attack points), POW (percentage of strength), and Available Attribute (element attributes that match the Drone). Drone card can be seen in Fig. 3.

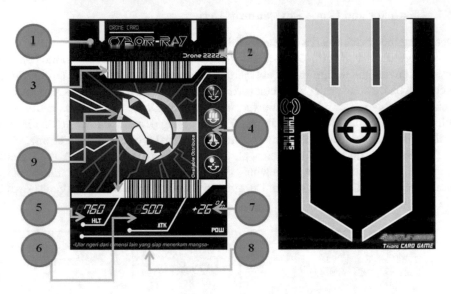

Fig 3. description :
1- Drone name
2- Drone code
3- Drone barcode
4- Available Attribute
5- HLT – Health Points

6- ATK – Attack Points
7- POW – Power Points
8- Drone Description
9- Drone display picture

Fig. 3.   Drone Card (face-up and face-down)

2. *Attribute Card* – Monster Element Attribute Card

Attribute Card is a card that can be input into computer/arcade machine by using barcode scanner. Its function is to increase the strength of drone that has been called to the arena (one of Attribute card can be seen in Fig. 4). If the inserted Attribute card is unmatched with the Drone, it will cause the Drone to die. There are four types of Attributes, namely:

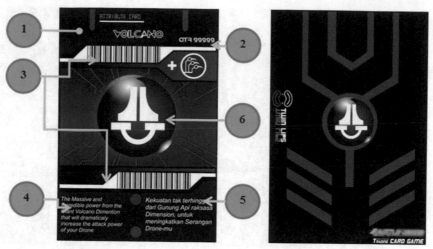

Fig 4. description :
1- Attribute Name
2- Attribute Code
3- Attribute Barcode

4- Description of Attribute in English
5- Description of Attribute in Bahasa
6- Attribute display picture

Fig. 4. Attribute Card (face-up and face-down)

a. ***Volcano*** is a fire element. An attribute card that can modify the Drone into an attacker monster, it will increase +10 HLT points, +150 ATK points, and +5 POW points
b. ***Deep*** is the depth of the sea element. An attribute card that can modify the Drone into a defensive monster, it will increase +300 HLT points and +2 POW points, but will decrease −10 ATK points
c. ***Sting*** is electric element. An attribute card that can modify the Drone into a fast attack monster, it will decrease −50 HLT points, but will increase +50 ATK points +20 POW points
d. ***Chaos*** is dark element. An attribute card that can increase all the strength of Drone in a balanced power, it will increase HLT, ATK, POW +30%

3. *Gear Card* – Monster Gear Card

Gear cards are cards that can be input into a computer/arcade game using a barcode scanner in the middle of a fight (one of the Gear cards can be seen in Fig. 5). Its function is to modify your or your opponent's Drone according to the effect written on each card. In one round of battles, each player can only activate three different Gear cards depending on the strategy used.

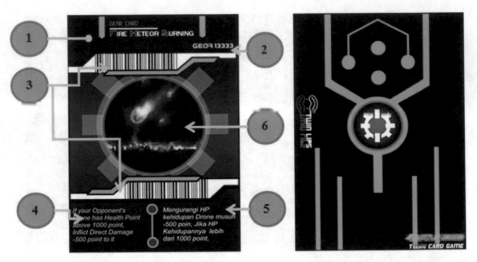

Fig 5. description :
1- Gear card name
2- Gear card code
3- Gear card barcode

4- Gear card description in English
5- Gear card description in Bahasa
6- Gear display picture

Fig. 5. Gear Card (face-up and face-down)

4.2   Game Art

The game art focuses on the theme and the character environment in the game that supports the story of the game.

4.2.1   Drone Character – Monster

There are four Drone positions in the battle (see in Fig. 6), namely: stand by, roar, attack mode and down.

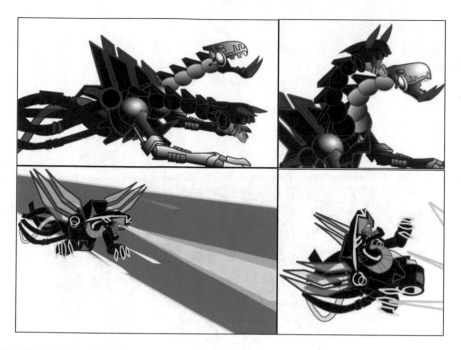

Fig. 6. (top) "D-CHIMERA" stand-by mode and roar; (bottom) "Cyborg-Ray" attack mode and down mode.

4.2.2   Battle Drone CTCG Interface

The following will describe some implementation results from the interface along with its description.

1. Title Scene

The initial scene of the Battle Drone CTCG is the title menu that shows the title of the game, and the Press Any Key to Continue command (see Fig. 7).

Fig. 7. Battle Drone CTCG opening scene

2. Battle Stage Scene

In the battle stage, both players must press Q & W for P1, and O & P for P2 to randomize power bonus and reward (see Fig. 8) which will improve and give effect directly to players' Drone in the arena. The player who has the most POW will attack first.

Fig 8. description :
1- Turn
2- Attribute of Drone P1
3- Attribute of Drone P2
4- HLT Drone P1
5- HLT Drone P2
6- Drone Name P1
7- Drone Name P2

8- Drone P1
9- Drone P2
10- Power Bonus P1
11- Power Bonus P2
12- Reward Bonus P1
13- Reward Bonus P2

Fig. 8. Battle stage scene

3. Battle Phase & Damage Scene

The player's Drone who has the most POW will attack first, and will change their turn with their opponent's Drone (see in Fig. 9 left). The attacked Drone will be down (see Fig. 9 right) and the HLT will be reduced by the amount of damage supplied from the percentage of ATK of their opponent's Drone.

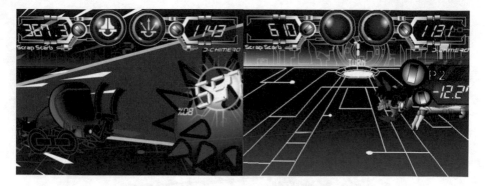

Fig. 9. Battle phase and battle damage scene

4. Result Battle Scene

After the battle is over, the game will display the battle result (the result of a match duel). Figure 10 left shows the battle result of single duel and Fig. 10 right shows result of match duel

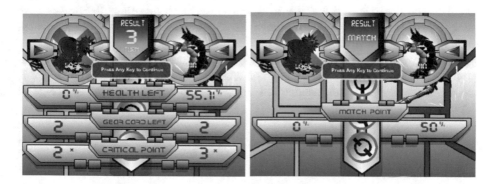

Fig. 10. Result battle scene

5 Game Testing

The barcode printed on each card is a one-dimensional barcode with Code 39. The test is performed on one of the "Fair Battle" Gear cards code: 83333 (see Fig. 11) using the ARGOX AS-8120 barcode scanner.

From the test as shown in Fig. 11, the barcodes printed on each card are identical to the code that is acceptable to the Notepad program, and it can be embodied in the game's animated characters. This proves that the barcode system is the right system to connect the collection card with the game application.

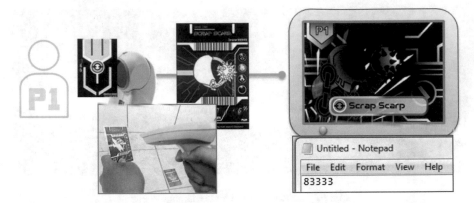

Fig. 11. Barcode test of "Scrap Scarp" card with Notepad recording software and Battle Drone CTCG

Beta Testing. Beta testing is a live test game application in an environment that cannot be controlled by the developer [4]. Trials were conducted in the form of a simple questionnaire filled by gamer and children from some game center. In this study, trials were conducted on 6 gamer and 4 childern, which presented 5 questions that refer to the gameplay and role of card game. Questionnaire questions are made as simple as possible so that childern can also fill them out.

From the results of beta testing in Table 1, we can determine the weight of the calculation for the answer "less" has a weight of 1, for the answer "enough" with weight 2, and the answer "good" with weight 3. Then we calculate the average percentage of respondents:

$$\bar{X} = \frac{34 \times 3 + 11 \times 2 + 5 \times 1}{50 \times 3} \times 100 = 86\% \tag{1}$$

Table 1. Result of beta testing

Question	Respondent's questions			Total respondents
	Good	Enough	Less	
How is character design?	7	3	0	10
How is card design?	7	2	1	10
How is the interface view of this game?	6	3	1	10
How is audio and music on this game?	6	2	2	10
How is gameplay of this game?	8	1	1	10
Total Answer	34	11	5	**50**

Based on the percentage acquisition the percentage is 86%, then the "Battle Drone" CTCG is acceptable because the presentation of the value obtained, above the minimum percentage of 50% (above enough), and close to 100% (very good).

6   Conclusion

Based on the result implementation, it can be concluded that, Computerized card strategy game, *Battle Drone*, is successfully created in game play with game development life cycle model system development.

The cards that are collected by the players can be connected to the *Battle Drone CTCG* software so that the cards can also be played in the software using the barcode system. It is advisable to the next research about computer-based trading card games, to use QR codes, it will make the card collection better suited for playing on smartphone.

References

1. Andrea, R.: Object position randomization technique "Find Me! – The Game". In: Proceeding SENAIK, 1 November 2013, Samarinda, Indonesia, pp. 301–303 (2013)
2. Andrea, R.: Game arena randomization technique "Find Me! – Bumi Etam" with the technique of shuffle random. In: Proceeding SENATKOM. 23 September 2015, Padang, Indonesia, pp. 823–827
3. Andrea, R., Akbar, R.I., Fitroni, M.: Developing battle of Etam earth game agent with finite state machine (FSM) and sugeno fuzzy. ICCS Proc. **1**(1), 184–187 (2014)
4. Hurd, D., dan Jenuings, E.: Standardized Educational Games Ratings: Suggested Criteria. Longman, London (2009)
5. Kyaw, A.S., Peters, C., Swe, T.N.: Unity 4.x Game AI Programming. Packt Publishing Ltd., Birmingham (2013)
6. Poole, D.L., Mackworth, A.K.: Artificial Intelligence Foundations of Computational Agents. Cambridge University Press, New York (2010)
7. Sommerville, I.: Software Engineering, 9th edn. Pearson Education Inc., United States of America (2011)

Wide Field of View Projection Using Rasterization

Grzegorz Muszyński, Krzysztof Guzek, and Piotr Napieralski(✉)

Institute of Information Technology, Lodz University of Technology,
ul. Wolczanska 215, 90-92, Łódź, Poland
{krzysztof.guzek,piotr.napieralski}@p.lodz.pl

Abstract. Perspective projection is a type of rendering that makes the images of three-dimensional objects approximate actual visual perception. The change of the field of view feature makes the camera switch to an either wider or narrower lens. The wide angle of virtual cameras in computer games, as well as in computer animation, can give a particular cinematic look. In this paper, we evaluate the possibility of using rasterization for the wide field of view. All methods have been implemented and verified experimentally, to be later compared with each other.

Keywords: Rendering · Rasterization · Field of view · FOV

1 Introduction

Rendering a wide field of view in computer generated images has been in the scope of researchers' interests for years. The reasoning behind creating such projections is to mimic the wide field of view of the human eye. 3D graphics have become very popular and intensively developed since the early eighties and currently has also reached standards in Web Technologies [1]. The problem of mapping the perspective for 3D space is still an important element of research. In this article, 'wide angle' is defined as a horizontal field of view of 100° or more. The limitations of rectilinear projection often do not allow for wide angle views. With the rectilinear perspective, a field of view above 100° produces noticeable artifacts in the form of geometry stretches. A field of view near 180° is virtually impossible to be perceived clearly with that projection. Overcoming these limitations is required to achieve a peripheral vision in VR environments, simulators, video games or other 3D rasterization software. The main issue with creating wide-angle images is to project the linear space, created using affine transformations, onto the non-linear space of the wide angle. For this reason, methods of non-linear projections were developed.

2 Related Works

2.1 Ray-Tracing Projection Methods

The idea of ray-trace rendering is to cast rays from the camera to surfaces that represent the reverse light travel. Basic implementation of ray-tracing makes it

© Springer Nature Switzerland AG 2019
K. Choroś et al. (Eds.): MISSI 2018, AISC 833, pp. 586–595, 2019.
https://doi.org/10.1007/978-3-319-98678-4_58

possible to emit rays with the angular step to cover more than 180° of the field of view, as described by Bourke [2]. This approach to wide-angle rendering is called angular fisheye projection. While ray-tracing can be used in real time to some extent, its rendering time remains inferior to the traditional rasterization approach [3]. A different method was developed by Brosz et al. to create non-linear ray-tracing, concerning the angle between the rays corresponding to the opposite edges of a perspective image [4].

2.2   Vertex Projection Methods

The methods of per-vertex projection evaluation can be used to project the required non-linear spaces [5]. These methods are found to work only for dense geometries (i.e. with a high concentration of triangles over a small area). The most basic vertex-based approach proposed by Lorenz is to render a scene n times for n sides of the geometric shape used for approximation [6]. However, this method is costly due to the n render passes. It can be minimized to 3 render passes, and optimization can be achieved with tessellation shaders. A different approach is to use vertex shaders to calculate per-vertex transformations. Such method was developed by Oros [7]. This approach, however, has its limits. A field of view near 180° can be achieved, but vertices near the camera can create artifacts. A high value for the near-clipping plane is recommended. It also creates a fisheye lens effect, which could be problematic in some cases, depending on the artistic goal. Alternatively, spherical mapping equations can be used to transform a scene into a sphere map. Such approach was used by Trapp et al. [8] This, however, produces geometric discontinuity. To overcome this, they propose a tesselation shader [5]. Recently, Perez et al. proposed two methods of non-linear projections adjusted for curved displays [9]. Methods use extensive use of tesselation shaders, and are ready to implement on current GPU pipelines. Pohl et al. tested various methods for distorting image for Head Mounted Displays [10], including geometry based distortion grid. Those test, however, excluded field of view. All of the previously mentioned methods have geometry artifacts with non-tessellated objects.

2.3   Image Projection Methods

A different approach is to render images and then transform them in the post-process stage. A common image-base method is to render the environment with a cube - or other geometric mesh - into multiple off-screen textures. It is then possible to recreate any wide-angle projection from a cube-map or other 360° map [11]. Another approach is to render a wide field of view with the rectilinear perspective. The image is then corrected in post-processing to adjust the rectilinear perspective stretches. However, this method restricts the obtainable field of view to less than 180° [12]. There are many different realtime barrel distortion methods developed, starting from Nijmeijer et al. [13]. Some other development in this method includes Park et al. [14] ideal image coordinates model and GPU implementation by Lee et al. [15]. While these methods can be used with the GPU acceleration, the cubemap approach requires changes in rendering the pipeline for multiple off-screen renders. Even then, they both produce certain parts of the images that are discarded in the process of projecting images onto the final image space. Another problem with image projection methods is filtering. Pixel information could be lost or blurred out in the process, which can lead to loss of sharp edges. This detremental effect is not present in hybrid projection, developed by Boustila et al. [16]. It uses both vertex and image approach to stitch two different type of projections: rectilinear and cylindrical, former used in centre of the image, while latter serves to improve peripheral vision. In the article by Pohl et al. [10] an image based barrel distortion model was also tested. Later, Toth et al. tested various plane projections [17]. In this paper, the common wide field of view rasterization projection methods are presented, comparing their maximum field of view, projection method, computing time and limitations.

3   Methods

3.1   Rectilinear Projection

The most basic type of perspective projection in rasterization is rectilinear projection. This projection is created using at least three affine transformations, described by 4×4 matrices: World (sometimes referred as Model), View and Projection. The World matrix W stores objects to scene transformations such as translation, rotation and scale, all basic affine transformations (Eqs. (1) and (2)).

$$
\begin{bmatrix} 1 & 0 & 0 & 0 \\ 0 & cos\phi & -sin\phi & 0 \\ 0 & sin\phi & cos\phi & 0 \\ 0 & 0 & 0 & 1 \end{bmatrix} \begin{bmatrix} cos\phi & -sin\phi & 0 & 0 \\ sin\phi & cos\phi & 0 & 0 \\ 0 & 0 & 1 & 0 \\ 0 & 0 & 0 & 1 \end{bmatrix} \begin{bmatrix} S_x & 0 & 0 & 0 \\ 0 & S_y & 0 & 0 \\ 0 & 0 & S_z & 0 \\ 0 & 0 & 0 & 1 \end{bmatrix} \begin{bmatrix} cos\phi & 0 & sin\phi & 0 \\ 0 & 1 & 0 & 0 \\ -sin\phi & 0 & cos\phi & 0 \\ 0 & 0 & 0 & 1 \end{bmatrix} \begin{bmatrix} 1 & 0 & 0 & T_x \\ 0 & 1 & 0 & T_y \\ 0 & 0 & 1 & T_z \\ 0 & 0 & 0 & 1 \end{bmatrix}
$$
(1)

The View matrix V is a description of the scene to camera transformations. The Projection matrix P is a description of the camera to projection transformation. To project a triangular mesh onto the screen, a position vector of each vertex must be multiplied by these matrices (Fig. 1).

$$
\begin{bmatrix} a & b & c & d \\ e & f & g & h \\ i & j & k & l \\ m & n & o & p \end{bmatrix} \begin{bmatrix} x \\ y \\ z \\ 1 \end{bmatrix} = \begin{bmatrix} x' \\ y' \\ z' \\ q \end{bmatrix}
$$
(2)

3.2   Curvilinear Vertex Projection

The first method capable of solving the problem in question is curvilinear vertex projection. In the traditional approach, the World, View and Projection transformations are often stored in a single WVP matrix. However, this method

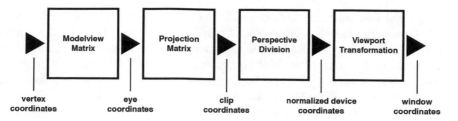

Fig. 1. Viewing transformations in rendering pipeline

requires two separate matrices: WV and P. Per-vertex transformation is additionally required between the World-View and View-Projection transformations. This transformation is described as substituting the z component of a vertex position vector. The new z component is calculated using Eq. (3):

$$z = (\frac{z_0 \pm d}{2})$$
(3)

where z_0 is the original z component, and d is the Euclidean distance from camera to vertex. Depending on whether the depth axis is positive or negative, the operation would include either addition or subtraction. Optimizing this formula includes storing a pre-divided fraction, by which the new depth is multiplied. As a result, vertices with the same original depth will be distributed curvilinearly after transformation, varying by camera offset. This allows for breaking the standard 180° limit of rectilinear projection. However, this projection can produce artefacts by drawing faces that should not be in front of the camera. These faces are mostly in the space near camera, and as such, can be marginalized by setting the near-clipping plane farther. Another problem with this projection method is operating on vertices. This means that lines will remain in the rectilinear space, whereas vertex positions will be in the curvilinear space. This discrepancy will cause visible geometry errors. To address this problem, the lines would need to be curvilinear as well. The recommended course of action is to operate on dense geometry or tessellate it directly before transformations, at the expense of rendering time.

3.3   Sphere Mapping with Pre-clip Tessellation

Another method for the high FOV projection is to use sphere mapping. In this type of projection, the only two transformation matrices used are World and View. The actual projection is carried out using Eq. (4):

$$x = \frac{2}{h_{fov}}\lambda$$
$$y = \frac{2}{v_{fov}}\phi$$
$$z = -1 + 2\frac{\rho - near}{far - near}$$
(4)

where $\lambda = arctan\frac{x}{-z}, \phi = arcsin\frac{y}{\rho}$ and $\rho = \sqrt{x^2 + y^2 + z^2}$.

Since the new coordinates are already in the screen-space, there is no need to divide by w component, as in the case of perspective projection in the traditional approach. This kind of mapping will produce artefacts of faces stretching horizontally along the image. For this very reason, the pre-clipping stage is required. Since the sphere mapping covers the angles $-\Pi$ to Π, every face that intersects the $-\Pi$, Π plane needs to be clipped. There are two case scenarios for this - one for faces on poles (intersecting the Y axis) and one for common intersection along $\pm Z$ (depending on the depth axis sign). This method, apart from adding an extra stage to the rasterization pipeline, comes across a similar problem of depending on geometric density. Without providing the dense scene or tessellation stage, this method will project lines instead of curves.

3.4   Barrel Distortion

The methods mentioned previously refer strictly to the vertex space. An image-space approach to this problem is to use distortion correction, or in this particular case, to apply distortion to the image. There are three main types of radially symmetric distortions: barrel, pincushion and handkerchief [13]. For the first two, a strong relationship can be observed. Barrel and pincushion are two opposites, and as such, they neutralize one another. An image processed with barrel and then pincushion distortion would remain the same as far as the perspective is concerned. Physical cameras with wide-angle lenses often produce images with barrel distortion. These types of lenses are typically referred to as fisheye lenses. To remove barrel distortion, pincushion distortion must be applied. This method produces the rectilinear perspective from a wide-angle image. For the problem presented in this article, the opposite effect is desired. A wide-angle image should be created from the existing rectilinear image. To do that, the opposite distortion - i.e. barrel distortion - should be applied [12]. Barrel distortion in rasterization is a post-process function that converts Cartesian coordinates of the input image to distorted, polar coordinates with the pole set in the centre of the image. This conversion is achieved with Eq. (5):

$$x = r \cdot cos(\theta) + 0.5$$
$$y = r \cdot sin(\theta) + 0.5 \tag{5}$$

where $\theta = arctan(\frac{x_0}{y_0})$, and r is a lens distance function. The two distance function tested in this article are $r = (\sqrt{x^2 + y^2})^{\alpha+1}$ and $r = tan(\sqrt{x^2 + y^2}) * \alpha$, where $\alpha = 0.5$ is a distortion strength parameter. With these new coordinates, the input image is then mapped onto the device. If new coordinates are mapped out of bounds, then an empty (black) pixel is used. This results in black spots outside the original image. Since this method relies on images projected rectilinearly, the FOV limit is not extended. However, this approach makes images with the FOV above 100° easier to read. This is achieved by scaling up the centre of the image while scaling down the peripheral objects.

3.5   Piecewise Perspective Projection

A basic method for approximating non-linear projections is to render n times [6]. For panoramas a cylindrical projection would be approximated. This includes dividing projection widthwise n times. Each stripe would have a width of $width/n$ and height would remain unchanged. Before piece projection scene should be aligned to the angle that this piece represents. This is done the easiest by repeatedly rotating camera Z vector by fov/n, given that this vector is aligned with the first piece. This algorithm can be described as follows:

```
for  i=0, i < n, i++:
    camera.rotate(fov/n);
    scene.transform();
    rasterize(w/n*i,0, w/n, h)
    //rasterize(x, y, width, height);
```

As this method needs scene to be reprojected n times, for each frame n different rasterization passes are needed.

3.6   Cube Mapping to Spherical Coordinates Projection

A popular method to store 360° images is cube mapping. It consists of 6 different images, each representing view aligned with axes in cartesian coordinate system, in both directions (see Fig. 2). Projection used is standard, rectilinear perspective, with both horizontal and vertical FOV equal to 90°. Then an equirectangular projection can be accomplished with Eq. (6):

$$x = -sin(s*2\pi)*sin(t*\pi)/a$$
$$y = cos(t*\pi)/a$$
$$z = -cos(s*2\pi)*sin(t*\pi)/a \tag{6}$$

where s and t are screen-space coordinates, $a = max(|x|, |y|, |z|)$ and x, y, z are mapped coordinates. When mapped coordinate is equal ±1, then according axis image should be used. The remaining coordinates are used to sample the texture.

Fig. 2. A cubemap images in flat, cross representation

4   Results

The test platform was equipped with the Intel Core I7-7700 HQ 2.8 GHz CPU
and 16 Gb memory. All methods were tested using the single-threaded software
rasterizer, written in C++ with libraries: QT 5.10, GLM 0.9.8.5. The test environment architecture was based on the OpenGL rendering pipeline. The testing
scene included a cube textured with the grid pattern, colored accordingly to axis
it is facing. Each square contains 64 smaller squares to simulate the pre-tessellated
scene (Fig. 3). A test cube was also created with 4 squares per face to simulate non-
tessellated scene (Fig. 4). To compare the methods, several factors were used, most
importantly the possible FOV values. A higher FOV means more peripheral vision
possible to achieve. The FOV is measured based on the maximum horizontal FOV
value that the algorithm can provide without visible errors. A valid method should
be able to produce images with FOV values above 100°, up to 180° and beyond.
Another factor is distortion type. Distortion intensity is measured by how objects
will behave when placed in certain areas. These areas include: central area, horizontal peripheral and vertical peripheral. The test environment operates based
solely on CPU, and thus the time values obtained do not refer to the conventional
rasterization process commonly operating on GPU. The values were used to determine the complexity of each method in relation to the basic rectilinear perspective.
The important factor is the limitation aspect of each of the methods, which may
be due to the FOV, dependency on other methods or a certain pipeline (Table 1).

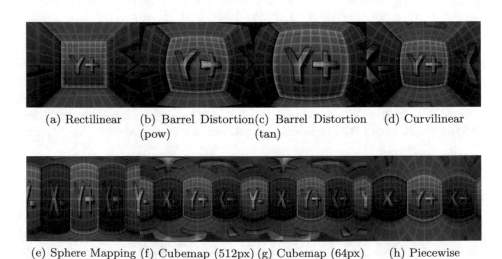

(a) Rectilinear   (b) Barrel Distortion(c) Barrel Distortion   (d) Curvilinear
                    (pow)   (tan)

(e) Sphere Mapping (f) Cubemap (512px) (g) Cubemap (64px)   (h) Piecewise

Fig. 3. Projection results for dense geometry

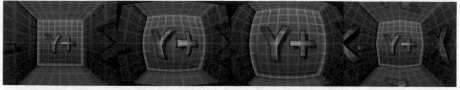

(a) Rectilinear   (b) Barrel Distortion (c) Barrel Distortion   (d) Curvilinear
                   (pow)   (tan)

(e) Sphere Mapping (f) Cubemap (512px) (g) Cubemap (64px)   (h) Piecewise

Fig. 4. Projection results for low-poly geometry

Table 1. Comparison of projection methods

Method	Frame time tes./no tes. (ms)	FOV test/max (°)	Central distortion	Horizontal peripheral distortion	Vertical peripheral distortion	Projection type
Rectilinear	2831/341	150/179	None	Heavy	Heavy	Rectilinear
Barrel distortion (Pow)	2863/374	150/179	Zoom	Radial	Radial	Fisheye
Barrel distortion (tan)	2846/371	150/179	Light	Radial	Radial	Fisheye
Curvilinear	2500/271	180/240	Light	Radial	Radial	Curvilinear
Sphere map	3016/265	360	No centre	Angular	Heavy	Equirect.
Cubemap (512px)	7571/1784	360	No centre	Angular	Heavy	Equirect.
Cubemap (64px)	146/51	360	No centre	Angular	Heavy	Equirect.
Piecewise	1250/373	360	No centre	Angular	Heavy	Equirect

5   Conclusions and Future Work

As expected, the vertex projection methods produced visible errors for the low-poly scene. The previously mentioned tessellation step would be necessary in production pipeline as it would increase the required resource significantly. Vertex projection methods can produce FOV higher than 180°, and they have stable framerate. They provide per-pixel results with framerates comparable to these of standard rectilinear projection. However, per-vertex calculations could be expensive on other platforms and pipelines. These are also methods that are most compatible with current GPU rendering pipelines, requiring only vertex and tesselation shaders. Texture space projections would be best suitable for artistic purposes. They have similar FOV restrictions as rectilinear perspective, however they resemble real life wide-angle lenses used in cinematography and photography. Compared to rectilinear perspective, these projections are easier to read, and do not present stretching effect. The render times for these is also similar to rectilinear. With hardware accelerated rasterization, these methods will produce stable, interactive framerate. The major drawback of these methods are pixelated central area. As they reproject rendered image, they are unable to produce new fragments on their own. With filtering algorithms this can be minimized, however, no new information would be introduced. Methods of multiple rendering passes are both the slowest and the fastest and are able to produce images with vertical FOV of 360°. Piecewise projection proved to be fastest than standard projection. This is most likely due to less geometry and fragment per stripe. However, this advantage would be negated when implemented on GPU pipeline. Sending and reading data to and from GPU could be expensive, especially with interactive framerates in mind. There are other problems, that storing multiple offscreen texture is resource demanding, and MRT may not be supported in some graphics pipelines. With cube mapping rasterization, the major problem is texture size. High resolution of cubemap required for accurate reprojection would mean long render times, due to computational complexity of $O(6N^2)$. However, with small texture size, frame time is 20 times faster than standard projection. The trade off is information loss during reprojection stage. These method would be best suited for dynamic creation of environmental maps instead of handling camera view. Each projection method gives different distortions that prove difficult to compare. Most methods produce equirectangular projection with some visible differences in vertical angle. Comparisons were made using a single-threaded CPU environment. Future works include comparison using the GPU software. In order to avoid tessellation, the vertex projection methods could be used alongside pixel projection methods. This would require changes in triangle intersection function, which is not a hardware-supported method of rasterization [17].

References

1. Dworak, D., Pietruszka, M.: PNG as fast transmission format for 3D computer graphics in the Web. In: Multimedia and Network Information Systems, January 2017
2. Bourke, P.: Computer Generated Angular Fisheye Projections (2001)
3. Gascuel, J.-D., Holzschuch, N., Fournier, G., Péroche, B.: Fast non-linear projections using graphics hardware. In: Proceedings of I3D, pp. 107–114 (2008)
4. Brosz, J., Samavati, F.F., Carpendale, M.S.T., Sousa, M.C.: Single camera flexible projection. In: Proceedings of NPAR, pp. 33–42 (2007)

5. Ardouin, J., Lécuyer, A., Marchal, M., Marchand, E.: Stereoscopic rendering of virtual environments with wide field-of-views up to 360°. In: 2014 IEEE Virtual Reality (VR), pp. 3–8 (2014)
6. Lorenz, H., Döllner, J.: Real-time piecewise perspective projections. In: Proceedings of GRAPP, pp. 147–155 (2009)
7. Oros, D.: A conceptual and practical look into spherical curvilinear projection. www.frost-art.com
8. Trapp, M., Lorenz, H., Döllner, J.: Interactive stereo rendering for non-planar projections of 3D virtual environments. In: Proceedings of GRAPP, pp. 199–204 (2009)
9. Perez, M., Rueda, S., Orudña, J.M.: Geometry-based methods for general non-planar perspective projections on curved displays. J. Supercomput., 1–15 (2018)
10. Pohl, D., Johnson, G.S., Bolkart, T.: Improved pre-warping for wide angle, head mounted displays. In: Proceedings of the 19th ACM Symposium on Virtual Reality Software and Technology, pp. 259–262 (2013)
11. Oortmerssen, W.: Fisheyequake. http://strlen.com/gfxengine/fisheyequake/
12. Carpentier, G.: Reducing stretch in high-FOV games using barrel distortion (2015). http://www.decarpentier.nl/lens-distortion
13. Nijmeijer, A.G.J., Boer, M.A., Slump, C.H., Samson, M.M., Bentum, M.J., Laanstra, G.J., Snijders, H., Smit, J., Herrmann, O.E.: Correction of lens-distortion for real-time image processing systems. In: VLSI Signal Processing VI (1993)
14. Park, J., Byun, S., Lee, B.: Lens distortion correction using ideal image coordinates. IEEE Trans. Consum. Electr. **55**(3) (2009)
15. Lee, T.Y., Wei, C.H., Lai, S.H., Lee, R.R.: Real-time correction of wide-angle lens distortion for images with GPU computing. In: 2012 IEEE Asia Pacific Conference on Circuits and Systems (APCCAS) (2012)
16. Boustila, S., Capobianco, A., Génevaux, O., Bechmann, D.: New hybrid projection to widen the vertical field of view with large screen to improve the perception of personal space in architectural project review. In: 2016 IEEE Virtual Reality (VR) (2016)
17. Toth, R., Nilsson, J., Akenine-Möller, T.: Comparison of projection methods for rendering virtual reality. In: Proceedings of High Performance Graphics, pp. 163–171 (2016)

Author Index

© Springer Nature Switzerland AG 2019
K. Choroś et al. (Eds.): MISSI 2018, AISC 833, pp. 597–599, 2019.
https://doi.org/10.1007/978-3-319-98678-4

Printed in the United States
By Bookmasters